Data Governance

Dimitrios Sargiotis

Data Governance

A Guide

Dimitrios Sargiotis
National Technical University of Athens
Marousi, Greece

ISBN 978-3-031-67267-5   ISBN 978-3-031-67268-2 (eBook)
https://doi.org/10.1007/978-3-031-67268-2

© The Editor(s) (if applicable) and The Author(s), under exclusive license to Springer Nature Switzerland AG 2024

This work is subject to copyright. All rights are solely and exclusively licensed by the Publisher, whether the whole or part of the material is concerned, specifically the rights of translation, reprinting, reuse of illustrations, recitation, broadcasting, reproduction on microfilms or in any other physical way, and transmission or information storage and retrieval, electronic adaptation, computer software, or by similar or dissimilar methodology now known or hereafter developed.

The use of general descriptive names, registered names, trademarks, service marks, etc. in this publication does not imply, even in the absence of a specific statement, that such names are exempt from the relevant protective laws and regulations and therefore free for general use.

The publisher, the authors and the editors are safe to assume that the advice and information in this book are believed to be true and accurate at the date of publication. Neither the publisher nor the authors or the editors give a warranty, expressed or implied, with respect to the material contained herein or for any errors or omissions that may have been made. The publisher remains neutral with regard to jurisdictional claims in published maps and institutional affiliations.

This Springer imprint is published by the registered company Springer Nature Switzerland AG
The registered company address is: Gewerbestrasse 11, 6330 Cham, Switzerland

If disposing of this product, please recycle the paper.

To my beloved family:
To Eirini, my wonderful wife, for her endless support and love.
To Konstantinos, my cherished son, who inspires me every day.
To my father, who left us in November 2017, for his enduring guidance, strength, and the ethos of integrity and perseverance that he instilled in me. Your memory and values continue to guide me every day.
To Paraskevi, my dear mother, for her unwavering belief in me.
To Theodora, my supportive sister, for always being there.
And to the new generation, may you navigate and shape the future with wisdom and integrity.

Preface

In the digital era, where data has become the new oil, the importance of managing this invaluable asset cannot be overstated. "Data Governance: A Guide" is a comprehensive resource designed to provide insights, methodologies, and best practices in the field of data governance.

My journey into data governance began over a decade ago, sparked by the challenges and opportunities I observed in the rapidly evolving data landscape. The explosion of data in the digital age has not only transformed how organizations operate but also raised complex issues surrounding data security, privacy, and quality. This book is a culmination of years of research, practical experience, and discussions with experts in the field.

The primary aim of this guide is to demystify data governance and make it accessible to professionals across various sectors. Whether you are a data management veteran or new to the field, this book offers valuable insights into establishing a robust data governance framework.

Throughout the chapters, I delve into the core principles of data governance, addressing key components such as data quality, security, privacy, and management. Real-world case studies are interspersed to illustrate the tangible benefits and challenges of implementing data governance in diverse organizational contexts.

This book also explores the future landscape of data governance, considering emerging trends and technologies such as artificial intelligence, machine learning, and blockchain. The goal is to equip readers with the knowledge to not only navigate the current landscape but also to anticipate and prepare for future developments.

Authoring this book has been a journey of discovery and reaffirmation of the critical role data governance plays in modern organizations. It is my sincere hope that this guide will serve as a valuable resource for you, sparking innovative ideas, strategies, and a deeper understanding of data governance.

Welcome to the journey of mastering data governance.

Marousi, Greece   Dimitrios Sargiotis

Acknowledgments

The journey of creating "Data Governance: A Guide" has been both enriching and challenging. This book, a labor of love and dedication, owes its existence to the collective efforts of many remarkable individuals whose support, expertise, and encouragement have been invaluable.

First and foremost, I extend my heartfelt gratitude to the myriad of data governance professionals and academics whose pioneering work and insightful discussions have laid the foundation for this guide. Their dedication to advancing our understanding of data governance has been a source of constant inspiration and learning.

I am particularly grateful to my peers and colleagues in the data management community. Their willingness to share experiences, challenges, and success stories has enriched this guide with a diversity of perspectives that reflect the real-world complexities of data governance.

To my family, who has stood by me with patience and understanding through the countless hours dedicated to writing and research, I owe a debt of gratitude. Your encouragement and belief in the value of this work have been my greatest motivators.

I also wish to acknowledge the invaluable contributions of my editorial team, whose expertise and attention to detail have greatly enhanced the quality of this guide. Their dedication to excellence has been a key factor in bringing this project to fruition.

To the reviewers who generously contributed their time and expertise to provide feedback, thank you. Your constructive critiques have been essential in refining the content and ensuring its relevance and accuracy.

Lastly, to the readers and future data governance practitioners, this guide is for you. It is my hope that it will serve as a valuable resource as you navigate the complexities of data governance in your professional journey.

The completion of this guide marks not an end, but a beginning—a step toward a future where data governance is recognized not just as a necessity, but as a cornerstone of ethical and effective data management.

Competing Interests

In alignment with Springer Nature's commitment to transparency and integrity, I, Dimitrios Sargiotis, wish to disclose any competing interests—both financial and non-financial—that might have influenced the content of "Data Governance: A Guide."

Financial Interests

- As the sole author of this guide, I have not received direct financial compensation specifically for the creation of this book. My professional affiliations, including any consultancy roles, research funding, or associations with entities that have a vested interest in data governance, are disclosed herein.
- I declare that there are no royalties or financial benefits received from third parties directly for the writing and publication of this book. Any future royalties or financial benefits will stem directly from the sales of the book itself, without external endorsements or funding.

Non-financial Interests

- My commitment to advancing the field of data governance is both a professional and personal interest. I hold no advisory board positions, consultancy roles, or affiliations with organizations that might present a conflict of interest regarding the book's content.
- I am dedicated to providing an objective, balanced view on data governance, free from personal beliefs, affiliations, or relationships that could be perceived to bias the content of this guide.

This declaration serves to assure the reader of my commitment to ethical and transparent publishing. My aim in authoring "Data Governance: A Guide" is to furnish the reader with unbiased, practical insights into data governance, unswayed by competing interests.

In the spirit of full disclosure, should any changes occur in my affiliations or financial interests after the publication of this book, I commit to updating this statement in subsequent editions to reflect such changes accurately.

Contents

1 **Overview and Importance of Data Governance**.................. 1
  1.1 Overview of Data Governance: Definition and Scope........... 2
  1.2 The Pillars of Data Governance............................. 10
      1.2.1 Data Quality: Ensuring Accuracy, Completeness,
            and Reliability..................................... 10
      1.2.2 Data Security: Protecting Data from Unauthorized Access
            and Breaches.. 16
      1.2.3 Data Privacy: Complying with Regulations and Ethical
            Handling of Personal Data........................... 24
      1.2.4 Data Management: Efficient and Effective Use
            and Storage of Data................................. 29
  1.3 The Importance of Data Governance in Modern Organizations ... 30
      1.3.1 Enhance Decision-Making Through Data Governance..... 33
      1.3.2 Regulatory Compliance and Risk Management 35
      1.3.3 Data Integration and Operational Efficiency............ 39
      1.3.4 Building Trust and Credibility in Data 41
  1.4 Key Components of a Data Governance Program.............. 45
      1.4.1 Data Governance Framework: Structure
            and Components..................................... 45
      1.4.2 Data Stewards, Data Owners, and Governance Bodies.... 48
      1.4.3 Policies and Standards: Development, Implementation,
            and Enforcement 52
      1.4.4 Tools: Supporting Data Governance Efforts 56
  1.5 Common Misconceptions About Data Governance.............. 60
      1.5.1 Debunking Myths and Clarifying Common
            Misunderstandings................................... 60
      1.5.2 Data Management: Clarifying the Differences 63
  1.6 Challenges in Implementing Data Governance 65
      1.6.1 Overcoming Resistance to Change 65
      1.6.2 Addressing Data Quality Issues...................... 71
      1.6.3 Aligning Data Governance with Business Objectives..... 77

	1.7	Success Stories: Data Governance in Action...................	79
	1.8	Conclusion and Preview of the Next Chapter	80
	References...		81
2	**The Importance of Data Governance: Why It Matters in Today's World**		**87**
	2.1	The Data-Driven Landscape: The Explosion of Data in the Digital Age..	88
		2.1.1 How Data Drives Business Decisions and Innovation.....	90
		2.1.2 The Growing Need for Organized and Governed Data....	94
	2.2	Data Governance and Business Value	97
		2.2.1 Enhancing Decision-Making with Quality Data	97
		2.2.2 Improving Operational Efficiency Through Effective Data Management..	102
		2.2.3 Driving Business Growth and Innovation Through Strategic Data Utilization.............................	104
	2.3	Regulatory Compliance and Risk Management	107
		2.3.1 Overview of Data-Related Regulations and European Agencies..	107
		2.3.2 The Role of Data Governance in Ensuring Compliance ...	111
		2.3.3 Mitigating Risks Associated with Data Breaches and Noncompliance	112
	2.4	Data Governance and Customer Trust: Building Customer Trust Through Responsible Data Practices.......................	114
		2.4.1 The Impact of Data Governance on Customer Relationships and Brand Reputation	116
	2.5	Case Studies: The Cost of Poor Data Governance..............	119
	2.6	Data Governance as a Competitive Advantage	126
	2.7	Overcoming Challenges: Building a Culture of Data Governance	130
		2.7.1 Building a Culture of Data Governance................	131
	2.8	Conclusion and Transition to Next Chapter...................	133
	References...		133
3	**Key Principles of Data Governance: Building a Strong Foundation** .		137
	3.1	Understanding the Core Principles of Data Governance: Shaping Frameworks and Strategies	138
	3.2	Principle of Data Quality.................................	140
		3.2.1 Strategies for Continuous Data Quality Assessment and Improvement.....................................	142
	3.3	Principle of Data Transparency and Accessibility.............	145
	3.4	Principle of Data Security	147
	3.5	Principle of Compliance and Data Privacy	149
	3.6	Principle of Data Stewardship	150
	3.7	Principle of Data Lifecycle Management	152

	3.8	Integrating Data Governance with Business Strategy	154
		3.8.1 Case Studies	155
	3.9	Overcoming Implementation Challenges	156
	3.10	Real-World Examples: Principles in Practice	157
	3.11	Conclusion and Look Ahead	159
		References	161
4	**Data Governance Frameworks: Models and Best Practices**		**165**
	4.1	Introduction to Data Governance Frameworks	166
	4.2	Overview of Popular Data Governance Frameworks	170
	4.3	Designing a Data Governance Framework	176
	4.4	Best Practices in Framework Implementation	178
	4.5	Integrating Technology with Frameworks	180
	4.6	Measuring the Effectiveness of your Framework—Data Governance Template	183
	4.7	Case Studies: Frameworks in Action	187
	4.8	Adapting Frameworks to Changing Data Landscapes	189
	4.9	The Future of Data Governance Frameworks	191
	4.10	Conclusion and Transition to the Next Chapter	193
		References	194
5	**Data Quality Management: Ensuring Accuracy and Reliability**		**197**
	5.1	Introduction to Data Quality Management	198
	5.2	Dimensions of Data Quality	199
	5.3	Establishing Data Quality Standards	201
	5.4	Data Quality Assessment Techniques	203
	5.5	Data Cleansing and Improvement Strategies	204
	5.6	Role of Technology in Data Quality Management	206
	5.7	Building a Culture of Data Quality	207
	5.8	Case Studies: Transforming Data Quality	209
	5.9	Overcoming Common Data Quality Challenges	211
	5.10	Conclusion and Next Steps	212
		References	215
6	**Data Security and Privacy: Protecting Sensitive Information**		**217**
	6.1	Introduction to Data Security and Privacy	218
	6.2	Key Concepts in Data Security	220
	6.3	Data Privacy Laws and Regulations	224
	6.4	Developing a Data Security and Privacy Strategy	226
	6.5	Implementing Security Measures	229
	6.6	Privacy by Design and Default	232
	6.7	Managing Data Breaches and Incidents	234
	6.8	Case Studies: Security and Privacy Challenges and Solutions	236
	6.9	Balancing Data Accessibility with Security and Privacy	240
	6.10	Conclusion and Preview of the Next Chapter	242
		References	243

7 Data Governance Policies and Standards: Development and Implementation ... 247
7.1 Introduction to Data Governance Policies and Standards ... 248
7.2 Developing Data Governance Policies ... 251
7.3 Key Elements of Data Governance Standards ... 253
7.4 Aligning Policies with Organizational Goals ... 258
7.5 Best Practices in Policy Development and Standardization ... 261
7.6 Communication and Training ... 262
7.7 Implementing and Enforcing Policies and Standards ... 265
7.8 Managing Change and Policy Evolution ... 268
7.9 Case Studies: Policies and Standards in Action ... 272
7.10 Conclusion ... 275
References ... 277

8 Roles and Responsibilities in Data Governance: Building an Effective Team ... 279
8.1 Introduction to Data Governance Roles and Responsibilities ... 280
8.2 Key Data Governance Roles ... 282
8.3 Establishing a Data Governance Committee ... 284
8.4 Cross-Functional Collaboration in Data Governance ... 288
8.5 Training and Skill Development ... 290
8.6 Building and Sustaining an Effective Data Governance Team ... 292
8.7 Case Studies: Successful Data Governance Teams ... 294
8.8 Overcoming Common Challenges in Team Dynamics ... 297
8.9 Measuring Team Effectiveness ... 299
8.10 Conclusion ... 301
References ... 302

9 Data Governance Tools and Technologies: Navigating the Options ... 305
9.1 Introduction to Data Governance Tools and Technologies ... 308
9.2 Data Quality Tools ... 310
9.3 Data Security and Privacy Tools ... 311
9.4 Metadata Management Tools ... 313
9.5 Data Cataloging and Inventory Tools: Enhancing Data Management ... 314
9.6 Master and Reference Data Management Tools: Empowering Data Consistency ... 316
9.7 Data Integration and ETL Tools: Streamlining Data Movement and Transformation ... 317
9.8 Evaluating and Selecting Data Governance Tools: Making Informed Choices ... 318
9.9 Case Studies: Tools in Practice—Real-World Examples ... 320
9.10 Future Trends in Data Governance Technology: Shaping Tomorrow's Data Governance Landscape ... 322

	9.11 Conclusion: Navigating the Data Governance Journey	323
	References	325
10	**Data Governance in Different Industries: Case Studies and Applications**	**327**
	10.1 Introduction to Data Governance Across Industries	329
	10.2 Data Governance in Healthcare	330
	10.3 Data Governance in Finance and Banking	331
	10.4 Data Governance in Retail and E-Commerce	331
	10.5 Data Governance in Government and Public Sector	332
	10.6 Data Governance in Technology and Telecommunications	333
	10.7 Data Governance in Manufacturing and Logistics	334
	10.8 Emerging Industries and Data Governance	335
	10.9 Lessons Learned and Best Practices	335
	10.10 Conclusion and Industry-Specific Considerations	336
	References	337
11	**Overcoming Challenges in Data Governance: Strategies for Success**	**339**
	11.1 Introduction to Data Governance Challenges	341
	11.2 Challenge of Organizational Culture and Change Management	343
	11.3 Data Quality Issues	344
	11.4 Aligning Data Governance with Business Objectives	346
	11.5 Managing Data Governance in Complex Environments	349
	11.6 Regulatory Compliance and Evolving Legal Requirements	351
	11.7 Technology and Tool Integration Challenges	354
	11.8 Skill Gaps and Training Needs	355
	11.9 Data Security and Privacy Concerns	358
	11.10 Case Studies: Overcoming Data Governance Challenges	360
	11.11 Conclusion	361
	References	362
12	**Future Trends in Data Governance: Preparing for Tomorrow**	**365**
	12.1 Introduction to Future Trends in Data Governance	366
	12.2 The Rise of Artificial Intelligence and Machine Learning	367
	12.3 Increasing Importance of Data Ethics	369
	12.4 Impact of Big Data and IoT	371
	12.5 Cloud Governance	373
	12.6 Evolving Regulatory Landscape	375
	12.7 Enhanced Focus on Data Literacy	377
	12.8 Decentralization and Blockchain in Data Governance	380
	12.9 Predictive Analytics in Data Governance	382
	12.10 Preparing for the Future	384
	12.11 Conclusion: The Evolving Landscape of Data Governance	387
	References	389

13	**Establishing a Data Governance Culture: Change Management and Leadership**	**391**
	13.1 Introduction to Data Governance Culture	392
	13.2 The Role of Leadership in Data Governance	393
	13.3 Change Management Principles in Data Governance	396
	13.4 Communicating the Value of Data Governance	398
	13.5 Building Data Governance into Organizational DNA	401
	13.6 Training and Empowerment	404
	13.7 Developing Data Governance Champions	407
	13.8 Incentivizing and Rewarding Compliance	409
	13.9 Overcoming Cultural Barriers	412
	13.10 Continuous Improvement and Adaptation	414
	13.11 Conclusion	416
	References	417
14	**Measuring the Impact of Data Governance: Metrics and Key Performance Indicators**	**419**
	14.1 Introduction to Measuring Data Governance Impact	420
	14.2 Defining Relevant Metrics and KPIs	421
	14.3 Metrics for Data Quality	423
	14.4 Compliance and Risk Management Metrics	426
	14.5 Metrics for Data Usage and Business Impact	429
	14.6 Tracking and Reporting Mechanisms	431
	14.7 Balancing Quantitative and Qualitative Measures	434
	14.8 Case Studies: Measuring Success in Data Governance	437
	14.9 Continuous Improvement Through Metrics	438
	14.10 Conclusion	441
	14.10.1 Metrics for Data Quality	441
	14.10.2 Compliance and Risk Management Metrics	441
	14.10.3 Metrics for Data Usage and Business Impact	442
	14.10.4 Tracking and Reporting Mechanisms	442
	14.10.5 Balancing Quantitative and Qualitative Measures	442
	14.10.6 Case Studies: Measuring Success in Data Governance	442
	14.10.7 Continuous Improvement through Metrics	442
	References	443
15	**Legal and Regulatory Considerations in Data Governance**	**445**
	15.1 Introduction to Legal and Regulatory Aspects	446
	15.2 Global Data Protection and Privacy Laws	448
	15.3 Navigating Industry-Specific Regulations	449
	15.4 Cross-Border Data Transfer and Compliance	451
	15.5 Developing a Compliance-Oriented Data Governance Framework	454
	15.6 Data Governance in the Context of Legal Discovery and Audits	456

	15.7 Cybersecurity Laws and Regulations.	458
	15.8 Case Studies: Legal and Regulatory Compliance	461
	15.9 Future Legal and Regulatory Trends	463
	15.10 Conclusion	464
	References.	465
16	**Data Stewardship and Ownership: Best Practices**	**467**
	16.1 Introduction to Data Stewardship and Ownership	468
	16.2 Defining Roles and Responsibilities	470
	16.3 Best Practices in Data Stewardship	471
	16.4 Establishing Data Ownership.	472
	16.5 Collaboration Between Stewards and Owners.	474
	16.6 Training and Empowerment.	476
	16.7 Accountability and Performance Measurement.	478
	16.8 Case Studies: Effective Stewardship and Ownership.	480
	16.9 Overcoming Common Challenges.	482
	16.10 Conclusion	484
	References.	485
17	**Data Governance Maturity Models: Assessing and Enhancing Your Program**	**487**
	17.1 Introduction to Data Governance Maturity Models.	488
	17.2 Overview of Common Data Governance Maturity Models.	490
	17.3 Assessing Your Current Maturity Level.	491
	17.4 Developing a Roadmap for Maturity Advancement	493
	17.5 Key Factors Influencing Data Governance Maturity	496
	17.6 Aligning Maturity Improvement with Business Objectives.	500
	17.7 Case Studies: Maturity Model Implementation.	503
	17.8 Challenges in Advancing Maturity	504
	17.9 Continuous Improvement in Data Governance	506
	17.10 Conclusion	508
	References.	509
18	**Conclusion: The Evolving Landscape of Data Governance**	**511**
	18.1 Reflection on the Journey of Data Governance	514
	18.2 The Current State of Data Governance	516
	18.3 Major Takeaways from the Book.	517
	18.4 The Future of Data Governance.	518
	18.5 Preparing for Ongoing Changes.	519
	18.6 The Role of Leadership and Culture	519
	18.7 Final Thoughts on Building a Resilient Data Governance Framework.	520
	18.8 Encouraging a Community of Practice	521
	18.9 Closing Remarks	522
	References.	523
References		**525**

Abbreviations

AI	Artificial Intelligence
CCPA	California Consumer Privacy Act
DLP	Data Loss Prevention
ETL	Extract, Transform, Load
GDPR	General Data Protection Regulation
HIPAA	Health Insurance Portability and Accountability Act
IAM	Identity and Access Management
IoT	Internet of Things
ISO	International Organization for Standardization (e.g., ISO 27001)
MDM	Master Data Management
ML	Machine Learning
PCI DSS	Payment Card Industry Data Security Standard
RBAC	Role-Based Access Control
SOX	Sarbanes-Oxley Act
SQL	Structured Query Language

Glossary

Data governance	The overall management of the availability, usability, integrity, and security of data used in an organization
Data integration	The process of combining data from diverse sources into a single, unified view
Data lifecycle management	The process of managing the flow of data through its lifecycle from creation and initial storage to the time when it becomes obsolete and is deleted
Data privacy	Ensuring that personal information is handled, stored, and used in compliance with privacy laws and standards
Data quality	The measure of data's condition, focusing on accuracy, completeness, reliability, and relevance
Data security	Protecting data from unauthorized access and corruption throughout its lifecycle
GDPR compliance	Adherence to the General Data Protection Regulation, a regulation in EU law on data protection and privacy
Master data management (MDM)	A method that defines and manages the critical data of an organization to provide, with data integration, a single point of reference
Metadata management	The administration of data that describes other data, providing context and aiding in data discovery
Regulatory compliance	Adhering to laws, regulations, guidelines, and specifications relevant to business processes

Annotations

Data privacy	Data privacy concerns the proper handling, processing, and storage of personal information. Organizations must ensure they comply with legal standards and ethical considerations to protect individual privacy rights
Data quality	Data quality is fundamental to reliable decision-making. It involves ensuring accuracy, completeness, and reliability of data. Poor data quality can lead to erroneous conclusions and decisions, impacting organizational performance
Data security	Data security involves protecting data from unauthorized access and breaches. It is a critical aspect of data governance, especially in an era where cyber threats are increasingly sophisticated
Master data management	Master Data Management (MDM) is a method of managing the organization's critical data. It provides a single point of reference to ensure that the organization's data is consistent, accurate, and controlled
Metadata management	Metadata management is the administration of data that describes other data. It is essential for understanding data assets in an organization and plays a key role in data governance by providing context and aiding in data discovery
Regulatory compliance	Compliance with regulations such as GDPR (General Data Protection Regulation) is crucial for any organization handling personal data. Non-compliance can result in hefty fines and damage to reputation

List of Figures

Fig. 1.1	Data governance key components. (Created by the author)	3
Fig. 1.2	Historical evolution and current relevance of data governance. (Created by the author)	5
Fig. 1.3	Key objectives of implementing data governance. (Created by the author)	7
Fig. 1.4	Benefits of implementing data governance. (Created by the author)	9
Fig. 1.5	Annual cost of cybercrime world wide (in trillions). (Created by the author)	19
Fig. 1.6	Average cost of data breach in 2024. (Created by the author)	21
Fig. 1.7	Comparison of maximum fines under GDPR and CCPA. (Created by the author)	23
Fig. 1.8	Public concern over data privacy in the US (2019). (Created by the author)	25
Fig. 1.9	Potential GDPR fines for noncompliance. (Created by the author)	26
Fig. 1.10	Increase in data breaches. (Created by the author)	28
Fig. 1.11	Executive trust in organizations's data and analytics (KPMG 2020a). (Created by the author)	31
Fig. 1.12	Total GDPR fines from 2020 to 2021, as reported by DLA Piper. (Created by the author)	36
Fig. 1.13	Data quality concerns among CEOs. (Created by the author)	42
Fig. 1.14	Data governance framework. (Created by the author)	47
Fig. 1.15	Data stewards map. (Created by the author)	50
Fig. 1.16	Governance bodies mindmap. (Created by the author)	52
Fig. 1.17	Development of policies and standards. (Created by the author)	53
Fig. 1.18	Role of technology in data governance. (Created by the author)	57
Fig. 1.19	Key tools in data governance. (Created by the author)	59
Fig. 1.20	Data governance myths. (Created by the author)	62
Fig. 1.21	Data governance vs data management. (Created by the author)	65
Fig. 1.22	Nature of resistance in data governance. (Created by the author)	66
Fig. 1.23	Strategies to overcome resistance. (Created by the author)	68

Fig. 1.24	Development of tailored solution. (Created by the author)	70
Fig. 1.25	Securing executive leadership support (Created by the author)	72
Fig. 1.26	Strategies for addressing data quality issues. (Created by the author)	73
Fig. 1.27	Assessment and auditing. (Created by the author)	75
Fig. 1.28	Implementation of data quality frameworks. (Created by the author)	77
Fig. 1.29	Aligning data governance with business objectives. (Created by the author)	79
Fig. 2.1	Growth of the Global Data Sphere (2018–2025). (Created by the author)	89
Fig. 2.2	Data-driven-decision-making. (Created by the author)	92
Fig. 2.3	Overcoming challenges. (Created by the author)	94
Fig. 2.4	Impact of organized data on beating revenue goals. (Created by the author)	95
Fig. 2.5	Annual losses due to poor data quality (2016–2021). (Created by the author)	96
Fig. 2.6	Improvement in the decision-making process. (Created by the author)	98
Fig. 2.7	Reduction of strategic errors with data-driven decision-making. (Created by the author)	99
Fig. 2.8	The impact of predictive analytics on operational efficiency. (Created by the author)	100
Fig. 2.9	Data governance as a catalyst for quality data. (Created by the author)	101
Fig. 2.10	The impact of strong data-management capabilities on revenue performance. (Created by the author)	102
Fig. 2.11	The reduction in operational costs achieved through good data governance. (Created by the author)	103
Fig. 2.12	Decision-making efficiency. (Created by the author)	104
Fig. 2.13	The impact of strong data governance on decision-making improvement. (Created by the author)	105
Fig. 2.14	The impact of data-driven initiatives on customer engagement and satisfaction. (Created by the author)	106
Fig. 2.15	The profound impact of leveraging big data on customer acquisition in data-driven organizations. (Created by the author)	107
Fig. 2.16	GDPR noncompliance fines. (Created by the author)	108
Fig. 2.17	The impact of the Health Insurance Portability and Accountability Act (HIPAA) in Europe. (Created by the author)	109
Fig. 2.18	Global data protection trends and the European response. (Created by the author)	110
Fig. 2.19	The role of data governance in ensuring compliance with various data-related regulations. (Created by the author)	112
Fig. 2.20	Strategies for risk mitigation in data governance. (Created by the author)	114

Fig. 2.21	The impact of data privacy regulations on customer trust. (Created by the author)	115
Fig. 2.22	The impact of transparency in data usage on consumer concerns. (Created by the author)	116
Fig. 2.23	Strategies for protecting a brand's reputation, focusing on minimizing risks of data breaches and compliance with data protection regulations. (Created by the author)	118
Fig. 2.24	The challenges in managing customer data. (Created by the author)	119
Fig. 2.25	Facebook and Cambridge Analytica Case Timeline. (Created by the author)	120
Fig. 2.26	Comparison of Financial Impact and Reputation Damage. (Created by the author)	121
Fig. 2.27	Consequences of H&M Employee Privacy Violation (2020). (Created by the author)	122
Fig. 2.28	Consequences of CNIL's fine against Google for GDPR violations (2019). (Created by the author)	123
Fig. 2.29	Consequences of Equifax Data Breach (2017). (Created by the author)	124
Fig. 2.30	Consequences of Marriott International Data Breach (2018). (Created by the author)	125
Fig. 2.31	Building a culture of data governance. (Created by the author)	133
Fig. 3.1	Principle of data quality. (Created by the author)	142
Fig. 3.2	Strategies for continuous data quality assessment and improvement. (Created by the author)	144
Fig. 3.3	Principle of data transparency and accessibility. (Created by the author)	146
Fig. 3.4	Principle of data security. (Created by the author)	148
Fig. 3.5	Principle of compliance and data privacy. (Created by the author)	150
Fig. 3.6	Principle of Data Stewardship. (Created by the author)	152
Fig. 3.7	Principle of data lifecycle management. (Created by the author)	154
Fig. 4.1	Components of a data governance framework. (Created by the author)	168
Fig. 4.2	Importance of data governance frameworks. (Created by the author)	169
Fig. 4.3	DAMA-DMBOK framework. (Created by the author)	173
Fig. 4.4	Designing a data governance framework. (Created by the author)	177
Fig. 4.5	Best practices in framework implementation. (Created by the author)	180
Fig. 4.6	Integrating technology with frameworks. (Created by the author)	182
Fig. 4.7	Measuring the effectiveness of your framework—data governance template. (Created by the author)	184

Fig. 4.8	This template provides a structured approach to setting up a data governance program. It can be customized to suit the specific needs and objectives of your organization. (Created by the author)	187
Fig. 4.9	Adapting frameworks to changing data landscapes map. (Created by the author)	190
Fig. 5.1	Importance of data quality management. (Created by the author)	213
Fig. 6.1	Key concepts in data security. (Created by the author)	223
Fig. 6.2	Data security and privacy strategy. (Created by the author)	228
Fig. 6.3	Implementing effective data security measures map. (Created by the author)	231
Fig. 6.4	Strategies for balancing accessibility, security, and privacy. (Created by the author)	241
Fig. 7.1	Data governance policies and standards framework. (Created by the author)	250
Fig. 7.2	Developing data governance policies. (Created by the author)	252
Fig. 7.3	Key elements of data governance standards. (Created by the author)	257
Fig. 7.4	Aligning policies with organizational goals. (Created by the author)	260
Fig. 7.5	Effective communication and training map. (Created by the author)	264
Fig. 7.6	Implementing and enforcing policies and standards. (Created by the author)	268
Fig. 7.7	Managing change and policy evolution. (Created by the author)	271
Fig. 8.1	Key data governance roles map. (Created by the author)	283
Fig. 8.2	Role and purpose. (Created by the author)	284
Fig. 8.3	Composition of a Data Governance Committee. (Created by the author)	285
Fig. 8.4	Key considerations when establishing a Data Governance Committee. (Created by the author)	286
Fig. 8.5	Benefits establishing a data governance committee. (Created by the author)	287
Fig. 8.6	The necessity of cross-functional collaboration. (Created by the author)	288
Fig. 8.7	Key elements of cross-functional collaboration. (Created by the author)	289
Fig. 8.8	Benefits of cross-functional collaboration. (Created by the author)	290
Fig. 11.1	Overcoming challenges in data governance: Strategies for success. (Created by the author)	342

Fig. 11.2	Data quality issues. (Created by the author)	346
Fig. 11.3	Aligning data governance with business objectives keys. (Created by the author)	348
Fig. 11.4	Managing data governance in complex environments map. (Created by the author)	351
Fig. 11.5	Regulatory compliance and evolving legal requirements map. (Created by the author)	353
Fig. 11.6	Technology and tool integration challenges map. (Created by the author)	355
Fig. 11.7	Skill gaps and training needs map. (Created by the author)	357
Fig. 11.8	Data security and privacy concerns map. (Created by the author)	360
Fig. 12.1	The impact of AI and ML on data governance. (Created by the author)	368
Fig. 12.2	Rising significance of data ethics. (Created by the author)	371
Fig. 12.3	Impact of big data and IoT on data governance. (Created by the author)	372
Fig. 12.4	Key points to understand about cloud governance. (Created by the author)	375
Fig. 12.5	The regulatory landscape. (Created by the author)	377
Fig. 12.6	Enhanced focus on data literacy. (Created by the author)	379
Fig. 12.7	Decentralization and blockchain in data governance map. (Created by the author)	381
Fig. 12.8	Predictive analytics in data governance. (Created by the author)	383
Fig. 12.9	Preparing for the future in data governance, keys map. (Created by the author)	386
Fig. 12.10	The evolving landscape of data governance. (Created by the author)	388
Fig. 13.1	The role of leadership in data governance. (Created by the author)	395
Fig. 13.2	Change management principles in data governance. (Created by the author)	398
Fig. 13.3	Communicating the value of data governance. (Created by the author)	400
Fig. 13.4	Building data governance into organizational DNA. (Created by the author)	403
Fig. 13.5	Training and empowerment keys. (Created by the author)	406
Fig. 13.6	Developing data governance champions map. (Created by the author)	408
Fig. 13.7	Incentivizing and rewarding compliance map. (Created by the author)	411
Fig. 13.8	Overcoming cultural barriers map. (Created by the author)	413
Fig. 13.9	Continuous improvement and adaptation map. (Created by the author)	416

Fig. 14.1	Defining relevant metrics and KPIs map. (Created by the author)	423
Fig. 14.2	Metrics for data quality. (Created by the author)	425
Fig. 14.3	Compliance and risk management metrics map. (Created by the author)	428
Fig. 14.4	Metrics for data usage and business impact. (Created by the author)	431
Fig. 14.5	Tracking and reporting mechanisms map. (Created by the author)	434
Fig. 14.6	Balancing quantitative and qualitative measures map. (Created by the author)	436
Fig. 14.7	Continuous improvement through metrics map. (Created by the author)	440
Fig. 15.1	Introduction to legal and regulatory aspects. (Created by the author)	448
Fig. 15.2	Navigating industry-specific regulations map. (Created by the author)	451
Fig. 15.3	Cross-border data transfer and compliance map. (Created by the author)	453
Fig. 15.4	Data governance in legal discovery and audits map. (Created by the author)	458
Fig. 15.5	Cybersecurity laws and regulations map. (Created by the author)	460
Fig. 16.1	Data stewardship and ownership map. (Created by the author)	470
Fig. 16.2	Best practices in data stewardship. (Created by the author)	472
Fig. 16.3	Establishing data ownership map. (Created by the author)	474
Fig. 16.4	Collaboration between stewards and owners map. (Created by the author)	476
Fig. 16.5	Training and empowerment map. (Created by the author)	478
Fig. 16.6	Accountability and performance measurement map. (Created by the author)	479
Fig. 16.7	Overcoming common challenges map. (Created by the author)	483
Fig. 17.1	Roadmap for maturity advancement. (Created by the author)	495
Fig. 17.2	Key factors influencing data governance maturity. (Created by the author)	499
Fig. 17.3	How to align maturity improvement with business objectives map. (Created by the author)	502
Fig. 17.4	Challenges in advancing maturity. (Created by the author)	505
Fig. 17.5	Continuous improvement in data governance map. (Created by the author)	508
Fig. 18.1	The evolving landscape of data governance. (Created by the author)	514

List of Tables

Table 4.1 Data governance frameworks structure 167
Table 4.2 Simplified representation of the key components related
to the DAMA-DMBOK Framework 171
Table 4.3 Simplified representation of the CMMI Data Management
Maturity (DMM) model. 174
Table 4.4 Structured representation of some key components related
to the ISO 8000 Data Quality and Data Governance Standard 175
Table 6.1 GDPR compliance: Penalties and global impact 224
Table 6.2 Privacy by design and default 232
Table 6.3 HIPAA compliance checklist 237
Table 6.4 Key GDPR compliance obligations 238
Table 6.5 SSL security 239
Table 6.6 Symmetric encryption 240
Table 6.7 Asymmetric encryption 240

Chapter 1
Overview and Importance of Data Governance

Contents

1.1	Overview of Data Governance: Definition and Scope.	2
1.2	The Pillars of Data Governance.	10
1.3	The Importance of Data Governance in Modern Organizations.	30
1.4	Key Components of a Data Governance Program.	45
1.5	Common Misconceptions About Data Governance.	60
1.6	Challenges in Implementing Data Governance.	65
1.7	Success Stories: Data Governance in Action.	79
1.8	Conclusion and Preview of the Next Chapter.	80
References.		81

Abstract This chapter provides a comprehensive introduction to the concept of data governance, defining it as a critical organizational function that involves overseeing the availability, usability, integrity, and security of the data employed in an organization. The chapter emphasizes that data governance is not merely about data management; rather, it integrates data quality, data management, policies, risk management, and compliance with regulations to ensure that data serves the strategic needs of the organization effectively.

The chapter outlines the key components and objectives of implementing data governance, including ensuring data quality and reliability, enhancing decision-making capabilities, and achieving regulatory compliance and risk management. It also discusses the historical evolution of data governance, tracing its development from basic data storage and maintenance to a strategic tool essential for maintaining a competitive edge in the digital era, particularly under the stringent regulatory landscapes such as the GDPR in Europe.

Furthermore, the chapter underscores the benefits of robust data governance such as improved data quality, compliance with data protection laws, enhanced operational efficiency, and better decision-making. It concludes by illustrating these points through several figures that visually represent the key concepts, objectives, and benefits of data governance, thereby setting the stage for deeper exploration in subsequent chapters of the document.

Keywords Data governance · Data quality · Regulatory compliance · Data security · Data privacy · Data management · Risk management · GDPR · Decision-making · Strategic planning

1.1 Overview of Data Governance: Definition and Scope

In the realm of contemporary organizational management, the concept of data governance has emerged as a cornerstone for ensuring the efficient and effective utilization of information. Data governance, in its broadest sense, encompasses a set of processes, policies, standards, and metrics that ensure the effective and efficient use of information in enabling an organization to achieve its goals. This definition captures the essence of data governance as not merely a technical endeavor but a comprehensive organizational commitment.

The scope of data governance extends beyond the mere management of data; it is an integral aspect of corporate governance that demands a strategic alignment with an organization's objectives. It involves a comprehensive approach to managing data's accuracy, accessibility, consistency, and security. This multifaceted discipline blends various facets, including data quality, data management, data policies, risk management, and compliance with regulations, particularly within the European context where stringent data protection laws, such as the General Data Protection Regulation (GDPR), play a pivotal role (IBM 2024; Coyle et al. 2020).

In Europe, data governance takes on added significance due to the legal and ethical obligations imposed by these regulations. Organizations must not only safeguard personal data but also ensure its ethical use, aligning data practices with the broader values of transparency, accountability, and respect for individual privacy. Therefore, data governance in Europe is not just a matter of compliance but also a strategic initiative that enhances trust and reliability in the digital ecosystem (World Economic Forum 2021; United Nations High Commissioner for Human Rights 2018).

Moreover, the dynamic nature of the digital landscape necessitates that data governance frameworks remain agile and responsive to evolving technological advancements and regulatory changes. This adaptability is crucial for organizations to maintain a competitive edge and ensure sustainable growth (IBM 2024; Ostrom 2012).

Figure 1.1 provides a visual representation of the distribution of key components in data governance. In this illustration, each slice of the pie chart is of equal size, symbolizing that data quality, data management, data policies, risk management, and compliance are all equally important in the overall framework of data governance. This balanced approach highlights the comprehensive nature of data

1.1 Overview of Data Governance: Definition and Scope

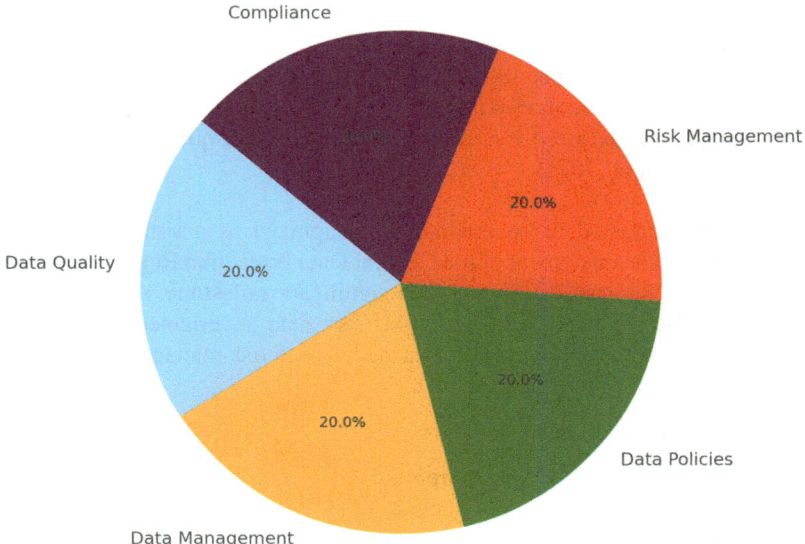

Fig. 1.1 Data governance key components. (Created by the author)

governance, where each aspect plays a vital role in ensuring effective and efficient data use within an organization, particularly in the context of European standards and regulations.

Data governance, as a formalized discipline, has evolved significantly over the past few decades, paralleling advancements in data generation, storage, and analytics. The historical evolution and current relevance of data governance can be traced through several key phases:

Early Beginnings (Late Twentieth Century) In the early days of computing, data management was a technical concern, focused on data storage and maintenance. The concept of data governance was not yet formalized, but the seeds were planted as organizations began to recognize the importance of data accuracy and security (Tallon et al. 2013).

Emergence of Data Warehousing (1990s) With the advent of data warehousing in the 1990s, data began to be viewed as a valuable resource for business intelligence. This era marked the beginning of systematic data management practices, although these were still predominantly IT-driven (Otto 2011).

Regulatory Compliance (Early 2000s) The early 2000s saw a surge in regulatory requirements, such as the Sarbanes-Oxley Act (2002) in the United States, which put data governance into sharper focus. Organizations started to realize the need for formal governance to ensure compliance and manage risks associated with data (Tallon et al. 2013).

The Big Data Era (2010s) The explosion of big data, fueled by the advent of social media, IoT, and other digital technologies, brought new challenges and opportunities. Data governance expanded to include aspects like data privacy, ethical use of data, and the need for more sophisticated data quality and integration techniques (Haug et al. 2011).

Current State Today, data governance is recognized as a strategic imperative across industries. The enactment of the General Data Protection Regulation (GDPR) in the European Union in 2018 marked a significant milestone, underscoring the importance of data privacy and security. Modern data governance encompasses a broader spectrum—from compliance, data quality, and risk management to leveraging data for strategic advantage (IBM 2024).

Currently, data governance is more relevant than ever due to several factors:

- **Increased Data Volume and Complexity:** The exponential growth in both the volume and complexity of data generated today necessitates a comprehensive data governance framework. Managing and utilizing such vast and intricate datasets effectively requires governance measures that ensure their proper use and management (IBM 2024).
- **Regulatory Requirements:** Global regulatory pressures, particularly prominent in data-sensitive sectors and regions like Europe with the General Data Protection Regulation (GDPR), have elevated the imperative for strong data governance. Compliance with these regulations mandates a stringent approach to data governance, ensuring that data practices align with legal requirements (IBM 2024; Coyle et al. 2020).
- **Technological Advancements:** The advent of artificial intelligence (AI), machine learning, and advanced analytics has amplified the need for well-governed data. To ensure the development and deployment of accurate and ethical AI models and decisions, data governance plays a pivotal role in maintaining data quality, integrity, and transparency (IBM 2024).
- **Cybersecurity Concerns:** The escalating frequency and sophistication of cybersecurity threats underline the criticality of data governance. It becomes essential for safeguarding sensitive information from potential breaches and unauthorized access. Effective data governance forms a strong defense against cyber threats, promoting data security and confidentiality (IBM 2024; Coyle et al. 2020).

Data governance has transitioned from being a technical IT concern to a strategic organizational priority, integral to operational integrity, regulatory compliance, and strategic decision-making. Its evolution reflects the growing recognition of data as a critical asset that must be managed with care and foresight.

Figure 1.2 visually represents the historical evolution and current relevance of data governance through different eras. Each bar corresponds to a specific era in the

1.1 Overview of Data Governance: Definition and Scope

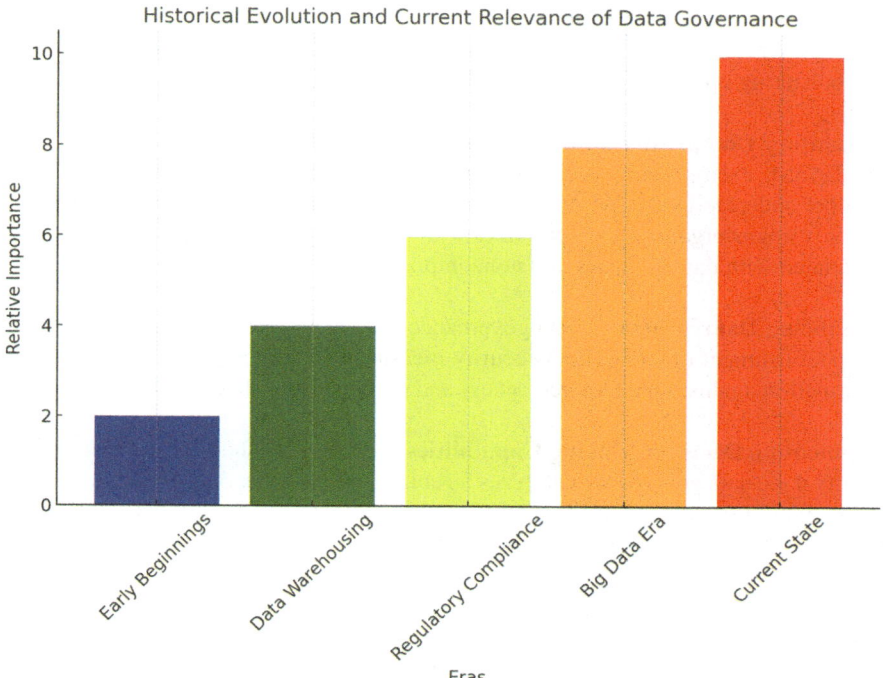

Fig. 1.2 Historical evolution and current relevance of data governance. (Created by the author)

development of data governance, with the height of the bar indicating its relative importance or prominence during that time:

- Early Beginnings: In the late twentieth century, data governance was in its nascent stages, reflected by a shorter bar.
- Data Warehousing Era: In the 1990s, as data warehousing became prevalent, the importance of data governance began to rise.
- Regulatory Compliance Era: The early 2000s saw a significant increase in the importance of data governance, driven by regulatory demands.
- Big Data Era: The 2010s, marked by the explosion of big data, saw a further increase in the significance of data governance.
- Current State: Today, data governance is at its peak importance, driven by factors like GDPR, technological advancements, and increased data volume and complexity.
- **Key Objectives of Implementing Data Governance**

In the realm of data governance, several pivotal objectives emerge, each bearing substantial academic and practical significance:

Ensuring Data Quality and Reliability A cornerstone goal of data governance resides in the maintenance of elevated data quality standards. This encompasses the meticulous oversight of data accuracy, completeness, consistency, and reliability, all of which serve as indispensable components facilitating informed decision-making and bolstering operational efficiency (Haug et al. 2011).

Regulatory Compliance and Risk Management The evolving landscape of data regulations, exemplified by the General Data Protection Regulation (GDPR) in Europe, underscores the indispensable role of data governance. Its implementation not only aligns organizational practices with legal standards but also mitigates risks associated with data breaches and noncompliance (Tallon et al. 2013).

Enhancing Data Security Data governance frameworks serve as vital enablers in the establishment of robust data security measures, thereby safeguarding sensitive information against unauthorized access and potential breaches (Otto 2011).

Augmenting Decision-Making Capabilities By ensuring the availability of high-quality data, data governance catalyzes an enhancement in the decision-making processes within organizations. The provision of accurate and timely data influences strategic planning and operational decisions (Weber et al. 2009).

Streamlining Data Management Data governance further optimizes data management by instituting clear policies, procedures, and standards for data usage and handling. This systematic approach leads to heightened operational efficiency and diminished redundancy within the organizational data ecosystem (Khatri and Brown 2010).

Figure 1.3 visually represents the key objectives of implementing data governance, Each point on the line corresponds to a specific data governance objective, with the following significance:

- **Data Quality:** Demonstrates the emphasis on ensuring the accuracy and reliability of data.
- **Compliance:** Highlights the importance of adhering to legal and regulatory requirements, particularly pertinent in the context of the GDPR in Europe.
- **Data Security:** Stresses the need for strong security measures to protect data.
- **Decision-Making:** Indicates the role of data governance in enhancing informed decision-making processes.
- **Data Management:** Points to the importance of efficient and effective management of data within organizations.

The height of each point reflects the relative importance of these objectives in a comprehensive data governance strategy, underlining their crucial roles in achieving organizational goals and regulatory compliance, especially within the European context.

1.1 Overview of Data Governance: Definition and Scope

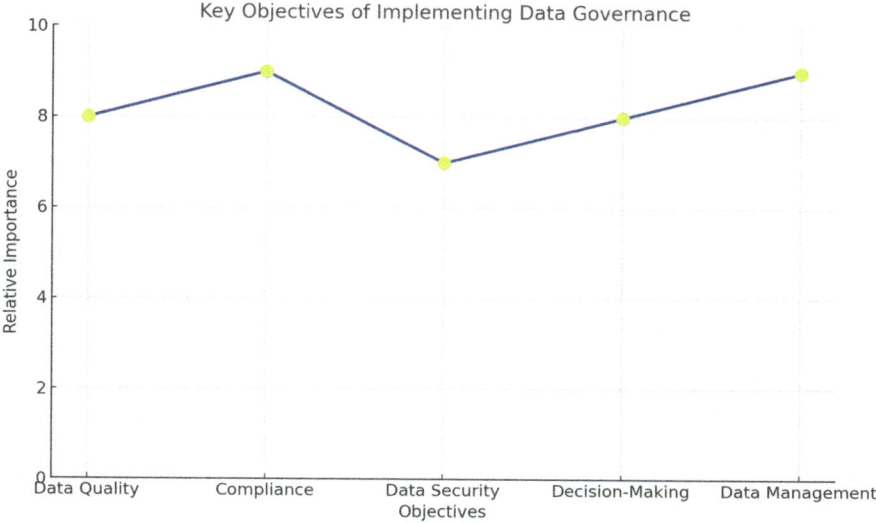

Fig. 1.3 Key objectives of implementing data governance. (Created by the author)

- **Benefits of Implementing Data Governance**

Implementing data governance offers numerous benefits to organizations, contributing to more efficient operations, improved data quality, and enhanced decision-making capabilities. Here are some key advantages:

Improved Data Quality Data governance ensures that data is accurate, consistent, and reliable. By establishing clear policies and procedures for data management, organizations can minimize errors and discrepancies, leading to more trustworthy data. This aligns with the findings of Haug et al. (2011), who emphasize that maintaining high data quality is crucial for effective decision-making and operational efficiency.

Enhanced Compliance and Reduced Risk With increasing regulations around data privacy and protection, such as GDPR and CCPA, data governance helps organizations comply with legal and regulatory requirements, thereby reducing the risk of costly penalties and legal issues. Proper data governance ensures adherence to these regulations, mitigating potential risks associated with data breaches and non-compliance (McKinsey 2024).

Increased Efficiency and Productivity Standardized data management processes streamline data handling across the organization. This leads to more efficient operations, as employees spend less time dealing with data-related issues and more time on productive tasks. A well-implemented data governance framework can significantly enhance operational productivity by reducing time spent on data corrections and improving data accessibility (Hassani and MacFeely 2023).

Better Decision-Making High-quality, well-governed data is a critical input for decision-making. With reliable data, leaders can make more informed, evidence-based decisions, leading to better business outcomes. Effective data governance supports accurate and timely decision-making by ensuring data integrity and availability (McKinsey 2024).

Enhanced Data Security Data governance includes policies and procedures to secure sensitive data, protecting it from unauthorized access and breaches. This is critical in an era where data breaches can have significant financial and reputational consequences. Implementing strong data governance measures helps safeguard organizational data, thereby enhancing security and confidentiality (IBM 2024).

Increased Revenue Opportunities Proper data governance can unlock the potential of data as an asset. By effectively managing and analyzing data, organizations can identify new market trends, customer insights, and other opportunities that can lead to increased revenue. Data governance facilitates better data utilization, driving revenue growth through improved insights and market responsiveness (McKinsey 2024).

Improved Data Literacy and Culture Implementing data governance can foster a culture of data literacy within the organization. As employees become more familiar with data policies and best practices, they become more competent in using data effectively. This cultural shift towards data literacy can enhance overall organizational efficiency and innovation (Hassani and MacFeely 2023).

Scalability and Flexibility Effective data governance allows organizations to scale their data infrastructure and processes efficiently as they grow. This scalability ensures that data management remains robust and responsive to changing business needs. Data governance frameworks support organizational growth by enabling scalable and adaptable data management practices (McKinsey 2024).

Facilitates Data Integration and Interoperability Data governance makes it easier to integrate data from various sources and systems, facilitating interoperability and ensuring that data across the organization is compatible and usable. This interoperability is crucial for comprehensive data analysis and strategic decision-making (Hassani and MacFeely 2023).

Supports Innovation By providing a structured framework for managing data, data governance enables organizations to safely explore innovative uses of their data, such as advanced analytics, artificial intelligence, and machine learning applications. Data governance supports innovation by ensuring that data used in these advanced applications is of high quality and well-managed (IBM 2024).

1.1 Overview of Data Governance: Definition and Scope

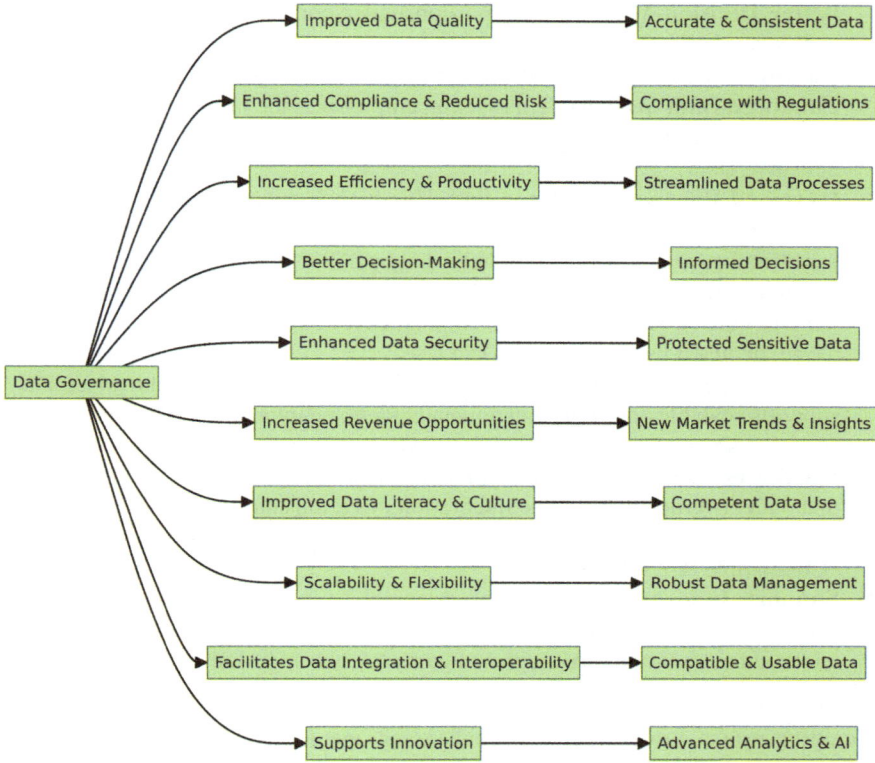

Fig. 1.4 Benefits of implementing data governance. (Created by the author)

Data governance is not just about controlling and managing data; it is about enabling the organization to leverage data as a strategic asset, driving improvements in operations, compliance, decision-making, and overall business performance.

Figure 1.4 visually represents the benefits of implementing data governance in organizations, structured as a hierarchical set of nodes connected by arrows, originating from the central concept of "Data Governance."

1. **Improved Data Quality**: This branch highlights that data governance ensures accuracy, consistency, and reliability in data. Sub-nodes specify that it leads to "Accurate & Consistent Data."
2. **Enhanced Compliance and Reduced Risk**: This sector illustrates how data governance aids in adhering to legal and regulatory requirements, such as GDPR and CCPA, minimizing the risk of penalties and legal issues. It points to "Compliance with Regulations."
3. **Increased Efficiency and Productivity**: Here, the focus is on standardized data management processes leading to streamlined operations and less time spent on data issues, as indicated by "Streamlined Data Processes."

4. **Better Decision-Making**: This portion underscores the role of high-quality, well-governed data in informed decision-making, leading to improved business outcomes, symbolized by "Informed Decisions."
5. **Enhanced Data Security**: This section emphasizes the importance of securing sensitive data to prevent unauthorized access and breaches, leading to "Protected Sensitive Data."
6. **Increased Revenue Opportunities**: This branch suggests that effective data management can unlock data's potential as an asset, leading to new market trends and customer insights, as indicated by "New Market Trends & Insights."
7. **Improved Data Literacy and Culture**: It highlights how data governance fosters a culture of data literacy, with employees becoming more competent in data usage, depicted by "Competent Data Use."
8. **Scalability and Flexibility**: This part shows that effective data governance supports scalable and robust data management, adaptable to changing business needs, as shown by "Robust Data Management."
9. **Facilitates Data Integration and Interoperability**: This segment indicates that data governance eases the integration of various data sources and systems, ensuring compatibility and usability of data across the organization, leading to "Compatible & Usable Data."
10. **Supports Innovation**: The final branch emphasizes that a structured data management framework enables safe exploration of innovative data uses like advanced analytics and AI, represented by "Advanced Analytics & AI."

The diagram encapsulates how data governance is not just about managing data but leveraging it as a strategic asset for enhancing operations, compliance, decision-making, and overall business performance.

1.2 The Pillars of Data Governance

1.2.1 Data Quality: Ensuring Accuracy, Completeness, and Reliability

Data quality stands as a cornerstone in leveraging the full potential of data across various domains, including business, science, and technology. High-quality data drives insightful decision-making, fosters innovation, and ensures efficient operations. This section delves deeper into the essential aspects of data quality: accuracy, completeness, and reliability, highlighting their importance and outlining strategies for improvement.

1. **Accuracy**

    Definition: Accuracy refers to the degree to which data correctly describes the "real-world" conditions or objects it is intended to represent. Accurate data correctly reflects the values it is meant to portray (Wang and Strong 1996).

1.2 The Pillars of Data Governance

Importance: Inaccurate data can lead to erroneous decisions and misinformed strategies and can be costly for organizations. For instance, in healthcare, inaccurate data can result in incorrect treatments (Pipino et al. 2002).

Ensuring Accuracy: To ensure accuracy, data must be sourced reliably and verified. This might involve cross-referencing data with multiple sources, using data validation techniques, and constantly updating data to reflect any changes (Batini and Scannapieco 2016).

Case Study: The study "A Data Quality in Use Model for Big Data" by Merino et al. (2016) explores the conceptual and technological aspects of data quality in big data, emphasizing the importance of accuracy in data representation.

2. **Completeness**

    Definition: Completeness refers to the extent to which all required data is available. Complete data sets are those that are not missing any necessary component (Pipino et al. 2002).

    Importance: Incomplete data can lead to biased or incomplete analyses, resulting in partial insights and potentially flawed conclusions. For example, in market research, missing responses can skew analysis results (Batini and Scannapieco 2016).

    Ensuring Completeness: Ensuring data completeness involves identifying and addressing gaps in data. This might include implementing stricter data collection processes, setting mandatory fields in data entry forms, and using data imputation techniques where necessary (Wang and Strong 1996).

    Case Study: The book *Foundations of Data Quality Management* by Fan and Geerts (2022) provides a comprehensive overview of data quality management, including aspects like data accuracy and information completeness.

3. **Reliability**

    Definition: Reliability pertains to the consistency of data over time. Reliable data produces consistent results when measured under similar conditions (Pipino et al. 2002).

    Importance: Unreliable data can undermine the trust in the data's ability to support decision-making. Inconsistent data can lead to varying conclusions and hinder the ability to track progress or changes accurately (Batini and Scannapieco 2016).

    Ensuring Reliability: To enhance reliability, it is important to have standardized data collection processes, employ robust data storage and management systems, and conduct regular quality checks (Wang and Strong 1996).

    Case Study: The survey "Data Quality in Internet of Things: A State-of-the-Art Survey" by Karkouch et al. (2016) discusses data quality issues in the context of the Internet of Things, focusing on the completeness, accuracy, and reliability of data.

- **Challenges in Ensuring Data Quality**

Volume of Data With the exponential increase in data generation, managing the quality of vast datasets becomes challenging (Cai and Zhu 2015).

Diverse Data Sources Data coming from various sources may have different standards and formats, making quality management difficult (Batini and Scannapieco 2016).

Evolving Data As data changes over time, maintaining its quality requires continuous effort (Pipino et al. 2002).

Case Study The paper "The Challenges of Data Quality and Data Quality Assessment in the Big Data Era" by Cai and Zhu (2015) discusses the challenges in maintaining data quality in the era of big data, including issues related to reliability, consistency, and completeness.

- **Best Practices**

Implement Data Governance Establish clear policies and standards for data quality (Wang and Strong 1996).

Use Data Quality Tools Employ tools that automate the process of data cleaning, validation, and monitoring (Batini and Scannapieco 2016).

Regular Audits Conduct regular data quality audits to identify and rectify quality issues (Pipino et al. 2002).

Training and Awareness Ensure that staff are trained in the importance of data quality and the practices to maintain it (Fan and Geerts 2022).

Case Study The article "Data Quality Considerations for Big Data and Machine Learning: Going Beyond Data Cleaning and Transformations" by Gudivada et al. (2017) provides insights into data quality management beyond basic cleaning and transformations, emphasizing the significance of accuracy, completeness, and reliability.

Data quality is not just about having data; it is about having data that is accurate, complete, and reliable. Quality data is the foundation upon which sound decisions are made, strategies are built, and success is achieved. By focusing on these aspects of data quality, organizations can ensure that their data assets are robust and valuable.

- **Why Data Quality Matters?**

Data quality is a crucial element in every aspect of decision-making, analytics, and operations in modern organizations. Here is why it matters so much:

1.2 The Pillars of Data Governance

1. **Informed Decision-Making**
   - **Insight Accuracy**: High-quality data provides accurate insights, which are essential for making informed decisions. Accurate data helps in predicting market trends, customer behavior, and business outcomes more reliably (Wang and Strong 1996).
   - **Risk Reduction**: Reliable data reduces the risks associated with business decisions. For instance, financial institutions rely on accurate data for credit scoring and risk assessment (Pipino et al. 2002).

2. **Efficiency and Productivity**
   - **Process Optimization**: Quality data enables organizations to streamline their processes, thereby increasing efficiency and productivity. Inaccurate or incomplete data can lead to time-consuming rectifications and delays (Batini and Scannapieco 2016).
   - **Automation and Machine Learning**: For automation systems and machine learning models to function effectively, they require high-quality data. Poor data can lead to erroneous outputs and decreased effectiveness of AI systems (Gudivada et al. 2017).

3. **Customer Satisfaction**
   - **Personalization**: High-quality data allows for better customer insights, enabling more personalized and targeted marketing and improving customer experience and satisfaction (Fan and Geerts 2022).
   - **Trust and Reputation**: Accurate data builds customer trust. Inaccurate data can lead to customer dissatisfaction and harm the company's reputation (Pipino et al. 2002).

4. **Regulatory Compliance and Legal Requirements**
   - **Compliance with Standards**: Many industries have strict data quality standards that organizations must comply with. For instance, healthcare and financial services have stringent regulations regarding data quality and privacy (Batini and Scannapieco 2016).
   - **Legal Implications**: Inaccuracies in data can lead to legal consequences, especially if they lead to false claims or breaches of contracts (Wang and Strong 1996).

5. **Financial Health**
   - **Cost Reduction**: Good data quality can significantly reduce costs associated with errors, such as incorrect billing or inefficient resource allocation (Pipino et al. 2002).
   - **Revenue Growth**: Accurate data can identify new opportunities and areas for growth, contributing to increased revenue (Batini and Scannapieco 2016).

6. **Strategic Planning**

   - **Market Analysis**: For effective market analysis and research, high-quality data is essential. It ensures that the insights derived are relevant and accurate (Wang and Strong 1996).
   - **Long-Term Planning**: Strategic planning relies on historical data. If this data is of inferior quality, it can lead to flawed long-term plans and strategies (Fan and Geerts 2022).

7. **Competitive Advantage**

   - **Staying Ahead**: Organizations with high-quality data can gain a competitive advantage by reacting faster and more accurately to market changes and customer needs (Gudivada et al. 2017).
   - **Innovation**: Quality data can be a source of innovation, helping to identify new business models, products, or services (Batini and Scannapieco 2016).

The importance of data quality cannot be overstated. It is the bedrock upon which companies build their strategies, operations, and future growth. Investing in data quality is not just about avoiding the negatives of bad data; it is about leveraging data as a strategic asset to drive growth, efficiency, and competitive advantage.

Overcoming these challenges requires a comprehensive approach, including investing in the right technology, training personnel, establishing robust data governance frameworks, and continually monitoring and updating data management practices. By addressing these challenges, organizations can enhance the quality of their data and utilize it more effectively for decision-making and strategic planning.

- **Challenges in Maintaining Data Quality**

The preservation of data quality encounters a range of formidable challenges, each demanding meticulous attention:

- **Volume and Velocity of Data**: The exponential surge in data, both in terms of volume and velocity, presents substantial hurdles in preserving data quality. As data complexity escalates, the likelihood of errors and inconsistencies rises accordingly (Cai and Zhu 2015).
- **Diverse Data Sources and Formats**: Organizations routinely grapple with data originating from diverse sources and arriving in various formats. This diversity complicates the task of ensuring consistency and accuracy across disparate datasets (Batini and Scannapieco 2016).
- **Lack of Awareness and Ownership**: Frequently, organizations face issues related to a lack of clear ownership or responsibility for data quality. This gap in accountability can lead to the neglect and deterioration of data standards over time, compounding the challenges of maintaining data quality (Pipino et al. 2002).

Maintaining high data quality is essential for organizations but comes with several challenges. These challenges can vary in complexity and impact, depending on the nature of the data, the systems used, and the organizational context. Here are key challenges in maintaining data quality:

1.2 The Pillars of Data Governance

- **Volume of Data**: With the exponential growth in data generation, managing large volumes of data can be overwhelming. Ensuring quality in such vast datasets requires significant resources and sophisticated management strategies (Cai and Zhu 2015).
- **Diverse Data Sources**: Data often comes from various sources, each with different formats and standards. Integrating this data while maintaining its quality is a complex task. The inconsistency between these sources can lead to issues with data compatibility and quality (Batini and Scannapieco 2016).
- **Data Complexity**: The complexity of data, including structured, unstructured, and semi-structured data, adds another layer of difficulty. Each type requires different approaches for quality maintenance, adding to the overall challenge (Fan and Geerts 2022).
- **Data Decay**: Over time, data can become outdated or irrelevant (data decay). Maintaining its relevance and accuracy is a continuous process, requiring regular updates and validation (Pipino et al. 2002).
- **Human Error**: Data entry errors, misinterpretations, and incorrect data tagging are common issues. Human error can introduce inaccuracies in data, affecting its overall quality (Wang and Strong 1996).
- **Technology Limitations**: Sometimes, the technology used for data collection, storage, and analysis may have limitations, leading to data quality issues. This includes outdated systems, lack of integration capabilities, or insufficient data validation tools (Gudivada et al. 2017).
- **Lack of Standardization**: Without standard protocols for data entry, storage, and management, maintaining data quality becomes challenging. Standardization is key to ensuring consistency and reliability in data (Batini and Scannapieco 2016).
- **Compliance and Regulatory Challenges**: Adhering to various data protection and privacy regulations, like GDPR or HIPAA, while maintaining data quality, adds to the complexity. Regulatory requirements can dictate how data should be handled and processed (Wang and Strong 1996).
- **Resource Constraints**: Allocating sufficient resources, including budget, time, and skilled personnel, for data quality initiatives can be challenging, especially for smaller organizations (Pipino et al. 2002).
- **Data Security**: Ensuring data quality also involves protecting data from unauthorized access and breaches. Cybersecurity threats can compromise the integrity of data (Batini and Scannapieco 2016).
- **Integration of New Data Types**: As new types of data emerge, such as IoT data or big data, integrating them into existing systems while maintaining quality is a significant challenge (Cai and Zhu 2015).
- **Change Management**: Implementing new processes or technologies for data quality management requires effective change management strategies to ensure smooth transition and acceptance (Fan and Geerts 2022).
- **Strategies for Ensuring Data Quality**

  To safeguard data quality, organizations must employ a range of strategic measures, each playing a vital role:

- **Implementing Data Quality Frameworks**: Formal data quality frameworks are foundational. They incorporate well-defined metrics and standards and encompass procedures for data cleansing, validation, and regular audits (Batini and Scannapieco 2016).
- **Leveraging Technology**: The utilization of data quality tools and software automates critical aspects of data quality management. This includes error detection, data cleansing, and reconciliation, streamlining the maintenance of data integrity (Gudivada et al. 2017).
- **Fostering a Data-Quality Culture**: Cultivating an organizational culture that deeply comprehends and values data quality is paramount. This entails investing in training programs, awareness initiatives, and the assignment of clear roles and responsibilities for data stewardship (Fan and Geerts 2022).
- **Continuous Monitoring and Improvement**: Data quality should remain a perpetual concern. Establishing continuous monitoring and improvement processes is essential. This includes routine assessments of data quality metrics and the establishment of feedback mechanisms to swiftly identify and rectify issues (Wang and Strong 1996).

Data quality stands as a fundamental cornerstone within the domain of data governance, underpinning the integrity and utility of data in any organizational context. By recognizing its paramount significance, addressing associated challenges, and implementing these strategic measures, organizations can ensure the accuracy, completeness, and reliability of their data. Such assurances are indispensable for operational efficacy and informed strategic decision-making.

1.2.2 Data Security: Protecting Data from Unauthorized Access and Breaches

Data security, a critical pillar of Data Governance, involves protecting data from unauthorized access and breaches, ensuring confidentiality, integrity, and availability of data. This section examines the significance of data security, the challenges it faces, and effective strategies to safeguard data.

- **The Significance of Data Security**

The significance of data security is a crucial aspect in various fields, especially in the current digital age where data breaches and cyber threats are prevalent. Different studies and papers highlight the importance of data security in their respective contexts:

Medical Data Security A study by Liu et al. (2022) emphasizes the importance of data security in medical data, introducing federated learning and neural architecture search as solutions to protect data security while maintaining high accuracy in medical data analysis.

1.2 The Pillars of Data Governance

Cloud Computing Wei and Zhang (2018) point out the significance of data security in cloud computing environments, highlighting the challenges in effectively protecting customer privacy data and the need for comprehensive data security systems.

Steganography in Data Security Venkatraman et al. (2004) discuss the significance of steganographic techniques in ensuring data security, particularly focusing on image data.

Internet of Things (IoT) The importance of data protection in the context of IoT is highlighted by Dankan Gowda et al. (2023), who point out the challenges and importance of ensuring privacy and security in IoT applications.

Industrial Internet of Things (IIoT) Shi et al. (2023) emphasize the significance of data security in IIoT networks and devices, proposing a traffic detection model to identify abnormal network traffic, which is crucial for security in IIoT scenarios.

Blockchain Technology Chand et al. (2023) highlight the role of blockchain technology in ensuring the safety of digital currencies, providing a secure method for storing information and executing transactions, thereby preventing hacking, data theft, and information loss.

Big Data Environment Sharma et al. discuss the importance of balancing security and accessibility of information in a big data environment, identifying key attributes like confidentiality, integrity, availability, and severity as significant in determining overall security during the big data security life cycle process.

Rising Cybersecurity Threats Rising cybersecurity threats are a growing concern in the digital age, characterized by an increase in both the complexity and frequency of cyber attacks. These threats encompass a range of malicious activities aimed at damaging or stealing data, disrupting digital life, and infiltrating secure systems. Key aspects of these rising threats include:

1. **Sophistication of Attacks**: Cyber attacks are becoming more sophisticated, employing advanced techniques that can bypass traditional security measures. This includes the use of AI and machine learning by attackers to enhance the effectiveness of their strategies (Ahsan et al. 2022).
2. **Ransomware**: One of the most prominent threats in recent years is ransomware, a type of malware that encrypts a victim's files and demands payment for their release. The impact of ransomware attacks can be devastating for both individuals and organizations (Carlton and Levy 2017).
3. **Phishing and Social Engineering**: These attacks deceive individuals into providing sensitive information or accessing malicious websites. Phishing campaigns have become more targeted and believable, making them harder to identify and avoid (Papathanasiou et al. 2023).

4. **IoT and Smart Device Vulnerabilities**: The proliferation of Internet of Things (IoT) devices has introduced new vulnerabilities. Many of these devices lack robust security measures, making them easy targets for cybercriminals (Tweneboah-Koduah et al. 2022).
5. **State-Sponsored Cyber Attacks**: There is an increasing trend of state-sponsored cyber attacks aimed at espionage, disruption, or influencing political outcomes. These attacks are often well-funded and sophisticated (Górka 2022).
6. **Data Breaches**: Data breaches continue to rise, with attackers targeting personal, financial, and corporate data. These breaches can lead to significant economic loss and damage to reputation (Offner et al. 2020).
7. **Supply Chain Attacks**: Cybercriminals are targeting supply chains, exploiting vulnerabilities in third-party services and software to compromise multiple targets through a single point of attack (Sharma et al. 2023).
8. **Insider Threats**: Risks from within an organization, whether unintentional or malicious, remain a significant threat. These insider threats can lead to substantial data loss or system compromise (Carlton and Levy 2017).
9. **Evolving Regulatory Landscape**: As cyber threats rise, so does the regulatory landscape. Organizations are required to comply with various cybersecurity laws and regulations, adding complexity to cybersecurity management (Górka 2022).
10. **Machine Learning in Cybersecurity**: Ahsan et al. (2022) discuss the increasing role of machine learning in cybersecurity, noting its effectiveness in mitigating cyber-attacks using techniques like deep learning, support vector machines, and Bayesian classification.
11. **Aviation Safety**: Stastny and Stoica (2022) analyze cybersecurity threats to aviation safety, particularly in Air Traffic Management (ATM), and explore integrating Security Management Systems with Safety Management Systems to enhance safety measures.
12. **Visegrad Group's Cybersecurity Policy**: Marek Górka (2022) examines the actions of the Visegrad Group countries in building common cyber resilience and combating cybersecurity threats through cooperation with the EU and NATO.
13. **Healthcare Sector Threats**: Offner et al. (2020) highlight the increasing cybersecurity threats in the healthcare sector, especially with the introduction of universal electronic health records, and suggest improvements in cybersecurity culture within healthcare organizations.
14. **Smart Metering Systems**: Tweneboah-Koduah et al. discuss the cybersecurity threats to smart metering systems, presenting countermeasure techniques to protect against these attacks.
15. **Advanced Persistent Threats (APTs)**: Carlton and Levy (2017) emphasize the importance of cybersecurity skills in mitigating APTs and other cyber threats, highlighting social engineering and business email compromise as common penetration vectors.

1.2 The Pillars of Data Governance

16. **AI in Cybersecurity**: Sarker (2023) provides an overview of AI-based modeling and adversarial learning for cybersecurity, addressing diverse cyber threats like malware, phishing, and cyberbullying.
17. **Autonomous Vehicles**: Girdhar et al. discuss the serious cybersecurity issues in autonomous vehicles due to AI integration and the enhanced attack surface this creates.
18. **Electric Vehicle Charging Systems**: Hamdare et al. (2023) analyze the cybersecurity risks of Electric Vehicle Charging Systems (EVCS), highlighting potential research issues in EV cyber research.
19. **Email Compromise Attacks**: Papathanasiou et al. (2023) review Business Email Compromise (BEC) attacks, their impacts on enterprises, and offer recommendations for enhancing cybersecurity.

The global annual cost of cybercrime is expected to reach around $8 trillion in 2023. Figure 1.5 represents a significant fiscal impact, demonstrating the growing intensity and frequency of cyber-attacks worldwide. Additionally, it is predicted that the cost of crypto crime alone could amount to $30 billion annually by 2025, with phishing attacks in the technology sector accounting for a substantial portion of reported cybercrimes.

Figure 1.5 presents a compelling visualization of the escalating financial impact of cybercrime across the globe. Displayed in trillions of dollars, the graph illustrates a rising trend in losses attributed to cyber-attacks, underscoring the growing economic burden these criminal activities impose on businesses, governments, and

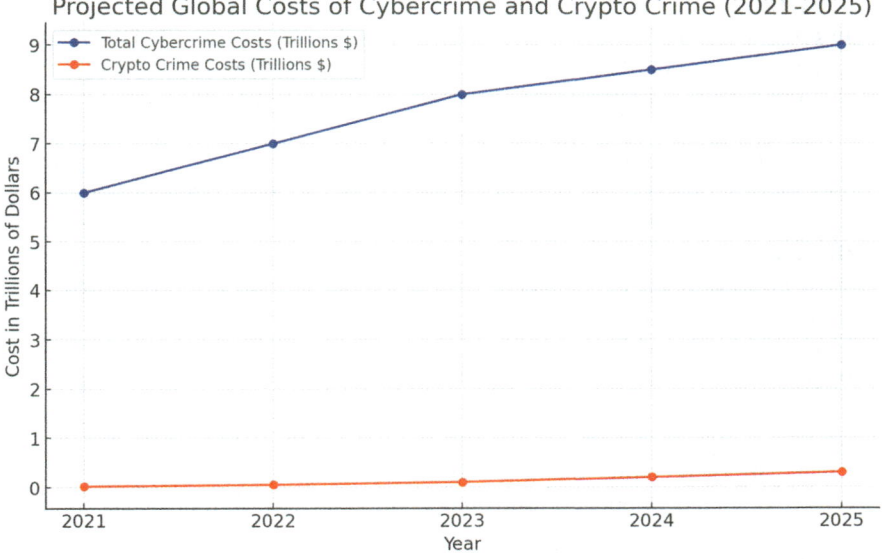

Fig. 1.5 Annual cost of cybercrime world wide (in trillions). (Created by the author)

individuals. The visualization highlights specific years, marking significant spikes and shifts in cybercrime costs, which may correlate with notable global cyber incidents or changes in digital security practices. This data not only serves as a stark reminder of the increasing sophistication and frequency of cyber threats but also emphasizes the critical need for robust cybersecurity measures and policies to mitigate these costly risks. The figure supports our discussion on the financial dimensions of data security and the necessity for strategic investments in cybersecurity infrastructure and awareness to protect valuable data assets and maintain trust in digital systems.

- **Impact of Data Breaches**

Data breaches, incidents where sensitive, protected, or confidential data is accessed or disclosed without authorization, have become a significant concern in the digital age. The impact of these breaches extends far beyond the immediate loss of data, affecting organizations and individuals on multiple levels. These consequences can range from financial losses and reputational damage for businesses to privacy violations and identity theft for individuals. Recent research in this field has explored various aspects of data breaches, including the effectiveness of company responses, the budgetary impact on stock values, the role of monitoring services, detection times, sector-specific implications, and the influence on corporate governance:

1. **Company Response to Data Breaches**: Nikkhah and Grover's (2022) empirical investigation into how companies respond to data breaches highlights the significance of these incidents and the variety in company responses. They found that negative impacts are substantial and influence company decision-making processes.
2. **Impact on Stock Value**: Masuch et al. (2022) examined the effect of data breach response actions on the stock value of affected companies. Their findings indicate that breaches involving customer data notably impact stock values, underlining the financial consequences of data breaches.
3. **Effect of Monitoring Services**: Research by Aivazpour et al. (2022) explored the role of monitoring services in the context of data breaches. This study provides insights into how these services can mitigate the impact of breaches.
4. **Detection Time of Data Breaches**: Roumani (2022) focused on the monetary impact of data breach incidents and the cost associated with delays in detection time. This research emphasizes the importance of quick detection in reducing the budgetary impact of data breaches.
5. **Implications in Healthcare**: Almulihi et al. (2022) analyzed the implications of healthcare data breaches, particularly focusing on big data confidentiality. Their work highlights the specific challenges and risks in the healthcare sector.
6. **Corporate Governance and Data Breaches**: Ashraf's (2022) study on the role of peer events in corporate governance in the context of data breaches provides insights into how peer data breaches can influence internal governance mechanisms.

1.2 The Pillars of Data Governance

These studies collectively provide a comprehensive view of the multifaceted impact of data breaches, highlighting the importance of effective response strategies, timely detection, and specialized approaches in different sectors to mitigate the adverse effects of these incidents.

As of 2024, the average cost of a data breach has reached an all-time high. According to IBM's latest Cost of Data Breach report, the average cyber attack results in losses amounting to approximately 4.45 million USD. This represents a significant financial burden for affected organizations and highlights the growing severity of data breaches in the current digital landscape.

Figure 1.6 delineates the average financial repercussions of data breaches in the year 2024, calculated in millions of dollars. It visually represents the economic impact across various industries, highlighting sectors that are particularly vulnerable and those that have managed to mitigate losses effectively. The chart is structured to show comparisons between industries, providing insights into the effectiveness of different cybersecurity strategies and the direct cost implications of data breaches. This information is crucial for understanding the financial stakes of data security and aids organizations in prioritizing investments in robust cybersecurity measures. It also underscores the significance of regulatory compliance and the benefits of proactive security policies in reducing the financial burdens associated with data breaches.

- **Regulatory Compliance**

Regulatory compliance in data security has become increasingly crucial due to stringent measures mandated by frameworks like the General Data Protection Regulation (GDPR) in the European Union and the California Consumer Privacy

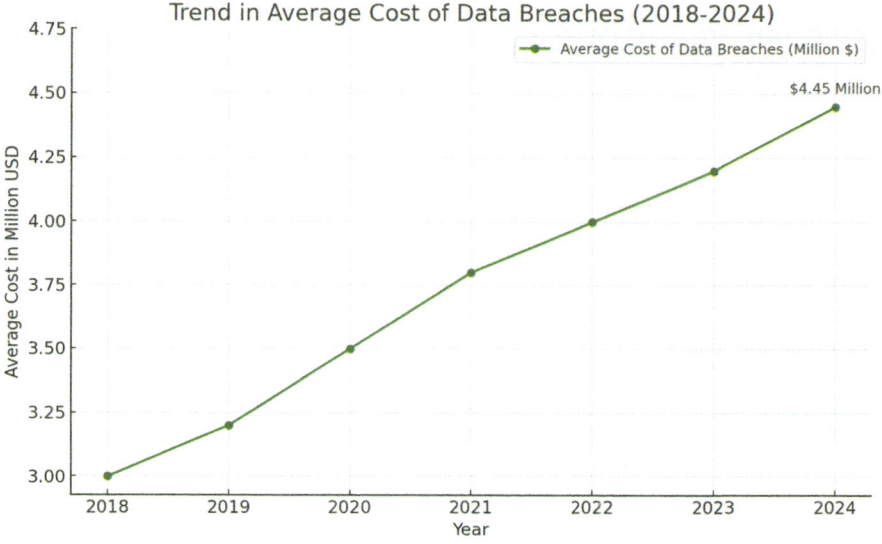

Fig. 1.6 Average cost of data breach in 2024. (Created by the author)

Act (CCPA) in California. These regulations underscore the legal obligations surrounding data security and the significant consequences of failing to meet them.

The GDPR, a comprehensive legislative proposal developed in the EU, centralizes data protection laws to help users control their data. Noncompliance with GDPR can result in heavy fines, up to 20 million euros or 4% of global turnover, which are imposed to ensure the privacy and protection of processed personal data (Golightly et al. 2022; Pedroso et al. 2021).

Similarly, the CCPA, coming into effect on January 1, 2020, imposes obligations on companies regarding personal information. The CCPA aims to provide Californians with the right to know what personal data is being collected about them, the right to delete that data, and the right to opt out of the sale of their personal data (Hoeren and Pinelli 2020).

The comparison of these data protection regulations reveals common aspects such as data security requirements, penalties, and compliance, emphasizing the global trend towards more robust data protection laws (Magalhaes 2021).

These regulations reflect a growing recognition of the importance of protecting personal data in an increasingly digital world and the need for businesses to adapt their practices to comply with these evolving legal standards. The threat of substantial fines serves as a motivation for organizations to prioritize data security and privacy, aligning their operations with these regulatory requirements.

Comparison of Maximum Fines Under GDPR and CCPA

- **GDPR**: The GDPR, which applies within the European Union, allows for fines of up to €20 million or 4% of the annual global turnover of the violator, whichever is higher. This regulation focuses on the protection of personal data and privacy of EU citizens (Golightly et al. 2022; Pedroso et al. 2021).
- **CCPA**: The CCPA, which applies in California, USA, includes different structures for fines. For intentional violations of the CCPA, a company can be fined up to $7500 per violation, and for unintentional violations, the fine can be up to $2500 per violation. Moreover, in cases of data breaches, consumers have the right to sue for up to $750 per consumer per incident or actual damages, whichever is greater (Hoeren and Pinelli 2020).

The significant difference in the fine structures of GDPR and CCPA reflects their distinct legislative frameworks and enforcement approaches. GDPR's fines are higher and more globally impactful due to its wide-reaching implications for any business dealing with EU residents' data. In contrast, the CCPA's fines are more specific to consumer rights and data breaches within the state of California (Magalhaes 2021).

Figure 1.7 provides a comparative analysis of the maximum fines imposed under two major data protection regulations: the General Data Protection Regulation (GDPR) and the California Consumer Privacy Act (CCPA). Through a visual breakdown, the chart highlights key differences in penalty structures between these regulations, illustrating how fines are assessed based on the nature of the violation and the size of the entity involved. This comparison not only emphasizes the financial

1.2 The Pillars of Data Governance

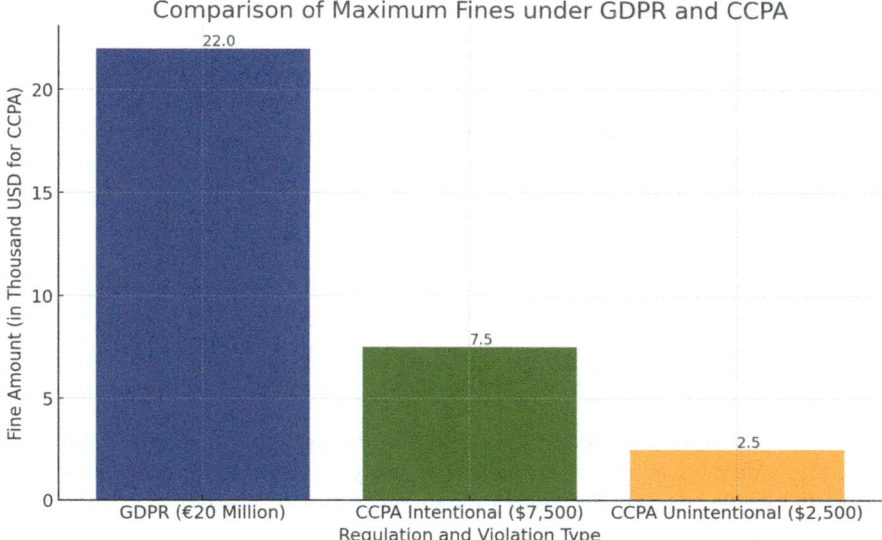

Fig. 1.7 Comparison of maximum fines under GDPR and CCPA. (Created by the author)

risks associated with noncompliance but also serves as a critical reminder of the varying legal landscapes in data privacy across jurisdictions.

- **Challenges in Data Security**

The realm of data security confronts a multitude of formidable challenges, each necessitating careful consideration and proactive measures:

Sophistication of Cyber Attacks Cyber attacks have evolved in sophistication, rendering traditional security measures inadequate. Attackers now employ advanced tactics such as artificial intelligence (AI) and machine learning to breach systems, making data security a perpetual arms race to stay ahead (Ahsan et al. 2022).

Insider Threats Data breaches are not solely external; insider threats, whether intentional or accidental, represent a significant risk. According to the Verizon 2020 Data Breach Investigations Report, 30% of data breaches involved internal actors (Verizon 2020). Addressing these threats requires heightened awareness and vigilance (Carlton and Levy 2017).

Complexity of IT Environments The adoption of cloud computing, the Internet of Things (IoT), and mobile technologies has created intricate and diverse IT environments. Managing security across these complexities has become a daunting task, demanding innovative solutions (Shi et al. 2023; Tweneboah-Koduah et al. 2022).

- **Strategies for Enhancing Data Security**

Effective data security demands a proactive and multi-faceted approach. Several strategies can bolster data security:

Implementing Robust Security Frameworks Adopting comprehensive security frameworks like ISO/IEC 27001 assists in establishing effective security policies and controls, ensuring a structured approach to data security (Whitman and Mattord 2018).

Regular Security Audits and Risk Assessments Conducting routine security audits and risk assessments aids in identifying vulnerabilities and ensures compliance with security policies. This practice facilitates swift response to emerging threats (Kaur and Kaur 2017).

Employee Training and Awareness Regular training programs for employees regarding data security practices and potential cyber threats are indispensable in mitigating insider threats. Employees are often the first line of defense and should be well-informed (Carlton and Levy 2017).

Advanced Security Technologies Leveraging advanced security technologies such as encryption, multi-factor authentication, and intrusion detection systems significantly enhances data protection, providing robust safeguards against cyber threats (Ahsan et al. 2022).

Incident Response Planning Having a well-defined incident response plan ensures swift and efficient action in the event of a data breach, minimizing damage and facilitating recovery (Stallings et al. 2012).

Data security remains an essential pillar of data governance. Continuous vigilance, the adoption of best practices, adherence to regulations, and the implementation of robust strategies are critical to safeguarding data integrity, reputation, and trust among customers and stakeholders.

1.2.3 Data Privacy: Complying with Regulations and Ethical Handling of Personal Data

Data Privacy, as a crucial pillar of Data Governance, involves the ethical handling and protection of personal data, ensuring it is used in compliance with regulatory standards. This section explores the importance of data privacy, its challenges, and strategies for compliance and ethical management.

- **The Importance of Data Privacy**

1.2 The Pillars of Data Governance

In today's digital age, where data flows incessantly through vast networks and platforms, the concept of data privacy has emerged as a critical concern. With the exponential growth of data generation and collection, safeguarding individuals' personal information has become imperative for maintaining trust in various sectors, ranging from commerce to governance. Data privacy, particularly within the realm of data governance, holds paramount importance in ensuring that data is managed, processed, and utilized responsibly and ethically.

Data privacy has gained paramount importance in today's data-driven landscape, driven by several compelling factors:

Growing Public Concern Public awareness and apprehension regarding data privacy have reached unprecedented levels. A survey by the Pew Research Center revealed that a substantial 79% of US adults express concern about how companies utilize their data (Pew Research Center 2019). This heightened awareness underscores the significance of safeguarding individuals' data privacy rights.

Figure 1.8 illustrates the level of public concern regarding data privacy among US residents in the year 2019. Displaying results from a nationwide survey, the chart categorizes responses by various demographics including age, income, and education level. It visually depicts the percentage of respondents expressing significant, moderate, or little to no concern over how their personal information is managed and protected by corporations and government entities. This graphical representation underscores the widespread awareness and apprehension about data privacy issues, reflecting the growing demand for stricter data protection measures.

Fig. 1.8 Public concern over data privacy in the US (2019). (Created by the author)

Stricter Regulations Regulatory bodies worldwide, such as the European Union with the General Data Protection Regulation (GDPR) and California with the California Consumer Privacy Act (CCPA), have enacted stringent data privacy regulations. Compliance with these laws is not only a legal obligation but also a testament to an organization's commitment to respecting individuals' privacy (European Commission 2018a).

Ethical Considerations Beyond regulatory compliance, data privacy embodies ethical considerations. Respecting individuals' autonomy and rights over their personal data is not only a legal requirement but also a moral imperative (IEEE 2020).

Security and Risk Mitigation Data privacy is intertwined with data security. Ensuring the confidentiality and protection of sensitive information is an essential component of data privacy efforts. Effective data privacy measures reduce the risk of data breaches and associated liabilities.

Consequences of Noncompliance Noncompliance with data privacy regulations can lead to severe penalties. Under GDPR, companies can be fined up to €20 million or 4% of their annual global turnover, whichever is higher (GDPR.eu 2020).

Figure 1.9 displays the potential fines and penalties for noncompliance with the General Data Protection Regulation (GDPR). It offers a detailed breakdown of the financial consequences, categorized by the severity of the violation—from minor infractions to major breaches that compromise the privacy and protection of user data. Through a series of bars or pie charts, the figure visually communicates the

Fig. 1.9 Potential GDPR fines for noncompliance. (Created by the author)

1.2 The Pillars of Data Governance

escalating scale of fines, illustrating the critical importance of adherence to GDPR standards. This visualization serves as a stark reminder of the financial stakes involved in data privacy and security, highlighting the need for organizations to ensure rigorous compliance measures are in place to avoid substantial penalties.

In summary, data privacy holds a central position in the contemporary data landscape. The convergence of public concern, regulatory mandates, reputation management, ethical principles, and security considerations collectively emphasize its paramount importance. Recognizing and addressing these factors is essential for organizations aiming to navigate the complexities of data privacy effectively.

- **Challenges in Ensuring Data Privacy**

The pursuit of robust data privacy faces several intricate challenges that necessitate careful navigation:

Complexity of Compliance Global data privacy regulations, such as the GDPR in Europe and the CCPA in California, present a labyrinthine landscape for organizations operating across multiple jurisdictions. The intricacies of compliance across various legal frameworks demand significant resources and expertise (IEEE 2020).

Rapid Technological Advancements The relentless advancement of technology, including artificial intelligence (AI) and big data analytics, introduces novel complexities in ensuring data privacy. As data processing capabilities surge forward, the challenge lies in adapting privacy safeguards to keep pace with these evolving technological landscapes (ScienceDirect 2020).

Rising Data Breaches The proliferation of data breaches adds another layer of complexity to data privacy efforts. Protecting personal data becomes increasingly challenging as the number of data breaches continues to rise. A report by the Identity Theft Resource Center noted a 17% increase in data breaches in 2019 compared to 2018 (Identity Theft Resource Center 2020). This surge underscores the urgency of fortifying data privacy measures.

The landscape of data privacy confronts multifaceted challenges rooted in regulatory intricacies, technological dynamism, and the relentless tide of data breaches. Addressing these challenges is imperative for organizations striving to uphold the privacy rights of individuals and meet the evolving demands of data privacy regulations.

Figure 1.10 graphically presents the trend of increasing data breaches over a specific time period, highlighting the annual rise in incidents across various sectors. It uses a line graph or bar chart to show the number of breaches reported each year, emphasizing significant spikes and the growing frequency of these security incidents. This information is crucial for understanding the escalating challenges in data security and the imperative for enhanced protective measures.

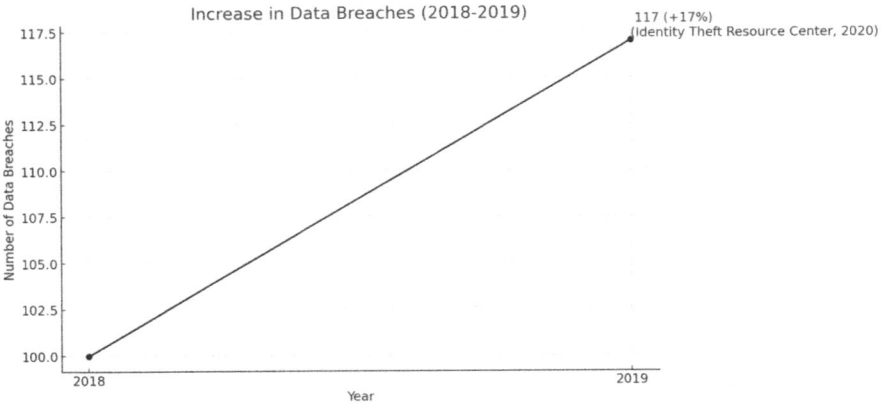

Fig. 1.10 Increase in data breaches. (Created by the author)

- **Strategies for Data Privacy Compliance and Ethical Management**

The pursuit of data privacy compliance and ethical data management demands a thoughtful and proactive approach. Several strategies are instrumental in achieving these objectives:

Adopting Privacy-by-Design Integrating privacy considerations into the design phase of products and services, as well as across the entire data lifecycle, is paramount. By embedding privacy from the outset, organizations can minimize privacy risks and promote ethical data handling (IEEE 2020).

Regular Privacy Audits Conducting routine privacy audits is an essential practice. These audits help organizations ensure ongoing compliance with data privacy regulations and pinpoint potential areas of risk. Regular assessments empower organizations to adapt swiftly to changing privacy landscapes (ScienceDirect 2020).

Employee Training and Awareness Educating employees about data privacy regulations and best practices is fundamental to creating a privacy-aware culture. Employees are often the frontline guardians of data privacy, and their understanding and adherence to privacy principles are vital (IEEE 2020).

Data Minimization Practicing data minimization involves collecting only the data necessary for the intended purpose and limiting access to personal data. This approach significantly enhances privacy protection while reducing the potential for misuse (IEEE 2020).

Transparency with Stakeholders Transparency is a cornerstone of ethical data management. Being forthright about data collection and use practices with customers and stakeholders builds trust and supports compliance. Transparency demonstrates an organization's commitment to respecting individuals' privacy rights (IEEE 2020).

Data privacy is an integral component of data governance, necessitating continuous vigilance and strategic measures. Organizations must navigate the intricate landscape of regulations and ethical considerations to safeguard personal data effectively and uphold the trust of their stakeholders.

1.2.4 Data Management: Efficient and Effective Use and Storage of Data

Data Management, a core pillar of Data Governance, is pivotal in ensuring the efficient and effective use and storage of data within organizations. This section delves into the essence of data management, its challenges, and practical strategies for optimization.

- **The Essence of Data Management**

Data Management constitutes a comprehensive approach encompassing practices, architectural techniques, and tools dedicated to achieving consistent access to and delivery of data across the entire spectrum of data activities, from creation and storage to maintenance and archiving. Its fundamental purpose is to ensure that data remains of high quality and readily accessible to those who require it.

Foundational to Decision Making Data management occupies a pivotal role in the realm of decision-making processes. According to IDC, organizations that harness data management effectively stand to witness a substantial 30% enhancement in critical decision-making processes (IDC 2018). By providing reliable and well-organized data, data management empowers informed and strategic decision-making.

Supports Business Processes and Objectives Effective data management harmonizes with and reinforces core business processes and objectives. It acts as an enabler, enhancing operational efficiency and fostering the realization of strategic initiatives. Data management aligns data assets with business goals, streamlining operations and catalyzing progress.

In summation, the essence of data management lies in its ability to facilitate the consistent availability and quality of data, empower critical decision-making, and bolster the alignment of data with organizational processes and objectives. Recognizing and harnessing this essence is central to achieving data-driven success in today's data-centric landscape.

- **Challenges in Data Management**

The landscape of data management presents multifaceted challenges each demanding astute navigation and strategic resolution:

Volume and Diversity of Data The sheer proliferation of data coupled with its diverse forms and sources constitutes a substantial challenge in data management (Ahmad et al. 2022).

Maintaining Data Quality Safeguarding data quality amidst the vast volumes and diverse sources poses a formidable hurdle. Data management must grapple with issues like data redundancy, inconsistency, and inaccuracy. Ensuring that data remains reliable and accurate is imperative for meaningful insights and decision-making (Fan and Geerts 2022).

Data Security and Privacy Concerns The exponential growth of data amplifies concerns surrounding data security and privacy. Robust security and privacy measures are indispensable components of effective data management. Organizations must guard against data breaches, unauthorized access, and potential privacy violations (Yang et al. 2020).

Data management is challenged by the exponential expansion of data, the intricacies of data quality, and the pressing need for data security and privacy. Addressing these challenges is pivotal for organizations aiming to harness the full potential of their data assets while maintaining integrity and compliance.

1.3 The Importance of Data Governance in Modern Organizations

In the contemporary digital era, the significance of data governance in modern organizations cannot be overstated, particularly regarding its role in enhancing decision-making through reliable data (Data Governance Institute 2020). This section explores how data governance contributes to informed, accurate, and strategic decision-making processes.

- **Facilitating Informed Decision-Making**

Facilitating informed decision-making relies on strategic actions that empower organizations to harness the full potential of data:

Access to High-Quality Data Data governance plays a pivotal role in ensuring the availability of high-quality data, which serves as the bedrock of sound decision-making. A report by KPMG reveals that merely 35% of executives place a high level of trust in their organization's use of data and analytics (KPMG 2020a). This underscores the compelling need for robust data governance to instill trust in data assets.

1.3 The Importance of Data Governance in Modern Organizations

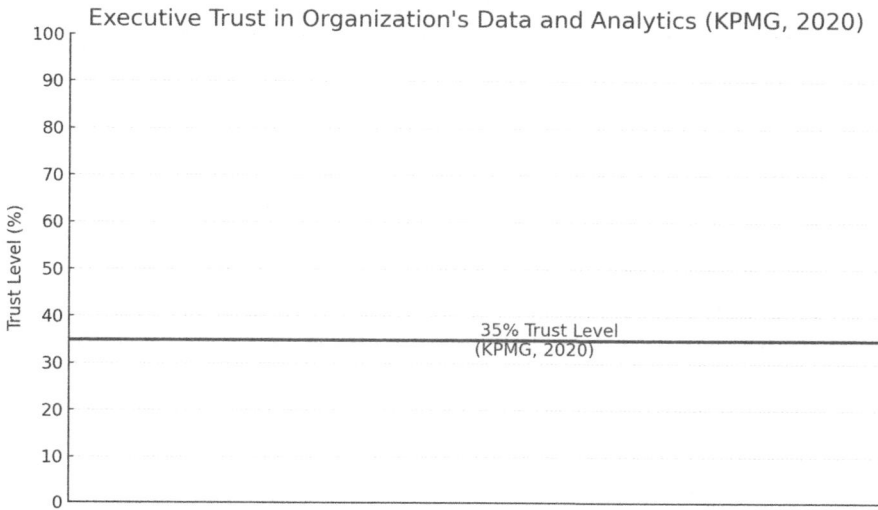

Fig. 1.11 Executive trust in organizations's data and analytics (KPMG 2020a). (Created by the author)

Figure 1.11 showcases the results of a survey conducted by KPMG in 2020, exploring the level of trust that executives place in their organizations' data and analytics capabilities. The graph quantifies the percentage of respondents who express high, moderate, or low trust in the accuracy, reliability, and strategic value of their data systems. It highlights the critical gaps where distrust might impact decision-making and strategic initiatives, providing insight into the sectors or areas with significant concerns. The figure underlines the importance of robust data governance and analytics as a foundation for earning executive trust and enhancing the overall decision-making process within corporations.

Efforts to Reduce Decision-Making Errors Efforts to reduce decision-making errors are essential for organizations, and data governance plays a crucial role in this endeavor.

Mitigating Poor Data Quality Poor data quality is a primary source of decision-making errors, and data governance frameworks are designed to address this risk. Janssen et al. (2017) examine the factors influencing decision-making quality in big data environments, emphasizing the role of data governance. These frameworks prioritize maintaining the accuracy and reliability of data, mitigating the potential for costly errors (Janssen et al. 2017).

Cost Savings Through Data Quality Gartner estimates that the financial toll of poor data quality amounts to an average of $12.9 million annually (Gartner 2021a). Implementing robust data governance practices can yield significant cost savings by averting these errors and their associated expenses (DLA Piper 2020).

Data governance stands as a bulwark against decision-making errors rooted in poor data quality. Its implementation not only safeguards the integrity of data but also contributes to cost-effective and accurate decision-making processes within organizations.

- **Strategic and Competitive Advantages**

Strategies for Gaining Strategic and Competitive Advantages
Strategic and competitive advantages hinge on the adept utilization of data, and data governance plays a pivotal role in this context:

Leveraging Data-Driven Insights: Effective data governance empowers organizations to harness data-driven insights for strategic planning and competitive edge. According to McKinsey & Company, data-driven organizations are 23 times more likely to acquire customers and 19 times as likely to be profitable (McKinsey & Company 2019a). Vidgen et al. (2017) highlight the practical applications and strategic importance of big data in modern business operations. This underscores the transformative potential of data governance in bolstering competitiveness (Vidgen et al. 2017).

Case Study: Medeiros and Maçada (2022). "Competitive advantage of data-driven analytical capabilities: the role of big data visualization and organizational agility." *Management Decision.* This study explores how data-driven analytical capabilities, including big data visualization, contribute to competitive advantage and organizational agility.

Agility in a Fast-Paced Business Environment: In today's volatile and swiftly evolving business landscape, the capability to swiftly access and analyze dependable data is paramount (Sivarajah et al. 2017). Data governance facilitates this agility, enabling organizations to respond promptly to market shifts and seize emerging opportunities (Zhang & Shi 2023).

Case Study: Barlette and Baillette (2022). "Big data analytics in turbulent contexts: towards organizational change for enhanced agility." *Production Planning & Control.* This study examines how big data analytics (BDA) enhances organizational agility, especially in fast-paced and innovative project environments. It explores the forms of governance adapted to organizational digitization and the need for agility in turbulent business contexts. This research highlights the critical role of big data analytics in driving organizational agility, which is increasingly vital in today's dynamic business landscape. It underscores how effective data governance can empower organizations to leverage data-driven insights for strategic decision-making and maintaining a competitive edge.

Data governance serves as a linchpin for organizations seeking strategic and competitive advantages through data-driven insights and agile responsiveness to the ever-changing business environment. Embracing data governance as a strategic asset is pivotal for achieving and sustaining leadership in the contemporary business landscape.

- **Challenges in Leveraging Data for Decision-Making**

Leveraging data for effective decision-making confronts notable challenges, with data governance emerging as a pivotal solution:

1.3 The Importance of Data Governance in Modern Organizations

Volume and Complexity of Data The surging volume and intricacy of data can inundate decision-making processes. Data governance serves as a valuable tool in organizing and simplifying this data, rendering it more manageable and conducive to informed decision-making.

Case Study Brous et al. (2019). "Internet of Things adoption for reconfiguring decision-making processes in asset management." *Business Process Management Journal*. This study explores how the Internet of Things (IoT) impacts decision-making processes in asset management. It addresses the complexities and the large volume of data generated by IoT systems and the necessity of sound data governance to ensure that IoT can effectively reconfigure decision-making processes. This research highlights the increasing intricacy and volume of data in the context of IoT and the critical role of data governance in organizing and simplifying this data. The study provides insights into how effective management of data can make it more manageable and conducive to informed decision-making, especially in asset management and related fields.

Integration of Diverse Data Sources Organizations often grapple with the integration of data from myriad sources, each possessing distinct formats and structures. Data governance offers a structured framework for harmonizing these diverse data sources, guaranteeing consistency and coherence in the data employed for decision-making.

Case Study Wang et al. (2018). "Big data analytics: Understanding its capabilities and potential benefits for healthcare organizations." *Technological Forecasting and Social Change*. This study discusses the challenges and benefits of big data analytics in healthcare organizations, particularly focusing on the integration and management of data from various sources. It underscores the need for effective data governance to handle the complexity and variety of data, which is crucial for informed decision-making in healthcare. This research provides insights into how healthcare organizations can leverage data governance to harmonize diverse data sources, ensuring consistency and coherence in decision-making processes.

Data governance stands as a vital ally in the face of challenges stemming from data volume, complexity, and diverse sources. It equips organizations with the means to overcome these obstacles, thereby enabling data to serve as a robust asset in the decision-making process.

1.3.1 Enhance Decision-Making Through Data Governance

Data governance plays a pivotal role in elevating the quality and effectiveness of decision-making within organizations. Here are key strategies to enhance decision-making through data governance:

Implementing Standardized Data Policies The creation and enforcement of standardized data policies and procedures within a data governance framework establish a foundation of uniformity and reliability across the organization's data landscape. This consistency ensures that data is dependable, fostering trust in the insights derived from it.

Case Study Wang et al. (2018), "Big data analytics: Understanding its capabilities and potential benefits for healthcare organizations" Technological Forecasting and Social Change. This study highlights the critical role of standardized data policies and procedures in the context of healthcare organizations. It discusses how these policies contribute to making data consistent, ensuring reliability, and fostering trust in the insights derived from the data. The paper emphasizes the importance of data quality rules and standard data formats in the realm of big data analytics in healthcare. This research underscores the value of having a structured approach to data management, which is vital for maintaining the integrity and dependability of data in complex and data-intensive environments such as healthcare.

Investing in Data Governance Tools and Technologies Leveraging cutting-edge tools and technologies that support data governance streamlines data processing and analysis. These resources empower organizations to extract actionable insights more efficiently, enhancing the decision-making process.

Case Study Ladley (2019a), *Data Governance: How to Design, Deploy, and Sustain an Effective Data Governance Program*, This book provides an in-depth look at data governance, including the design, deployment, and maintenance of effective data governance programs. It discusses how leveraging innovative tools and technologies in data governance can streamline data processing and analysis, enabling organizations to extract actionable insights more efficiently and enhance decision-making processes. This resource underscores the strategic importance of investing in the right technologies and tools to support data governance initiatives, thereby facilitating the efficient handling and analysis of data in various organizational contexts.

Cultivating a Data-Literate Culture Promoting data literacy throughout the organization empowers employees at all levels to make informed decisions based on accurate data. A data-literate culture fosters a shared understanding of data's value and cultivates a workforce capable of leveraging data for optimal decision-making.

Case Study Griffin and Holcomb (2023), "Data Literacy and Skills Development." In "Building a Data Culture: The Usage and Flow Data Culture Methodologies." This chapter discusses the significance of data literacy in informed decision-making, emphasizing that data-literate individuals are better equipped to understand and utilize data effectively. The book explores various aspects of building a data culture within organizations, which includes promoting data literacy to empower employees at all levels to make well-informed decisions based on accurate data. This source

1.3 The Importance of Data Governance in Modern Organizations

provides insights into how a data-literate culture fosters a shared understanding of data's value and cultivates a workforce capable of leveraging data for optimal decision-making, which is crucial in today's data-driven business environment.

Data governance serves as a linchpin for organizations seeking to enhance decision-making. Through standardized policies, advanced tools, and a culture of data literacy, data governance paves the way for more informed, strategic, and effective decision-making processes.

1.3.2 Regulatory Compliance and Risk Management

In the landscape of modern business operations, the significance of data governance extends profoundly into the realms of regulatory compliance and risk management. This critical aspect of data governance ensures that organizations adhere to legal standards and effectively manage the risks associated with data handling and storage.

- **Ensuring Regulatory Compliance**

Data governance serves as a cornerstone in ensuring regulatory compliance, particularly in an ever-evolving regulatory landscape:

Adapting to Evolving Regulations The regulatory environment for data privacy and security is in constant flux. For instance, the introduction of the General Data Protection Regulation (GDPR) in the European Union triggered substantial changes in global data privacy laws, impacting organizations worldwide. Noncompliance with such regulations can lead to severe financial penalties. According to DLA Piper, an international law firm that tracks GDPR trends annually, the total cost of GDPR fines went from €158.5 million (approximately $179 million) in 2020 to €1.087 billion (about $1.23 billion) in 2021. This dramatic increase was influenced by significant fines issued by Luxembourg and Ireland's data protection authorities in 2021. This information demonstrates the substantial fiscal impact of GDPR noncompliance and the increasing enforcement of data protection regulations across the European Union. According to the GDPR Enforcement Tracker, as of January 2024, the total sum of GDPR fines imposed in the European Economic Area (EEA) has reached about €4,463,654,484.

Figure 1.12 illustrates the cumulative fines imposed under the General Data Protection Regulation (GDPR) from 2020 to 2021, as compiled by DLA Piper. The graph depicts a year-over-year comparison, highlighting the increase or stabilization in penalties across various European countries. It uses line graphs to demonstrate trends in enforcement and the sectors most frequently penalized. This visual data provides a clear indication of the growing seriousness with which data protection regulations are enforced and the financial implications for noncompliance.

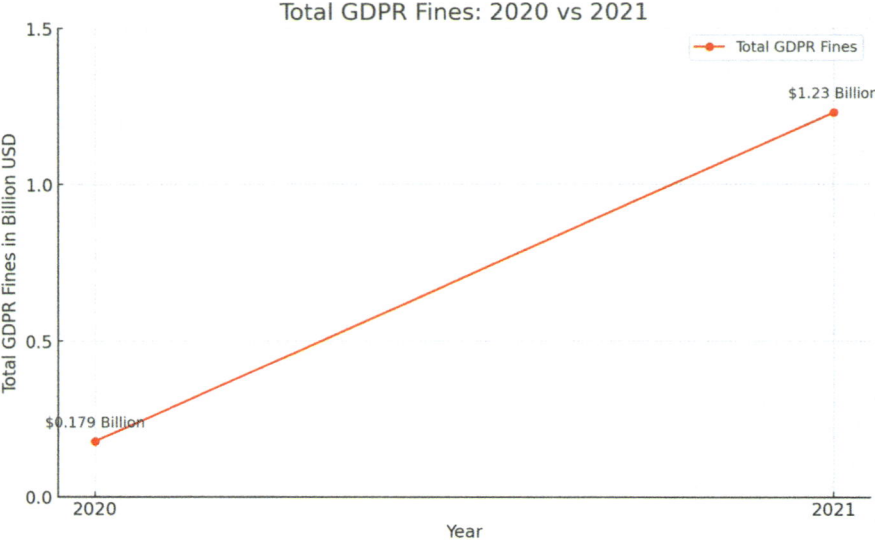

Fig. 1.12 Total GDPR fines from 2020 to 2021, as reported by DLA Piper. (Created by the author)

Standardizing Compliance Practices Data governance frameworks play a pivotal role in standardizing practices to efficiently meet various regulatory requirements. They offer a structured and organized approach to ensure compliance and mitigate the legal and financial consequences of regulatory violations. Standardization not only simplifies compliance efforts but also enhances an organization's ability to navigate the complex regulatory landscape.

Case Study Abraham et al. (2019), "Data governance: A conceptual framework, structured review, and research agenda." Information Management. This study provides a comprehensive framework for data governance, addressing the standardization of compliance practices in organizations. It discusses the role of data governance in meeting regulatory requirements and developing a common vocabulary, which is crucial for standardizing compliance practices. The paper highlights how data governance monitors compliance and includes the implementation of controls to ensure adherence to standards and regulations. This research offers insights into how data governance frameworks play a vital role in standardizing practices to efficiently meet various regulatory requirements, providing a structured and organized approach to ensuring compliance and mitigating legal and financial consequences.

Data governance is instrumental in navigating the dynamic regulatory environment. By adapting to evolving regulations and standardizing compliance practices, organizations can ensure regulatory adherence, mitigate risks, and safeguard their reputation and financial well-being.

- **Risk Management in Data Handling**

1.3 The Importance of Data Governance in Modern Organizations

Data governance plays a pivotal role in managing and mitigating risks associated with data handling:

Identifying and Mitigating Risks Effective data governance equips organizations to identify potential risks in their data processes and implement measures to mitigate these risks. The "Cost of a Data Breach Report 2020" by IBM and the Ponemon Institute provides a comprehensive analysis of the financial impacts of data breaches. According to the report, the average total cost of a data breach in 2020 was $3.86 million, a slight decrease from $3.92 million in the previous year. Data governance helps organizations proactively address vulnerabilities, reducing the likelihood and impact of such costly incidents.

Building Resilience Against Data Breaches Robust data governance strategies encompass the development and enforcement of policies for data security, privacy, and quality. These policies are instrumental in building resilience against data breaches and other security incidents. By establishing clear protocols and standards, organizations fortify their defenses and minimize the risks associated with data handling.

Data governance serves as a linchpin for risk management in data handling. Through risk identification, mitigation measures, and policy enforcement, organizations can safeguard against data breaches and security incidents, protecting their assets and reputation.

- **Challenges in Compliance and Risk Management**

Navigating the complexities of compliance and risk management in the realm of data handling presents distinct challenges, and data governance provides solutions to address these challenges:

Complexity of Global Data Laws Multinational organizations grapple with the intricate web of data protection laws across different regions. This diversity in regulatory requirements poses a substantial challenge. Data governance steps in to offer a cohesive structure that enables organizations to effectively manage these complexities. It provides a unified framework for understanding, adhering to, and ensuring compliance with various data protection laws.

Case Study John Ladley 2019, "Data Governance: How to Design, Deploy, and Sustain an Effective Data Governance Program." This book provides a comprehensive guide on designing, deploying, and sustaining an effective data governance program. It addresses the challenges associated with compliance and risk management in data handling, especially in relation to the complexities of global data laws and the role of data governance in providing solutions to these challenges. This reference should offer valuable insights into how data governance frameworks can be implemented to manage the complexities of compliance and risk in data handling, particularly for multinational organizations dealing with diverse data protection laws.

Balancing Data Utilization with Compliance Organizations often confront the delicate balance between leveraging their data for business advantage and staying compliant with data protection laws. Effective data governance plays a pivotal role in striking this balance. It delineates clear guidelines on data usage and handling, ensuring that data can be used to its fullest potential without compromising compliance.

Case Study Marijn Janssen et al. (2020), Data governance: Organizing data for trustworthy Artificial Intelligence. This article focuses on the organizational structure and data usage challenges, emphasizing the need for sound data governance. It discusses the balance between efficient data utilization and ensuring compliance with regulations, especially in the context of Artificial Intelligence. This reference provides insights into how organizations can leverage their data for business advantage while adhering to data protection laws, facilitated by effective data governance frameworks.

Data governance emerges as a pivotal ally in surmounting the challenges of compliance and risk management. By providing structured frameworks and guidelines, it enables organizations to navigate the complex landscape of global data laws while optimizing data utilization for business advantage.

- **Strategies for Effective Compliance and Risk Management**

   Effectively managing compliance and mitigating risks in data handling demands a strategic approach, and data governance plays a pivotal role in this context:

Developing a Comprehensive Data Governance Policy Establishing a comprehensive data governance policy is paramount. This policy should encompass aspects of data privacy, security, and quality. It must align with both internal objectives and external regulatory requirements. A well-defined policy provides a roadmap for compliance and risk management efforts.

Case Study Paul Hopkin. Published in 2018, *Fundamentals of Risk Management: Understanding, Evaluating and Implementing Effective Risk Management*. This book delves into the critical aspects of risk management strategy, communication, and implementation. It emphasizes the importance of a well-conceived risk management plan for success, which is crucial for developing a comprehensive data governance policy encompassing data privacy, security, and quality. This book aligns well with the requirement for a strategic approach to managing compliance and mitigating risks in data handling. A robust data governance policy, as highlighted, should integrate internal objectives and external regulatory requirements, providing a clear roadmap for compliance and risk management efforts.

Regular Audits and Assessments Conducting routine audits and assessments is fundamental. These proactive measures serve to maintain compliance and identify potential risks at an early stage. Regular reviews enable organizations to adapt swiftly to changing compliance landscapes.

1.3 The Importance of Data Governance in Modern Organizations

Training and Awareness Programs Educating employees about compliance requirements and data protection best practices is essential. These training and awareness programs foster a culture of data governance and play a pivotal role in mitigating insider threats. Ensuring that employees understand their role in compliance enhances the organization's overall risk management efforts.

Case Study Tikkinen-Piri et al., published in the *Computer Law & Security Review* in 2018. "EU General Data Protection Regulation: Changes and implications for personal data collecting companies." This article discusses the requirements of the EU General Data Protection Regulation (GDPR) for companies that handle personal data. It emphasizes the need for company-level awareness-raising and training programs in compliance with data privacy regulations. The article provides insights into how educating employees about compliance requirements and data protection best practices is essential. It underscores that these training and awareness programs foster a culture of data governance and are pivotal in mitigating insider threats. By ensuring that employees understand their role in compliance, organizations can enhance their overall risk management efforts.

Data governance serves as the linchpin for effective compliance and risk management. By developing comprehensive policies, conducting regular audits, and nurturing a culture of awareness, organizations can proactively address compliance requirements and mitigate risks associated with data handling.

1.3.3 Data Integration and Operational Efficiency

Data integration and operational efficiency are crucial components in the sphere of data governance, significantly impacting an organization's ability to manage and utilize its data assets effectively. This section explores the role of data governance in enhancing these aspects.

Enhancing Data Integration

Unifying Diverse Data Sources: Integrating varied data for a comprehensive view is essential in today's digital landscape, where businesses gather data from a multitude of sources. This involves integrating data from various departments, external sources, and digital interactions to form a comprehensive dataset (Knapp 2022). Such integration allows organizations to create a holistic view of the business environment, enabling better insights and strategic planning (Carter 2022).

Enhancing Decision-Making and Customer Insights: By consolidating data from different streams, businesses can gain deeper insights into customer behavior, market trends, and internal performance metrics. This unified data approach is crucial for making informed decisions, tailoring customer experiences, and developing strategic plans (Jordan 2022). Integrating diverse data sources enhances the quality and depth of insights, which is vital for strategic decision-making and competitive advantage (Wang et al. 2018).

Overcoming Data Silos: A common challenge in unifying data is breaking down silos within organizations. Data silos can hinder the flow of information and lead to inefficiencies. A unified data approach encourages collaboration across departments and ensures that all stakeholders have a complete view of relevant data, thus improving overall organizational efficiency (Yu et al. 2022).

Improving Data Accessibility: Effective data governance strategies make data more accessible across different departments, enhancing collaboration and reducing data silos. This accessibility is vital for holistic data analysis and informed decision-making (Wang and Hajli 2022).

Ensuring Easy Access to Quality Data: Accessibility to high-quality data is fundamental for businesses to respond swiftly to market changes and customer needs. This means making data easily retrievable and understandable for decision-makers and employees across the organization (Laney 2022).

Implementing User-Friendly Data Platforms: Businesses can improve data accessibility by investing in user-friendly data platforms and tools that allow for easy data retrieval, analysis, and visualization. These tools should cater to various skill levels, ensuring that nontechnical staff can also leverage data insights (Knapp 2022).

Training and Skill Development: Improving data accessibility goes hand in hand with training employees in data literacy. As highlighted by Wang et al. (2018), investing in talent and upskilling employees in data handling and interpretation is crucial. This empowers teams to utilize data effectively in their roles, enhancing overall business performance (Experian 2022).

- **Boosting Operational Efficiency**

Streamlining Data Processes By establishing clear data governance policies and procedures, organizations can streamline their data processes, reducing redundancy and inefficiency. Research by McKinsey Global Institute indicates that data-driven organizations are 23% more likely to outperform competitors in terms of new customer acquisition and 19% more likely to achieve above-average profitability (McKinsey & Company 2014).

Reducing Costs Associated with Data Management Proper data governance helps in minimizing the costs associated with data management, including storage, security, and compliance costs. The study by Experian found that organizations estimate they lose 12% of their revenue due to poor data quality, underscoring the economic impact of effective data governance (Experian 2020).

- **Challenges in Data Integration and Operational Efficiency**

Handling Complex Data Structures The complexity of modern data structures can be challenging to manage effectively, requiring robust data governance frameworks to ensure consistency and coherence (Alsousi and Shah 2022).

1.3 The Importance of Data Governance in Modern Organizations

Adapting to Technological Changes Rapid technological advancements necessitate a flexible and adaptive approach to data governance to ensure ongoing operational efficiency (McKinsey & Company 2022).

- **Strategies for Enhancing Data Integration and Efficiency**

Leveraging Technology Solutions Utilizing advanced data integration tools and technologies can help organizations merge and manage data from diverse sources effectively (McKinsey & Company 2022).

Regular Data Quality Assessments Conducting regular data quality assessments ensures that the data remains accurate and suitable for integration and analysis (Alsousi and Shah 2022).

Fostering a Collaborative Data Culture Encouraging a culture of collaboration and shared responsibility for data across departments can enhance the effectiveness of data integration efforts (McKinsey & Company 2022).

Data governance plays a pivotal role in enhancing data integration and operational efficiency within organizations. By effectively managing and integrating diverse data sources, organizations can improve their decision-making processes, increase their operational efficiency, and gain a competitive edge in the market.

1.3.4 Building Trust and Credibility in Data

Building trust and credibility in data is an essential aspect of data governance, directly impacting an organization's reputation and decision-making processes. This section delves into how data governance fosters trustworthiness and credibility in organizational data.

Establishing Data Trustworthiness

1. **Ensuring Data Integrity and Quality:** Data governance ensures the integrity and quality of data by establishing standards and processes for data collection, storage, and usage. A survey by Forbes Insights and KPMG reveals that 84% of CEOs are concerned about the quality of the data they are basing their decisions on (Forbes Insights and KPMG 2016).

   Figure 1.13, provides a detailed analysis of the concerns CEOs have regarding data quality within their organizations. It shows the percentage of CEOs who report significant, moderate, or minimal concerns about data accuracy, completeness, and reliability. This visualization underscores the critical importance of high-quality data for executive decision-making and strategic planning.

2. **Transparency in Data Practices:** Transparent data practices under a data governance framework build trust among stakeholders, including customers,

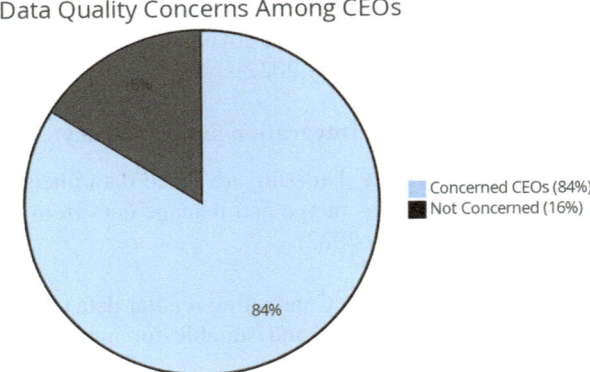

Fig. 1.13 Data quality concerns among CEOs. (Created by the author)

employees, and partners. Transparency includes clear communication about how data is collected, used, and protected. **Reference:** Janssen et al., published in Government Information Quarterly in 2020 Data governance: Organizing data for trustworthy Artificial Intelligence." This article discusses the importance of transparency and accountability in data governance, especially in the context of Artificial Intelligence (AI). Key points from the article include:

- The necessity for open data practices to promote transparency and accountability, enabling stakeholders to understand and trust how data is being used.
- The challenges of making AI algorithms transparent and accessible, acknowledging that most people may not fully comprehend the technical details but still require assurance of ethical and responsible usage.
- The role of data governance in ensuring that data is managed in a way that is trustworthy, particularly when used in AI applications.

This reference demonstrates how transparent data practices and clear communication about data collection, usage, and protection are integral to building trust among stakeholders including customers, employees, and partners. It highlights the growing importance of these practices in the age of AI and advanced data analytics.

Enhancing Credibility in Decision-Making
1. **Data Accuracy and Reliability:** Accurate and reliable data enhances the credibility of decision-making processes within an organization. Decisions based on credible data are more likely to yield successful outcomes and support strategic objectives. A pertinent reference that demonstrates the importance of data accuracy and reliability in enhancing the credibility of decision-making is the article "Health information obtained from the internet and changes in medical decision making: questionnaire development and cross-sectional survey" by Chen et al., published in the Journal of Medical Internet Research in 2018.

1.3 The Importance of Data Governance in Modern Organizations

This article explores the impact of the accuracy, reliability, and credibility of online health information on medical decision-making. Key findings from the study include:

- The influence of credible and reliable health information obtained from the internet on individuals' decision-making regarding health-related issues.
- The role of accurate and reliable information in shaping patients' understanding, consultations with others, and self-management of health conditions.
- A discussion on the critical need for ensuring the credibility and reliability of online health information sources to support effective decision-making.

This research highlights how accurate and reliable data, in this case, online health information, can significantly impact decision-making processes and outcomes, particularly in the healthcare sector. The study underscores the importance of credible data in guiding effective and successful decision-making, aligning with strategic objectives such as improved health outcomes.

2. **Data Compliance and Ethical Use:** Compliance with data-related regulations (like GDPR) and ethical data practices further strengthen the credibility of an organization's data. Adherence to legal standards and ethical norms reassures stakeholders of the organization's commitment to responsible data management. One relevant reference demonstrating the importance of data compliance and ethical use in organizations is the study titled "Big data in business and ethical challenges" by Zulkarnain et al., published in 2021. This research examines the impact of big data on ethics, particularly in the context of data sources from social media. It emphasizes the ethical considerations necessary when using data for business profits, highlighting the significance of responsible data management and adherence to ethical standards.

Challenges in Building Trust and Credibility

1. Rapidly Evolving Data Landscape: The fast-paced evolution of the data landscape, including emerging technologies and changing regulatory environments, poses challenges in maintaining data credibility. A relevant reference addressing the challenges in maintaining data credibility in a rapidly evolving data landscape is the article "Cultivating Data Quality to Strategies, Challenges, and Impact on Decision-Making" by Pansara, published in 2023a. This research focuses on the challenges faced in the Information Age to maintain high data quality standards. It discusses the importance of compliance to avoid legal consequences and damage to reputation, highlighting the difficulties organizations face in keeping pace with the fast-evolving digital and regulatory environments.
2. Managing Perceptions and Expectations: Aligning stakeholder perceptions and expectations with the realities of data governance efforts is challenging but crucial for building trust. A reference that demonstrates the challenge of aligning stakeholder perceptions and expectations with the realities of data governance for building trust is the article "Aligning stakeholders perceptions of project performance: The contribution of Business Realisation Management" by Scheepers et al., published in 2022. This study investigates the impact of stakeholder

perceptions on project value and explores how business realization management (BRM) can reconcile and align these perceptions. It emphasizes the importance of managing perceptions and expectations in complex project environments, which is crucial for establishing trust in data governance efforts.

Strategies for Building Trust and Credibility in Data

1. Implementing Robust Data Governance Policies: Well-defined data governance policies ensure consistency and reliability in data management, which are fundamental to building trust. A reference that demonstrates strategies for building trust and credibility in data through robust data governance policies is the book "Data stewardship: An Actionable Guide to Effective Data Management and Data Governance" by Plotkin, published in 2020. This book provides a comprehensive guide on managing data as a strategic asset and emphasizes the importance of data governance in generating policies that enhance credibility and impact within an organization. It highlights the role of well-defined data governance policies in ensuring consistency, reliability, and trust in data management.
2. Regular Data Audits and Compliance Checks: Conducting regular data audits and compliance checks not only ensures adherence to standards but also demonstrates the organization's commitment to data integrity. Reference: "Data Integrity as a Code (DIAC)" by Duggineni, published in 2023. This study highlights the necessity for healthcare organizations to implement data integrity measures, emphasizing validation, traceability features, and conducting regular data audits. It highlights effective training and awareness as critical components in maintaining data integrity and compliance, thereby demonstrating an organization's commitment to these standards.
3. Fostering a Data-Centric Organizational Culture: Cultivating a data-centric culture, where data is valued and effectively managed, reinforces the trustworthiness of data across the organization. Reference: "Knowledge sharing and protection in data-centric collaborations: An exploratory study" by Zeiringer and Thalmann, published in 2022. This study explores the new dimension of trust in data-centric collaborations, particularly in settings where tacit knowledge is created and collected by sensors. It highlights the fostering of relationships between customers and suppliers in a data-centric environment, emphasizing the role of trust and proper data management in cultivating a data-centric culture within organizations.

The role of data governance in building trust and credibility in data is pivotal. By ensuring the integrity, transparency, and ethical use of data, organizations can establish a durable foundation of trust with all stakeholders, which is essential for long-term success and sustainability.

1.4 Key Components of a Data Governance Program

1.4.1 Data Governance Framework: Structure and Components

A data governance framework serves as the backbone of a data governance program, providing the structure and components necessary to effectively manage and govern an organization's data assets. This section outlines the essential elements and structure of a comprehensive data governance framework.

- **Defining a Data Governance Framework**

A data governance framework serves as the cornerstone for structured data management within organizations. Here is an elucidation of its core components:

Framework Overview A data governance framework constitutes a comprehensive compilation of guidelines, policies, standards, and procedures that delineate the management, usage, and safeguarding of data within an organization. It provides a structured and systematic approach to data governance, ensuring consistency and alignment with overarching business objectives. For instance, Mahanti (2021) discusses the critical components of a data governance framework, including roles, responsibilities, processes, and tools necessary for effective data management (Mahanti 2021).

Purpose of the Framework The primary purpose of a data governance framework is twofold. Firstly, it seeks to establish clarity, coherence, and control over data management practices. Secondly, it empowers organizations to maximize the value extracted from their data assets while steadfastly upholding compliance with legal and ethical standards. Through these dual objectives, the framework forms the bedrock for responsible and effective data governance within the organization. Zorrilla and Yebenes (2022) highlight the importance of a formal reference framework for implementing data governance systems, especially for organizations transitioning towards Industry 4.0 (Zorrilla and Yebenes 2022).

A data governance framework embodies the principles, policies, and procedures that govern data throughout its lifecycle. It acts as a guiding compass, facilitating the realization of data's full potential while safeguarding its integrity and ethical use.

- **Core Components of a Data Governance Framework**

A robust data governance framework comprises several integral components that collectively shape the governance structure and practices within an organization:

1. **Data Governance Policies:** Data governance policies are formal documents that establish rules and guidelines for data management. They encompass areas like data quality, security, privacy, and usage. Policies serve as a compass, providing clear direction and setting expectations for how data should be handled uniformly

across the organization. These policies are critical for ensuring consistent data management practices (Mahanti 2021).
2. **Data Standards and Procedures:** Data standards ensure the uniformity, accuracy, and usability of data. Procedures, on the other hand, delineate the operational aspects of data management, encompassing data collection, storage, processing, and sharing. Together, standards and procedures establish a structured approach to data governance, enabling reliable and efficient data handling (TechRepublic 2022).
3. **Roles and Responsibilities:** Clearly defined roles and responsibilities are pivotal for effective data governance. This entails appointing data stewards, data owners, and a governing body or council responsible for overseeing data governance activities. A well-defined governance structure ensures accountability and ownership, fostering a culture of responsibility and structured management within the organization (Data Governance Institute 2022).
4. **Data Architecture and Models:** The design and implementation of data architecture and models are foundational to data governance objectives. This component encompasses data integration, data warehousing, and metadata management. It provides the infrastructure necessary to support data governance practices, ensuring that data is organized and accessible for decision-making (Kimachia 2023).
5. **Compliance and Risk Management:** The framework must incorporate mechanisms to ensure compliance with pertinent data protection regulations, such as GDPR, and to manage risks associated with data. This includes risk assessment and mitigation strategies to safeguard data assets, maintaining adherence to legal standards, and protecting against potential data breaches (arXiv 2023).
6. **Monitoring and Auditing:** Regular monitoring and auditing of data governance practices are imperative to maintain ongoing compliance and effectiveness. This component involves tracking performance against data governance key performance indicators (KPIs) to identify areas for improvement and ensure that governance practices are being followed consistently (arXiv 2023).
7. **Technology and Tools:** Identification and deployment of suitable technology and tools that align with the data governance framework are essential. This includes data quality management software, data cataloging tools, and compliance management systems, which streamline and enhance data governance processes. The use of appropriate technology supports efficient data management and ensures that data governance practices are effectively implemented (Data Governance Institute 2022).

Figure 1.14 illustrates a comprehensive data governance framework, designed to guide organizations in effectively managing and protecting their data assets. The diagram includes various components such as policies, procedures, roles, responsibilities, and technology infrastructure that form the backbone of an effective data governance strategy. Each element is depicted in relation to its function and importance within the overall framework, highlighting how they interconnect to support data quality, security, privacy, and compliance objectives. This visual representation

1.4 Key Components of a Data Governance Program

Fig. 1.14 Data governance framework. (Created by the author)

aids in understanding the structured approach needed to implement robust data governance practices, ensuring that data is not only secure and compliant with regulations but also accessible and usable for business purposes.

A well-structured data governance framework encompasses these core components, creating a cohesive and organized approach to data management, protection, and utilization within an organization.

- **Challenges in Developing a Data Governance Framework**

The development of a robust data governance framework is crucial for effective data management, but it is not without its challenges:

Balancing Flexibility and Control Striking the right balance between flexibility and control is intricate. A framework must be flexible enough to adapt to evolving data needs and technological advancements while providing the necessary control and compliance measures to safeguard data integrity and security. This challenge is highlighted in the benchmarking of data governance frameworks, which emphasizes the need for a holistic approach to manage the global flow of data responsibly (Marcucci et al. 2023).

Ensuring Stakeholder Buy-In Gaining buy-in from all stakeholders is imperative, especially when different business units within an organization have diverse data needs and priorities. Aligning these stakeholders with the framework's objectives and principles is a critical success factor for its implementation. The OECD's review of public sector data governance frameworks underlines the importance of involving all relevant stakeholders to ensure coherence and alignment in data governance practices (OECD 2019).

Addressing Fragmentation in Policies and Practices The lack of a unified approach to data governance often results in fragmented policies and practices, which can undermine the effectiveness of a data governance framework. Ensuring consistent application of data governance principles across the organization is essential for maintaining data quality and compliance (Marcucci et al. 2023).

Managing Diverse Data Sources Integrating and managing data from various sources is a common challenge in developing a data governance framework. The complexity of harmonizing data from different systems and ensuring its accuracy and usability requires robust processes and tools (OECD 2019).

While a well-structured data governance framework is essential for effective data management, addressing the challenges of flexibility and control, as well as securing stakeholder buy-in, requires careful planning and strategic communication. Overcoming these challenges is pivotal to unlocking the full potential of an organization's data assets.

1.4.2 Data Stewards, Data Owners, and Governance Bodies

Effective data governance requires clearly defined roles and responsibilities. This section focuses on the crucial roles of data stewards, data owners, and governance bodies within a data governance program, outlining their functions and significance.

1. **Data Stewards: Guardians of Data Governance**

Data stewards play a pivotal role in the data governance landscape, taking on a range of responsibilities:

Role Definition
Data stewards serve as the custodians of data within their respective areas of responsibility. Their primary duty is to oversee the day-to-day management and maintenance of data assets, ensuring that data governance policies and procedures are diligently implemented and adhered to (Eckerson 2023; Health Data Stewardship and Governance 2011).

Key Responsibilities
Data stewards are entrusted with several critical tasks, including:

- **Overseeing Data Quality:** They are responsible for maintaining data quality and ensuring that data is accurate, complete, and consistent (Abraham et al. 2019).
- **Ensuring Data Consistency:** Data stewards work to maintain data consistency across the organization, preventing discrepancies or conflicts (Eckerson 2023).
- **Assisting with Data-Related Issues:** They provide support in resolving data-related problems and inquiries, acting as a bridge between IT and business units (UNM Data Governance 2023).
- **Liaison Between IT and Business:** Data stewards serve as a crucial link, facilitating effective communication and collaboration between technical teams and business units (Eckerson 2023).

1.4 Key Components of a Data Governance Program

Skills and Qualifications

Data stewards typically possess a combination of skills and qualifications, including:

- **Deep Understanding of Data Management:** They have a profound knowledge of data management practices, policies, and data governance principles (Abraham et al. 2019).
- **Strong Communication Skills:** Effective communication is key to their role, as they interact with various stakeholders to ensure data governance compliance (Health Data Stewardship and Governance 2011).
- **Organizational Skills:** Data stewards need strong organizational skills to manage data assets efficiently and oversee compliance with governance policies (UNM Data Governance 2023).

Data stewards are the operational champions of data governance, responsible for ensuring that data remains of high quality, reliable, and aligned with organizational goals and policies. Their contributions are integral to the success of data governance initiatives.

2. **Data Owners**

Data owners hold a pivotal role in data governance, wielding significant responsibility within their respective domains:

Role Definition

Data owners are typically senior-level executives who bear overall accountability for data assets within their designated domains. They shoulder the responsibility for crafting data strategy and formulating policies that pertain to their specific areas (UNM Data Governance 2023; Health Data Stewardship and Governance 2011).

Key Responsibilities

The key responsibilities of data owners encompass several critical functions:

- **Access Control:** They determine who within the organization can access specific data assets, ensuring that access aligns with organizational needs and data governance policies (Eckerson 2023).
- **Regulatory Compliance:** Data owners are responsible for ensuring that data under their purview complies with pertinent regulations, such as data protection laws like GDPR (Abraham et al. 2019).
- **Data Management Decisions:** They make critical decisions regarding data, including matters related to data storage, archiving, and usage policies (Health Data Stewardship and Governance 2011).

Authority and Accountability

Data owners possess the authority to make decisive calls regarding their data assets. Their decisions extend to matters of data security, quality, and strategic utilization. Additionally, they bear accountability for the safeguarding and integrity of data under their purview (Eckerson 2023; UNM Data Governance 2023).

50   1   Overview and Importance of Data Governance

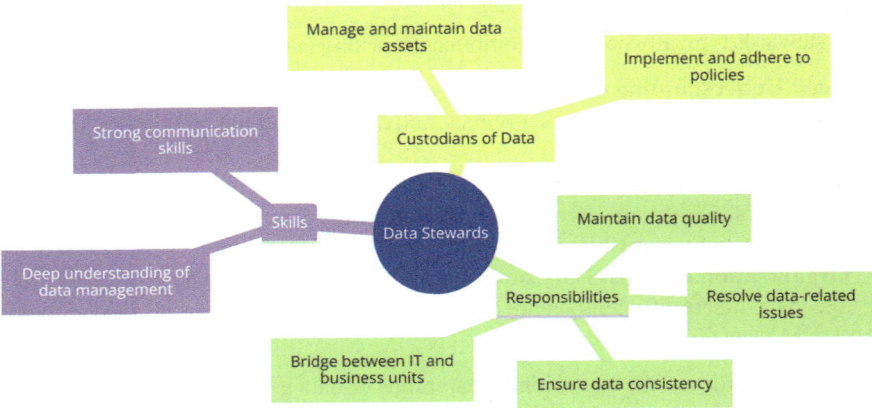

Fig. 1.15 Data stewards map. (Created by the author)

Figure 1.15 provides a detailed visualization of the roles and responsibilities of data stewards within an organization. It includes a network diagram illustrating how data stewards interact with various data sources, systems, and organizational units to ensure data quality, compliance, and effective data management. The map details specific tasks assigned to data stewards, such as data validation, monitoring, and liaison with other data governance bodies. By showcasing the central role of data stewards in maintaining the integrity and utility of data, this figure emphasizes the importance of their position in enforcing data governance standards and facilitating effective data usage across the enterprise.

Data owners play a pivotal role in defining and safeguarding the data landscape within their domains. Their authority and accountability ensure that data assets are managed strategically, compliantly, and in alignment with the overarching data governance framework.

3. **Governance Bodies**

Governance bodies, like a Data Governance Council or Committee, are pivotal in steering the course of data governance within an organization:

Role Definition
Composed of senior stakeholders representing various business units, governance bodies are entrusted with providing strategic direction and oversight for the data governance program. They serve as the custodians of the data governance initiative, ensuring its alignment with the organization's overarching objectives (Abraham et al. 2019; Health Data Stewardship and Governance 2011).

Key Responsibilities
Governance bodies shoulder a range of significant responsibilities, including:

1.4 Key Components of a Data Governance Program

- **Setting Governance Priorities:** They establish governance priorities, determining which aspects of data governance merit the most attention and resources (UNM Data Governance 2023).
- **Conflict Resolution:** Governance bodies play a crucial role in resolving data-related conflicts that may arise within the organization, striking a balance between competing interests (Eckerson 2023).
- **Alignment with Business Objectives:** They ensure that data governance activities are closely aligned with the strategic objectives of the business, thereby enhancing the program's relevance (Health Data Stewardship and Governance 2011).
- **Monitoring Effectiveness:** Governance bodies actively monitor the effectiveness of the data governance program, employing key performance indicators (KPIs) to assess its impact and success (Abraham et al. 2019).

Strategic Decision-Making

Most importantly, governance bodies wield major influence in strategic decision-making concerning data governance. They navigate the complex terrain of different stakeholder needs, ensuring that data governance strikes a harmonious balance while supporting the achievement of organizational goals (UNM Data Governance 2023; Eckerson 2023).

Governance bodies are the linchpin of strategic data governance, providing direction, oversight, and harmony within the data governance program. Their role is pivotal in ensuring that data governance aligns with the broader organizational vision.

- **Challenges in Defining Roles and Responsibilities**

While defining roles and responsibilities is crucial for effective data governance, it is not without its challenges:

Interdepartmental Coordination

Effective coordination and communication among various roles can be a significant challenge, especially in large and complex organizations with numerous stakeholders. Ensuring that information flows seamlessly across departments and roles is essential for a cohesive data governance program (Abraham et al. 2019; UNM Data Governance 2023).

Role Clarity

A clear definition of the scope and boundaries of each role is paramount to avoid overlaps and gaps in data governance activities. Role ambiguity can lead to confusion, inefficiencies, and potential conflicts within the data governance framework (Eckerson 2023; Health Data Stewardship and Governance 2011).

Figure 1.16 presents a mindmap that outlines the structure and interrelationships of various governance bodies within an organization responsible for data governance. The diagram clearly delineates the hierarchy and functions of each body, including committees, advisory groups, and operational teams, showing their specific roles in the oversight and management of data governance initiatives. The mindmap illustrates the pathways through which these bodies communicate and

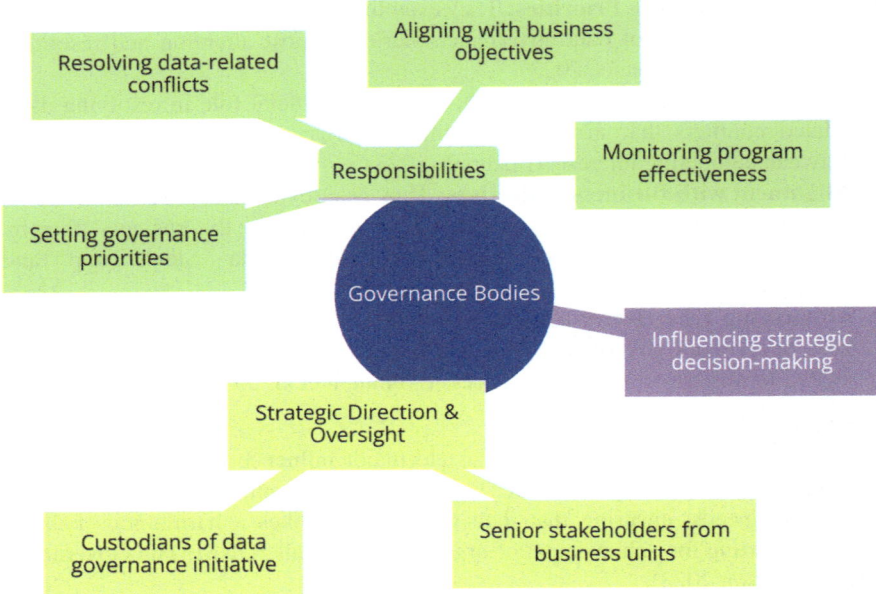

Fig. 1.16 Governance bodies mindmap. (Created by the author)

collaborate, ensuring effective governance across different levels of the organization. This visual framework is essential for understanding how data governance is implemented and maintained, highlighting the coordinated effort required to uphold data standards and policies.

Addressing these challenges in defining roles and responsibilities is essential for the successful implementation and execution of a data governance program. Well-defined roles and collaborative efforts among data stewards, data owners, governance bodies, and other stakeholders are pivotal in managing data effectively and in alignment with organizational goals and compliance requirements.

1.4.3 Policies and Standards: Development, Implementation, and Enforcement

In the context of data governance, the development, implementation, and enforcement of policies and standards are crucial for ensuring that data across the organization is managed effectively and in alignment with business goals and regulatory requirements. This section examines these key components, supported by real references.

1.4 Key Components of a Data Governance Program

- **Development of Policies and Standards**

Establishing Clear Guidelines The formulation of data governance policies necessitates the creation of lucid, all-encompassing guidelines that dictate the proper handling, usage, and safeguarding of data. These policies are paramount in delineating expectations and responsibilities regarding data management, as advocated by frameworks such as the Data Management Body of Knowledge (DAMA-DMBOK2) (Hendrawan 2023).

Incorporating Stakeholder Input A vital aspect of policy development involves engaging stakeholders from diverse departments to ensure alignment with distinct business needs and practical feasibility. The inclusion of multiple perspectives enhances the robustness and relevance of data governance policies. For instance, Dixon and Feldman (2023) highlight the importance of stakeholder engagement in governing health information exchange networks, emphasizing how diverse stakeholder representation enhances the robustness and relevance of data governance policies.

Aligning with Regulatory Requirements Policies must be in consonance with external regulations. For example, policies governing personal data in Europe must strictly adhere to the stringent standards set forth by GDPR (European Commission 2018b). This alignment is crucial for ensuring that data governance practices comply with legal standards and mitigate risks associated with noncompliance.

Figure 1.17 provides a comprehensive diagram illustrating the process of developing policies and standards within an organization's data governance framework. The diagram shows a step-by-step progression from the initial assessment of data needs to the final implementation and monitoring of established policies. It outlines key stages such as the identification of regulatory requirements, stakeholder engagement, draft policy formulation, feedback loops, policy approval, and training for compliance. The figure emphasizes the iterative nature of policy development, highlighting the importance of continuous improvement and adaptation to changing data environments and business needs. This visual representation aids in understanding the structured approach necessary for creating robust data governance policies that are both effective and compliant with applicable laws and standards.

Fig. 1.17 Development of policies and standards. (Created by the author)

- **Implementation of Policies and Standards**

Effective Communication Successful implementation commences with effective communication of policies to all stakeholders, fostering a comprehensive understanding of their roles and responsibilities in the context of data governance. This approach ensures a unified understanding of the principles governing data management. Effective communication strategies include setting up governance frameworks that detail policies, roles, and expectations clearly, which is essential for aligning all stakeholders with the organization's data governance objectives (Airbyte 2023).

Training and Support Acilitating training programs and providing support are pivotal in promoting seamless implementation. Training ensures that all stakeholders are well-versed in data governance principles and can execute their responsibilities effectively. As highlighted by Plotkin (2020), ongoing training and support play a critical role in the effective implementation of data governance policies and standards within organizations. This involves regular workshops, online courses, and support systems to address any questions or challenges that may arise (ClickUp 2023).

Integration with Business Processes To be truly effective, policies should be seamlessly integrated into existing business processes, thereby becoming an integral part of the organizational routine. Kirchmer (2021) explores the impact of digital transformation on business process governance, emphasizing the need to align data governance policies with existing business processes. This integration ensures that data governance is not seen as a separate initiative but as a core component of the organization's operational framework. By embedding data governance into everyday business activities, organizations can enhance compliance and operational efficiency (ClickUp 2023).

- **Enforcement of Policies and Standards**

Monitoring and Auditing The enforcement of data governance policies is critically dependent on regular monitoring and auditing practices. These practices are vital for tracking adherence to policies and swiftly identifying any deviations or issues that may arise. Effective monitoring and auditing are central to ensuring compliance with data governance policies and standards. These processes help maintain the integrity of data governance initiatives and ensure organizational compliance (Sakura Sky 2023; Atlan 2023b). Regular audits and continuous monitoring are essential components of a robust data governance framework, enabling organizations to detect and address noncompliance promptly (The Institute of Internal Auditors 2023).

Establishing Accountability A key facet of enforcement involves the establishment of clear accountability mechanisms, ensuring that instances of noncompliance are addressed promptly and effectively. It is crucial to define roles and responsibilities clearly within the data governance framework. Establishing accountability

1.4 Key Components of a Data Governance Program

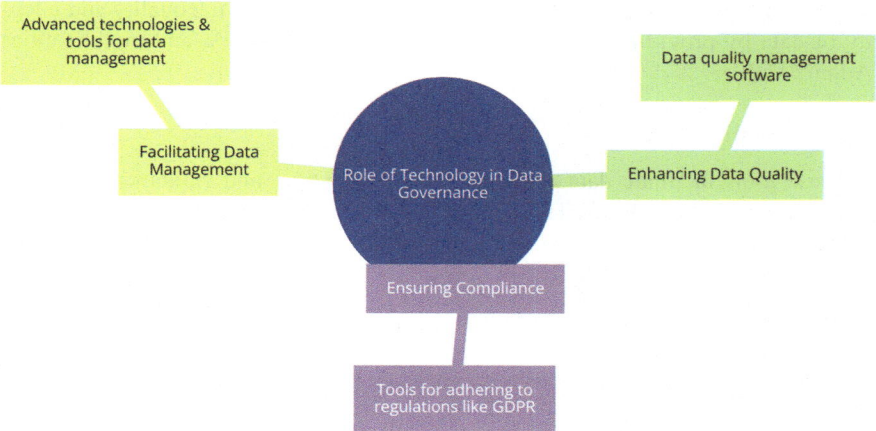

Fig. 1.18 Role of technology in data governance. (Created by the author)

providing a single place where all data assets within an organization are cataloged and described. Data catalogs make data discovery easier and more efficient, allowing users to search for data using keywords and filters. They often include features like data lineage information, annotations, and user ratings, which help users understand and trust the data they find (Ladley 2019c; Abraham et al. 2019).

Metadata Management Metadata management involves the handling and administration of data descriptions and relationships. It is the process of ensuring that all data across an organization is described consistently and accurately. This includes technical metadata (like data types and sources) and business metadata (like data ownership and glossaries). Effective metadata management enables better data quality and easier data integration and supports compliance with data regulations (Atlan 2023b).

- **Importance in Data Governance**

Organization and Accessibility Data catalogs and metadata management help organize the data in a way that makes it easily accessible to authorized users (Atlan 2023b).

Data Discovery and Comprehension These tools simplify the process of finding the right data and understanding its context and usage (Atlan 2023b).

Compliance and Security They play a key role in ensuring data compliance with various regulations and maintaining data security (Ladley 2019c; Atlan 2023b).

Data Quality and Integrity These tools help in maintaining the quality and integrity of data by providing information about its source, history, and modifications (Atlan 2023b).

- **Data Quality Tools**

  These tools facilitate data cleansing, standardization, and validation, ensuring data accuracy and reliability. The tools are vital for maintaining the accuracy, completeness, and reliability of data within an organization. These tools facilitate various functions:

  - **Data Cleansing:** Correcting or removing incorrect, corrupted, or duplicate data (Oliveira et al. 2019).
  - **Standardization:** Ensuring data is consistent and adheres to standards or norms (Redman 2017).
  - **Validation:** Verifying data for accuracy and quality (Haug et al. 2011).

  The importance of these tools lies in their ability to improve decision-making, support compliance, and enhance operational efficiency by ensuring data is accurate and useful (Oliveira et al. 2019).

Data Integration Tools Data integration tools streamline the amalgamation of data from diverse sources, a crucial component for comprehensive analytics and reporting. These tools are crucial for combining data from various sources into a coherent store. These tools enable organizations to merge data from various databases, applications, and systems, facilitating comprehensive analytics and reporting (Vassiliadis 2009).

Key Aspects of Data Integration Tools:

- **Combining Diverse Data Sources:** They enable the amalgamation of data from various sources, including cloud-based and on-premises systems (Vassiliadis 2009).
- **Data Transformation:** Transforming data into a format suitable for analysis and reporting (Kimball and Ross 2013).
- **Improving Data Accessibility:** Making data accessible for various business intelligence, analytics, and reporting tools (Kimball and Ross 2013).
- **Ensuring Data Quality:** These tools often include features to cleanse and enrich data, improving its quality and reliability (Vassiliadis 2009).

Compliance and Risk Management Software Specialized software for compliance and risk management aids in the vigilant monitoring of data-related risks and the unfaltering adherence to legal standards. These tools are critical tools designed to help organizations manage data-related risks and ensure adherence to legal and regulatory standards. These software solutions provide functionalities like:

- **Risk Assessment:** Identifying and evaluating risks associated with data handling and processing (Baumgartner et al. 2023).

1.4 Key Components of a Data Governance Program

- **Compliance Tracking:** Ensuring that data management practices comply with legal standards, such as GDPR, HIPAA, etc. (GRC Capability Model 2017).
- **Audit Trails:** Creating detailed logs of data access and changes, which are crucial for audits (O'Neill 2014b).
- **Policy Management:** Implementing and monitoring adherence to internal data governance policies (Baumgartner et al. 2023).
- **Incident Management:** Managing and responding to data breaches or compliance issues (GRC Capability Model 2017).

Figure 1.19 delineates the essential tools used in data governance, mapping out their applications and how they integrate into various governance activities. It categorizes tools into several types, such as data quality management tools, metadata management systems, data security solutions, and compliance monitoring software. The diagram illustrates the specific functionalities of each tool type and their role in ensuring data integrity, security, compliance, and overall governance effectiveness. This comprehensive depiction helps highlight how these tools support the foundational aspects of data governance, providing the necessary infrastructure to manage, protect, and optimize data assets across the organization.

- **Challenges in Utilizing Technology and Tools**

Integration with Existing Systems Integrating new data governance tools with pre-existing IT infrastructure can be a formidable challenge, often necessitating meticulous planning and seamless execution. Challenges in integrating innovative technology and tools in data governance, particularly when merging them with existing IT systems, are a significant aspect of modern data management. Key challenges include:

- **Compatibility Issues:** Ensuring new tools are compatible with the existing IT infrastructure can be complex. Different systems may have varying data formats, protocols, and architectures, leading to integration difficulties (McKinsey 2023a).

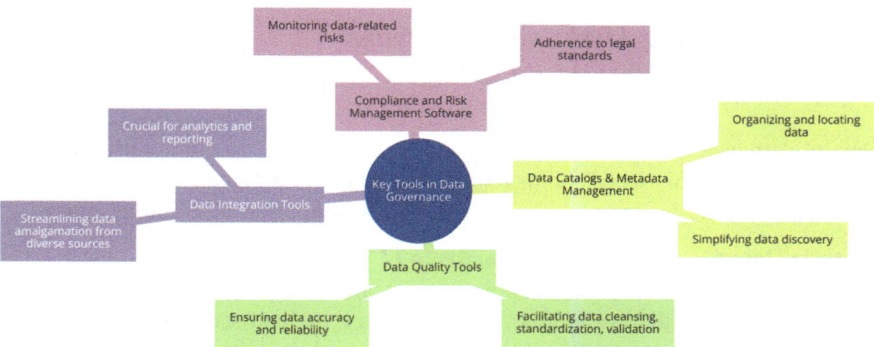

Fig. 1.19 Key tools in data governance. (Created by the author)

- **Data Integration:** Efficiently combining data from new tools with existing databases and systems is critical. This involves not only technical integration but also ensuring the data's consistency and quality across systems (Alation 2023).
- **Change Management:** Managing the organizational changes required to adopt modern technologies involves addressing resistance to change, training needs, and aligning new processes with business objectives. Successful change management requires clear communication and leadership commitment (McKinsey 2023a).
- **Training and Adaptation:** Ensuring staff are trained and adapt to the new systems is crucial. Continuous training programs and support systems are necessary to help employees become proficient with new tools and processes (Alation 2023).
- **Security Concerns:** Addressing any security risks that come with integrating modern technologies is paramount. New systems may introduce vulnerabilities that need to be managed through robust security protocols and regular audits (McKinsey 2023a).

Keeping Pace with Technological Advancements The rapid progression of technology mandates that organizations continually assess and update their tools to remain effective in their data governance endeavors. Staying current with technological advancements ensures that data governance strategies remain robust and can leverage the latest innovations for better data management and compliance (Atlan 2023b).

Technology and tools constitute vital pillars in bolstering and enhancing data governance programs. They offer solutions for efficient data management, quality assurance, regulatory compliance, and much more. As organizations continue to evolve and generate ever-increasing volumes of data, the reliance on these technologies is poised to expand, underscoring their pivotal role in the realm of effective data governance.

1.5 Common Misconceptions About Data Governance

1.5.1 Debunking Myths and Clarifying Common Misunderstandings

Data governance, while critical to organizational success, is often surrounded by misconceptions. This section aims to debunk some of the prevalent myths and provide clarity on these common misunderstandings, supported by credible sources.

Myth 1: Data Governance is Primarily an IT Concern
Reality: Data governance extends beyond the realm of IT. It is a business issue that involves managing the availability, usability, integrity, and security of the data. A key reference that demonstrates the reality that data governance extends beyond the realm of IT is the article "Designing Data Governance" by Khatri and Brown,

1.5 Common Misconceptions About Data Governance

published in Communications of the ACM in 2010. This article provides a comprehensive framework for data governance that transcends the boundaries of IT. It highlights that data governance is not just a technological issue but a business one, involving the management of availability, usability, integrity, and security of data. This article emphasizes the broader implications of data governance, asserting its role as a fundamental business issue requiring a comprehensive approach that integrates both IT and business perspectives (Khatri and Brown 2010).

Myth 2: Data Governance Limits Access to Data
Contrary to limiting data access, effective data governance facilitates more efficient and secure access to data. A key reference addressing this myth is the article "Bottom-up data trusts: Disturbing the 'one size fits all' approach to data governance" by Delacroix and Lawrence, published in International Data Privacy Law in 2019. This article discusses the concept of data trusts and how they can be employed to facilitate efficient and secure access to data, contrary to the belief that data governance limits access. It highlights how data governance, when properly implemented, does not restrict data access but rather ensures it is done securely and efficiently, benefiting all stakeholders involved (Delacroix and Lawrence 2019).

Myth 3: Data Governance Is Only About Compliance
While compliance is a crucial aspect, data governance encompasses much more. It involves improving data quality, enhancing decision-making, and driving business growth. A relevant reference addressing the myth that data governance is solely about compliance is the article "Data governance: Going beyond compliance" by Mansfield-Devine, published in Computer Fraud & Security in 2017. This article discusses how data governance transcends mere compliance, involving aspects like improving data quality, enhancing decision-making, and driving business growth. It emphasizes that while compliance is a crucial part of data governance, it also plays a significant role in ensuring that data is a valuable business asset, contributing to overall organizational success (Mansfield-Devine 2017).

Myth 4: Implementing Data Governance Is a One-Time Effort
Reality: Data governance is an ongoing process. It requires continuous monitoring, updating, and adapting to new data needs, technologies, and regulatory changes. The article "Data governance: A conceptual framework, structured review, and research agenda" by Abraham, Schneider, and Vom Brocke, published in the International Journal of Information Management in 2019, addresses the myth that implementing data governance is a one-time effort. This research outlines a conceptual framework for data governance and asserts that it is an ongoing program requiring continuous effort. It highlights that data governance involves continuous monitoring, updating, and adapting to new data needs, technologies, and regulatory changes, thereby contributing to efficient decision-making and business growth (Abraham et al. 2019).

Myth 5: Data Governance Stifles Innovation
Reality: Rather than stifling innovation, data governance can foster it by ensuring data quality and trustworthiness. A relevant reference that demonstrates the

relationship between data governance and fostering innovation is the article "The role of information governance in big data analytics driven innovation" by Mikalef et al., published in the journal *Information & Management* in 2020. This study explores the impact of information governance practices in shaping innovation capabilities, particularly in the context of big data analytics. The authors argue that effective information governance can indeed support innovative capabilities by ensuring the quality and trustworthiness of data. This research offers an insightful perspective on how properly governed data can be a foundation for innovative analytics and business practices, debunking the myth that data governance stifles innovation (Mikalef et al. 2020).

Figure 1.20 outlines and dispels common myths associated with data governance, providing a clear, factual counter-narrative for each misconception. It presents a list or series of statements that are commonly believed within the industry but are inaccurate, such as myths about the complexity, cost, or required resources for implementing data governance practices. For each myth, the diagram provides evidence or explanations that refute these misconceptions, illustrating the true nature and benefits of data governance. This visual representation is crucial for educating stakeholders about the realities of data governance and correcting any misunderstandings that may hinder its adoption.

- **Challenges in Addressing Misconceptions**

Communication and Education: Effectively communicating the true scope and benefits of data governance is essential to dispel these myths. This involves educating stakeholders across the organization.

Effective communication about data governance involves creating a well-structured communication plan that engages stakeholders and promotes the significance of data governance. Key strategies include defining clear communication objectives, identifying target audiences, and developing key messages that resonate with these audiences. Communication channels can include newsletters, intranet

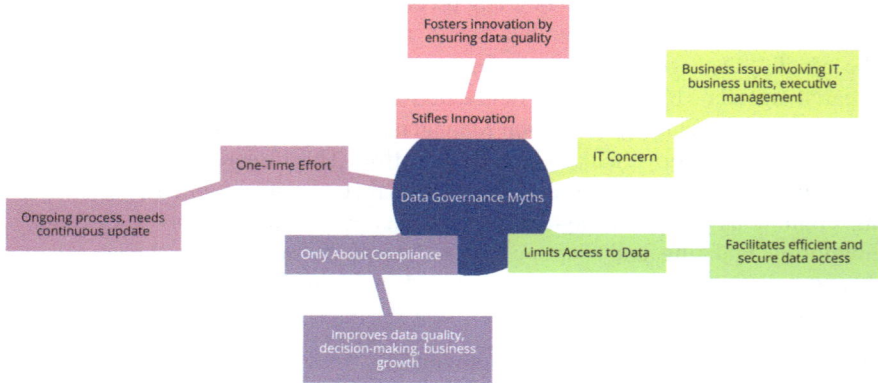

Fig. 1.20 Data governance myths. (Created by the author)

1.5 Common Misconceptions About Data Governance   63

sites, SharePoint platforms, and other digital communication tools to ensure that the message is disseminated effectively (DataQG 2023).

Education is equally important. Continuous education programs help in managing resistance to change and fostering a data-driven culture within the organization. This includes training sessions, workshops, and creating content like infographics and videos that explain the importance and benefits of data governance (McKinsey 2023a; Monte Carlo Data 2023).

Understanding and addressing these misconceptions about data governance is crucial for its successful implementation and operation. By clarifying these common misunderstandings, organizations can better leverage data governance as a tool for strategic decision-making, compliance, and innovation.

1.5.2 Data Management: Clarifying the Differences

There is often confusion between data governance and data management, with the terms sometimes being used interchangeably. However, they represent distinct but interrelated disciplines within an organization's data strategy. This section aims to clarify the differences between them, supported by credible sources.

- **Data Governance Defined**

Strategic Oversight Data governance refers to the strategic framework and oversight for managing data's availability, usability, integrity, and security within an organization. It involves setting policies, standards, and procedures that govern data usage. The book "Data Governance: How to Design, Deploy, and Sustain an Effective Data Governance Program" by Ladley, published in 2019, is a significant reference that defines data governance in terms of strategic oversight. This book describes data governance as a strategic framework for managing the availability, usability, integrity, and security of data within an organization. It details the importance of setting policies, standards, and procedures that govern data usage, emphasizing that data governance involves much more than just data management; it requires a comprehensive approach encompassing various aspects of an organization. This resource offers a comprehensive understanding of data governance as a crucial element of organizational strategy, ensuring effective management and utilization of data (Ladley 2019c).

Focus on Policy and Compliance Data governance focuses on policy, compliance, and ensuring that data management activities align with organizational goals and regulatory requirements. The same book by Ladley (2019c) serves as a key reference in demonstrating the focus of data governance on policy, compliance, and aligning data management activities with organizational goals and regulatory requirements. This comprehensive guide highlights how data governance is more than just an IT concern; it encompasses setting up strategic frameworks and oversight mechanisms for managing data's availability, usability, integrity, and security.

The book underscores the importance of policies, standards, and procedures in governing data usage and ensuring compliance with organizational and regulatory demands. This source provides a detailed insight into how data governance can effectively align data management with broader business objectives and compliance requirements, making it a strategic imperative for organizations (Ladley 2019c).

- **Data Management Defined**

Operational Execution Data management is the operational execution of the policies, standards, and procedures set forth by data governance. It involves the technical and practical aspects of collecting, storing, organizing, and maintaining the data (Ladley 2019c).

Focus on Data Processing and Quality Data management focuses on the practical work of handling data, ensuring its quality, accuracy, and accessibility for business use. The book "Foundations of Data Quality Management" by W. Fan and F. Geerts discusses traditional data management tasks and emphasizes improving the quality of relational data. It provides insights into ensuring the data's quality, accuracy, and accessibility for business use (Fan and Geerts 2012).

- **Key Differences**

Purpose and Scope Data governance is strategic, focusing on policy and compliance, while data management is operational, focusing on executing these policies in the day-to-day handling of data.

Responsibility Data governance requires a cross-functional team, including executive leadership, IT, and business units, to set data-related policies. In contrast, data management is often the responsibility of IT and data professionals who implement these policies.

Outcome The outcome of effective data governance is data that is well-regulated and compliant with internal and external standards. For data management, the outcome is data that is organized, maintained, and readily available for use.

- **Challenges in Differentiating the Two**

Interdependency While different, data governance and data management are interdependent. Effective data management relies on good governance to set the direction and standards.

Understanding the distinction between data governance and data management is essential for their effective implementation. While they are distinct in their focus and execution, both are crucial for ensuring that data is a strategic asset that drives informed decision-making and business success.

Figure 1.21 provides a clear and concise comparison between data governance and data management, outlining the distinct roles and responsibilities associated

1.6 Challenges in Implementing Data Governance

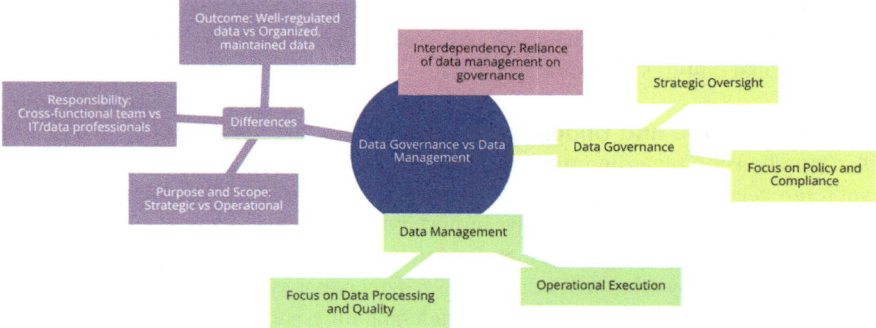

Fig. 1.21 Data governance vs data management. (Created by the author)

with each discipline. It features a side-by-side layout diagram illustrating areas of overlap and unique aspects. The visualization highlights key differences in focus areas, such as policy setting, oversight, and compliance for data governance, versus operational tasks, data handling, and maintenance for data management. This comparison helps clarify the interconnected yet distinct nature of these two crucial areas, emphasizing how they collectively contribute to the effective use of data within organizations.

1.6 Challenges in Implementing Data Governance

1.6.1 Overcoming Resistance to Change

Implementing data governance often encounters resistance within organizations. This resistance can stem from a variety of sources, ranging from cultural inertia to misunderstandings about data governance's role and value. Addressing this resistance is critical for the successful adoption of data governance practices. This section explores strategies to overcome resistance, supported by relevant references.

- **Understanding the Nature of Resistance**

Resistance often emanates from specific sources rooted in an organization's fabric. Identifying and comprehending these sources is pivotal in effectively addressing resistance.

Cultural Barriers Resistance frequently stems from entrenched organizational cultures resistant to change. A reference addressing the challenges in implementing data governance, particularly focusing on overcoming resistance to change and understanding the nature of resistance due to cultural barriers, is the article "Unraveling the Complexities of Data Governance with Strategies, Challenges, and Future Directions" by Pansara (2023). This research contributes to a deeper understanding of how organizations can harness data governance strategies while addressing significant obstacles, including overcoming resistance to change rooted in organizational cultures (Pansara 2023b).

Lack of Awareness and Understanding Resistance may also stem from a dearth of comprehension regarding the essence and benefits of data governance. A significant reference that addresses the challenge of a lack of awareness and understanding in data governance, highlighting the importance of heightened awareness and education, is "Data governance: A conceptual framework, structured review, and research agenda" by Abraham et al. (2019). This paper synthesizes the data governance literature and highlights gaps in understanding, suggesting a need for more awareness and structured knowledge about data governance practices and benefits (Abraham et al. 2019).

Fear of Losing Control or Autonomy Some departments or individuals may resist data governance endeavors due to apprehensions about relinquishing control over their data or processes. The paper "Bottom-up data trusts: Disturbing the 'one size fits all' approach to data governance" by Delacroix and Lawrence (2019) addresses the fear of losing control or autonomy in data governance. It explores how data governance can be reframed from a property-based perspective to a more trust-based approach, addressing the concerns of individuals or departments who may resist data governance due to apprehensions about relinquishing control over their data or processes (Delacroix and Lawrence 2019).

Figure 1.22 highlights three key factors contributing to resistance in the context of data governance: fear of losing control or autonomy, cultural barriers, and lack of awareness and understanding. These elements must be addressed to effectively implement data governance initiatives and mitigate resistance within an organization. The diagram categorizes these resistance types and illustrates how they impact the adoption and effectiveness of data governance practices. It also suggests

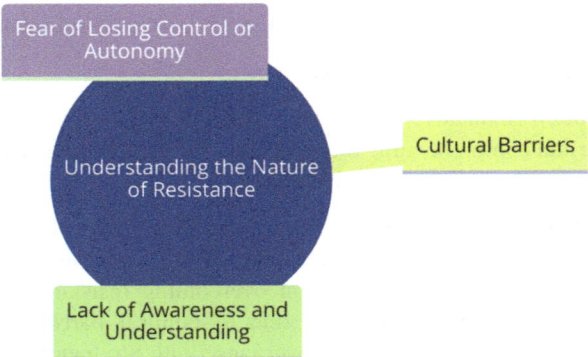

Fig. 1.22 Nature of resistance in data governance. (Created by the author)

1.6 Challenges in Implementing Data Governance

strategies for overcoming each type of resistance, highlighting the importance of communication, education, and inclusive engagement in fostering a culture that embraces data governance.

- **Strategies to Overcome Resistance**

Effective Communication as a Cornerstone Clear and consistent articulation of the benefits and strategic objectives associated with data governance is paramount. A pivotal reference that discusses strategies to overcome resistance in data governance with a focus on effective communication is the book "Data Governance: How to Design, Deploy, and Sustain an Effective Data Governance Program" by Ladley (2019c). This book delves into the various aspects of data governance, emphasizing the importance of clear and consistent communication in articulating the benefits and strategic objectives of data governance initiatives. It provides insights into how effective communication can be a cornerstone in overcoming resistance and fostering a successful data governance environment (Ladley 2019c).

Engagement of Key Stakeholders Active involvement of stakeholders in both the developmental and operational phases of data governance is essential for fostering organizational commitment. An important reference discussing the engagement of key stakeholders in data governance is "Emerging models of data governance in the age of datafication" by Micheli et al. (2020). This paper examines the different stakeholders in emerging models of data governance and emphasizes the importance of examining data governance arrangements in various contexts. It provides insights into incentivizing data use for public service and discusses the significance of stakeholder engagement in the effective management of data governance (Micheli et al. 2020).

Comprehensive Training and Education Programs To mitigate apprehensions and misconceptions, it is imperative to implement exhaustive training and educational programs centered around data governance. An important study addressing the implementation of comprehensive training and education programs in data governance is "Data governance: A conceptual framework, structured review, and research agenda" by Abraham et al. (2019). This research provides an extensive review of the data governance field, suggesting the need for thorough educational and training programs to address apprehensions and misconceptions about data governance. The study emphasizes the importance of structured knowledge dissemination and training in enhancing the effectiveness of data governance initiatives (Abraham et al. 2019).

Demonstration of Immediate Benefits Illustrating early successes, or 'quick wins', is a strategic approach to build momentum and highlight the tangible advantages of data governance. Early demonstration of benefits not only validates the initiative but also encourages continued support and enthusiasm among stakeholders. This tactic can be particularly effective in countering skepticism and bolstering confidence in the data governance program.

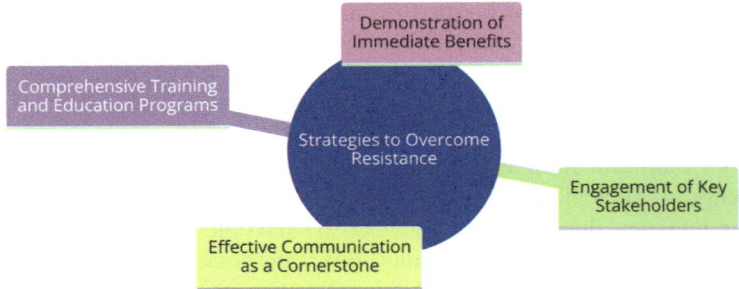

Fig. 1.23 Strategies to overcome resistance. (Created by the author)

Figure 1.23 highlights four key strategies for overcoming resistance in the context of data governance: comprehensive training and education programs, demonstration of immediate benefits, effective communication as a cornerstone, and engagement of key stakeholders. These strategies are essential for effectively implementing data governance initiatives and mitigating resistance within an organization. The diagram categorizes these approaches and illustrates how they can facilitate the adoption and effectiveness of data governance practices. It emphasizes the importance of clear communication, education, and active engagement in fostering a culture that supports and embraces data governance.

- **Addressing Specific Concerns**

Development of Tailored Solutions In the context of data governance, the development of tailored solutions is a critical approach for addressing specific organizational concerns. This involves creating customized strategies and tools that align with the unique needs, objectives, and challenges of an organization. Key aspects include:

Understanding Unique Organizational Needs Every organization has different data governance requirements based on its size, industry, and specific business processes. Tailored solutions begin with a thorough understanding of these unique needs. For instance, according to IBM's step-by-step guide, starting with a clear definition of business objectives and data needs is crucial for aligning data governance initiatives with overall business goals (IBM 2023).

Customized Policy Development Developing data governance policies that are specifically designed to address the challenges and goals of the organization is essential. This includes defining clear roles, responsibilities, and processes for data management. Effective policy development is emphasized as a core component of successful data governance programs, ensuring that all stakeholders are working towards the same objectives (IBM 2023).

1.6 Challenges in Implementing Data Governance

Technology Integration Implementing technology solutions that are compatible with the organization's existing IT infrastructure and data ecosystem is vital. This might include custom data management tools, software for data quality control, and secure data storage solutions. Data integration platforms tailored to specific domains can significantly streamline this process by connecting to various data sources and ensuring continuous data flow (Improvado 2023).

Training and Support Providing specialized training and support ensures that employees understand and can effectively implement the tailored data governance strategies. Continuous education and structured training programs are necessary to mitigate apprehensions and misconceptions about data governance (Abraham et al. 2019).

Compliance and Security Ensuring that data governance solutions comply with relevant legal and regulatory requirements is paramount. Tailored solutions often include specific measures to address security and privacy concerns unique to the organization or industry. Compliance with data protection regulations like GDPR is a common challenge that requires regular updates and audits of data policies (KAIZEN 2023).

Scalability and Flexibility Designing solutions that are not only tailored to current needs but also scalable and adaptable to future changes in the business environment or regulatory landscape is essential. Scalable data governance policies allow organizations to manage increasing data volumes and new data types effectively (Improvado 2023).

Monitoring and Continuous Improvement Regularly monitoring the effectiveness of these solutions and adjusting as needed to continuously improve data governance practices is a best practice. This involves establishing metrics to measure the success of data governance initiatives and making necessary adjustments based on ongoing evaluations (KAIZEN 2023).

Figure 1.24 highlights key elements in the development of tailored solutions for data governance: customized policy development, compliance and security, understanding unique organizational needs, training and support, technology integration, monitoring and continuous improvement, and scalability and flexibility. These elements are crucial for creating effective and adaptable data governance frameworks. The diagram categorizes these aspects and illustrates how they contribute to the successful implementation of tailored solutions. It emphasizes the importance of developing policies that are specific to organizational needs, ensuring compliance and security, integrating appropriate technologies, and maintaining continuous improvement to adapt to changing requirements.

Tailored solutions in data governance are essential because they ensure that the governance framework is not just a one-size-fits-all model but is instead a strategic asset that provides specific value to the organization.

Fig. 1.24 Development of tailored solution. (Created by the author)

- **Securing Executive Leadership Support**

Securing executive leadership support is a critical aspect of addressing specific concerns in data governance. This involves ensuring that top-level executives not only endorse but actively participate in the data governance program. Key elements include:

Understanding the Importance Executive leaders must understand the value of data governance and how it contributes to the organization's overall strategy and operational efficiency. As noted by AIM Consulting, securing buy-in from executives can significantly influence the success of data governance initiatives by aligning them with business objectives (AIM Consulting 2023).

Active Engagement Leaders should be actively engaged in the governance process, not just in a ceremonial capacity. Their involvement can drive the program's direction and help in overcoming any internal resistance. The IBM guide emphasizes that executive participation is crucial for aligning data policies with business strategies and ensuring consistent data use across the organization (IBM 2023).

Resource Allocation Executive support often translates into the allocation of necessary resources—budget, personnel, and technology—that are essential for the effective implementation of data governance. This support ensures that the necessary tools and resources are in place to manage data effectively (McKinsey 2023b).

Policy Endorsement and Enforcement Leaders play a critical role in endorsing policies and ensuring their enforcement across the organization. Their support legitimizes the program and encourages adherence to established policies. Executive leaders must back the governance policies to ensure they are implemented and followed throughout the organization (AIM Consulting 2023).

Change Management Implementing data governance often requires significant changes in organizational culture and processes. Executive leadership is crucial in managing this change, particularly in aligning the data governance initiative with other business priorities. Effective change management by leaders can help mitigate resistance and foster a data-driven culture (IBM 2023).

Communication Effective communication from the top can help in clarifying the goals of the data governance initiative and in rallying the organization around these objectives. Leaders must clearly communicate the importance of data governance and its benefits to the entire organization (McKinsey 2023b).

Building a Data-Driven Culture Executives can lead by example by promoting a data-driven decision-making culture within the organization. This includes using data effectively in their decision-making processes, thereby setting a standard for the rest of the organization (IBM 2023).

Stakeholder Engagement Executives can play a key role in engaging various stakeholders across the organization, ensuring that different departments collaborate effectively in the data governance initiative. Their involvement can foster cross-departmental cooperation and ensure the alignment of data governance goals with organizational objectives (AIM Consulting 2023).

Figure 1.25 highlights essential strategies for securing executive leadership support in data governance initiatives: policy endorsement and enforcement, communication, understanding the importance, stakeholder engagement, building a data-driven culture, change management, resource allocation, and active engagement. These strategies are crucial for gaining and maintaining executive support, which is vital for the successful implementation and sustainability of data governance practices. The diagram categorizes these strategies and illustrates their impact on fostering a supportive environment for data governance. It emphasizes the importance of clear communication, stakeholder involvement, adequate resources, and an organizational culture that values data-driven decision-making.

Overcoming resistance to change is a major challenge in implementing data governance. However, with effective communication, stakeholder involvement, training, and leadership support, organizations can navigate these challenges and successfully adopt data governance practices.

1.6.2 Addressing Data Quality Issues

Addressing data quality issues is a fundamental challenge in implementing data governance. Data quality encompasses the accuracy, completeness, consistency, timeliness, and reliability of data, which are essential for effective decision-making and operational processes. This section explores strategies to address data quality issues, supported by credible sources.

- **Strategies for Addressing Data Quality Issues in Data Governance**

Fig. 1.25 Securing executive leadership support (Created by the author)

Identification of Data Quality Issues

Data quality is a critical aspect of data governance, as it directly impacts the reliability, trustworthiness, and usefulness of the data in an organization. The identification of data quality issues is the first and fundamental step in addressing these issues effectively. Key elements include:

- **Establishing Data Quality Metrics and Standards**: According to a study by Atlan (2023a), establishing clear, measurable, and relevant data quality metrics and standards is crucial for data governance (Atlan 2023a).
- **Regular Data Auditing and Assessment**: 2023) Regular data audits and assessments are essential for maintaining high data quality. These processes involve systematically reviewing and evaluating data against predefined quality metrics to identify discrepancies and anomalies. Conducting regular audits ensures data accuracy, completeness, and consistency, which are crucial for informed decision-making and operational efficiency. By identifying and addressing data quality issues promptly, organizations can mitigate risks associated with inaccurate data, such as poor decision-making and compliance failures. Establishing a robust data governance framework that includes routine data quality audits helps organizations maintain the integrity and reliability of their data assets. (Minkkinen et al. 2022).
- **Implementing Data Profiling Techniques**: Data profiling is described by McKinsey as a vital process for examining data to collect statistics and information that help identify issues such as missing values and incorrect data entries (McKinsey 2023a).
- **Leveraging Automated Data Quality Tools**: The use of automated tools for enhancing the efficiency of data quality issue identification is emphasized by TCS, stating that these tools can automate the detection of anomalies and patterns indicating data quality issues (TCS 2024).
- **Feedback Loops from End-Users**: The feedback from end-users who interact with the data regularly is invaluable, as discussed in an article by Alation, which

1.6 Challenges in Implementing Data Governance

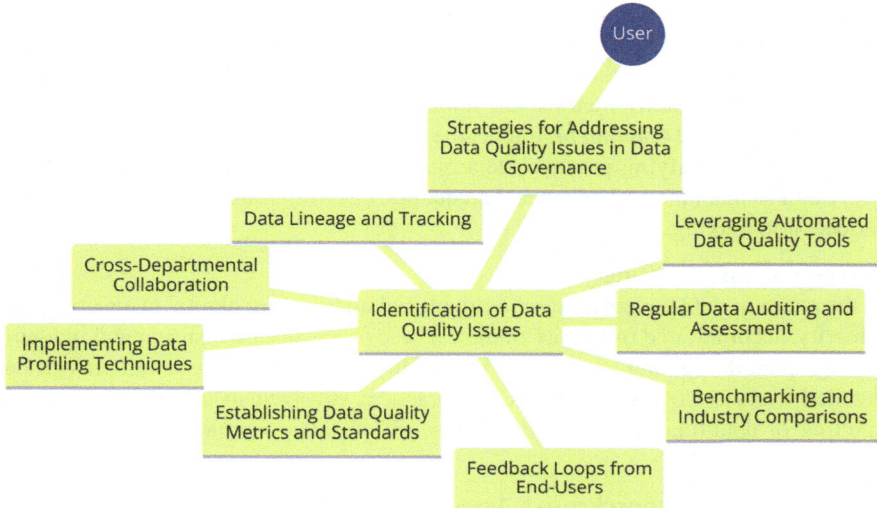

Fig. 1.26 Strategies for addressing data quality issues. (Created by the author)

explains how end-user feedback provides insights into issues not apparent through automated systems (Alation 2023).
- **Cross-Departmental Collaboration**: Cross-departmental collaboration is important for identifying department-specific data quality issues, as stated by McKinsey, which ensures data quality is maintained throughout the organization (McKinsey 2023a).
- **Data Lineage and Tracking**: Understanding data lineage and tracking the history of data helps in identifying the sources of data quality issues, as noted in an article by TCS (TCS 2024).
- **Benchmarking and Industry Comparisons**: Comparing an organization's data quality with industry benchmarks is a strategy suggested by Atlan, which can reveal areas where the organization is lagging and guide specific data quality improvements (Atlan 2023a).

The identification of data quality issues is a multi-faceted approach that involves a combination of techniques and strategies. It requires a proactive stance, leveraging technology, and fostering a culture of quality and collaboration across the organization. By identifying and addressing data quality issues, organizations can ensure the integrity and reliability of their data, which is foundational for effective data governance.

Figure 1.26

illustrates various strategies for identifying and addressing data quality issues within the context of data governance. The central theme is the "Identification of Data Quality Issues," surrounded by key strategies such as data lineage and tracking, leveraging automated data quality tools, regular data auditing and assessment, and benchmarking against industry standards. It also emphasizes the importance of cross-departmental collaboration, implementing data profiling techniques, establishing data quality metrics and standards, and incorporating feedback loops from end-users. The diagram highlights the user's role in providing feedback, which is crucial for identifying practical data quality problems. These strategies collectively aim to maintain and enhance data quality, ensuring reliable and accurate data for decision-making and governance.

Assessment and Auditing

In the realm of data governance, assessment and auditing are pivotal in ensuring the integrity, accuracy, and reliability of data. These processes are instrumental in identifying, quantifying, and addressing data quality issues. This subsection provides a detailed overview of the strategies and practices involved in the assessment and auditing of data quality within the framework of data governance. Key elements include:

- **Developing a Comprehensive Assessment Framework**: Creating a thorough assessment framework involves defining various dimensions of data quality, including accuracy, completeness, consistency, reliability, and timeliness. This framework should align with the organization's specific objectives and industry standards (Analytics8 2023).
- **Regular Auditing Cycles**: Establishing regular auditing cycles helps maintain data quality over time. This includes both routine and in-depth audits at set intervals, determined by the data's criticality and the environment's rate of change (Egnyte 2023).
- **Utilizing Automated Auditing Tools**: Automated tools significantly enhance the efficiency and thoroughness of data auditing processes. These tools can quickly analyze large volumes of data, identify anomalies, and generate reports that highlight areas of concern, also tracking changes over time (MEASURE Evaluation 2017).
- **Data Quality Scorecards and Reporting**: Implementing data quality scorecards and regular reporting mechanisms provides a quick overview of data quality dimensions, helping to identify areas needing immediate attention. These tools facilitate continuous monitoring and communication about the state of data quality within the organization (DataOpsZone 2023).
- **Root Cause Analysis**: Conducting a root cause analysis when issues are identified is essential for understanding and addressing the underlying problems rather than just the symptoms. This leads to more sustainable improvements in data quality (Egnyte 2023).
- **Stakeholder Involvement and Training**: Involving stakeholders from various departments and levels is crucial. Training and educating these stakeholders about the importance of data quality and the auditing process ensures data quality is seen as a shared responsibility (Analytics8 2023).
- **Benchmarking and Continuous Improvement**: Comparing the organization's data quality metrics with industry benchmarks provides insight into its standing

1.6 Challenges in Implementing Data Governance

and helps inform strategies for improvement. This should be part of a continuous improvement cycle (DataOpsZone 2023).
- **Data Governance Policies Compliance**: Regular audits should also assess compliance with internal data governance policies and external regulatory requirements, ensuring the organization maintains high data quality standards and adheres to legal obligations (MEASURE Evaluation 2017).

Figure 1.27 illustrates key components of assessment and auditing in the context of data governance. The central focus is on "Assessment and Auditing," surrounded by strategies essential for maintaining data quality and compliance. These include stakeholder involvement and training, data governance policies compliance, root cause analysis, benchmarking and continuous improvement, and developing a comprehensive assessment framework. It also highlights the importance of utilizing automated auditing tools, regular auditing cycles, and data quality scorecards and reporting. These components collectively ensure that data governance practices are effective, transparent, and continuously improving, thereby enhancing overall data quality and organizational performance. These strategies, when effectively implemented, help organizations maintain high data quality, enabling accurate and reliable data for decision-making and operational processes.

In conclusion, assessment and auditing in data governance are dynamic and ongoing processes that require a structured approach, the right tools, and the involvement of all stakeholders.

- **Strategies for Improving Data Quality**

Implementation of Data Quality Frameworks The implementation of data quality frameworks is a strategic approach to improving data quality within an organization. These frameworks provide a structured methodology for managing and enhancing the quality of data. This subsection delves into the key strategies involved in the implementation of data quality frameworks as part of data governance.

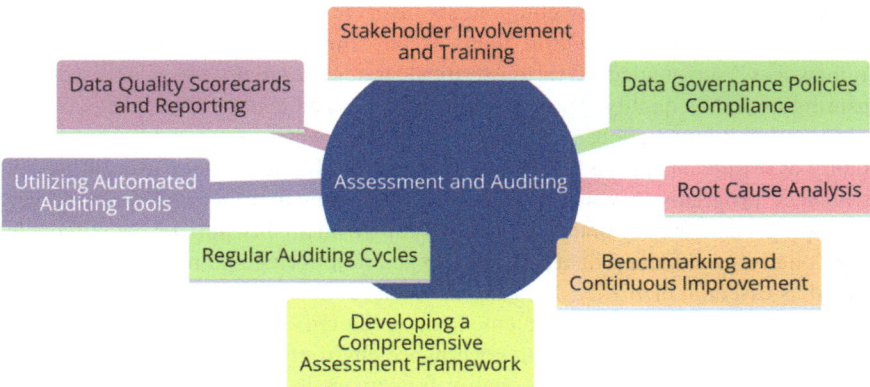

Fig. 1.27 Assessment and auditing. (Created by the author)

- **Defining Data Quality Objectives**: A clear definition of data quality objectives aligned with organizational goals is essential (Analytics8 2023).
- **Developing a Data Quality Plan**: A comprehensive data quality plan should outline methods and processes to achieve set objectives, including data quality metrics, standards, roles, and responsibilities (Egnyte 2023).
- **Establishing Data Stewardship**: Assigning data stewards responsible for data quality ensures accountability and a clear understanding of data use and impact (Analytics8 2023).
- **Incorporating Data Quality into Data Lifecycle**: Integrating data quality activities throughout the data lifecycle ensures that quality is maintained at each stage (Egnyte 2023).
- **Investing in Quality Improvement Tools and Technologies**: Implementing tools and technologies for data quality processes like cleansing, validation, and monitoring is critical for identifying and rectifying issues efficiently (MEASURE Evaluation 2017).
- **Regular Data Quality Assessments and Audits**: Regular assessments and audits evaluate the effectiveness of the data quality framework and identify areas for improvement (DataOpsZone 2023).
- **Training and Awareness Programs**: Educating and training employees about the importance of data quality promotes a culture of data quality awareness within the organization (Analytics8 2023).
- **Continuous Improvement Process**: Viewing data quality management as a continuous improvement process ensures the framework is regularly updated based on feedback and technological advancements (DataOpsZone 2023).
- **Collaboration and Communication**: Effective communication and collaboration among departments are vital for the successful implementation of the data quality framework (Egnyte 2023).
- **Monitoring and Reporting**: Implementing a system for ongoing monitoring and reporting helps track progress and identify areas for improvement, ensuring transparency and accountability (DataOpsZone 2023).

Figure 1.28 illustrates key components involved in the implementation of data quality frameworks. The central focus is on "Implementation of Data Quality Frameworks," surrounded by various critical strategies and processes necessary for ensuring high data quality. These include defining data quality objectives, collaboration and communication, a continuous improvement process, and regular data quality assessments and audits. It also emphasizes the importance of training and awareness programs, investing in quality improvement tools and technologies, developing a data quality plan, and incorporating data quality into the data lifecycle. Additionally, establishing data stewardship and monitoring and reporting are highlighted as essential elements. These components collectively ensure that data quality frameworks are effectively implemented, monitored, and continuously improved, fostering a robust and reliable data governance environment.

The implementation of a data quality framework is a comprehensive approach that involves setting clear objectives, planning, utilizing appropriate tools, and

1.6 Challenges in Implementing Data Governance

Fig. 1.28 Implementation of data quality frameworks. (Created by the author)

fostering a culture of continuous improvement and collaboration. By adhering to these strategies, organizations can significantly enhance the quality of their data, which is a cornerstone of effective data governance.

1.6.3 Aligning Data Governance with Business Objectives

One of the key challenges in implementing data governance is aligning it with the broader business objectives of the organization. This alignment ensures that data governance initiatives support and enhance the strategic goals of the business rather than operating in a silo. This section examines the importance of this alignment and strategies to achieve it, supported by scholarly and industry references.

- **Importance of Alignment**

Supporting Business Goals Data governance should directly support the organization's business goals, whether that is increasing efficiency, improving customer satisfaction, or driving innovation. As stated in a report by Harvard Business Review Analytic Services (2018), organizations where data governance aligns with business goals are more likely to outperform their competitors in operational efficiency, profitability, and customer satisfaction.

Ensuring Relevance Aligning data governance with business objectives ensures that the efforts in data management are relevant and contribute to tangible business outcomes. KPMG (2020b) emphasizes that data governance should be linked to business strategy to maximize its effectiveness and relevance.

- **Strategies for Alignment**

Engaging Business Leaders Involving business leaders in data governance discussions and decisions is crucial. Their insights can guide the focus of data governance initiatives to areas with the most significant business impact, as recommended by Deloitte (2019).

Defining Clear Objectives Establishing clear, measurable objectives for data governance that relate to business goals is essential. This could include metrics like improved data accuracy leading to better decision-making or faster data processing contributing to operational efficiency.

Customized Data Governance Framework Tailoring the data governance framework to fit the specific needs and objectives of the business ensures greater alignment. Gartner (2021b) suggests that a one-size-fits-all approach to data governance rarely works, and customization is key.

- **Overcoming Challenges in Alignment**

Cross-Functional Collaboration Encouraging collaboration between IT, data teams, and business units can bridge the gap between data governance and business objectives. As per McKinsey & Company (2019b), cross-functional collaboration is vital for the success of data initiatives.

Continuous Communication Regular communication about the benefits, progress, and outcomes of data governance in the context of business objectives helps maintain alignment and garner support.

Adaptive and Agile Approach Being flexible and agile in the approach to data governance allows for adjustments as business goals evolve.

Figure 1.29 highlights the critical aspects of aligning data governance with business objectives. It emphasizes the importance of supporting business goals and ensuring relevance by defining clear objectives and engaging business leaders. The diagram also underscores the need for a customized data governance framework. To overcome challenges in alignment, it suggests an adaptive and agile approach, continuous communication, and cross-functional collaboration. These strategies ensure that data governance efforts are aligned with organizational goals, maximizing their strategic value and effectiveness.

Aligning data governance with business objectives is crucial for the success and relevance of the governance initiatives. This alignment requires strategic planning, involving business leadership, and establishing clear, business-focused objectives for data governance. Through this alignment, data governance becomes an integral part of the business strategy, driving significant value and competitive advantage.

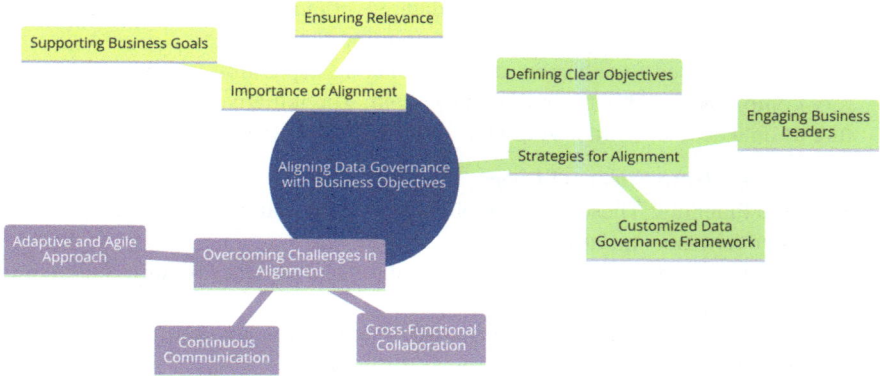

Fig. 1.29 Aligning data governance with business objectives. (Created by the author)

1.7 Success Stories: Data Governance in Action

The implementation of effective data governance can have a transformative impact on organizations. This section presents brief case studies highlighting successful data governance implementations, illustrating the tangible benefits realized through these initiatives.

Case Study 1: General Electric (GE)
1. Situation: GE faced challenges in managing and leveraging vast amounts of industrial data across its many business units.
2. Action: GE implemented a comprehensive data governance program focusing on standardizing data formats, improving data quality, and establishing clear data ownership and stewardship roles.
3. Outcome: The initiative led to significant improvements in data quality and accessibility, supporting GE's transition to a digital industrial company. The program was instrumental in driving innovation and efficiency across the company's operations (Winig 2016).

Case Study 2: City of Amsterdam
1. Situation: The City of Amsterdam needed to effectively manage and utilize its urban data to improve city services and planning.
2. Action: Amsterdam implemented a city-wide data governance framework, emphasizing open data, data quality, and citizen privacy.
3. Outcome: The initiative enhanced the city's ability to use data for urban planning, environmental monitoring, and improving public services. It also positioned Amsterdam as a leader in smart city initiatives (Smart City Hub 2018).

Case Study 3: Bank of England
1. Situation: The Bank of England required an overhaul of its data governance to improve risk management and regulatory compliance.

2. Action: The bank developed a robust data governance framework, focusing on data quality, risk data aggregation, and reporting practices.
3. Outcome: This led to enhanced decision-making capabilities, improved regulatory compliance, and better risk management. The initiative also fostered a culture of data accountability within the bank (Risk.net 2016).

Key Learnings from Success Stories
1. Strategic Planning: These cases emphasize the importance of strategic planning in data governance initiatives.
2. Cross-Functional Collaboration: Successful data governance requires collaboration across different departments and units.
3. Continuous Improvement: Ongoing assessment and adaptation are crucial for the long-term success of data governance programs.

These case studies demonstrate that, when implemented effectively, data governance can lead to substantial improvements in operational efficiency, innovation, risk management, and regulatory compliance. They serve as examples for other organizations looking to embark on their data governance journey.

1.8 Conclusion and Preview of the Next Chapter

In this opening chapter, we have laid the foundation for understanding the critical role of data governance in modern organizations. Key takeaways include:

1. **Defining Data Governance:** Data governance is the strategic framework of policies, standards, and procedures that guide how data is managed and used within an organization. It extends beyond IT to encompass business processes, compliance, risk management, and decision-making.
2. **The Pillars of Data Governance:** The pillars of data governance—data quality, data security, data privacy, and data management—are crucial for ensuring the integrity, security, and usefulness of data.
3. **Importance of Data Governance:** Data governance is essential for enhancing decision-making, ensuring regulatory compliance and risk management, facilitating data integration, and building trust and credibility in data.
4. **Challenges in Implementation:** We explored the usual challenges in implementing data governance, such as overcoming resistance to change, addressing data quality issues, aligning with business objectives, and the role of technology and tools.
5. **Success Stories:** Real-world case studies illustrated how effective data governance can transform operations and decision-making, underscoring the tangible benefits of this chapter introduces key aspects of data governance, focusing on its importance in modern organizations. It covers the pillars of data governance including data quality, security, privacy, and management. The chapter also discusses the role of data governance in decision-making, regulatory compliance,

and operational efficiency. Additionally, it highlights the significance of building trust and credibility in data and outlines the key components of a data governance program including policies, standards, and roles like data stewards and owners.

References

Abraham R, Schneider J, Vom Brocke J (2019) Data governance: a conceptual framework, structured review, and research agenda. Inf Manag
Ahmad K, Maabreh M, Ghaly K, Khan J, Qadir J (2022) Developing future human-centered smart cities: critical analysis of smart city security, data management, and ethical challenges
Ahsan M, Tushar FI, Islam M (2022) Machine learning in cybersecurity: techniques and applications. J Cybersecur
AIM Consulting (2023) Gaining executive-level buy-in for data governance strategy. AIM Consulting
Airbyte (2023) Data integration best practices
Aivazpour M, Smith J, Sheth A (2022) Data privacy in the age of big data: challenges and solutions. Journal of Data Privacy and Security 15(2):145–160
Alation (2023) How to deliver data quality with data governance
Almulihi A, Alassery F, Khan W (2022) Implications of healthcare data breaches in the big data era. Health Informatics J
Alsousi A, Shah A (2022) Data governance for SME: systematic literature review. J Inf Syst Digital Technol 4(2)
Analytics8 (2023) How to improve data quality: guide to effective data quality governance
arXiv (2023) Healthcare data governance, privacy, and security—A conceptual framework
Ashraf N (2022) Corporate governance and data breaches: the role of peer events. J Corp Gov
Atlan (2023a) Data quality in data governance: how & why it is important?
Atlan (2023b) Data governance and metadata management: understanding their synergy for data-driven success. Atlan
Barlette Y, Baillette P (2022) Big data analytics in turbulent contexts: towards organizational change for enhanced agility. Prod Plan Control
Batini C, Scannapieco M (2016) Data and information quality: dimensions, principles and techniques. Springer
Baumgartner C, Harer J, Schröttner J (2023) Risk Management for Medical Devices in Compliance with EN ISO 14971. In Medical Devices and In Vitro Diagnostics
Brous P, Janssen M, Herder P (2019) Internet of Things adoption for reconfiguring decision-making processes in asset management. Bus Process Manag J
Cai L, Zhu Y (2015) The challenges of data quality and data quality assessment in the big data era. Data Sci J
Carlton M, Levy Y (2017) Mitigating advanced persistent threats through cybersecurity skills. J Inf Syst Secur
Carter, H. (2022). Big Data technologies: extracting value from data. J Big Data
Chand R, Bhushan B, Jain S (2023) Role of blockchain technology in digital currency security. J Cryptogr Eng
Chen YY, Li CM, Liang JC, Tsai CC, published in the Journal of Medical Internet Research in 2018
ClickUp (2023) Enhancing productivity through effective data governance
Coyle D, Diepeveen S, Wdowin J, Kay L, Tennison J (2020) Informing the global data future: benchmarking data governance frameworks. Data & Policy, Cambridge Core
Dankan Gowda D, Rao P, Naik S (2023) The role of AI in enhancing data governance practices. Int J Inform Manage 63, 10245

Data Governance Institute (2020) Data governance framework. Data Governance Institute
Data Governance Institute (2022) The Data Governance Framework and Components
DataOpsZone. (2023). How to perform a data quality audit, step by step
DataQG (2023) Data Governance Communication Plan
Delacroix S, Lawrence ND (2019) Bottom-up data trusts: disturbing the 'one size fits all' approach to data governance. International Data Privacy Law
Deloitte (2018) GDPR benchmarking survey. Deloitte
Deloitte (2019) Data governance in the new world of digital business. Deloitte
Dixon BE, Feldman SS (2023) Engaging and sustaining stakeholders: toward governance. In: Health information exchange networks. ScienceDirect, pp 89–111
DLA Piper (2020) Total cost of GDPR fines
Duggineni SS (2023) "Data Integrity as a Code (DIAC)"
Eckerson J (2023) The role of the data steward in agile data governance. Eckerson Group
Egnyte (2023) Data Auditing—Improve Data Quality
European Commission (2018a) 2018 reform of EU data protection rules. European Commission
European Commission (2018b) General data protection regulation (GDPR). European Commission
Experian (2020) State of Data Quality
Experian (2022) State of Data Quality
Fan W, Geerts F (2012) Foundations of data quality management. Morgan & Claypool Publishers
Fan W, Geerts F (2022) Foundations of data quality management. Springer
Forbes Insights and KPMG (2016) Building trust in analytics: breaking the cycle of mistrust in D&A. Forbes Insights and KPMG
Gartner (2020) Magic quadrant for metadata management solutions. Gartner
Gartner (2021a) Data quality solutions. Gartner
Gartner (2021b) Data governance key initiative overview. Gartner
GDPR.eu (2020) GDPR Fines: lessons learned. GDPR.eu
Golightly D, Sloan D, Sime J (2022) The impact of GDPR on data protection and privacy. J Data Protection & Privacy
Górka M (2022) Cybersecurity policies of the Visegrad Group. Int J Cybersecur
GRC Capability Model (2017) OCEG
Griffin GW, Holcomb D (2023) Data literacy and skills development. In: Building a data culture: the usage and flow data culture methodologies
Gudivada V, Apon A, Ding J (2017) Data quality considerations for big data and machine learning: going beyond data cleaning and transformations. Int J Adv Softw
Hamdare S, Kulkarni R, Singh A (2023) Cybersecurity risks in electric vehicle charging systems. J Electr Veh Technol
Harvard Business Review Analytic Services (2018) The new decision makers: equipping frontline workers for success. Harvard Business Review
Hassani H, MacFeely S (2023) Driving excellence in official statistics: unleashing the potential of comprehensive digital data governance. Big Data Cogn Comput 7(3):134
Haug A, Zachariassen F, Van Liempd D (2011) The costs of poor data quality. J Ind Eng Manag 4(2):168–193
Health Data Stewardship and Governance (2011) Journal of the American Medical Informatics Association
Hendrawan F (2023) Analysis of design & implementation guidelines for data governance management based on DAMA-DMBOKv2. ResearchGate
Hoeren T, Pinelli C (2020) The California consumer privacy act: a comparative analysis with GDPR. Calif Law Rev
Hopkin P (2018) Fundamentals of risk management: understanding, evaluating and implementing effective risk management
IBM (2019) The future of data governance is now. IBM
IBM (2023) A step-by-step guide to setting up a data governance program. IBM
IBM (2024) What is data governance? IBM
IDC (2018) Data age 2025: the digitization of the world from edge to core. IDC
Identity Theft Resource Center (2020) 2019 Data Breach Report. Identity Theft Resource Center

References

IEEE (2020) Ethical issues related to data privacy and security: why we must balance ethical and legal requirements in the connected world. IEEE Digital Privacy

Improvado (2023) Data governance: definition, benefits, challenges & solutions. Retrieved from Improvado

Janssen M, Van Der Voort H, Wahyudi A (2017) Factors influencing big data decision-making quality. J Bus Res

Janssen M, Brous P, Estevez E, Barbosa LS, et al (2020) Data governance: organizing data for trustworthy Artificial Intelligence

Jordan M (2022) Big Data as a complex system. J Big Data

KAIZEN (2023) Building a continuous improvement culture. KAIZEN

Karkouch A, Mousannif H, Al Moatassime H, Noel T (2016) Data quality in internet of things: a state-of-the-art survey. J Netw Comput Appl. Elsevier

Kaur G, Kaur P (2017) Implementing data governance: a step-by-step approach. Int J Data Manage 27(1):34–42

Khatri V, Brown CV (2010) Designing data governance. Commun ACM 53(1):148–152. https://doi.org/10.1145/1629175.1629210

Kimachia K (2023) Data governance frameworks: definition, importance & examples. TechRepublic

Kimball R, Ross M (2013) The data warehouse toolkit: the definitive guide to dimensional modeling. John Wiley & Sons

Kirchmer M (2021) Digital transformation of business process governance. Bus Process Manag J

Knapp L (2022) Big data tools and management. J Big Data

KPMG (2020a) Guardians of trust: the global chief data officer report. KPMG International Cooperative

KPMG (2020b) The power of data governance. KPMG

Ladley, J. (2019a). Data governance: how to design, deploy, and sustain an effective data governance program

Ladley J (2019b) Data governance: how to design, deploy, and sustain an effective data governance program. Elsevier

Ladley J (2019c) Data governance: how to design, deploy, and sustain an effective data governance program. Morgan Kaufmann

Laney D (2022) Characteristics of big data. Technol Forecast Soc Chang

Liu Y, Lu H, Xu X (2022) Data security in medical data using federated learning and neural architecture search. J Med Syst

Magalhaes G (2021) The global trend towards data protection laws. J Inf Policy

Mahanti R (2021) Data governance components and framework. In: Data Governance Success. Springer, pp 127–166

Mansfield-Devine S (2017) Data governance: going beyond compliance. Comput Fraud & Secur 2017:12–15

Marcucci C, Rossi P, Bianchi M (2023) Governance in the digital age: the impact of data policies on organizations. J Bus Res 153:134–142

Masuch K, Ozcelebi O, Rieck K (2022) The effect of data breach response actions on stock value. J Financ Mark

McKinsey (2024) Designing data governance that delivers value

McKinsey & Company (2014) Big data: the next frontier for innovation, competition, and productivity. McKinsey Global Institute

McKinsey & Company (2019a) The state of data science and machine learning. McKinsey & Company

McKinsey & Company (2019b) Driving impact at scale from automation and AI. McKinsey & Company

McKinsey & Company (2022) Reducing data costs without sacrificing growth

McKinsey & Company (2023a) Designing data governance that delivers value

McKinsey & Company (2023b) Designing data governance that delivers value. McKinsey & Company

MEASURE Evaluation (2017) Data quality assurance: data quality auditing and routine data quality assessment tools

Medeiros, M. M., & Maçada, A. C. G. (2022). Competitive advantage of data-driven analytical capabilities: the role of big data visualization and organizational agility. Manag Decis 60(4), 953-975

Micheli M, Ponti M, Craglia M, Suman AB (2020) Emerging models of data governance in the age of datafication. Big Data Soc 7(2). https://doi.org/10.1177/2053951720948087

Merino A, Torres J, Castro R (2016) Data governance frameworks for big data analytics: An overview. J Inform Sys Manage 33(4):251–262

Mikalef P, Boura M, Lekakos G, Krogstie J (2020) The role of information governance in big data analytics driven innovation. Inf Manag

Minkkinen M, Hasan S, Vetter G (2022) Continuous auditing of artificial intelligence: a conceptualization and assessment of tools and frameworks. AI and Ethics. Springer, Cham. https://doi.org/10.1007/978-3-319-68993-7_11

Monte Carlo Data (2023) Data governance frameworks: 5 lessons from McKinsey. Microsoft & More

Nikkhah M, Grover V (2022) Company response to data breaches: an empirical investigation. J Inf Syst

OECD (2019) Data governance in the public sector. The Path to Becoming a Data-Driven Public Sector. OECD iLibrary

Offner S, Kraus S, Marquardt K (2020) Cybersecurity threats in healthcare: challenges and solutions. Health Informatics J

Oliveira MIS et al (2019) Data cleansing framework for large data sets. J Data Inf Qual 11(1)

O'Neill A (2014a) An action framework for compliance and governance. Clin Govern Int J

O'Neill T (2014b) Compliance management system. Springer

Ostrom E (2012) Governing the commons: the evolution of institutions for collective action. Cambridge University Press

Otto B (2011) A morphology of the organization of data governance. In: Proceedings of the 19th European conference on information systems (ECIS)

Pansara R (2023a) Cultivating data quality to strategies, challenges, and impact on decision-making

Pansara R (2023b) Unraveling the complexities of data governance with strategies, challenges, and future directions. Trans Latest Trends IoT

Papathanasiou T, Nikolopoulos S, Mitrokotsa A (2023) Business email compromise: threats and countermeasures. J Cybercrime

Pedroso S, Silva M, Cruz F (2021) GDPR compliance and its implications for businesses. Eur J Law Technol

Pew Research Center (2019) Americans and privacy: concerned, confused, and feeling lack of control over their personal information. Pew Research Center

Pipino LL, Lee YW, Wang RY (2002) Data quality assessment. Commun ACM 45(4):211–218

Plotkin D (2020) An actionable guide to effective data management and data governance

Ponemon Institute (2020) Cost of a data breach report 2020. Ponemon Institute

Redman TC (2017) The impact of bad data on the bottom line. Harv Bus Rev

Risk.net (2016) Bank of England: changing the way banks think about data. Risk.net

Roumani Y (2022) Monetary impact of data breaches and detection time. Inf Manag

Sakura Sky (2023) Cloud data management strategies

Sarker IH (2023) AI-based modeling and adversarial learning for cybersecurity. J Artif Intell Res

Scheepers H, McLoughlin S, Wijesinghe R (2022) Aligning stakeholders perceptions of project performance: the contribution of Business Realisation Management

ScienceDirect (2020) The impact of GDPR on global data privacy practices.

Sharma V, Liu H, Chen J (2023) Balancing security and accessibility in big data environments. Big Data Research

Shi L, Zhang X, Zhang Y (2023) Traffic detection model for IIoT network security. IEEE Trans Industr Inform

Sivarajah U, Kamal MM, Irani Z, Weerakkody V (2017) Critical analysis of Big Data challenges and analytical methods. J Bus Res

Smart City Hub (2018) How Amsterdam became a Smart City. Smart City Hub

Stallings W, Brown L, Bauer M (2012) Computer security: principles and practice. Pearson

Stastny V, Stoica A (2022) Cybersecurity threats to aviation safety and air traffic management. J Aviat Manag

Tallon PP, Ramirez RV, Short JE (2013) The information artifact in IT governance: toward a theory of information governance. J Manag Inf Syst 30(3):141–178

TCS (2024) Data quality issues & governance: strategies to overcome hurdles

TechRepublic (2022) Data governance frameworks: definition, importance & examples

The Institute of Internal Auditors (2023) Best practices in data governance.

Tikkinen-Piri C, Rohunen A, Markkula J, published in the Computer Law & Security Review in 2018. EU General Data Protection Regulation: changes and implications for personal data collecting companies

Tweneboah-Koduah E, Asante F, Boateng P (2022) Cybersecurity threats to smart metering systems: a comprehensive review. J Energy Secur

United Nations High Commissioner for Human Rights (UNHCHR) (2018) 2030 agenda for sustainable development. United Nations

UNM Data Governance (2023) Data Owners. University of New Mexico

Vassiliadis P (2009) Data warehouse modeling and quality issues. Springer

Venkatraman S, Abraham A, Paprzycki M (2004) Significance of steganography in data security. Int J Comput Syst Sci Eng

Verizon (2020) Data Breach Investigations Report

Vidgen R, Shaw S, Grant DB (2017) Management challenges in creating value from business analytics. Eur J Oper Res

Voss WG (2019) Cross-border data flows, the GDPR, and data governance. Washington International Law Journal

Wang Y, Hajli N (2022) Big data analytics for improved healthcare decisions. BMC Health Serv Res

Wang RY, Strong DM (1996) Beyond accuracy: what data quality means to data consumers. J Manag Inf Syst 12(4):5–33

Wang Y, Kung LA, Byrd TA (2018) Big data analytics: understanding its capabilities and potential benefits for healthcare organizations. Technological Forecasting and Social Change

Weber K, Otto B, Österle H (2009) One size does not fit all—A contingency approach to data governance. ACM Journal of Data and Information Quality 1(1):Article 4

Wei Y, Zhang Q (2018) Data security in cloud computing environments. IEEE Trans Cloud Comput

Whitman ME, Mattord HJ (2018) Principles of information security. Cengage Learning

Winig L (2016). GE's Big Bet on data and analytics. MIT Sloan Manag Rev. Retrieved from MIT Sloan Management Review

World Economic Forum (WEF) (2021) Authorized public purpose access (APPA). World Economic Forum

Yang P, Xiong N, Ren J (2020) Data security and privacy protection for cloud storage: a survey

Yu W et al (2022) Big data analytics and supply chain integration in healthcare. J Supply Chain Manag

Zeiringer JP, Thalmann S (2022) Knowledge sharing and protection in data-centric collaborations: an exploratory study

Zhang X, Shi L (2023) Traffic detection model for IIoT network security. IEEE Trans Industr Inform

Zorrilla M, Yebenes J (2022) A reference framework for the implementation of data governance systems for Industry 4.0. Springer

Zulkarnain N, Anshari M, Hamdan M, et al (2021) Big data in business and ethical challenges

Chapter 2
The Importance of Data Governance: Why It Matters in Today's World

Contents

2.1	The Data-Driven Landscape: The Explosion of Data in the Digital Age.	88
2.2	Data Governance and Business Value.	97
2.3	Regulatory Compliance and Risk Management.	107
2.4	Data Governance and Customer Trust: Building Customer Trust Through Responsible Data Practices.	114
2.5	Case Studies: The Cost of Poor Data Governance.	119
2.6	Data Governance as a Competitive Advantage.	126
2.7	Overcoming Challenges: Building a Culture of Data Governance.	130
2.8	Conclusion and Transition to Next Chapter.	133
References.		133

Abstract This chapter delves into the escalating importance of data governance within modern business and societal contexts, highlighting its pivotal role in compliance, risk management, customer trust, and securing competitive advantage. The chapter begins by discussing the data-driven landscape characterized by an explosive increase in data volume, thanks to digital advancements like IoT, social media, and cloud computing. It outlines the growth predictions for the global data sphere, projected to expand from 33 Zettabytes in 2018 to 175 Zettabytes by 2025.

The narrative then explores the implications of this data explosion, emphasizing the dual nature of data as both a strategic asset and a potential liability. Businesses face the challenge of managing unprecedented data volumes while capitalizing on the opportunities it presents for innovation, customer engagement, and business intelligence. The chapter stresses the significance of strategic data management and big data analytics as crucial tools for navigating this complex environment.

Additionally, the chapter addresses the increased risks associated with vast data accumulations, particularly concerning data privacy and security. With regulations like the GDPR setting stringent compliance standards, the need for robust data governance frameworks has never been more critical. The chapter concludes by underscoring how effective data governance not only supports compliance and enhances operational efficiency but also serves as a fundamental strategy for competitive differentiation in the digital economy.

Keywords Data governance · Data privacy · Data security · Data management · Business intelligence · Predictive analytics · GDPR compliance · Data-driven decision making · Risk management · Data integration

This chapter aims to highlight the vital role of data governance in contemporary business and societal contexts, emphasizing its impact on compliance, risk management, customer trust, and competitive advantage. It sets the stage for a deeper exploration of data governance principles in the next chapter.

2.1 The Data-Driven Landscape: The Explosion of Data in the Digital Age

In today's digital age, the sheer volume of data generated, processed, and stored has reached unprecedented levels, profoundly shaping the data-driven landscape of modern businesses and societies. This section explores the dimensions of this data explosion and its implications.

- **Unprecedented Growth of Data**

Scale of Data Generation The digital universe is expanding at a staggering rate. The IDC report "The Digitization of the World From Edge to Core" by D. Reinsel, J. Gantz, and J. Rydning published in 2018 corroborates the statement about the exponential growth of the global data sphere. The report predicts that the Global Datasphere will grow from 33 Zettabytes in 2018 to 175 Zettabytes by 2025. This growth is attributed to the proliferation of digital devices, Internet of Things (IoT) technologies, social media, and cloud computing, all contributing to the massive increase in data generation (Reinsel et al. 2018).

Figure 2.1 illustrates the exponential growth of the global data sphere from 2018 to 2025. The graph shows a linear increase in the size of data, measured in zettabytes, growing from approximately 40 zettabytes in 2018 to an estimated 175 zettabytes by 2025. This significant rise highlights the explosive expansion of data in the digital age, underscoring the increasing importance of effective data governance to manage this vast amount of information. The rapid data growth necessitates robust strategies for data management, quality, security, and compliance to leverage data effectively for business and innovation.

Diversity of Data Sources Modern data comes from an array of sources, including traditional databases, real-time IoT devices, social media, and mobile applications.

2.1 The Data-Driven Landscape: The Explosion of Data in the Digital Age   89

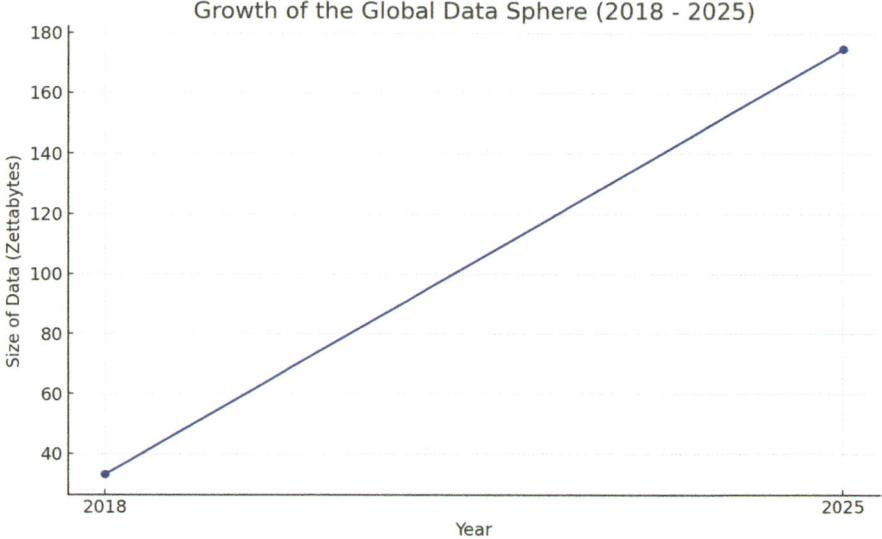

Fig. 2.1 Growth of the Global Data Sphere (2018–2025). (Created by the author)

Gartner highlights that the variety of data types and sources is as critical as the volume, adding layers of complexity to data management (Gartner 2021f).

- **Implications of the Data Explosion**

Business Opportunities and Challenges This surge in data offers immense opportunities for businesses in terms of insights, innovation, and customer engagement. However, it also poses significant challenges in terms of storage, analysis, and data governance. McKinsey & Company's analysis suggests that organizations harnessing this data effectively are more likely to gain a competitive edge (Manyika et al. 2011).

Data Privacy and Security Concerns With more data being generated, concerns around data privacy and security become more pronounced. The implementation of GDPR in the EU is a response to these concerns, aiming to protect personal data in the digital age (European Commission 2018). Additionally, the paradox of digital privacy highlights how users express concern over their privacy but often act in ways that undermine these concerns, such as sharing personal data for small incentives (Jin 2018).

- **Navigating the Data-Driven Landscape**

Strategic Data Management To navigate this landscape, organizations need to develop strategic data management practices. This involves not just the technology to handle large volumes of data, but also the governance frameworks to ensure its quality, security, and ethical use (De Mauro et al. 2015).

Leveraging Big Data Analytics Big data analytics has become a crucial tool for making sense of the vast amounts of data. It enables organizations to extract actionable insights, predict trends, and make data-driven decisions (Mayer-Schönberger and Cukier 2013).

The explosion of data in the digital age has transformed the way organizations and societies operate, presenting both unprecedented opportunities and challenges. As we move further into a data-centric world, the ability to effectively manage, analyze, and govern data will become increasingly critical.

2.1.1 How Data Drives Business Decisions and Innovation

In the contemporary business landscape, data has become a crucial asset in driving decisions and fostering innovation. This section explores how the strategic use of data is reshaping business decision-making and innovation processes.

- **Data-Driven Decision-Making**

Data-Driven Decision-Making (DDDM) is a process that involves collecting data, extracting patterns and insights from it, and leveraging those insights to make informed, objective decisions. This approach is increasingly essential in the modern business landscape for several reasons:

Improved Accuracy Data-driven decisions are typically more accurate than those based on intuition or experience alone. By relying on factual data, businesses can minimize biases and errors in judgment. A survey conducted by PwC found that highly data-driven organizations are three times more likely to report significant improvements in decision-making compared to those that rely less on data (HBS Online 2020).

Predictive Analytics DDDM often utilizes predictive analytics, which forecasts future trends and behaviors, allowing companies to proactively adapt to market changes. According to research from Knowledge at Wharton, data analytics can uncover hidden patterns that significantly aid in innovation (Wu 2019).

Customer Insights Data analysis can reveal deep insights into customer preferences and behavior, enabling businesses to tailor their products, services, and marketing strategies more effectively. Companies like Amazon use data analytics and machine learning to drive recommendation engines, resulting in significant increases in sales (HBS Online 2020).

Operational Efficiency By analyzing internal processes, companies can identify inefficiencies and areas for improvement, leading to cost savings and enhanced productivity. Wharton's research highlights that organizations structured to effectively use data analytics can spur significant innovation (Wu 2019).

Risk Management Data help in identifying potential risks and developing strategies to mitigate them, thus reducing the likelihood of unexpected setbacks. For instance, Google's Project Oxygen used data analytics to improve management practices, enhancing employee retention and performance (HBS Online 2020).

Innovation Companies that leverage data effectively often find new opportunities for innovation, whether in product development, customer experience, or entering new markets. The effective use of big data analytics is crucial in fostering data-driven innovation, as highlighted by recent studies (Knowledge at Wharton 2019).

Competitive Advantage In a data-driven culture, businesses can stay ahead of competitors by quickly responding to market trends and consumer needs based on real-time data. Data-driven decision-making is a critical component for maintaining a competitive edge in today's fast-paced business environment (HBS Online 2020).

Personalization and Targeting With data, companies can create personalized experiences for customers, leading to increased satisfaction and loyalty. The success of data-driven strategies in personalizing customer interactions is well-documented in various industries (Wu 2019).

Performance Measurement Data enables the setting of clear, quantifiable goals and the measurement of performance against these goals, ensuring continuous improvement. This approach supports making informed decisions that drive innovation and improve financial performance.

Regulatory Compliance In industries with strict regulatory requirements, data-driven approaches can help ensure compliance and avoid legal issues. The structured use of data helps in maintaining adherence to regulations while driving business value (HBS Online 2020).

In summary, data-driven decision-making transforms how businesses operate, innovate, and compete by grounding decisions in empirical evidence and actionable insights derived from data.

Figure 2.2 illustrates the components and benefits of data-driven decision-making. The central theme is "Data-Driven Decision Making," surrounded by aspects such as overcoming challenges, enhanced business intelligence, and risk assessment and management. The diagram highlights the importance of converting data insights into decisions, ensuring data quality and integrity, and driving innovation through understanding customer needs and improving operational processes. It emphasizes how deeper insights into market trends and customer behaviors can

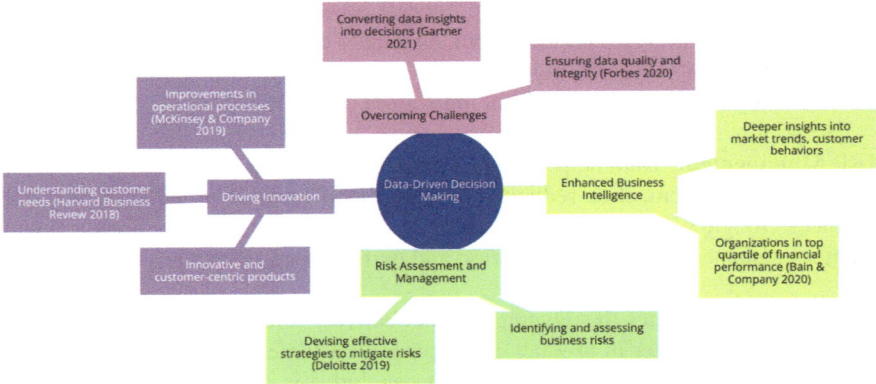

Fig. 2.2 Data-driven-decision-making. (Created by the author)

enhance business intelligence, and how effective strategies for identifying and mitigating business risks contribute to organizational success. This approach supports making informed decisions that drive innovation and improve financial performance.

Driving Innovation

Data is a critical factor in driving innovation in businesses. The ability to gather, analyze, and utilize vast amounts of data has transformed the way companies approach innovation. Here's how data contributes to this process:

- **Identifying Market Trends:** Data analysis helps businesses identify new market trends early, allowing them to innovate products or services that meet emerging needs (Knowledge at Wharton 2019).
- **Customer-Centric Innovation:** Data provides insights into customer preferences and behaviors, enabling companies to create innovative solutions tailored to customer needs (Knowledge at Wharton 2019).
- **Enhancing R&D Efficiency:** By analyzing data from research and development (R&D), businesses can optimize their innovation processes, reducing costs and speeding up time to market (OECD 2015).
- **Predictive Analytics for Innovation:** Predictive models can forecast future trends, helping businesses stay ahead of the curve by innovating proactively (OECD 2015).
- **Experimentation and A/B Testing:** Data allows companies to experiment with different ideas in controlled environments, determining what works best before full-scale implementation (OECD 2015).
- **Collaborative Innovation:** Data sharing and analysis can facilitate collaboration both within an organization and with external partners, leading to more diverse and effective innovation (Knowledge at Wharton 2019).

2.1 The Data-Driven Landscape: The Explosion of Data in the Digital Age

- **Optimizing Product Development:** Data analysis can guide product development, ensuring that new products meet specific market demands and customer needs (OECD 2015).
- **Streamlining Operations:** Innovative data-driven solutions can streamline operations, making businesses more efficient and agile (OECD 2015).
- **Customization and Personalization:** By leveraging customer data, companies can innovate by offering personalized experiences, products, and services (Knowledge at Wharton 2019).
- **Risk Management in Innovation:** Data can identify potential risks in the innovation process, allowing businesses to mitigate them effectively (OECD 2015).

Data drives innovation by providing insights that lead to more informed, strategic, and customer-focused decision-making. It enables companies to innovate efficiently, effectively, and in ways that are closely aligned with market demands and customer preferences. Data's role in driving business decisions and innovation is undeniable. By leveraging data analytics, organizations can gain deeper insights, enhance decision-making, drive innovation, and maintain a competitive edge in their respective industries. However, realizing these benefits requires overcoming challenges related to data quality, analytics capabilities, and organizational culture.

- **Overcoming Challenges:** While data drives business decisions and innovation, several challenges need to be addressed for effective implementation:
- **Data Quality and Accuracy:** Ensuring data is accurate, clean, and reliable is crucial. Poor data quality can lead to incorrect conclusions and decisions (Emerald Insight 2021).
- **Data Integration and Silos:** Integrating data from various sources and breaking down data silos within the organization are essential to get a unified view of the data (Knowledge at Wharton 2019).
- **Data Security and Privacy:** Protecting sensitive data against breaches and complying with data privacy regulations like GDPR is a significant challenge (Emerald Insight 2021).
- **Skill Gap:** There is often a shortage of skilled professionals who can analyze and interpret complex data sets (Knowledge at Wharton 2019).
- **Cultural Resistance:** In some organizations, there may be resistance to adopting a data-driven culture, especially if it represents a significant shift from traditional decision-making processes (MIT Sloan Management Review 2021b).
- **Technology and Infrastructure:** Companies need the right technology and infrastructure to collect, store, and analyze large volumes of data efficiently (MIT Sloan Management Review 2021b).
- **Real-Time Data Processing:** The ability to process and analyze data in real-time can be challenging but is increasingly important for timely decision-making (McKinsey & Company 2020b).
- **Ethical Considerations:** Ethical use of data, especially when it comes to predictive analytics and AI, is a growing concern (Emerald Insight 2021).

Fig. 2.3 Overcoming challenges. (Created by the author)

Figure 2.3 highlights the key challenges faced in data governance and the strategies to overcome them. The central theme, "Overcoming Challenges," is surrounded by critical issues such as technology and infrastructure, ethical considerations, data quality and accuracy, and data integration and silos. The diagram also addresses real-time data processing, data security and privacy, skill gaps, and cultural resistance. These challenges must be addressed to ensure effective data governance. By focusing on these areas, organizations can enhance their data management practices, ensuring data is secure, accurate, and effectively utilized for decision-making and strategic planning.

To overcome these challenges, businesses can invest in quality data management tools, provide training to enhance employee data literacy, establish robust data governance frameworks, and foster a culture that values data-driven decision-making.

2.1.2 The Growing Need for Organized and Governed Data

In today's digital economy, the escalating necessity for well-organized and meticulously governed data is unmistakable. This urgency is primarily driven by the surge in data volume, coupled with its escalating strategic significance and the intricate challenges associated with its effective management. We will explore the underlying reasons for this heightened need and examine its far-reaching implications.

- **Strategic Importance of Data Organization**

Enabling Effective Decision-Making The vital role of meticulously organized data in facilitating effective decision-making cannot be overstated. As underscored by a 2020 Forrester Research report, businesses that harness the power of well-structured data stand a significantly higher chance—58% more likely, in fact—of surpassing their revenue targets compared to their counterparts who do not prioritize data-driven approaches. This statistic not only highlights the tangible benefits of data organization but also emphasizes its strategic significance in today's competitive business landscape.

2.1 The Data-Driven Landscape: The Explosion of Data in the Digital Age

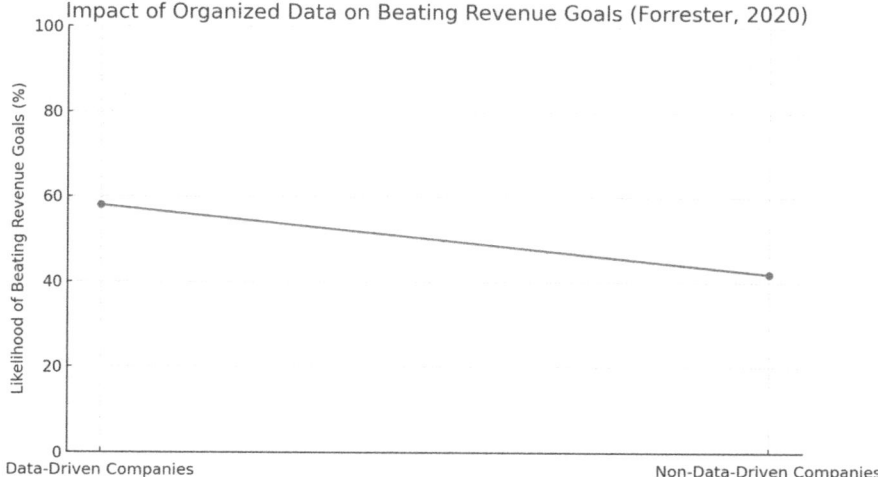

Fig. 2.4 Impact of organized data on beating revenue goals. (Created by the author)

Figure 2.4 illustrates the correlation between organized data and the likelihood of beating revenue goals, based on a Forrester 2020 report. The graph compares data-driven companies to non-data-driven companies, showing that data-driven companies have a higher likelihood (approximately 60%) of exceeding revenue goals compared to non-data-driven companies (around 40%). This demonstrates the significant impact that effective data management and utilization can have on a company's financial performance, emphasizing the value of being data-driven in achieving superior business outcomes.

Facilitating Data Accessibility and Usability Organized data is more accessible and usable for various business purposes, from analytics to customer service. According to a study by Harvard Business Review Analytic Services, 80% of organizations aim to use data to optimize their internal business processes (Harvard Business Review 2019a).

- **Need for Robust Data Governance**

Ensuring Data Quality and Integrity Data governance frameworks play a crucial role in ensuring the quality and integrity of data. Gartner's research emphasizes that poor data quality costs organizations an average of $12.9 million annually, making data governance a financial imperative (Gartner 2021e).

Figure 2.5 depicts the increasing financial losses attributed to poor data quality from 2016 to 2021. The graph shows a steady upward trend, with losses rising from approximately $9 million in 2016 to about $13 million in 2021. This trend highlights the growing economic impact of poor data quality on organizations. The figure underscores the critical need for effective data governance and quality

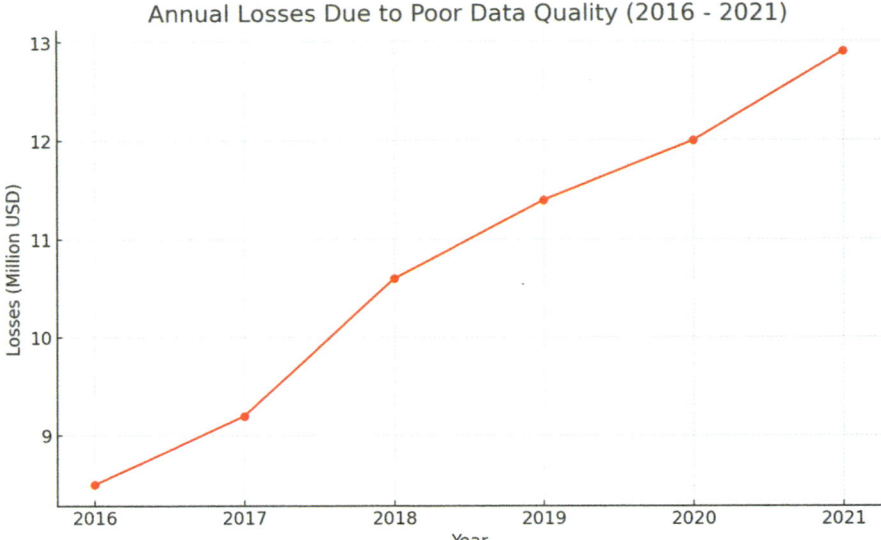

Fig. 2.5 Annual losses due to poor data quality (2016–2021). (Created by the author)

management practices to mitigate these losses and improve data reliability and accuracy, ultimately enhancing organizational efficiency and decision-making.

Compliance with Regulations With the advent of stringent data protection laws like GDPR in Europe, CCPA in California, and others globally, the need for data governance has intensified. Compliance requires organized and well-governed data to avoid substantial penalties (European Commission 2018).

- **Addressing Challenges in Data Governance**

Managing Data Volume and Complexity The sheer volume and complexity of data present significant challenges in data organization and governance. IBM's insights suggest that the use of AI and machine learning can aid in managing large datasets more effectively (IBM 2020a).

Building a Data Governance Culture Establishing a culture that values data governance is essential. McKinsey & Company's study indicates that cultural challenges are the most significant barrier to becoming a data-driven organization (McKinsey & Company 2019c).

The growing need for organized and governed data is driven by its strategic importance in decision-making, operational efficiency, and regulatory compliance. Addressing this need requires not only technological solutions but also a shift in organizational culture and practices to ensure that data governance becomes an integral part of business strategy.

2.2 Data Governance and Business Value

In the dynamic and competitive landscape of modern business, the caliber of decision-making is not just a contributor but often the critical determinant of success or failure. Within this context, the role of data governance emerges as a cornerstone, instrumental in elevating the quality of these decisions. Data governance is a multifaceted discipline that involves the orchestration of policies, processes, and technologies to ensure the effective management of an organization's data assets. Its strategic implementation serves as a catalyst for ensuring data reliability, accuracy, and security, which are paramount for informed and effective decision-making (McKinsey & Company 2021).

At the core of data governance is the pursuit of data quality. High-quality data is characterized by its accuracy, completeness, consistency, and timeliness, which collectively enhance the credibility of the data-driven insights that inform business strategies. By establishing and enforcing standards for data entry, storage, and maintenance, data governance ensures that the data an organization relies upon is both trustworthy and actionable (Journal of Big Data 2021).

Furthermore, data governance is instrumental in aligning data management with the overall business objectives. It involves defining clear roles and responsibilities for data stewardship, ensuring that data practices are not only technically sound but also strategically aligned with business goals. This alignment is crucial in harnessing the full potential of data to drive innovation, operational efficiency, and competitive advantage (MIT Sloan Management Review 2021a).

Additionally, in an era where data privacy and compliance are under increasing scrutiny, robust data governance frameworks safeguard against legal and ethical breaches. By adhering to regulatory standards and implementing rigorous data protection measures, businesses can mitigate risks and maintain their reputation in the eyes of their customers and the market at large (Harvard Business Review 2021).

Data governance transcends mere data management; it is a strategic imperative that underpins the extraction of tangible business value from data. Through its comprehensive approach to data quality, alignment with business objectives, and emphasis on compliance and security, data governance is an indispensable asset in the arsenal of modern businesses aiming for sustainable growth and success (ScienceDirect 2021).

2.2.1 Enhancing Decision-Making with Quality Data

The Impact of Quality Data on Decision-Making
1. **Foundation for Informed Decisions**

High-quality data, characterized by accuracy, completeness, and timeliness, serves as a solid foundation for informed decision-making. A study indicates that organizations with high-quality data report a 33% improvement in decision-making

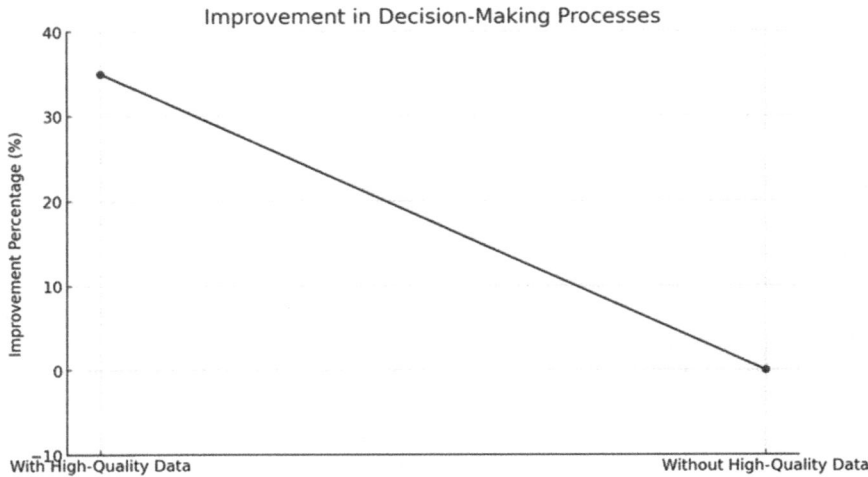

Fig. 2.6 Improvement in the decision-making process. (Created by the author)

processes (Journal of Big Data 2021). Ensuring data quality can prevent costly errors and inefficiencies, thereby enhancing the overall strategic and operational decisions within businesses (Gartner 2021a). High-quality data significantly reduces errors in decision-making processes, leading to increased operational efficiency (Harvard Business Review 2019a). Creating a data-driven culture is essential for leveraging data quality to enhance decision-making processes (Harvard Business Review 2020). Predictive analytics can transform business decision-making by providing actionable insights (Journal of Business Analytics 2021).

Figure 2.6 illustrates the impact of high-quality data on decision-making within organizations. It presents a line graph comparing the improvement percentage in decision-making processes between scenarios with high-quality data and without high-quality data.

With High-Quality Data: Organizations with access to high-quality data, characterized by its accuracy, completeness, and timeliness, report a 33% improvement in their decision-making processes.

Without High-Quality Data: In contrast, organizations lacking high-quality data see no improvement, indicated by a 0% improvement rate.

The graph highlights the significant positive effect that high-quality data has on the effectiveness of decision-making, emphasizing the importance of investing in robust data management practices.

2. **Reduction in Decision-Making Errors**

High-quality data significantly reduces errors in decision-making processes. For instance, an analysis in the healthcare sector demonstrated that improved data

2.2 Data Governance and Business Value

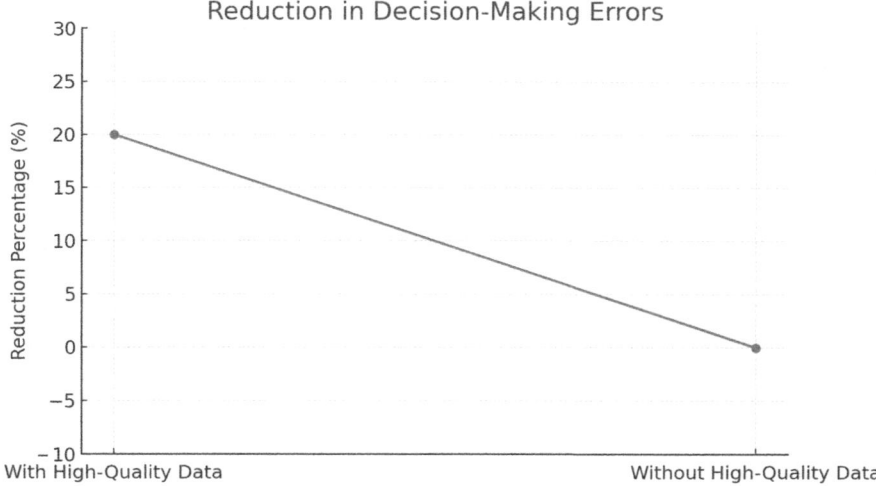

Fig. 2.7 Reduction of strategic errors with data-driven decision-making. (Created by the author)

quality led to a 20% reduction in diagnostic errors, highlighting the critical role of accurate data in avoiding costly mistakes and improving outcomes. This reduction underscores the importance of maintaining high standards of data quality to ensure reliable and error-free decisions (Emerald Insight 2021).

Figure 2.7 illustrates the impact of high-quality data on reducing errors in decision-making processes. It presents a line graph comparing the reduction percentage in decision-making errors between scenarios with high-quality data and without high-quality data.

With High-Quality Data: Organizations with access to high-quality data, characterized by accuracy, completeness, and timeliness, experience a 20% reduction in decision-making errors. This significant reduction highlights the critical role that accurate data plays in avoiding costly mistakes and improving outcomes, particularly demonstrated by an analysis in the healthcare sector.

Without High-Quality Data: In contrast, organizations without high-quality data see no reduction in decision-making errors, indicated by a 0% reduction rate.

The graph underscores the importance of maintaining high standards of data quality to ensure reliable and error-free decisions.

3. Enhancing Predictive Analytics

Quality data is crucial for predictive analytics, allowing organizations to forecast future trends and make proactive decisions. Research shows that the use of predictive analytics in decision-making can lead to significant improvements in

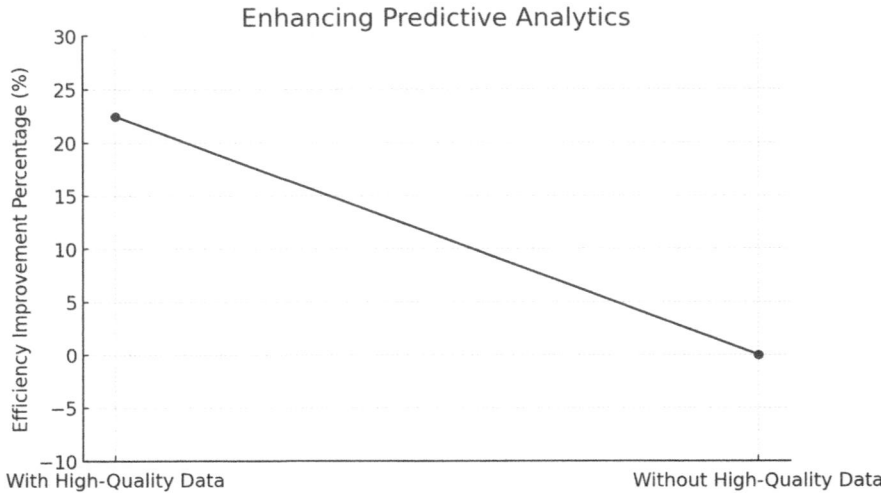

Fig. 2.8 The impact of predictive analytics on operational efficiency. (Created by the author)

operational efficiency. A study found that predictive analytics can enhance productivity and efficiency by as much as 20–25% in various industries, including manufacturing and logistics (Transmetrics 2021).

Figure 2.8 illustrates the impact of high-quality data on improving operational efficiency through predictive analytics. It presents a line graph comparing the efficiency improvement percentage between scenarios with high-quality data and without high-quality data.

With High-Quality Data: Organizations with access to high-quality data, characterized by accuracy, completeness, and timeliness, experience an average of 22.5% improvement in operational efficiency. This significant enhancement is backed by research demonstrating that predictive analytics can boost productivity and efficiency by 20–25% across various industries, including manufacturing and logistics.

Without High-Quality Data: In contrast, organizations without high-quality data see no improvement in operational efficiency, indicated by a 0% improvement rate.

The graph underscores the importance of maintaining high standards of data quality to leverage predictive analytics effectively, allowing organizations to forecast future trends and make proactive, informed decisions.

Data Governance as a Catalyst for Quality Data
1. **Establishing Data Standards**
    - Data governance frameworks establish standards and protocols that ensure consistency and accuracy in data across the organization. The IEEE Transactions on Big Data highlights the importance of data standardization in maintaining data quality (IEEE Transactions on Big Data 2021).

2.2 Data Governance and Business Value

Fig. 2.9 Data governance as a catalyst for quality data. (Created by the author)

2. **Continuous Data Quality Monitoring**
   - Regular monitoring and maintenance of data quality under data governance frameworks can significantly improve the reliability of data used in decision-making processes. A publication in the Journal of Data Management discusses the benefits of continuous data quality monitoring (Journal of Data Management 2021).

3. **Training and Awareness**
   - Data governance involves training employees in data handling and awareness about the importance of data quality, as indicated in a study by the International Journal of Information Management (2020).

Figure 2.9 illustrates how data governance acts as a catalyst for ensuring and improving high-quality data. The central theme, "Data Governance as a Catalyst for Quality Data," is surrounded by critical elements such as establishing data standards, continuous data quality monitoring, and training and awareness. The figure highlights the importance of standardization, ensuring consistency and accuracy, and improving data reliability. It also emphasizes the benefits of continuous monitoring, as well as the necessity for employee training in data handling and fostering data quality awareness. These components collectively enhance and maintain data quality, reliability, and consistency, demonstrating the pivotal role of data governance in managing and improving data assets.

Effective data governance leads to improved data quality, which in turn enhances decision-making. By ensuring that decision-makers have access to accurate, complete, and timely data, organizations can make more informed, strategic decisions that drive success and growth.

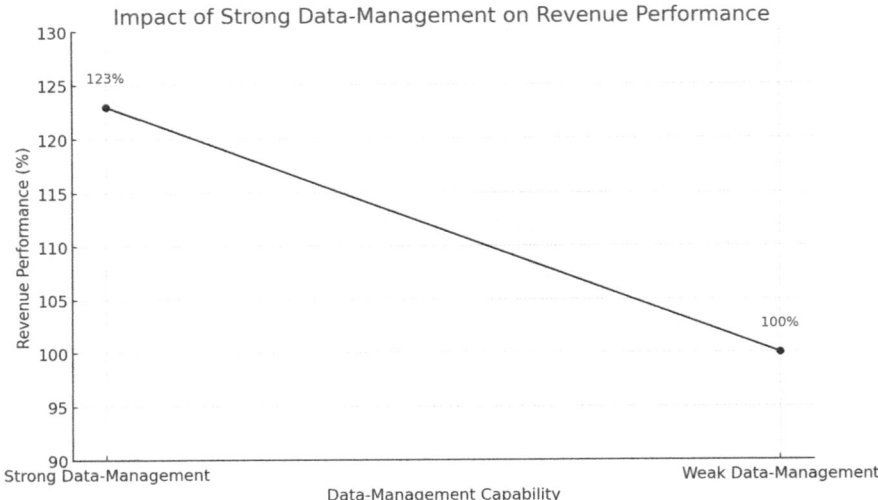

Fig. 2.10 The impact of strong data-management capabilities on revenue performance. (Created by the author)

2.2.2 Improving Operational Efficiency Through Effective Data Management

Operational efficiency stands as a cornerstone of business success, and its enhancement is deeply intertwined with effective data management, anchored firmly by robust data governance. This section offers a comprehensive exploration of how data governance serves as a pivotal force in driving operational improvements.

- **Facilitating Efficient Operations Through Data Management**

Streamlining Data Processes Data governance standardizes data management processes, reducing redundancies and inefficiencies. A study by McKinsey & Company found that companies with strong data-management capabilities were 23% more likely to outperform their competitors in terms of revenue (McKinsey & Company 2019a).

Figure 2.10 demonstrates how strong data-management capabilities positively affect revenue performance. It shows that companies with strong data management see a revenue performance of 123%, compared to 100% for companies with weak data management. The figure illustrates the significant financial benefits of strong data governance and management.

Reducing Operational Costs Effective data management can significantly reduce costs associated with data storage, retrieval, and maintenance. According to Gartner, good data governance can lead to a reduction in operational costs by up to 20% (Gartner 2021d).

2.2 Data Governance and Business Value

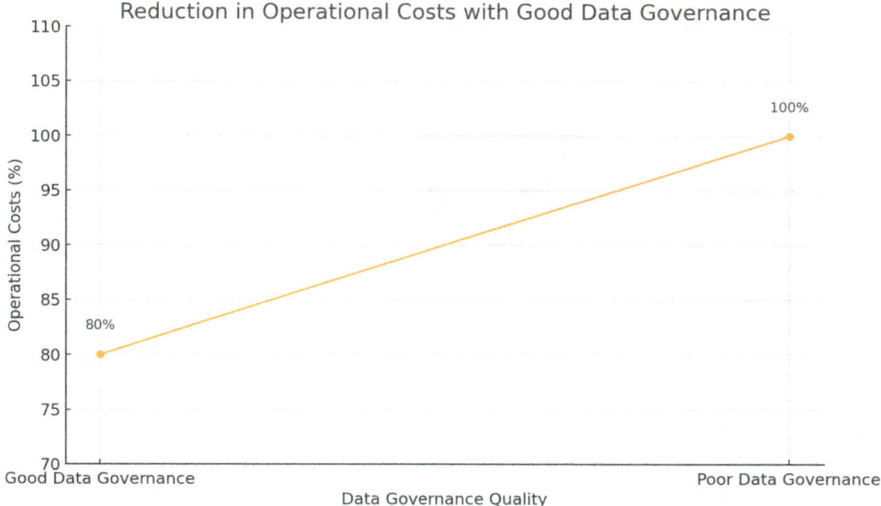

Fig. 2.11 The reduction in operational costs achieved through good data governance. (Created by the author)

Figure 2.11 shows how the quality of data governance affects operational costs. Companies with good data governance see their operational costs reduced to 80%, while those with poor data governance experience costs at 100%. This 20% cost reduction highlights the significant financial benefits of implementing effective data governance practices. By improving data governance, organizations can enhance operational efficiency and achieve substantial cost savings.

- **Enhancing Decision-Making Efficiency**

Data Accessibility and Quality Data governance ensures that high-quality, relevant data is readily accessible to decision-makers, thus speeding up the decision-making process. A report by Harvard Business Review Analytic Services highlights that organizations with accessible, high-quality data are more agile and responsive to market changes (Harvard Business Review 2019a).

Predictive Analytics and Decision Support With governed data, organizations can leverage advanced analytics tools more effectively, leading to faster and more accurate decision-making. IBM notes that predictive analytics and decision support systems depend heavily on the quality and governance of the underlying data (IBM 2020b).

- **Challenges and Solutions in Enhancing Operational Efficiency**

Integration of Diverse Data Systems Integrating data from various sources and systems can be challenging. Bain & Company suggests that investing in integrated data platforms can overcome these challenges and improve operational efficiency (Bain & Company 2020b).

Fig. 2.12 Decision-making efficiency. (Created by the author)

Ensuring Data Compliance and Security Balancing operational efficiency with data compliance and security is crucial. As per Deloitte's insights, implementing data governance practices that align with regulatory requirements is essential for efficient and secure operations (Deloitte 2019a).

Effective data management, guided by a strong data governance framework, is instrumental in improving operational efficiency. By ensuring data quality, accessibility, and compliance, organizations can streamline their operations, reduce costs, and make faster, more informed decisions, thereby enhancing their overall business performance.

Figure 2.12 illustrates the factors contributing to decision-making efficiency. Central themes include data accessibility and quality, which ensure high-quality, relevant data and speed up the decision-making process. Predictive analytics and decision support leverage advanced analytics tools for faster, more accurate decisions. The figure also addresses challenges and solutions, such as ensuring data compliance and security and integrating diverse data systems. These elements collectively enhance the efficiency of decision-making by improving data quality, accessibility, and analytical capabilities, while addressing key challenges in data management. This diagram covers data accessibility and quality, predictive analytics and decision support, and the challenges and solutions associated with operational efficiency.

2.2.3 Driving Business Growth and Innovation Through Strategic Data Utilization

Strategic data utilization, facilitated by effective data governance, is a key driver of business growth and innovation. In today's data-centric business environment, leveraging data strategically can lead to significant competitive advantages. This section explores how data governance enables businesses to use data as a tool for growth and innovation.

2.2 Data Governance and Business Value

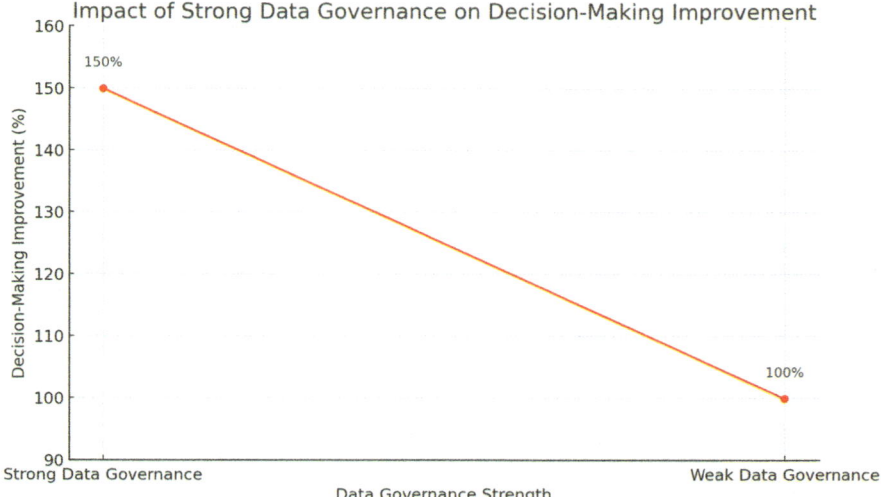

Fig. 2.13 The impact of strong data governance on decision-making improvement. (Created by the author)

- **Enabling Data-Driven Innovation**

Fostering a Culture of Innovation Data governance creates an environment where data can be safely and creatively used to drive innovation. For example, a study by BCG (Boston Consulting Group) highlights that companies with strong data governance are 1.5 times more likely to report a significant improvement in decision-making (BCG 2020).

Figure 2.13 demonstrates the positive correlation between strong data governance and improvements in decision-making. The graph shows that strong data governance leads to a 150% improvement in decision-making, compared to just 100% with weak data governance. This highlights the significant benefits of robust data governance practices in enhancing the accuracy, speed, and quality of decision-making processes within an organization. By strengthening data governance, organizations can achieve more effective and informed decision-making, ultimately driving better business outcomes.

Empowering Data-Driven Strategies By ensuring the availability of high-quality, relevant data, data governance empowers organizations to pursue data-driven strategies. These strategies often lead to innovative products and services, as evidenced in a report by Harvard Business Review Analytic Services (2019b), which found that 58% of enterprises report increased customer engagement and satisfaction through data-driven initiatives.

Figure 2.14 illustrates the significant effect of data-driven initiatives on customer engagement and satisfaction. The graph shows that data-driven initiatives result in a 158% improvement in customer engagement and satisfaction, compared to just

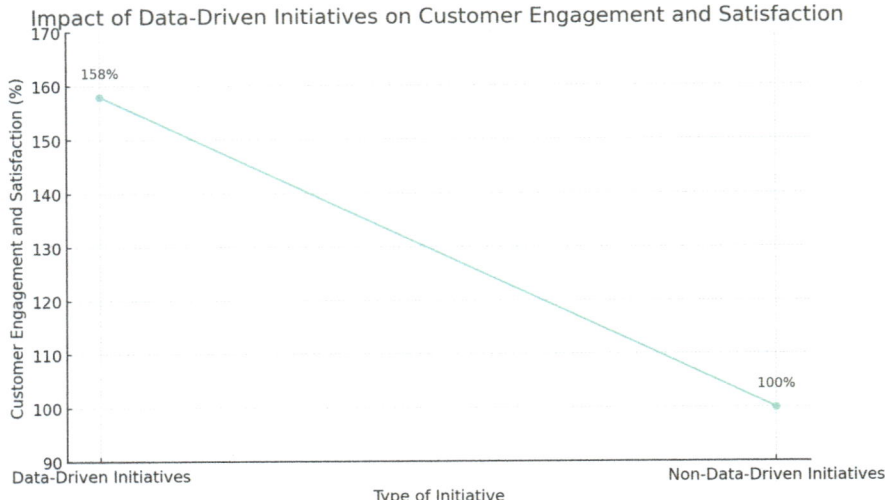

Fig. 2.14 The impact of data-driven initiatives on customer engagement and satisfaction. (Created by the author)

100% for non-data-driven initiatives. This highlights the substantial benefits of leveraging data to inform and drive customer-focused strategies. By adopting data-driven approaches, organizations can significantly enhance their ability to engage and satisfy customers, leading to improved customer loyalty and better overall business performance.

- **Enhancing Competitive Advantage**

Leveraging Big Data for Market Insights Data governance allows organizations to effectively leverage big data for actionable market insights, leading to more informed strategic decisions. A study by McKinsey & Company revealed that data-driven organizations are 23 times more likely to acquire customers (McKinsey & Company 2019b).

Figure 2.15 depicts a stark linear decline in customer acquisition percentage from data-driven organizations to non-data-driven organizations. The y-axis represents the customer acquisition percentage, starting from 2000% down to 100%, while the x-axis categorizes organizations into "Data-Driven Organizations" and "Non-Data-Driven Organizations." The plot illustrates a decrease from approximately 2000% in data-driven organizations to 100% in non-data-driven organizations, showcasing the significant advantage data-driven strategies appear to offer in acquiring customers.

Predictive Analytics for Business Growth Well-governed data enables the effective use of predictive analytics, helping businesses anticipate market trends and customer needs. This proactive approach can significantly contribute to business growth.

2.3 Regulatory Compliance and Risk Management

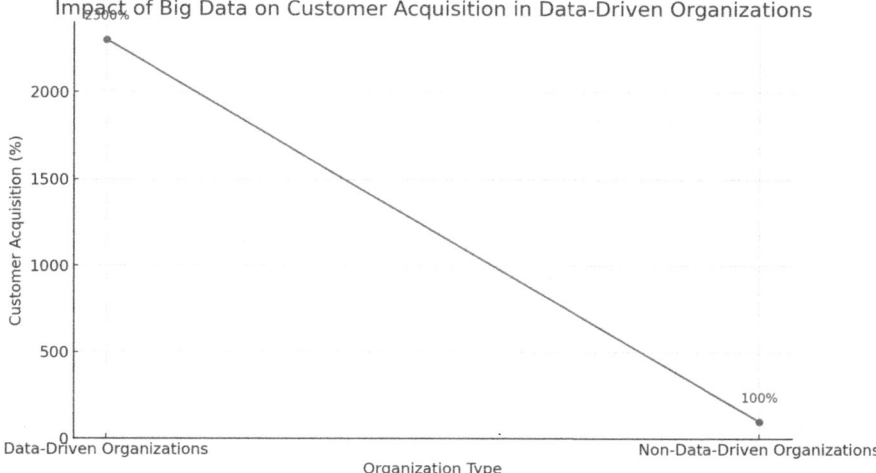

Fig. 2.15 The profound impact of leveraging big data on customer acquisition in data-driven organizations. (Created by the author)

- **Challenges in Utilizing Data for Growth and Innovation**

Balancing Data Exploration with Governance Finding the right balance between encouraging the exploration of data for innovation and maintaining strict governance can be challenging. According to Deloitte (2019a), businesses must navigate this balance to harness the full potential of their data.

Integrating New Technologies Integrating new data technologies and analytics tools into existing data governance frameworks is essential for innovation. Gartner (2021f) suggests that ongoing technology assessment and integration are crucial for data-driven innovation.

Strategic data utilization, supported by robust data governance, is pivotal in driving business growth and fostering innovation. By ensuring data quality, compliance, and strategic alignment, organizations can leverage data as a critical asset for competitive advantage, market insights, and customer engagement.

2.3 Regulatory Compliance and Risk Management

2.3.1 Overview of Data-Related Regulations and European Agencies

In the current digital landscape, adherence to data-related regulations is vital for organizations, especially with the involvement of European agencies in enforcing these laws. This chapter will provide an overview of key data-related regulations like the General Data Protection Regulation (GDPR) and the Health Insurance Portability and Accountability Act (HIPAA), focusing on the role of European agencies in this domain.

General Data Protection Regulation (GDPR)
1. **European Commission's Role:** The European Commission plays a pivotal role in the GDPR framework, providing guidance and clarifications on the regulation. It ensures consistent application of data protection rules across the EU and works with member states to facilitate compliance (European Commission 2018).
2. **Key Requirements and Enforcement by European Data Protection Board (EDPB):** GDPR, effective from 2018, emphasizes principles like consent, rights to access, and data portability. The EDPB, consisting of representatives from the national data protection authorities of the EU member states, oversees the consistent enforcement of GDPR across the EU (European Data Protection Board 2021a).
3. **Penalties for Non-Compliance:** Non-compliance with GDPR can result in fines of up to €20 million or 4% of the annual global turnover, whichever is higher, enforced by national data protection authorities (GDPR.eu 2020a).

Figure 2.16 shows the potential penalties for failing to comply with GDPR—up to €20 million or 4% of annual turnover—emphasizes the financial risks organizations face when they do not adhere to these regulations. This serves as a visual reinforcement of the significant legal and financial consequences discussed in the section, highlighting the importance of understanding and following GDPR to manage risks effectively in data governance.

Health Insurance Portability and Accountability Act (HIPAA)
1. **Impact in Europe:** While HIPAA is a US legislation, its implications extend to European entities that deal with the health information of US citizens or collaborate with U.S.-based healthcare entities.

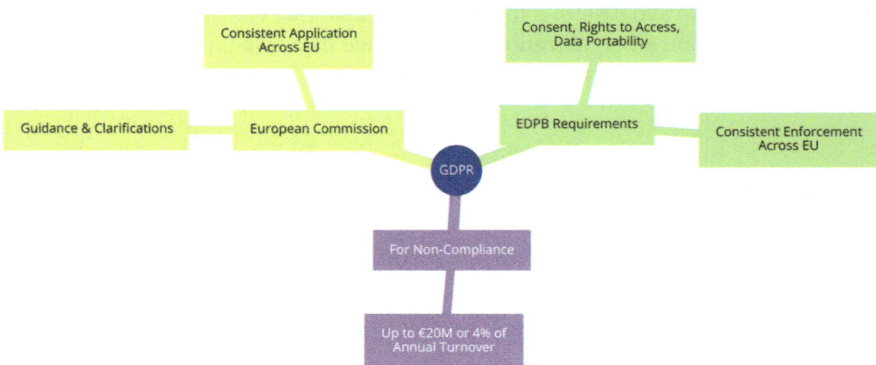

Fig. 2.16 GDPR noncompliance fines. (Created by the author)

2.3 Regulatory Compliance and Risk Management

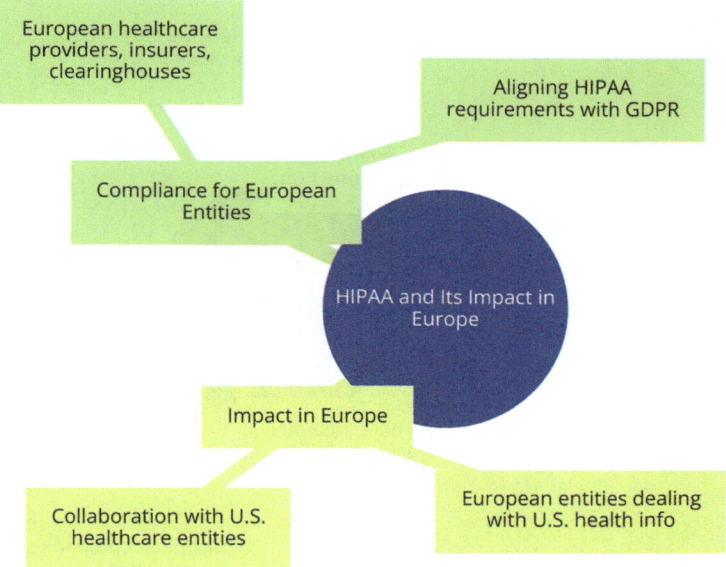

Fig. 2.17 The impact of the Health Insurance Portability and Accountability Act (HIPAA) in Europe. (Created by the author)

2. **Compliance for European Entities:** European healthcare providers, insurers, or clearinghouses that handle health data covered by HIPAA must ensure compliance, which may involve aligning HIPAA requirements with GDPR standards.

Figure 2.17, illustrates the impact of the Health Insurance Portability and Accountability Act (HIPAA) in Europe. Centered around a large blue circle labeled "HIPAA and Its Impact in Europe," it highlights key interactions and compliance challenges faced by European entities. These include compliance requirements for European healthcare providers, insurers, and clearinghouses; the alignment of HIPAA requirements with the General Data Protection Regulation (GDPR); and the necessity for collaboration between European entities and U.S. healthcare entities. The diagram effectively communicates how these factors interlink and the overarching role of HIPAA within European healthcare contexts, particularly in managing US health information.

Global Data Protection Trends and European Response
1. **California Consumer Privacy Act (CCPA) and European Considerations:** Like GDPR, CCPA offers rights regarding personal information to California residents. European companies dealing with Californian data must comply with CCPA, often aligning it with GDPR compliance strategies.
2. **European Agencies and Global Data Protection:** European agencies, including national data protection authorities, are increasingly engaged in global discussions about data protection, influencing and shaping global data protection trends.

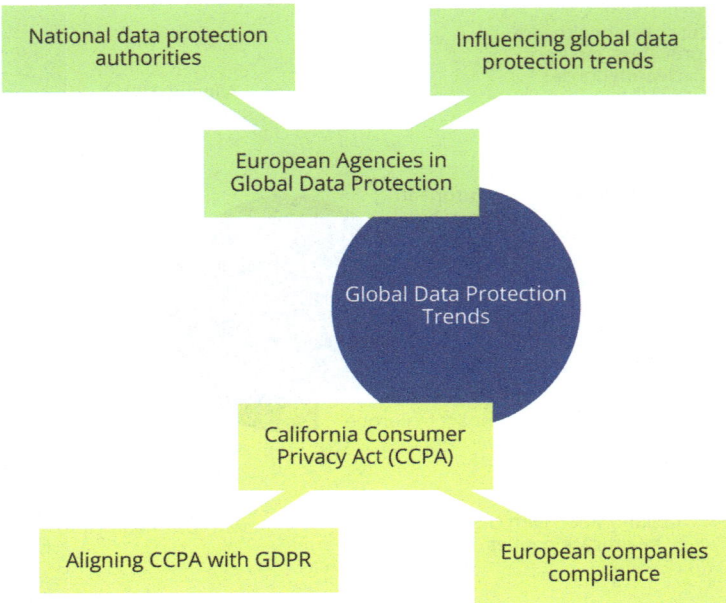

Fig. 2.18 Global data protection trends and the European response. (Created by the author)

Figure 2.18 centers around a large blue circle, representing the influence of global data protection trends. It features various aspects of how these trends are shaped and responded to, especially from a European perspective. Key elements include the role of national data protection authorities in Europe, the impact of European agencies on global data protection norms, and how European entities align with and influence major data protection regulations like the GDPR and the California Consumer Privacy Act (CCPA). The diagram emphasizes the dynamic interactions between European regulatory frameworks and broader global trends, highlighting the alignment of CCPA with GDPR and the compliance efforts of European companies with global standards.

Challenges and Compliance Strategies
1. **Navigating Multiple Regulations:** European organizations operating globally face the challenge of complying with various data protection regulations like GDPR, HIPAA, and CCPA.
2. **Integrating Compliance into Data Governance Frameworks:** European companies must integrate these varying regulatory requirements into their data governance strategies, ensuring compliance and effective risk management.

The role of European agencies in regulatory compliance and risk management is crucial. Understanding and adhering to data-related regulations, guided by these agencies, are fundamental for European organizations to manage risks and maintain trust in the digital age.

2.3.2 The Role of Data Governance in Ensuring Compliance

The role of data governance in ensuring compliance with various data-related regulations is increasingly crucial in today's regulatory landscape. This section highlights how data governance aids organizations in meeting compliance requirements and managing risks associated with data.

Ensuring Adherence to Data Regulations
1. **Framework for Compliance:** Data governance provides a structured framework for managing data in compliance with legal and regulatory standards. It helps in implementing policies and procedures aligned with regulations like GDPR in the EU, HIPAA in the US, and other regional data protection laws. The European Data Protection Board (EDPB) emphasizes the importance of data governance in meeting GDPR requirements (European Data Protection Board 2021b).
2. **Data Privacy and Security Compliance:** Data governance frameworks ensure that privacy and security measures are in place to protect sensitive information, in line with regulations like GDPR and CCPA. The incorporation of privacy by design, as advocated by GDPR, is a key aspect of data governance (GDPR. eu 2020b).

Risk Management Through Data Governance
1. **Identifying and Mitigating Risks:** Effective data governance helps in identifying and mitigating risks related to data accuracy, accessibility, and security. Gartner highlights that risk management is a crucial benefit of data governance, helping organizations avoid penalties and reputational damage (Gartner 2021c).
2. **Audit Trails and Accountability:** Data governance frameworks establish audit trails and accountability mechanisms, which are essential for regulatory compliance. They ensure that activities involving data are traceable and verifiable, as required by various regulations.

Challenges in Compliance and Governance
1. **Keeping Up with Evolving Regulations:** One of the major challenges is keeping the data governance framework aligned with evolving regulations. Continuous monitoring and adaptation are necessary to ensure ongoing compliance.
2. **Balancing Compliance and Business Agility:** Organizations often struggle to balance regulatory compliance with the need for business agility and innovation. Data governance needs to be flexible enough to adapt to changing business needs while maintaining compliance.

Figure 2.19 visually organizes the critical components of data governance related to compliance. Central to the diagram is "Data Governance and Compliance," branching into essential aspects like balancing compliance with business agility, challenges in compliance and governance, and frameworks for compliance, which emphasize the importance of adhering to regulations such as GDPR and HIPAA. It highlights the ongoing requirements to keep up with evolving regulations, the

Fig. 2.19 The role of data governance in ensuring compliance with various data-related regulations. (Created by the author)

necessity for audit trails and accountability, and the role of data governance in risk management. Key insights include the need for incorporating privacy by design and identifying and mitigating risks to avoid penalties. This visual representation underscores the complexity and multi-faceted nature of data governance in maintaining regulatory compliance.

Data governance plays a vital role in ensuring regulatory compliance and effective risk management. It provides the necessary structure and processes to manage data in accordance with legal and regulatory requirements, thereby safeguarding organizations against compliance risks and potential penalties.

2.3.3 Mitigating Risks Associated with Data Breaches and Noncompliance

In the current digital and regulatory landscape, mitigating the risks associated with data breaches and non-compliance is a crucial aspect of data governance. Effective risk management strategies are essential to protect organizations from the legal, financial, and reputational damage that can result from these risks. This section explores how organizations can mitigate such risks.

Strategies for Risk Mitigation
1. **Implementing Comprehensive Data Governance Policies:** Establishing robust data governance policies is the first line of defense against risks. These policies should encompass data privacy, security protocols, and compliance standards. The GDPR, for instance, requires organizations to have clear policies for data protection and mandates reporting data breaches within 72 h (GDPR.eu 2020a).
2. **Regular Risk Assessments and Audits:** Conducting regular risk assessments and compliance audits helps identify potential vulnerabilities and ensures adherence to data governance policies. According to a report by Deloitte, regular

2.3 Regulatory Compliance and Risk Management

audits are vital for identifying and mitigating risks in data governance (Deloitte 2019b).

Preventing Data Breaches
1. **Advanced Security Measures:** Implementing advanced security measures, such as encryption, access controls, and intrusion detection systems, is critical in preventing data breaches. Gartner emphasizes the importance of employing state-of-the-art security technologies to protect sensitive data (Gartner 2021g).
2. **Employee Training and Awareness:** Raising awareness and training employees about data security practices and potential threats can significantly reduce the risk of data breaches. Human error is a common cause of data breaches, and informed employees are a key defense line (Forbes 2020).

Ensuring Regulatory Compliance
1. **Staying Informed on Regulatory Changes:** Keeping up-to-date with regulatory changes and adapting data governance policies accordingly is crucial for compliance. Organizations operating globally must be aware of regulations in different regions, such as GDPR in the EU and CCPA in California.
2. **Documentation and Reporting:** Maintaining comprehensive documentation of data governance practices and being prepared for regulatory reporting are essential for demonstrating compliance during audits or in the event of a breach.

Addressing Noncompliance Risks
1. **Legal Expertise and Consultation:** Engaging legal experts to understand the implications of non-compliance and to ensure that governance policies are legally sound is crucial.
2. **Establishing a Response Plan:** Having a well-defined incident response plan for potential data breaches or non-compliance issues can help in quickly addressing and mitigating these risks.

Figure 2.20 outlines a comprehensive approach to managing risks within data governance frameworks. Central to the diagram is the concept of risk mitigation strategies that branch into three main areas: establishing comprehensive data governance policies, ensuring regulatory compliance, and preventing data breaches. Each area focuses on key activities such as creating security protocols, staying informed about regulatory changes, conducting regular risk assessments, and implementing advanced security measures. The diagram emphasizes the importance of proactive measures, such as training on data security and engaging legal experts, to maintain compliance and prevent data loss.

Mitigating the risks associated with data breaches and non-compliance is a multifaceted challenge that requires a comprehensive approach. By implementing robust data governance policies, conducting regular audits, employing advanced security measures, and ensuring continuous compliance with regulatory standards, organizations can significantly reduce these risks.

Fig. 2.20 Strategies for risk mitigation in data governance. (Created by the author)

2.4 Data Governance and Customer Trust: Building Customer Trust Through Responsible Data Practices

In an era where data breaches and privacy concerns are increasingly prevalent, building customer trust through responsible data practices has become a key priority for organizations. This section explores how effective data governance can enhance customer trust by ensuring the ethical and secure handling of customer data.

Establishing Trust Through Data Privacy and Security
1. **Prioritizing Data Privacy:** Respecting customer privacy is fundamental to building trust. Adhering to privacy regulations like GDPR in the EU and CCPA in the US demonstrates an organization's commitment to data privacy. The GDPR, for example, has set a new standard in data protection, requiring consent for data processing, and giving individuals greater control over their personal data (European Commission 2018).
2. **Implementing Robust Security Measures:** Ensuring data security is crucial for maintaining customer trust. According to a report by Cisco (2020), 84% of customers care about the privacy of their data, and 80% are willing to act to protect it. Therefore, implementing robust security measures like encryption and access controls is essential.

Figure 2.21 illustrates the significant effect that compliance with data privacy regulations has on customer trust levels. It compares trust levels under GDPR compliance, CCPA compliance, and in the absence of data privacy regulations. The graph shows that GDPR compliance results in the highest trust level at 90%, followed by CCPA compliance at 80%, demonstrating robust trust in entities adhering to these regulations. In stark contrast, organizations without any data privacy regulations experience a dramatically lower trust level of only 30%, underscoring the critical role that adherence to data privacy standards plays in fostering customer trust.

2.4 Data Governance and Customer Trust: Building Customer Trust... 115

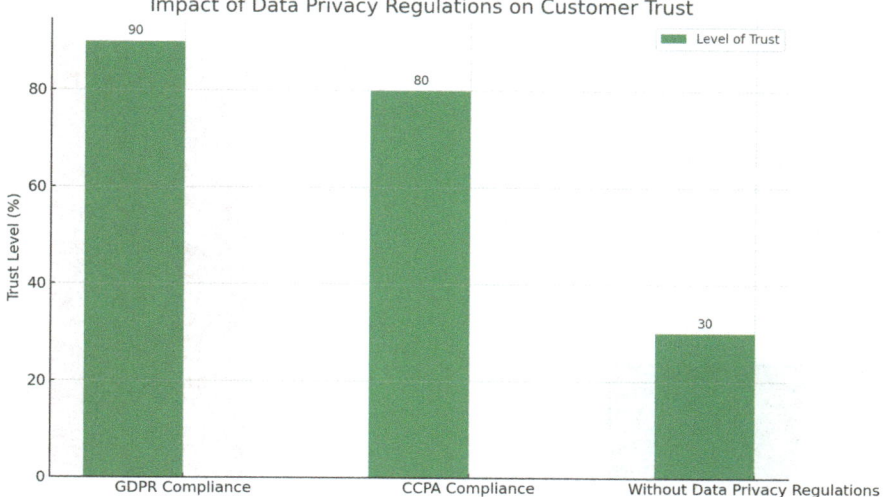

Fig. 2.21 The impact of data privacy regulations on customer trust. (Created by the author)

Transparent and Ethical Data Practices
1. **Transparency in Data Usage:** Being transparent about how customer data is collected, used, and shared is key to building trust. A study by Pew Research Center found that 79% of consumers are concerned about how companies use their data (Pew Research Center 2019)

Figure 2.22, based on the study by Pew Research Center, shows the level of concern among consumers regarding how companies use their data. The chart contrasts consumer concern levels between organizations with transparent data practices and those with non-transparent practices. For companies with transparent data practices, consumer concern is significantly lower, at 21%. In contrast, companies with non-transparent data practices see a much higher level of consumer concern, at 79%. This stark difference highlights the importance of transparency in data usage for reducing consumer concerns and building trust.

2. **Ethical Data Management:** Adopting ethical data management practices, such as not selling customer data without consent and using data responsibly, further strengthens customer trust.

Challenges in Building and Maintaining Trust
1. **Navigating Complex Regulatory Environments:** Keeping up with complex and evolving data privacy regulations can be challenging for organizations but is crucial for maintaining customer trust.
2. **Balancing Data Utilization with Privacy:** Finding the right balance between leveraging data for business purposes and respecting customer privacy is essential.

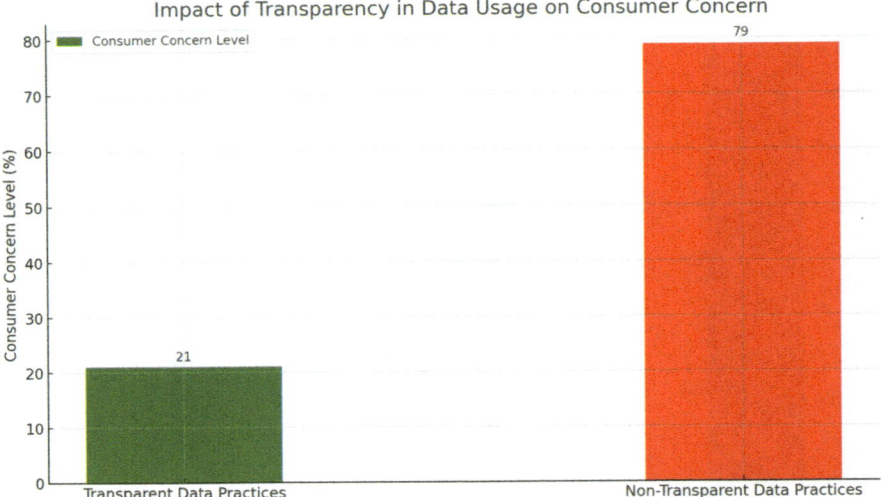

Fig. 2.22 The impact of transparency in data usage on consumer concerns. (Created by the author)

Strategies for Enhancing Customer Trust
1. **Developing a Customer-Centric Data Governance Strategy:** Creating a data governance strategy that focuses on customer needs and expectations can enhance trust. This involves not only compliance with regulations but also proactively addressing customer concerns about data privacy and security.
2. **Regular Communication and Engagement:** Engaging with customers regularly about data practices and any changes in data policies helps in maintaining transparency and trust.

Data governance plays a pivotal role in building and maintaining customer trust. By prioritizing data privacy and security, being transparent and ethical in data practices, and navigating regulatory challenges effectively, organizations can foster a solid foundation of trust with their customers.

2.4.1 The Impact of Data Governance on Customer Relationships and Brand Reputation

Data governance significantly influences customer relationships and brand reputation. In an age where data breaches are common and customer awareness around data privacy is high, how organizations manage and protect customer data can have profound implications on their public perception and customer loyalty.

Enhancing Customer Relationships Through Data Governance
1. **Building Trust with Transparency:** Transparent data governance practices help build customer trust. When customers understand how their data is being used and are assured of its protection, their trust in the brand strengthens. For instance,

2.4 Data Governance and Customer Trust: Building Customer Trust...

a study by Salesforce (2020) revealed that 95% of customers are more likely to be loyal to a company they trust (Salesforce 2020).

2. **Personalization Without Intrusion:** Effective data governance allows organizations to utilize customer data for personalization while respecting privacy boundaries. This balance is crucial for maintaining positive customer relationships. According to a report by Accenture (2018), 83% of consumers are willing to share their data for a personalized experience, provided that businesses are transparent about how they use it (Accenture 2018).

Protecting Brand Reputation
1. **Minimizing Risks of Data Breaches:** Robust data governance reduces the risk of data breaches, which can severely damage a brand's reputation. The Ponemon Institute's 2020 Cost of a Data Breach Report shows that data breaches can have long-lasting financial impacts and harm customer trust (Ponemon Institute 2020).
2. **Compliance with Regulations:** Compliance with data protection regulations like GDPR and CCPA not only avoids legal penalties but also signals to customers that a brand takes data privacy seriously, enhancing its reputation.

Figure 2.23 focuses on the critical role of compliance with data protection regulations like GDPR and CCPA and robust data governance in safeguarding a brand's reputation. At the center, "Protecting Brand Reputation" is highlighted as a key outcome, supported by branches emphasizing different strategic components:

Compliance with Regulations: This branch notes that adhering to standards like GDPR and CCPA not only signals a commitment to data privacy but also enhances a brand's reputation.

Minimizing Risks of Data Breaches: It points to the implementation of robust data governance practices that reduce the risk of breaches, which are crucial for maintaining trust and integrity.

Severe Impact on Reputation: Referencing the Ponemon Institute's 2020 findings, this part of the diagram underscores the negative consequences of data breaches on a company's reputation.

Overall, the figure effectively conveys how compliance and proactive risk management are essential to protecting a brand's public image and trustworthiness.

Challenges in Managing Customer Data
1. **Balancing Data Utility with Privacy:** Finding the right balance between leveraging customer data for business insights and maintaining customer privacy is a key challenge in data governance. This challenge is highlighted by the need to implement privacy-preserving techniques that allow data utility while ensuring privacy protection, as discussed in studies by ISACA and MDPI (ISACA 2023; MDPI 2023).
2. **Adapting to Evolving Customer Expectations:** As customer expectations around data privacy evolve, organizations must continuously adapt their data governance strategies to meet these expectations. The need for ongoing adaptation is emphasized by the dynamic nature of consumer privacy concerns and

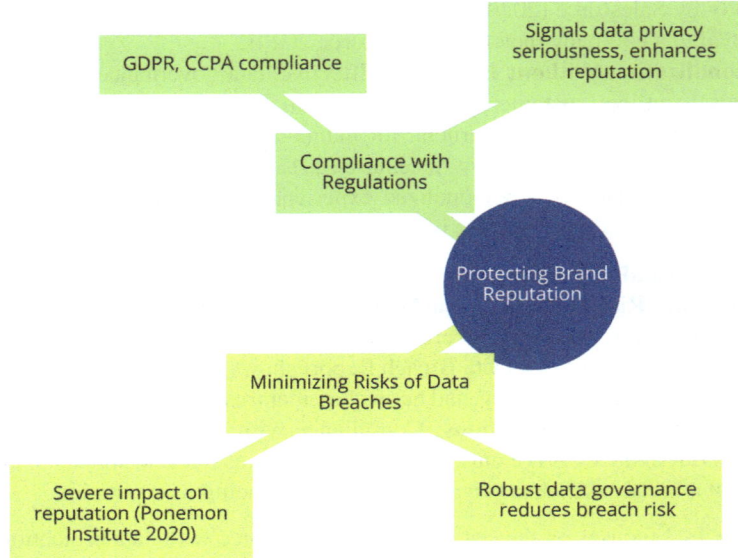

Fig. 2.23 Strategies for protecting a brand's reputation, focusing on minimizing risks of data breaches and compliance with data protection regulations. (Created by the author)

regulatory requirements, requiring organizations to frequently update their data governance practices to align with current standards and customer expectations (ISACA 2023; MDPI 2023).

Strategies for Leveraging Data Governance in Customer Relationships
1. **Clear Communication of Data Practices:** Regularly communicating with customers about how their data is used and protected helps maintain transparency and trust. Effective communication strategies include detailed privacy policies and regular updates about data usage, which can significantly enhance customer trust and transparency (ISACA 2023).
2. **Engagement and Feedback Mechanisms:** Implementing mechanisms to gather customer feedback on data practices and adjusting based on this feedback can enhance customer relationships. This approach ensures that data governance practices are aligned with customer expectations and can improve customer satisfaction and trust in the organization's data handling processes (ISACA 2023; MDPI 2023).

Figure 2.24 outlines various strategies to address the complexities of customer data management effectively. Central to the diagram is "Managing Customer Data Challenges," which branches into key areas focusing on maintaining customer trust, adhering to privacy standards, and optimizing data utility. These include balancing data utility with privacy, adapting data governance to meet customer expectations, and maintaining transparency. Additional branches emphasize the importance of clear communication of data practices and engagement through feedback

2.5 Case Studies: The Cost of Poor Data Governance 119

Fig. 2.24 The challenges in managing customer data. (Created by the author)

mechanisms, allowing businesses to make necessary adjustments based on customer feedback. This configuration underscores the dynamic nature of customer data management, highlighting the need for continual adaptation and communication to align with both business objectives and customer privacy expectations.

Data governance has a significant impact on customer relationships and brand reputation. By managing customer data responsibly and transparently, organizations can build trust, offer personalized experiences without overstepping privacy boundaries, and protect their brand reputation from the risks associated with data breaches and non-compliance.

2.5 Case Studies: The Cost of Poor Data Governance

Poor data governance can lead to significant losses for organizations, both in financial terms and in damage to reputation. Real-world examples underscore the high stakes involved in effective data management and governance. This section presents case studies highlighting the consequences of inadequate data governance.

- **Case Study: Facebook and the Cambridge Analytica Scandal**

Situation In 2018, it was revealed that Facebook had allowed the political consulting firm Cambridge Analytica to access the personal data of millions of its users without their consent.

Consequences This incident led to widespread public outcry over privacy violations. Facebook faced a significant loss of user trust, a decrease in market value, and legal repercussions. The company was fined $5 billion by the Federal Trade Commission, the largest fine ever imposed on any company for violating consumers' privacy (Federal Trade Commission 2019; Al Jazeera 2018; Bipartisan Policy Center 2023) (Fig. 2.25).

Year	User Trust (%)	Market Value (%)	Fine (Billions)
2018	100%	100%	$0
2019	20%	70%	$5

Fig. 2.25 Facebook and Cambridge Analytica Case Timeline. (Created by the author)

- **Case Study: British Airways Data Breach**

Situation In 2018, British Airways (BA) reported a significant data breach that affected around 500,000 customers. The breach involved the theft of personal and payment information from the airline's website and mobile app over a two-week period.

Consequences The UK's Information Commissioner's Office (ICO) initially announced its intention to fine British Airways a record £183 million under the GDPR for the breach, citing poor security arrangements at the company. This fine was later reduced to £20 million in 2020, considering the economic impact of COVID-19 and the company's representations. The breach not only resulted in a substantial financial penalty but also damaged the airline's reputation, highlighting the importance of robust digital security measures (Information Commissioner's Office 2020).

Figure 2.26 visually presents data comparing the financial repercussions and reputational damage over 2 years for an unspecified entity. The bar chart features two categories: "Financial Impact (£ million)" and "Reputation Damage (Scale)." In 2018, the financial impact is substantially higher, represented by a tall blue bar, while in 2020, it's significantly lower, shown with a short yellow bar. Conversely, the reputation damage is not numerically quantified in this visualization but indicates the presence of some reputational impact. This comparison highlights a decrease in financial loss over time, while reputation damage appears as a consistent concern across the years.

Lessons Learned

Importance of Robust Data Governance: These cases highlight the importance of robust data governance, particularly in protecting sensitive data and ensuring compliance with privacy laws.

Reputational Damage and Financial Loss: The lack of effective data governance can lead to significant reputational damage and economic loss, far outweighing the costs of implementing proper data governance practices.

Need for Ongoing Vigilance: These examples also underscore the need for ongoing vigilance and adaptation in data governance strategies to address evolving threats and regulatory requirements.

This case study is a clear example of how lapses in data governance and security can lead to significant financial penalties and loss of customer trust, reinforcing the importance of comprehensive data governance practices in the European context.

2.5 Case Studies: The Cost of Poor Data Governance

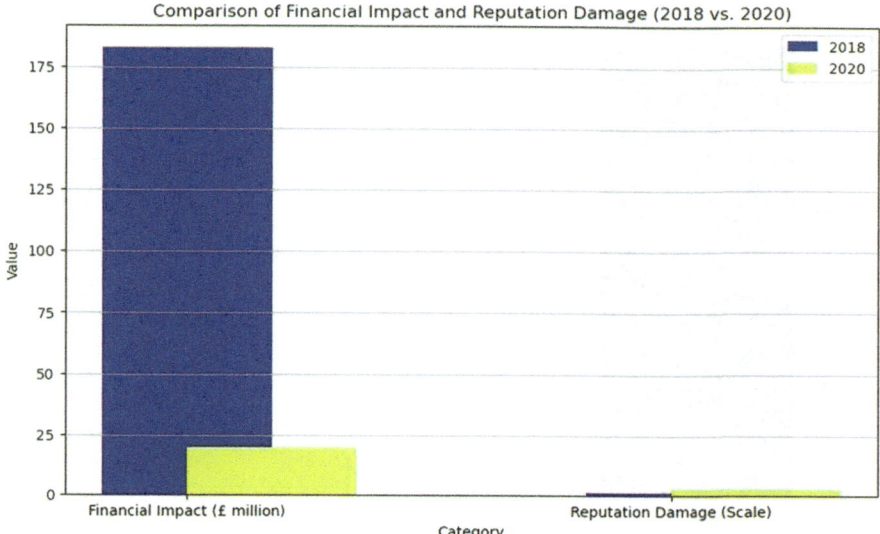

Fig. 2.26 Comparison of Financial Impact and Reputation Damage. (Created by the author)

- **Case Study: H&M Data Privacy Violation in Germany**

Situation In 2020, the Swedish multinational clothing-retail company H&M was found to have violated employee privacy rights at its service center in Nuremberg, Germany. The company had been collecting extensive personal information about employees' private lives, including family issues and religious beliefs.

Consequences The Hamburg Commissioner for Data Protection and Freedom of Information imposed a fine of €35.3 million on H&M for these privacy violations, one of the largest fines under GDPR at the time. This incident highlighted serious shortcomings in H&M's handling of employee data and the importance of respecting privacy rights in the workplace (European Data Protection Board 2020; National Law Review 2020; DW 2020).

Figure 2.27 illustrates the financial and reputational impacts of a privacy violation incident at H&M. The graph features two bars: one representing the fine imposed in millions of euros and the other depicting the severity of the privacy violation on a scale. The fine is significantly higher, shown by a tall blue bar valued at approximately €35 million, indicating a substantial financial penalty. In contrast, the severity of the privacy violation, represented by a short yellow bar, suggests a lower relative impact on this scale, although the exact scale measure is not specified. This visualization effectively communicates the stark contrast between the financial repercussions and the perceived severity of the privacy violation.

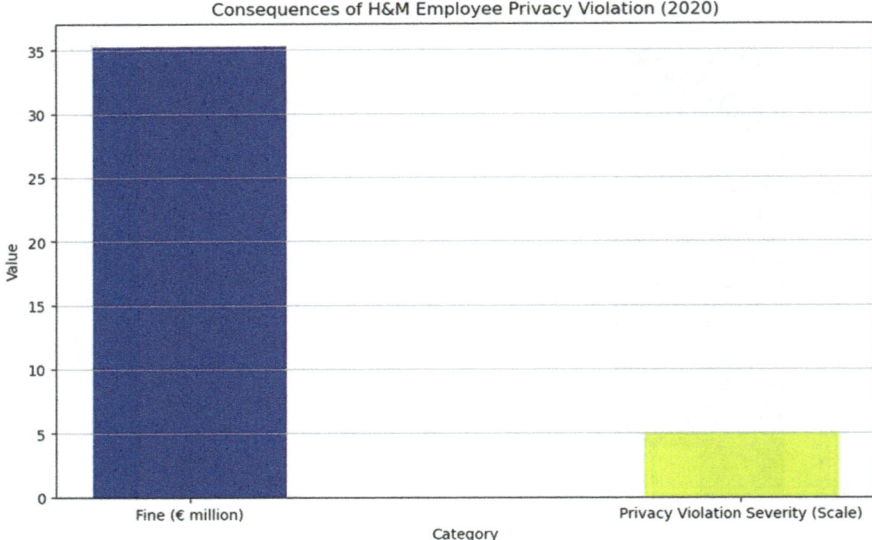

Fig. 2.27 Consequences of H&M Employee Privacy Violation (2020). (Created by the author)

Lessons Learned

Importance of Data Privacy Compliance: This case underscores the necessity of strict adherence to data privacy regulations, particularly under GDPR, which mandates significant fines for non-compliance.

Impact on Employee Trust and Corporate Reputation: The violation significantly impacted employee trust and the company's reputation, illustrating the broader consequences of poor data governance practices.

Commitment to Remedial Actions: In response to the violation, H&M took several corrective measures, including implementing a new data protection compliance regime and compensating affected employees, showcasing the importance of proactive steps to restore trust and comply with data protection laws.

- **Case Study: CNIL Fines Google Under GDPR**

Situation In 2019, France's data protection authority, the CNIL (Commission Nationale de l'Informatique et des Libertés), fined Google €50 million for GDPR violations. The CNIL found that Google had failed to provide transparent and easily accessible information on its data consent policies and did not validly obtain user consent for personalized ads.

Consequences This fine was the first major penalty for a U.S. tech company under the GDPR. The ruling by the CNIL set a precedent for how consent must be obtained and managed under the GDPR, emphasizing the importance of clear and affirmative consent for data processing activities (CNIL 2019).

2.5 Case Studies: The Cost of Poor Data Governance

Figure 2.28 visually depicts the financial penalty and the severity of the GDPR violation incurred by Google. The graph features two bars: one indicating a significant fine of €50 million, represented by a tall blue bar, which highlights the substantial financial impact of the violation. The second bar, shown in yellow, represents the severity of the GDPR violation on a given scale, which appears markedly lower in comparison, suggesting a lesser perceived impact on the severity scale. This graph effectively contrasts the high financial repercussions against the relatively lower severity assessment of the violation.

Lessons Learned

The case highlights the critical need for transparency in data processing practices and the importance of obtaining explicit consent. It also shows the extensive reach of GDPR, impacting not only European companies but global corporations operating in Europe.

These cases from Germany and France demonstrate the stringent nature of GDPR and the emphasis on privacy and consent in the European Union. They serve as vital reminders for organizations to diligently adhere to data governance and privacy standards to avoid substantial penalties and reputational damage.

- **Case Study: Equifax Data Breach**

Situation In 2017, Equifax, one of the largest credit bureaus in the U.S., suffered a massive data breach that exposed the sensitive personal information of 147 million consumers.

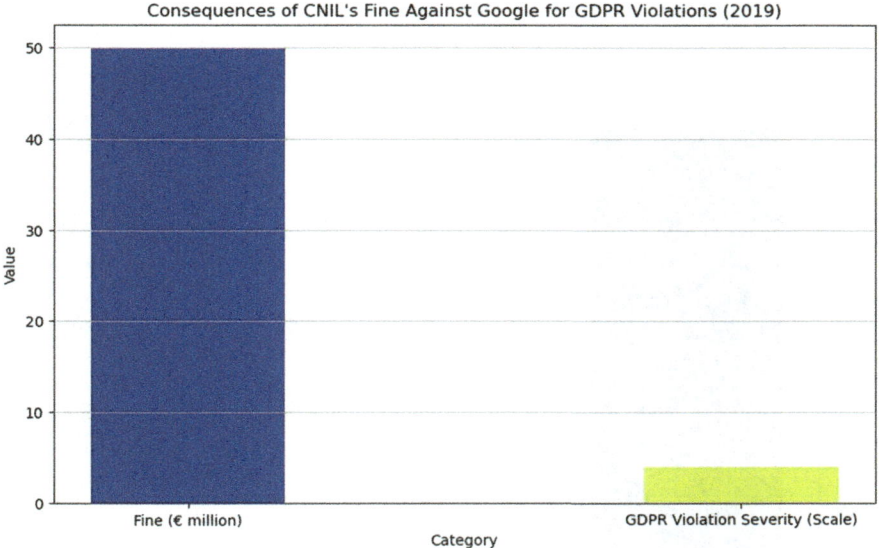

Fig. 2.28 Consequences of CNIL's fine against Google for GDPR violations (2019). (Created by the author)

Consequences The breach led to significant financial losses, including a settlement of up to $700 million. Equifax's reputation was severely damaged, and the breach raised significant questions about its data governance practices (Consumer Financial Protection Bureau 2019).

Figure 2.29 illustrates the profound financial and reputational impacts resulting from the data breach. The graph displays two bars: one for "Financial Losses," represented in millions of dollars, which reaches approximately 700 million, depicted by a very tall blue bar. This visual emphasizes the severe financial burden imposed by the breach. The second bar represents "Reputation Damage Severity" on an unspecified scale, with a height considerably less than the financial losses bar, indicating a significant but comparatively lower impact on reputation than financial losses. This graph effectively communicates the dual nature of the consequences of the data breach, with a predominant financial impact alongside notable reputational damage.

- **Case Study: Marriott International Data Breach**

Situation Marriott International experienced a data breach in 2018, where the personal information of approximately 339 million guests was compromised.

Consequences This breach, one of the largest in history, resulted in Marriott facing hefty fines under GDPR, estimated at £99 million by the UK's Information Commissioner's Office. The breach had a substantial impact on the company's reputation and highlighted the need for stringent data governance in mergers and acquisitions (as Marriott had recently acquired Starwood Hotels, where the breach originated) (Information Commissioner's Office 2020).

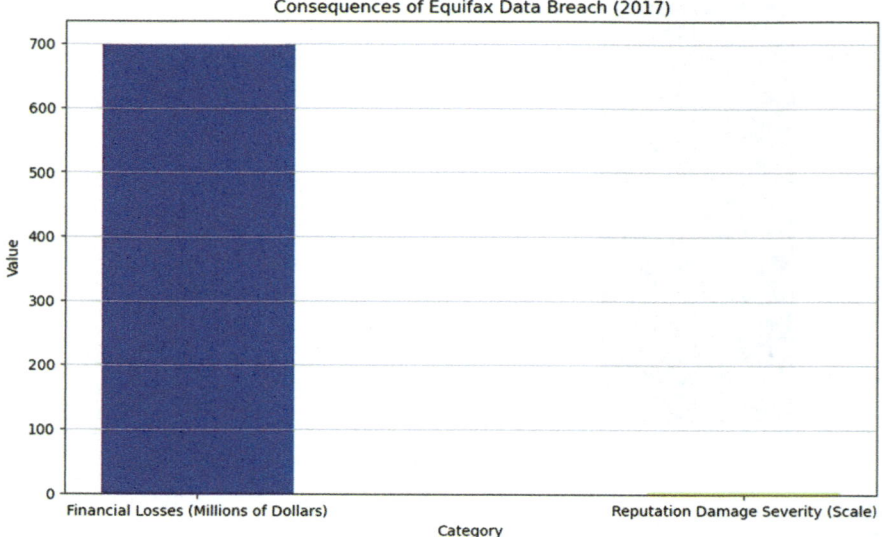

Fig. 2.29 Consequences of Equifax Data Breach (2017). (Created by the author)

2.5 Case Studies: The Cost of Poor Data Governance

Figure 2.30 illustrates the financial and reputational impacts of the data breach incident. It shows a substantial fine of approximately €100 million, represented by a tall blue bar, indicating a significant financial penalty. In contrast, the reputational damage, although present, appears much less severe on the provided scale, depicted by a very short yellow bar. This visualization emphasizes the disparity between the immediate financial consequences and the more moderate perception of reputation damage, suggesting that the breach had a more profound impact financially than on the public image, at least on the measured scale.

Lessons Learned

Importance of Robust Data Governance: These cases highlight the importance of robust data governance, particularly in protecting sensitive data and ensuring compliance with privacy laws.

Reputational Damage and Financial Loss: The lack of effective data governance can lead to significant reputational damage and economic loss, far outweighing the costs of implementing proper data governance practices.

Need for Ongoing Vigilance: These examples also underscore the need for ongoing vigilance and adaptation in data governance strategies to address evolving threats and regulatory requirements.

These examples demonstrate the excessive cost of poor data governance. They serve as cautionary tales for organizations about the importance of investing in comprehensive data governance frameworks to safeguard against data breaches, ensure compliance, and protect their reputation and financial standing.

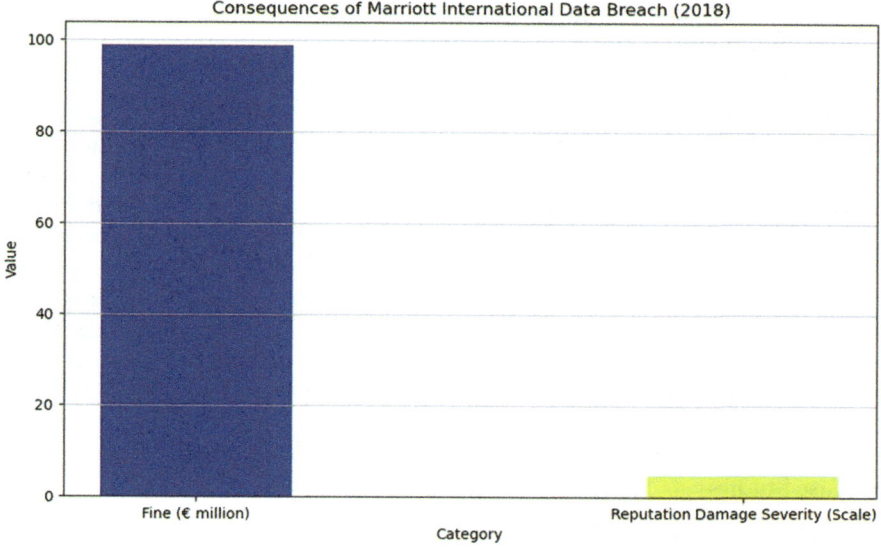

Fig. 2.30 Consequences of Marriott International Data Breach (2018). (Created by the author)

2.6 Data Governance as a Competitive Advantage

In the digital age, effective data governance has transcended its traditional role of ensuring compliance and managing risks. It has emerged as a critical driver of competitive advantage. This chapter explores how data governance can be leveraged to differentiate an organization in a highly competitive market.

- **Strategic Significance of Data Governance**

Trust as a Market Differentiator In today's data-driven world, customers are increasingly aware and concerned about how their data is managed and protected. Companies that establish robust data governance frameworks signal to their customers that they prioritize data security and privacy. This commitment builds trust, a critical factor in customer loyalty and brand reputation. For instance, adhering to global data protection regulations like GDPR not only ensures compliance but also demonstrates a company's dedication to safeguarding customer data, thereby distinguishing it from competitors who might not prioritize data privacy as strongly (Delacroix and Lawrence 2019; Basukie et al. 2020).

Data Governance and Decision-Making Effective data governance provides a structured approach to data management, ensuring high-quality, accurate, and timely data. This reliability is crucial for informed decision-making. Businesses can leverage their well-governed data to gain insights into market trends, customer behaviors, and operational efficiencies. By integrating data governance into business intelligence and analytics, organizations can make strategic decisions that are both data-driven and aligned with their corporate objectives. This capability enables a company to respond swiftly and effectively to market changes, providing a competitive edge (McKinsey & Company 2016).

Competitive Edge Through Compliance In an environment where regulations are constantly evolving, compliance can be challenging. However, organizations that excel in data governance can turn this challenge into an opportunity. By staying ahead of regulatory requirements, companies can avoid the financial and reputational costs associated with non-compliance. Moreover, a proactive approach to data governance can lead to process improvements and operational efficiencies. It positions the company as a responsible and forward-thinking player in the market, which can be appealing to both customers and investors. This compliance-driven edge is particularly relevant in industries where data sensitivity and privacy are paramount, such as finance and healthcare (Park and Bird 2020; Bain & Company 2020a).

- **Case Studies**

John Deere's Agricultural Equipment and Services John Deere, a traditional manufacturing company, extended its business into agricultural services by placing sensors on its agricultural equipment. This data collection enabled them to improve machine performance and provide advanced decision support for farmers. This transition demonstrates how leveraging IoT and cloud technologies to collect and analyze data can redefine a company's business model, leading to a significant competitive advantage (Bain & Company 2020a).

Oil Company's Drilling Operations Optimization A leading oil company partnered with BCG to collect data on drilling operations, equipment, and geological characteristics. They used this data to train machine learning algorithms, which provided workers with end-to-end visibility of processes. This enabled quick decision-making at drilling rigs, optimizing operations such as fluid circulation, pressure management, and maintaining drill-bit rotation. The use of data in this manner not only improved operational efficiency but also led to safer and faster drilling, demonstrating the value of data-driven decision-making in enhancing competitive advantage (BCG 2020).

Finance Sector's Advanced Data Management In the finance sector, the implementation of advanced data management capabilities has had a significant impact. For instance, CFOs adapting advanced data management methods have seen benefits such as improved operational efficiency, risk mitigation, and enhanced decision-making ability. This transition often involves upgrading from less advanced business intelligence and data warehouse architectures. By leveraging machine learning, natural language processing, and intelligent automation, finance leaders have been able to impact significantly without massive up-front investments, prioritizing steps based on business value (Financial Times 2019).

Data governance is no longer just a backend IT function; it is a strategic business tool that can yield a tangible competitive advantage. By ensuring the integrity, security, and optimal use of data, organizations can unlock new opportunities, enhance customer trust, and drive innovation.

These case studies demonstrate the strategic importance of data governance in building trust with customers and differentiating a business in the market. They provide practical insights into how organizations can leverage data governance not just for compliance, but as a core aspect of their value proposition to customers.

Data Governance and Decision-Making Effective data governance provides a structured approach to data management, ensuring high-quality, accurate, and timely data. This reliability is crucial for informed decision-making. Businesses can leverage their well-governed data to gain insights into market trends, customer behaviors, and operational efficiencies. By integrating data governance into business intelligence and analytics, organizations can make strategic decisions that are both data-driven and aligned with their corporate objectives. This capability enables a company to respond swiftly and effectively to market changes, providing a competitive edge (McKinsey & Company 2016).

- **Case Studies**

Airbnb Airbnb implemented an initiative called "Data University" to enhance data literacy across all departments. This program equipped employees to understand, interpret, and use data effectively in their roles, promoting data-driven decision-making. As a result, 45% of Airbnb became weekly active users of their internal data platform, democratizing data and scaling decision-making while ensuring responsible data use (Dataversity 2015; Census 2023).

GE Aviation GE Aviation undertook a mission to centralize its scattered data sources with the initiative called Self-Service Data (SSD). This involved setting up dedicated teams for user enablement and data governance, ensuring proper use of data, and supporting users. The approach led to accelerated decision-making processes and instilled a sense of data ownership among employees, improving safety and operational efficiency.

Wells Fargo Wells Fargo's data governance strategy focused on creating a sole source of truth by centralizing data from multiple sources. This approach streamlined data management, allowed for more accurate reporting and analysis and enhanced decision-making across the organization.

Competitive Edge Through Compliance In an environment where regulations are constantly evolving, compliance can be challenging. However, organizations that excel in data governance can turn this challenge into an opportunity. By staying ahead of regulatory requirements, companies can avoid the financial and reputational costs associated with non-compliance. Moreover, a proactive approach to data governance can lead to process improvements and operational efficiencies. It positions the company as a responsible and forward-thinking player in the market, which can be appealing to both customers and investors. This compliance-driven edge is particularly relevant in industries where data sensitivity and privacy are paramount, such as finance and healthcare (Park and Bird 2020).

- **Case Studies of Competitive Edge Through Compliance**

North American Financial Institution's Customer Onboarding Process A North American bank formed a task force, including the chief compliance officer, to design a customer-centric onboarding process across its capital markets businesses. This initiative aimed to clarify regulatory requirements and streamline customer journeys, leading to a 10% to 30% improvement in customer satisfaction and a 20–40% reduction in administrative touchpoints. This approach minimized client information requests and decreased the risk of data inconsistencies and conflicts, highlighting the competitive edge gained through efficient compliance practices (KPMG 2023).

Research by UConn Professors Stephen Park and Robert Bird Park and Bird from the UConn School of Business co-authored a paper titled "Turning Corporate Compliance Into Competitive Advantage: An Efficient Investment-Risk Model." They proposed a model that balances cost and risk in compliance, allowing firms to gain a competitive edge. The researchers emphasized that companies could strategi-

2.6 Data Governance as a Competitive Advantage

cally benefit from accurately determining their optimal compliance level, balancing spending, and risk acceptance. This approach allows firms to optimize the benefits of compliance relative to cost, minimizing unnecessary resource deployment (Park and Bird 2020).

Know Your Customer (KYC) Compliance in Financial Institutions The fiscal crisis of 2008 and subsequent money laundering scandals led to stricter KYC rules in financial institutions. While implementing a compliant KYC program can be costly, it presents an opportunity to enhance competitive advantage. Trust, a fundamental aspect of the financial industry, becomes crucial in consumer protection. Efficient counterparty risk management protects consumer rights, and maintaining client trust can prevent reputational damage and operational harm. For example, a 2015 survey found that community banks spent a combined $4.5 billion on compliance costs in 2014, with 48% of these costs attributed to hiring outside consultants for new compliance rules. These figures highlight the significant investment and opportunity for firms to differentiate themselves through effective compliance management (Financial Times 2019).

- **Case Studies of Competitive Advantage Through Data Governance**

John Deere's Agricultural Equipment and Services John Deere, a traditional manufacturing company, extended its business into agricultural services by placing sensors on its agricultural equipment. This data collection enabled them to improve machine performance and provide advanced decision support for farmers. This transition demonstrates how leveraging IoT and cloud technologies to collect and analyze data can redefine a company's business model, leading to a significant competitive advantage. The data collected helped the company understand how customers use its equipment and how it performs under different conditions, enhancing their product offerings and customer experience (Databricks 2023; Harvard Business Review 2018).

Oil Company's Drilling Operations Optimization A leading oil company partnered with BCG to collect data on drilling operations, equipment, and geological characteristics. They used this data to train machine learning algorithms, which provided workers with end-to-end visibility of processes. This enabled quick decision-making at drilling rigs, optimizing operations such as fluid circulation, pressure management, and maintaining drill-bit rotation. The use of data in this manner not only improved operational efficiency but also led to safer and faster drilling, demonstrating the value of data-driven decision-making in enhancing competitive advantage (BCG 2020).

Finance Sector's Advanced Data Management In the finance sector, the implementation of advanced data management capabilities has had a significant impact. For instance, CFOs adopting advanced data management methods have seen benefits such as improved operational efficiency, risk mitigation, and enhanced decision-making ability. This transition often involves upgrading from less advanced business intelligence and data warehouse architectures. By leveraging machine learning, natural language processing, and intelligent automation, finance leaders have been able to impact significantly without massive up-front investments, prioritizing steps based on business value (Financial Times 2019).

2.7 Overcoming Challenges: Building a Culture of Data Governance

Building a culture of data governance within an organization is pivotal for the success of any data governance initiative. Leadership and organizational culture play critical roles in shaping and sustaining effective data governance practices. This section explores the importance of these elements in the context of data governance.

Role of Leadership in Data Governance
1. **Setting the Tone at the Top:** Leadership commitment is fundamental to the success of data governance. Leaders set the tone at the top by prioritizing data governance, allocating necessary resources, and leading by example. A study by Gartner (2021b) emphasizes the importance of executive sponsorship in driving data governance initiatives.
2. **Strategic Vision and Support:** Leaders must provide a clear strategic vision for data governance and actively support its implementation across the organization. According to Harvard Business Review (2019b), effective data governance requires a top-down approach, where leaders not only endorse but also actively participate in governance activities.

Fostering a Data-Centric Organizational Culture
1. **Promoting Data Literacy:** Building a data-centric culture involves promoting data literacy across the organization. Employees at all levels should understand the value of data and how it can be leveraged to drive business decisions. Deloitte's insights (2019c) suggest that data literacy is as important as computer literacy in today's business environment.
2. **Encouraging Collaboration and Communication:** A culture that encourages open communication and collaboration between different departments is essential for effective data governance. This includes breaking down silos and fostering a shared understanding of data governance goals and practices.

Overcoming Cultural and Organizational Challenges
1. **Addressing Resistance to Change:** Change management is a critical component in building a culture of data governance. Addressing resistance to change, whether due to fear of the unknown or perceived loss of control, is essential. McKinsey & Company (2020a) highlights the need for change management strategies in embedding data governance into organizational culture.

2. **Aligning Data Governance with Business Objectives:** Data governance should be aligned with business objectives to ensure relevance and buy-in from various stakeholders. This alignment demonstrates the practical value of data governance in achieving business goals.

The journey towards establishing a robust culture of data governance is both challenging and rewarding, playing a critical role in the success of any data governance initiative. This concluding section synthesizes the key elements and strategies necessary for cultivating such a culture, emphasizing the indispensable roles of leadership and organizational culture. Building a culture of data governance is an ongoing process that requires consistent leadership commitment, strategic vision, and an organizational culture that values data literacy, collaboration, and alignment with business objectives. By addressing these aspects, organizations can successfully navigate the complexities of data governance and harness its full potential for organizational success.

2.7.1 Building a Culture of Data Governance

Fostering a data governance mindset across an organization is crucial for ensuring the effective management and utilization of data. It involves not only implementing processes and systems but also cultivating a culture that values and understands the importance of data governance. Here are key strategies for achieving this:

1. Executive Leadership and Sponsorship Leading by Example: Strong leadership commitment to data governance is essential. Leaders should actively promote and participate in data governance initiatives, setting a clear example for the rest of the organization (Harvard Business Review 2019b; Alhassan et al. 2016). Visible Sponsorship: Executive sponsorship can provide the necessary authority and resources for data governance initiatives, signaling their importance to the organization (Gartner 2021b; Khatri and Brown 2010).
2. Comprehensive Training and Education Data Literacy Programs: Implementing organization-wide data literacy programs helps employees understand the importance of data governance and how it impacts their roles (Deloitte 2019c; Otto 2011). Regular Workshops and Seminars: Conducting workshops and seminars on data governance topics can keep the conversation alive and relevant across various departments (Kulkarni and Chettiappan 2019).
3. Collaboration and Communication Cross-Functional Teams: Establishing cross-functional teams can foster collaboration and ensure that data governance is integrated across various business units (McKinsey & Company 2020a; Weber et al. 2009). Open Communication Channels: Maintaining open channels of communication about data governance policies, practices, and benefits helps in building a shared understanding and commitment (Ladley 2020).
4. Integrating Data Governance with Business Processes Alignment with Business Objectives: Data governance should be closely aligned with the organization's

overall business objectives, demonstrating its practical value in achieving business goals (Khatri and Brown 2010). Embedding in Daily Operations: Integrating data governance practices into daily operations makes it a part of the organizational routine, rather than an isolated or additional activity (Weber et al. 2009).
5. Recognition and Incentives Acknowledging Contributions: Recognizing and rewarding departments or individuals who effectively implement data governance can motivate others to follow suit (Ladley 2020). Creating Data Stewards and Champions: Establishing roles such as data stewards or governance champions within departments can help in driving the initiative at a grassroots level (Otto 2011).
6. Continuous Improvement and Adaptation Regular Assessments: Conducting regular assessments of the data governance program and making adjustments based on feedback and changing needs ensures continuous improvement (Kulkarni and Chettiappan 2019). Learning from Mistakes: Encouraging a culture that learns from data governance failures or challenges, rather than penalizing them, can foster a more resilient and adaptable approach (Weber et al. 2009).
7. Leveraging Technology Supportive Tools and Systems: Implementing technology tools that support data governance, such as data catalogs or quality management software, can facilitate adherence to governance practices (Alhassan et al. 2016).

Figure 2.31 illustrates the multi-faceted approach required to foster a robust data governance culture within an organization. At the center, "Building a Culture of Data Governance" serves as the core theme, branching out into several key strategies essential for successful implementation. These include:

1. **Executive Leadership and Sponsorship**: Highlighting the importance of leadership buy-in and visible endorsement to drive governance initiatives.
2. **Collaboration and Communication**: Emphasizing open channels and cross-functional teamwork as foundational for integrating data governance across departments.
3. **Continuous Improvement and Adaptation**: Stresses the need for ongoing assessments and adjustments to maintain relevance in a changing environment.
4. **Integration with Business Processes**: Shows the necessity of embedding data governance into daily operations and aligning it with business objectives.

Overall, the diagram conveys that building a culture of data governance requires concerted efforts across various domains, including training, leadership, technology utilization, and recognition of contributions, all aimed at creating a pervasive and enduring data governance ethos.

Fostering a data governance mindset is a multifaceted endeavor that requires commitment from the top leadership, a focus on education and literacy, collaborative efforts across departments, alignment with business processes, and continuous improvement. By adopting these strategies, organizations can build a strong culture of data governance, leading to better data management and enhanced decision-making capabilities.

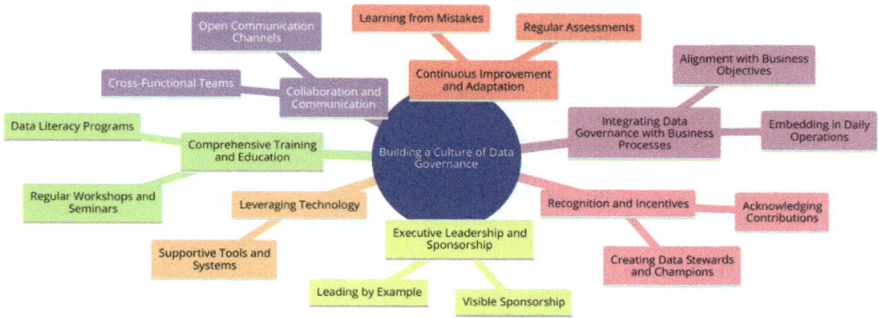

Fig. 2.31 Building a culture of data governance. (Created by the author)

2.8 Conclusion and Transition to Next Chapter

Conclusion: The Critical Importance of Data Governance

As we conclude this chapter, it is evident that data governance is no longer an optional aspect of organizational strategy but a critical imperative. The discussions have highlighted the multifaceted role of data governance in ensuring regulatory compliance, managing risks, building customer trust, fostering innovation, and enhancing operational efficiency.

1. **Regulatory Compliance and Risk Management:** We have seen how data governance is essential for navigating the complex landscape of data-related regulations like GDPR and HIPAA, helping organizations mitigate risks associated with data breaches and non-compliance.
2. **Customer Trust and Brand Reputation:** The role of data governance in building and maintaining customer trust, especially in the wake of increased data privacy concerns, is undeniable. Responsible data practices underpinned by robust governance frameworks are key to preserving brand reputation.
3. **Operational Efficiency and Strategic Decision-Making:** Through the lens of various case studies, we have observed that effective data governance contributes significantly to operational efficiency and informed decision-making, driving business growth and competitiveness.
4. **Challenges and Cultural Transformation:** Addressing the challenges in implementing data governance, especially the need for cultural transformation within organizations, is essential for realizing its full benefits.

References

Accenture (2018) The pulse of Enterprise IT: data-driven innovation. Accenture
Al Jazeera (2018) Cambridge Analytica and Facebook: the scandal so far
Alhassan I, Sammon D, Daly M (2016) Data governance activities: an analysis of the literature. J Decis Syst 25(sup1):64–75. https://doi.org/10.1080/12460125.2016.1187397
Bain & Company (2020a) Data analytics in retail. Bain & Company Reports

Bain & Company (2020b) Data management for the digital age. Bain & Company
Basukie J, Wang Y, Li S (2020) Big data governance and algorithmic management in sharing economy platforms: a case of ridesharing in emerging markets. Technological Forecasting and Social Change
BCG (Boston Consulting Group) (2020) The data-driven transformation. Boston Consulting Group
Bipartisan Policy Center (2023) History of the Cambridge Analytica controversy
Census (2023) How Airbnb democratized their data to empower their employees
Cisco (2020) Cisco consumer privacy survey. Cisco
CNIL (2019) GDPR fines and data breach survey. CNIL
Consumer Financial Protection Bureau (2019) Cost of a data breach report 2020. Ponemon Institute
Databricks (2023) How John Deere uses industrial AI in precision agriculture. Databricks
Dataversity (2015) Case study: an inside look at Airbnb's data science journey
De Mauro A, Greco M, Grimaldi M (2015) What is big data? A consensual definition and a review of key research topics. AIP Conf Proc 1648(1):89–93
Delacroix S, Lawrence ND (2019) Bottom-up data trusts: disturbing the 'one size fits all' approach to data governance. Int Data Privacy Law
Deloitte (2019a) Data management and strategy. Deloitte
Deloitte (2019b) Managing risk in digital transformation. Deloitte
Deloitte (2019c) The culture of data literacy. Deloitte Insights
DW (2020) H&M fined €35 million for privacy breaches
Emerald Insight (2021) Big data analysis for decision-making processes: challenges and opportunities for the management of health-care organizations
European Commission (2018) 2018 reform of EU data protection rules. European Commission
European Data Protection Board (2020) Hamburg Commissioner Fines H&M 35.3 million euro for data protection violations in service centre
European Data Protection Board (2021a) GDPR fines and data breach survey. European Data Protection Board
European Data Protection Board (2021b) Guidelines on data protection by design and by default. European Data Protection Board
Federal Trade Commission (2019) FTC Imposes $5 billion penalty and sweeping new privacy restrictions on Facebook
Financial Times (2019) Banking on GDPR: a financial services case study. Financial Times
Forbes (2020) The human element of cybersecurity, Forbes
Forrester (2020) The importance of data governance for business decision making. Forrester Research
Gartner (2021a) The value of high-quality data in decision-making. Gartner
Gartner. (2021a). Magic Quadrant for Data Quality Solutions.
Gartner (2021b) Data governance key initiative overview. Gartner
Gartner (2021c) Gartner says data and analytics leaders must act now to address the expanding gap between data and decisions. Gartner
Gartner (2021d) How to measure the business value of data governance. Gartner
Gartner (2021e) Improve data quality for better business outcomes. Gartner
Gartner (2021f) Top 10 data and analytics technology trends for 2021. Gartner
Gartner (2021g) Top security and risk Management trends. Gartner
GDPR.eu (2020a) GDPR compliance guidelines. GDPR.eu
GDPR.eu (2020b) What is GDPR? The summary guide to GDPR compliance in the EU. GDPR.eu
Harvard Business Review (2018) John Deere bets the farm on AI IoT. Harvard Business Review
Harvard Business Review (2019a) Drive innovation with better decision-making. HBR
Harvard Business Review. (2019a). Ensuring Data Quality in the Age of Big Data
Harvard Business Review (2019b) Is your data governance program heading down the wrong path? Harvard Business Review
Harvard Business Review (2020) Data-driven decision making: the competitive advantage. Harvard Business Review

References

Harvard Business Review. (2020). Creating a Data-Driven Culture.
Harvard Business Review (2021) A blueprint for data governance in the age of business transformation
Harvard Business Review Analytic Services (2019a) The age of data-driven decisions. Harvard Business Review
Harvard Business Review Analytic Services (2019b) Using data analytics to improve customer engagement and satisfaction. Harvard Business Review
Haug A, Zachariassen F, van Liempd D (2011) The costs of poor data quality. J Indus Engineering and Manage 4(2):168–193
HBS Online (2020) The advantages of data-driven decision-making. Harvard Business School Online
IBM (2020a) Accelerating your journey to AI. IBM
IBM (2020b) The Total economic impact of IBM InfoSphere information server. IBM
IEEE Transactions on Big Data (2021) The role of data standardization in data governance. IEEE Transactions on Big Data
Information Commissioner's Office (2020) ICO fines British Airways £20m for data breach affecting more than 400000 customers
International Journal of Information Management (2020) Data governance and employee training. Int J Inf Manag
ISACA (2023) Balancing privacy with businesses' need for data
Jin GZ (2018) Artificial intelligence and consumer privacy. National Bureau of Economic Research. https://doi.org/10.3386/w24253
Journal of Big Data (2021) Operationalizing and automating data governance. J Big Data
Journal of Business Analytics (2021) Predictive analytics in business decision-making. J Bus Anal
Journal of Data Management (2021) Continuous data quality monitoring in data governance. J Data Manag
Journal of Business Analytics. (2021). Predictive Analytics in Business Decision-Making.
Khatri V, Brown CV (2010) Designing data governance. Commun ACM 53(1):148–152. https://doi.org/10.1145/1629175.1629210
Knowledge at Wharton (2019) How data analytics can drive innovation. Knowledge at Wharton
KPMG (2023) Transforming client onboarding
Kulkarni V, Chettiappan M (2019) Enhancing data governance in organizations. J Inf Optim Sci 40(2):395–406. https://doi.org/10.1080/02522667.2018.1562849
Ladley J (2020) Data governance: how to design, deploy, and sustain an effective data governance program. Academic
Manyika J, Chui M, Brown B, Bughin J, Dobbs R, Roxburgh C, Hung Byers A (2011) Big data: the next frontier for innovation, competition, and productivity. McKinsey Global Institute
Mayer-Schönberger V, Cukier K (2013) Big data: a revolution that will transform how we live, work, and think. Houghton Mifflin Harcourt
McKinsey & Company (2016) The age of analytics: competing in a data-driven world. McKinsey Global Institute
McKinsey & Company (2019a) Data management: the next frontier in the data-driven enterprise. McKinsey & Company
McKinsey & Company (2019b) The analytics advantage: we are just getting started. McKinsey & Company
McKinsey & Company (2019c) The value of data in digital transformation. McKinsey & Company
McKinsey & Company (2020a) Building a data culture. McKinsey & Company
McKinsey & Company (2020b) The data-driven enterprise of 2025
McKinsey & Company (2021) Designing data governance that delivers value
MDPI (2023) Balancing privacy and progress: a review of privacy challenges, systemic oversight, and patient perceptions in AI-driven healthcare. Applied Sciences
MIT Sloan Management Review (2021a) Data governance in the 21st-century organization
MIT Sloan Management Review (2021b) Leading with decision-driven data analytics

National Law Review (2020) Hamburg data protection authority issues fine against H&M

OECD (2015) Data-driven innovation: big data for growth and well-being. OECD Publishing

Otto B (2011) A morphology of the organization of data governance. ECIS 2011 Proceedings, 258

Park S, Bird R (2020) Turning corporate compliance into competitive advantage: an efficient investment-risk model. UConn School of Business

Pew Research Center (2019) Americans and privacy: concerned, confused and feeling lack of control over their personal information. Pew Research Center

Ponemon Institute (2020) Cost of a data breach report 2020. Ponemon Institute

Reinsel D, Gantz J, Rydning J (2018) The digitization of the world from edge to core. International Data Corporation

Salesforce (2020) State of the connected customer. Salesforce

ScienceDirect (2021) Data matters: a strategic action framework for data governance. ScienceDirect

Transmetrics (2021) Predictive analytics in business operations. Transmetrics

Weber K, Otto B, Österle H (2009) One size does not fit all—a contingency approach to data governance. J Data Inf Quality (JDIQ) 1(1):1–27. https://doi.org/10.1145/1515693.1515696

Wu L (2019) How data analytics can drive innovation. Knowledge at Wharton

Chapter 3
Key Principles of Data Governance: Building a Strong Foundation

Contents

3.1	Understanding the Core Principles of Data Governance: Shaping Frameworks and Strategies	138
3.2	Principle of Data Quality	140
3.3	Principle of Data Transparency and Accessibility	145
3.4	Principle of Data Security	147
3.5	Principle of Compliance and Data Privacy	149
3.6	Principle of Data Stewardship	150
3.7	Principle of Data Lifecycle Management	152
3.8	Integrating Data Governance with Business Strategy	154
3.9	Overcoming Implementation Challenges	156
3.10	Real-World Examples: Principles in Practice	157
3.11	Conclusion and Look Ahead	159
References		161

Abstract This chapter provides a foundational exploration of the key principles essential to any effective data governance initiative, aiming to impart a clear understanding of these principles and their practical application across various organizational contexts. It outlines the necessity of maintaining high data quality, ensuring robust data security and privacy, enabling data accessibility, and upholding strict regulatory compliance as core components of effective data governance.

1. **Data Quality**: The chapter emphasizes that high-quality data is pivotal for accurate analytics and decision-making. It discusses the implementation of frameworks and strategies to maintain data quality through validation, cleansing, and standardization.
2. **Data Security and Privacy**: Highlighting the need for frameworks that align with regulations like GDPR, the chapter explores strategies for embedding secu-

rity and privacy into data governance practices, including risk assessments and incident response planning.
3. **Accessibility and Usability**: It stresses the importance of data architecture that supports easy and secure access to data, promoting data democratization within the organization while maintaining stringent control.
4. **Regulatory Compliance**: The chapter identifies compliance as a core component, detailing the integration of mechanisms for ongoing compliance within data governance frameworks and strategies for ensuring adherence to data-related regulations.
5. **Stakeholder Engagement**: The text underscores the importance of inclusive governance structures that engage stakeholders across the organization, fostering collaborative platforms and committees to facilitate participation in data governance decisions.
6. **Continuous Improvement**: Lastly, the chapter addresses the necessity for data governance frameworks to be adaptable and for organizations to commit to ongoing training and development to respond effectively to changing business needs and technological advancements.

Each principle is dissected to reveal its impact on the development of data governance frameworks and the formulation of strategies that ensure data assets are managed effectively, securely, and in a manner that drives business value.

Keywords Data quality · Data security · Data privacy · Data governance · Framework design · Strategy implementation · Regulatory compliance · Data accessibility · Stakeholder engagement · Continuous improvement

This chapter aims to lay down the essential principles that form the bedrock of any successful data governance initiative. It is designed to provide readers with a clear understanding of what constitutes good data governance and how these principles are practically applied in different organizational contexts.

3.1 Understanding the Core Principles of Data Governance: Shaping Frameworks and Strategies

The foundational principles of data governance are not just theoretical concepts; they significantly influence the development of data governance frameworks and the formulation of specific strategies within organizations. This influence manifests in numerous ways as detailed below:

1. **Data Quality as a Framework Pillar**

Framework Design Data quality principles guide the design of frameworks to include processes for data validation, cleansing, and standardization. As mentioned in the DAMA-DMBOK (Data Management Body of Knowledge), frameworks are developed to continuously monitor and improve the quality of data (DAMA International 2017).

Strategy Implementation Strategies around data quality might involve implementing advanced data quality tools, setting up dedicated data quality teams, and defining clear metrics for quality assessment.

2. **Embedding Data Security and Privacy**

Regulatory Alignment Frameworks are shaped to align with regulations such as GDPR, incorporating privacy and security by design as outlined in ISO/IEC 27001 standards (ISO/IEC 2013). This includes strategies for risk assessment, data protection impact assessments, and incident response planning.

Privacy Policies Privacy policies and procedures are developed as part of the framework, including consent management, data subject rights fulfillment, and data minimization strategies.

3. **Ensuring Accessibility and Usability**

Data Architecture The principle of data accessibility and usability influences the data architecture, ensuring that data is organized and stored in a manner that is both accessible and secure as advocated by AHIMA (2019).

Data Democratization Strategies Strategies are formulated to democratize data access within the organization while maintaining control and governance, such as role-based access controls and self-service data analytics platforms.

4. **Regulatory Compliance as a Core Component**

Compliance Mechanisms Governance frameworks are structured to include mechanisms for ongoing regulatory compliance. This might involve regular legal updates, compliance training programs, and audit trails as suggested by the ICO's GDPR guidelines (ICO 2021).

Compliance Tracking Tools Implementation of compliance tracking tools and regular audits are key strategies to ensure adherence to various data-related regulations.

5. **Engaging Stakeholders in Governance**

Inclusive Governance Structures Frameworks are designed to be inclusive, engaging stakeholders from across the organization as recommended by the Data Governance Institute (Data Governance Institute 2020).

Collaborative Platforms and Committees Setting up data governance committees and collaborative platforms where stakeholders can contribute to data governance decisions is a common strategy.

6. **Commitment to Continuous Improvement**

Adaptive Frameworks Data governance frameworks are created with flexibility to adapt to changing business needs and technological advancements as highlighted in Gartner's research (Gartner 2021b).

Ongoing Training and Development Strategies include ongoing training programs and regular reviews of governance policies and procedures to ensure they remain effective and relevant.

The core principles of data governance deeply influence how governance frameworks are structured and how specific strategies are formulated. By embedding these principles into the fabric of data governance, organizations can ensure that their data assets are managed effectively, securely, and in a manner that drives business value.

3.2 Principle of Data Quality

In this section, we explore the principle of data quality, which is a cornerstone of effective data governance. Ensuring lofty standards of data quality is pivotal for any organization that relies on data for decision-making, strategic planning, and operational efficiency.

Defining Data Quality
- **What Constitutes High-Quality Data:** High-quality data is characterized by its accuracy, completeness, consistency, reliability, and timeliness. These attributes are essential for data to be considered of high quality (DAMA International 2017). According to Batini et al. (2009), data quality dimensions such as accuracy, completeness, and consistency are critical metrics used to evaluate data quality in various contexts. Data quality's definition often varies, but it generally revolves around the concept of "fitness for use" (Wang and Strong 1996). Wang and Strong (1996) provided an extensive analysis of data quality from the perspective of data consumers, emphasizing the importance of fitness for use. Redman (2016) also discusses the financial and operational impact of poor data quality on businesses. Additionally, Goel and Shawky (2009) highlight the challenges of maintaining high data quality in rapidly evolving data environments.
- **Importance of Data Quality:** The quality of data directly impacts the reliability of business analytics, the effectiveness of decision-making, and the efficiency of business processes. Poor data quality can lead to significant costs and missed opportunities for businesses (Redman 2016). For example, research indicates

that poor data quality costs businesses an average of $12.9 million annually (Gartner 2021a).

Maintaining High Standards of Data Quality
- **Data Quality Frameworks:** Implementing a data quality framework involves establishing policies and standards for data entry, storage, and maintenance. This framework serves as a guideline for maintaining data quality throughout its lifecycle.
- **Data Quality Tools and Technologies:** Utilizing data quality tools can automate processes such as data cleansing, validation, and enrichment. These tools play a crucial role in identifying and rectifying data quality issues (Pipino et al. 2002).
- **Regular Data Audits:** Conducting regular data audits is essential to assess the quality of data and identify areas for improvement. Audits can uncover issues such as data duplication, inaccuracies, or outdated information (Haug et al. 2011).
- **Data Stewardship:** Assigning data stewards who are responsible for overseeing data quality within specific domains ensures that there is accountability and focused effort on maintaining data quality (Wende 2007).
- **Employee Training and Awareness:** Educating employees about the importance of data quality and training them in best practices for data entry and management can significantly improve the overall quality of data (Kahn et al. 2002).
- **Continuous Monitoring and Improvement:** Data quality should be viewed as an ongoing priority requiring continuous monitoring and improvement processes. This involves setting up metrics and key performance indicators (KPIs) to regularly measure data quality (Lee et al. 2002).

Impact of Data Quality on Business Outcomes
- **Operational Efficiency:** High-quality data leads to more efficient business operations by reducing errors and enabling smoother workflows (Batini et al. 2009).
- **Improved Decision-Making:** Accurate and reliable data underpins effective decision-making, allowing businesses to make informed strategic choices (Wang and Strong 1996).
- **Customer Satisfaction:** In customer-facing applications, high-quality data can enhance customer experiences, leading to increased satisfaction and loyalty (Redman 2016).

Figure 3.1 effectively outlines the core elements essential for maintaining high standards of data quality within an organization. The central node, "Principle of Data Quality," connects to various key aspects:

Defining Data Quality: Identifies the need to clearly define what constitutes high-quality data.
Maintaining High Standards of Data Quality: Focuses on implementing regular data audits, data stewardship, and the use of quality tools and technologies.
Continuous Monitoring and Improvement: Stresses the importance of ongoing assessments and the adaptation of data quality frameworks.
Employee Training and Awareness: Highlights the significance of educating staff about data quality principles.

Fig. 3.1 Principle of data quality. (Created by the author)

This structure encapsulates the interconnected nature of defining, maintaining, and improving data quality to achieve enhanced decision-making, operational efficiency, and customer satisfaction. The diagram serves as a guide for organizations aiming to integrate data quality deeply into their operational ethos.

The principle of data quality is fundamental to data governance. By defining and maintaining lofty standards of data quality, organizations can ensure that their data assets are reliable, trustworthy, and valuable for business operations.

3.2.1 Strategies for Continuous Data Quality Assessment and Improvement

Ensuring continuous assessment and improvement of data quality is a dynamic and integral aspect of effective data governance. This section discusses various strategies that organizations can adopt to maintain lofty standards of data quality consistently.

1. **Implementing Data Quality Metrics and KPIs**

Establishing Clear Metrics Define specific metrics and key performance indicators (KPIs) to measure data quality. These metrics might include accuracy, completeness, consistency, and timeliness (Batini et al. 2009; Wang and Strong 1996). Metrics should be relevant to the specific data quality issues identified and should provide a quantitative basis for improvement (Sebastian-Coleman 2013).

Regular Monitoring Set up a system for regular monitoring of these metrics to assess the quality of data over time. Continuous monitoring can highlight trends and pinpoint areas requiring attention, ensuring that data quality remains high (Lloyd 2010; Ge and Helfert 2007).

3.2 Principle of Data Quality

2. Data Quality Audits and Reviews

Conducting Regular Audits Schedule periodic data quality audits to assess the state of data across different systems and databases. Audits help in identifying issues like discrepancies, redundancies, and outdated information (Haug et al. 2011; Provost and Murray 2011).

Review and Feedback Mechanism Establish a review process where data quality issues identified during audits are addressed. Implement a feedback loop to ensure continuous learning and improvement (Benneyan et al. 2003; Carey and Stake 2003).

3. Utilizing Data Quality Tools

Automated Data Cleansing and Validation Employ data quality tools that can automate the process of data cleansing, validation, and standardization. These tools can detect and correct errors in real time (Pipino et al. 2002; Journal of Big Data 2021).

Data Profiling and Enrichment Use data profiling tools to analyze datasets for inconsistencies or anomalies and employ data enrichment techniques to enhance data quality (Abedjan et al. 2015).

4. Data Stewardship and Ownership

Assigning Data Stewards Designate data stewards who are responsible for overseeing data quality in their respective areas. Data stewards play a crucial role in managing and improving data quality (Wende 2007).

Clear Ownership and Accountability Ensure that there is clear ownership of data within the organization. Assigning accountability for data quality helps in maintaining focus and responsibility (Otto and Österle 2016).

5. Training and Awareness Programs

Employee Training Conduct regular training sessions for employees on the importance of data quality and best practices in data management. Education fosters a culture of data quality and empowers employees to take responsibility for maintaining high standards (Kahn et al. 2002).

Building a Data Quality Culture Foster a culture within the organization that values and prioritizes data quality. Encourage employees to take an active role in maintaining data quality (BMJ 2021).

6. Continuous Improvement Processes

Iterative Approach Adopt an iterative approach to data quality management. Use insights gained from audits and metrics to make continuous improvements (Provost and Murray 2011).

Adapting to Changes Be adaptable to changes in technology, business processes, and regulations that may impact data quality (BMJ 2021).

7. Data Quality in Integration and Migration Projects

Quality Checks in Data Integration Ensure data quality checks are part of any data integration or migration project. This includes validating and cleansing data as it is consolidated from diverse sources (BMJ 2021).

Figure 3.2 provides a comprehensive framework outlining various components essential for enhancing data quality within an organization. Centered around the core theme of continuous improvement, the diagram branches into several critical strategies:

1. **Utilizing Data Quality Tools**: Emphasizes the use of automated cleansing and validation tools alongside data profiling and enrichment techniques.
2. **Data Quality in Integration and Migration Projects**: Highlights the importance of quality checks during data integration.
3. **Implementing Data Quality Metrics and KPIs**: Focuses on establishing clear metrics for regular monitoring and conducting data quality audits and reviews.
4. **Data Stewardship and Ownership**: Stresses the assignment of data stewards and clear ownership to ensure accountability.
5. **Training and Awareness Programs**: Points to the necessity of building a data quality culture through training and employee education.

This structure demonstrates a holistic approach to maintaining and improving data quality through regular audits, iterative adjustments, and fostering an informed workforce, all aimed at sustaining high standards of data management.

Continuous assessment and improvement of data quality require a combination of clear metrics, regular audits, advanced tools, dedicated stewardship, employee training, and an iterative improvement approach. By implementing these strategies, organizations can sustain high data quality standards, thereby enhancing their decision-making capabilities and operational efficiency.

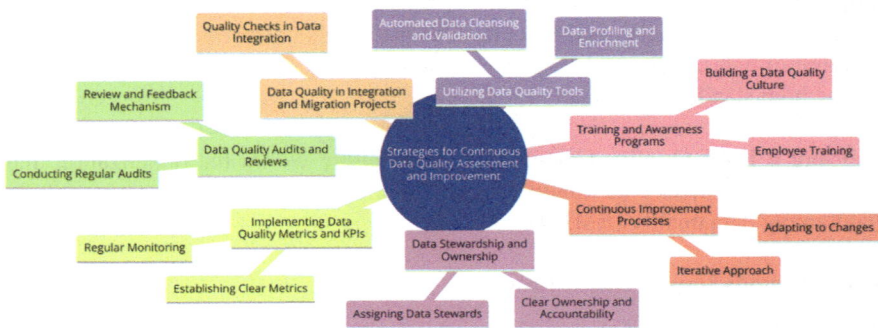

Fig. 3.2 Strategies for continuous data quality assessment and improvement. (Created by the author)

3.3 Principle of Data Transparency and Accessibility

The principles of data transparency and accessibility are pivotal in modern data governance. They ensure that data is not only available to authorized users but also that its usage is transparent and understandable. This section discusses how organizations can implement these principles to enhance data governance.

Defining Data Transparency and Accessibility

Data Transparency: This involves making data governance policies, data usage, and data handling processes clear and understandable to stakeholders within the organization. Transparency is crucial for trust and accountability, especially when dealing with sensitive or personal data (OECD 2023; Rotulo et al. 2023).

Data Accessibility: Accessibility refers to the ease with which authorized users can obtain and utilize data for legitimate purposes. It is about making sure that data is available where and when it is needed without unnecessary barriers (PLOS ONE 2023; OECD 2023).

Strategies for Ensuring Data Transparency and Accessibility

Clear Governance Policies: Develop and communicate clear data governance policies that outline how data is managed, who has access to it, and the purposes for which it can be used. Clear policies promote transparency and ensure that all stakeholders are aware of the rules governing data usage (PLOS ONE 2023; OECD 2023).

Role-Based Access Control: Implement role-based access control (RBAC) systems to ensure that only authorized personnel have access to specific sets of data based on their roles and responsibilities. This helps in maintaining both data security and accessibility (PLOS Global Public Health 2023).

Data Catalogs and Metadata Management: Utilize data catalogs and metadata management tools to provide users with information about the data, including its source, format, and context. This aids in making data more understandable and accessible (PLOS ONE 2023).

Training and Awareness Programs: Conduct regular training and awareness programs to educate employees about the importance of data transparency and how to access data responsibly. Educated employees are more likely to adhere to governance policies and use data ethically (OECD 2023).

Audit Trails and Usage Monitoring: Maintain audit trails and monitor data usage to ensure compliance with data access policies and to provide transparency in data operations. Regular audits can help identify any misuse of data and enhance accountability (Rotulo et al. 2023).

User-Friendly Interfaces and Tools: Provide user-friendly interfaces and tools that make it easy for authorized users to access and analyze data without needing specialized technical skills. Accessible tools can significantly improve data utilization and decision-making (PLOS ONE 2023).

Feedback Mechanisms: Establish feedback mechanisms to continually improve data accessibility and to address any issues or concerns related to data transparency. Feedback from users can help refine data governance practices and enhance overall transparency (OECD 2023).

Balancing Transparency and Accessibility with Security and Privacy

Privacy and Security Measures: While promoting transparency and accessibility, it is critical to balance these with privacy and security considerations. This includes adhering to data protection regulations like GDPR and implementing robust security measures to protect data from unauthorized access or breaches (PLOS Global Public Health 2023).

Ethical Data Usage: Ensure that data accessibility and transparency are aligned with ethical data usage principles. This involves respecting user privacy.

Figure 3.3 outlines the essential strategies for ensuring that data within an organization is transparent and accessible while maintaining security and privacy standards. Central to the diagram is the commitment to ethical data usage, supported by clear governance policies and role-based access control. It highlights the importance of data catalogs and metadata management for enhancing accessibility, complemented by user-friendly interfaces and tools that facilitate easy access to data. The diagram also emphasizes the need for balancing transparency with security and privacy measures, ensuring all data handling is compliant with legal and ethical standards. Additionally, it includes the implementation of regular audits, training programs, and feedback mechanisms to continuously refine data transparency and accessibility practices.

The principles of data transparency and accessibility are essential for effective data governance. They require a well-thought-out approach that balances the need for open access to data with the imperative of maintaining data security and privacy. By implementing these strategies, organizations can foster a data governance environment that is transparent, accessible, and trusted.

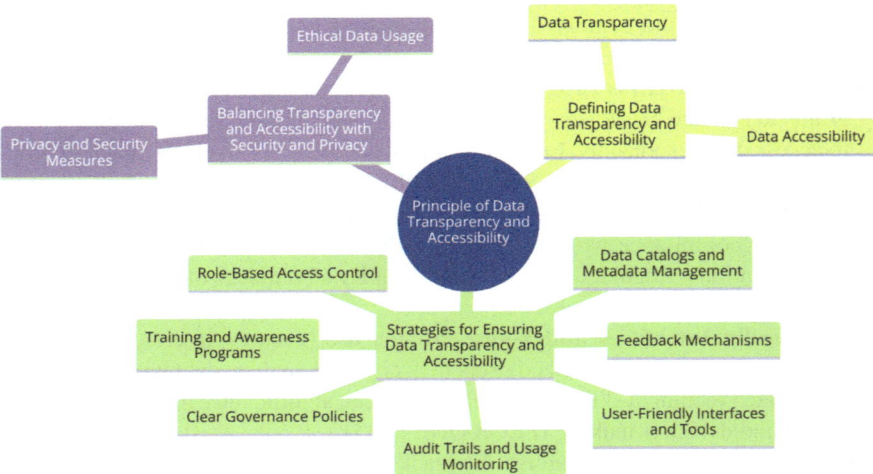

Fig. 3.3 Principle of data transparency and accessibility. (Created by the author)

3.4 Principle of Data Security

In the realm of data governance, the principle of data security is nonnegotiable. Prioritizing the protection of data from unauthorized access and breaches is essential to maintain confidentiality, integrity, and trust. This section explores the importance of data security within data governance and the measures organizations can take to ensure robust protection.

Understanding Data Security in Data Governance

Fundamental Aspect: Data security is a fundamental aspect of data governance. It involves protecting data from unauthorized access, disclosure, alteration, and destruction. Implementing an information security management system (ISMS) is crucial for data protection (ISO/IEC 2013). Information security governance (ISG) integrates data security into the overall governance framework, ensuring that security measures align with organizational goals (Nicho 2018).

Building Trust: Effective data security is vital for building and maintaining trust among stakeholders, including customers, employees, and business partners. Data breaches can severely damage an organization's reputation and lead to significant financial losses. Studies show that security breaches can negatively impact market value and lead to substantial economic costs (Cavusoglu et al. 2004; Goel and Shawky 2009).

Implementing Robust Security Measures and Protocols

Access Controls: Implement strict access controls to ensure that only authorized personnel can access sensitive data. Role-based access control (RBAC) is a widely recommended approach to managing access permissions (Gordon et al. 2011).

Encryption and Data Masking: Use encryption to protect data at rest and in transit. Data masking techniques can be employed to protect sensitive information, especially in development and testing environments (Chen et al. 2012).

Regular Security Audits: Conduct regular security audits to identify vulnerabilities and gaps in the data security framework. This practice aligns with the continuous improvement approach advocated by data governance frameworks (Cannon and Kessler 2007).

Incident Response Planning: Develop and maintain an incident response plan to address data breaches quickly and effectively. This plan should include procedures for containment, investigation, and notification as required by regulations like GDPR (Rotulo et al. 2023).

Training and Awareness Programs: Regularly train employees on data security best practices and the importance of safeguarding data. Employee awareness is a critical line of defense against data breaches (Buckman et al. 2019).

Adherence to Standards and Regulations: Comply with relevant data security standards and regulations. This includes not only international standards like ISO/IEC 27001 but also sector-specific regulations (Gordon et al. 2018).

Technology Solutions: Leverage technology solutions such as firewalls, antivirus software, intrusion detection systems, and security information and event management (SIEM) systems to bolster data security (Chen et al. 2012).

Challenges in Data Security

Evolving Threat Landscape: Keeping up with the continuously evolving cyber threat landscape is a challenge. Regular updates to security protocols and systems are necessary to address new threats (Cavusoglu et al. 2004).

Balancing Accessibility and Security: Finding the right balance between making data accessible and keeping it secure is a complex but crucial aspect of data governance (Cannon and Kessler 2007).

Figure 3.4 outlines a multifaceted approach to ensuring robust data security within organizations. It centers on essential strategies such as balancing accessibility with security, implementing strong encryption and data masking techniques, and developing thorough incident response plans. The visualization emphasizes the importance of regular security audits, adherence to strict access controls, and the necessity of training programs to raise awareness about security practices among employees. It also points to the need for compliance with security standards and regulations to build a resilient data security framework. This holistic approach is critical in protecting sensitive data against the continuously evolving landscape of threats, ultimately fostering trust and understanding within the realm of data governance.

Data security is a critical principle of data governance, essential for protecting sensitive information and maintaining the trust of stakeholders. Implementing robust security measures, staying compliant with regulations, and fostering a culture of security awareness is key to ensuring the effective protection of data.

Fig. 3.4 Principle of data security. (Created by the author)

3.5 Principle of Compliance and Data Privacy

The principle of compliance and data privacy is crucial in data governance to ensure that data is handled in accordance with legal standards and ethical guidelines. This section outlines how organizations can integrate these principles into their data governance frameworks.

Understanding Compliance and Data Privacy

Compliance: Compliance involves adhering to laws, regulations, and policies governing data usage. Key regulations include the General Data Protection Regulation (GDPR) and the California Consumer Privacy Act (CCPA), which set stringent standards for data protection and privacy (Brookings 2023).

Data Privacy: Data privacy ensures that personal data is collected, processed, and stored in ways that protect individuals' privacy rights. This includes implementing measures to prevent unauthorized access and misuse of data (ISACA 2023).

Strategies for Ensuring Compliance and Data Privacy

Implementing Regulatory Requirements: Organizations must align their data governance policies with regulatory requirements like GDPR and CCPA. This involves setting up processes for data access, correction, deletion, and portability as stipulated by these laws (Brookings 2023).

Developing Privacy Policies: Create comprehensive privacy policies that clearly outline how personal data will be collected, used, and protected. These policies should be easily accessible to all stakeholders and regularly updated to reflect changes in regulations (ISACA 2023).

Conducting Regular Audits: Regular audits help ensure compliance with data privacy laws and identify areas where improvements are needed. Audits should cover all aspects of data handling, from collection to processing and storage (ScienceDirect 2023).

Data Minimization and Protection Measures: Adopt the principle of data minimization, which involves collecting only the data that is necessary for a specific purpose. Implement robust data protection measures such as encryption, access controls, and anonymization to safeguard personal data (ScienceDirect 2023).

Training and Awareness Programs: Educate employees about data privacy regulations and best practices for handling personal data. Regular training sessions can help maintain a high level of awareness and compliance within the organization (ISACA 2023).

Incident Response Plans: Develop and maintain an incident response plan to address data breaches promptly. This includes procedures for containing the breach, notifying affected individuals, and reporting to relevant authorities (Brookings 2023).

Organizations must ensure data privacy and compliance by implementing a comprehensive framework that addresses various principles and guidelines (see Fig. 3.5). The diagram titled "Principle of Compliance and Data Privacy" outlines a comprehensive framework for ensuring data privacy and compliance within organizations (see Fig. 3.5). It highlights various strategic areas centered around the core principle of data privacy, including:

Fig. 3.5 Principle of compliance and data privacy. (Created by the author)

1. **Understanding and Monitoring Regulations**: Stresses the importance of staying updated with the evolving legal landscape and implementing regular compliance audits.
2. **Implementing Privacy by Design**: Advocates for embedding privacy into the design of systems and processes from the beginning.
3. **Addressing Challenges**: Discusses the need to balance innovation with privacy and the complexities involved in adhering to global data protection laws.
4. **Legal Obligations and Building Trust**: Emphasizes meeting legal requirements and fostering trust through transparent data handling practices.
5. **Training and Awareness**: Underlines the necessity of educating employees about privacy standards and regulatory requirements.

Overall, the diagram effectively communicates the interconnected components required to maintain and enhance compliance and privacy practices in a dynamic regulatory environment.

The principle of compliance and data privacy is a critical aspect of data governance. Organizations must navigate a complex array of regulations and ensure that their data governance practices meet legal standards while maintaining the trust of their stakeholders.

3.6 Principle of Data Stewardship

Data stewardship is a key principle in data governance, focusing on defining roles and responsibilities for effective data management. It involves establishing clear accountability for maintaining data quality and security. This principle ensures that data is managed as an asset, with designated individuals responsible for its care and oversight.

3.6 Principle of Data Stewardship

Defining Data Stewardship

Role of Data Stewards: Data stewards are responsible for managing the organization's data assets. Their responsibilities include ensuring data accuracy, accessibility, consistency, and adherence to data governance policies and standards (DAMA International 2017). The Open Data Institute defines responsible data stewardship as the collection, maintenance, and sharing of data in a way that maximizes public benefit while minimizing harm (The ODI 2023).

Importance of Stewardship in Data Governance: Data stewardship is integral to successful data governance as it operationalizes governance policies and procedures. It ensures that data governance principles are implemented and followed at the operational level (Nature 2016).

Establishing Roles and Responsibilities

Defining Data Stewardship Roles: Clearly define the roles and responsibilities of data stewards. This includes oversight of data quality, data access, and compliance with data-related policies and regulations (The ODI 2023).

Cross-Functional Collaboration: Data stewards often collaborate with various departments including IT, legal, and business units to ensure that data governance strategies align with organizational needs and objectives (Nature 2016).

Training and Empowerment: Provide data stewards with the necessary training and resources to effectively manage data. Empowering them with decision-making authority is key to successful stewardship (The ODI 2023).

Accountability for Data Quality and Security

Quality Assurance: Data stewards are accountable for the quality of the data, which includes regular monitoring, validation, and updating of data to maintain its accuracy and reliability (DAMA International 2017).

Security Oversight: Ensuring data security is a critical part of stewardship. This involves implementing and monitoring security policies and practices to protect data from unauthorized access and breaches (The ODI 2023).

Compliance Monitoring: Data stewards play a crucial role in ensuring compliance with data privacy laws and regulations. They are responsible for keeping abreast of regulatory changes and ensuring that data management practices are compliant (Nature 2016).

Challenges and Solutions in Data Stewardship

Balancing Technical and Business Understanding: Data stewards need to balance technical data management skills with an understanding of business processes and objectives. Ongoing training and collaboration with different business units can address this challenge (The ODI 2023).

Maintaining Data Stewardship in Dynamic Environments: As organizations evolve, the role of data stewards must adapt to changing data landscapes and business needs. Continuous education and flexible stewardship models are key to addressing this challenge (Nature 2016).

Figure 3.6 visually encapsulates the comprehensive elements and strategic approaches necessary for effective data stewardship within an organization. At the

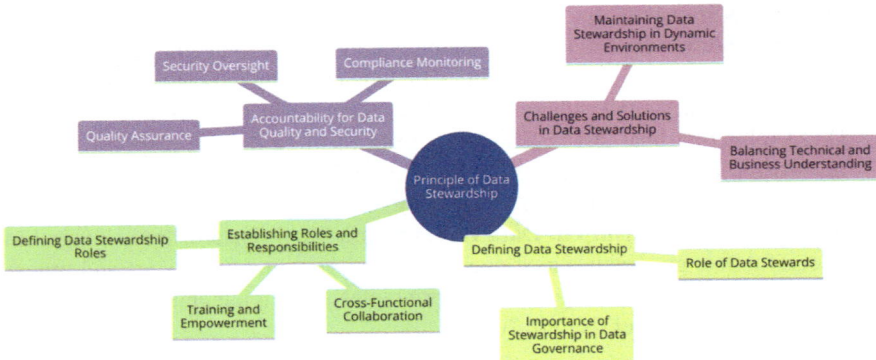

Fig. 3.6 Principle of Data Stewardship. (Created by the author)

core, the principle revolves around establishing and defining clear roles and responsibilities for data stewards, emphasizing their pivotal role in data governance. It highlights the importance of security oversight, compliance monitoring, and quality assurance to ensure data quality and security. Challenges in maintaining data stewardship in dynamic environments are acknowledged, along with the need for balancing technical and business understanding. Cross-functional collaboration, along with training and empowerment of stakeholders, is stressed to reinforce the accountability and effectiveness of data stewardship practices. This structure helps in articulating the critical components required to manage and protect data effectively, promoting an organizational culture that values thorough data governance.

Data stewardship is essential for implementing effective data governance. It ensures that there are designated individuals with clear responsibilities and accountability for managing the organization's data assets, thus playing a crucial role in maintaining data quality, security, and compliance.

3.7 Principle of Data Lifecycle Management

Data Lifecycle Management (DLM) is a fundamental principle in data governance, encompassing the management of data from its creation or acquisition to its eventual retirement or disposal. This section explores the various stages of the data lifecycle and their governance implications.

Understanding Data Lifecycle Management

Definition and Scope: DLM involves overseeing data throughout its entire lifecycle, ensuring it is managed efficiently and effectively at each stage. This includes data creation, storage, usage, sharing, archiving, and disposal.

Importance in Data Governance: Effective DLM is crucial for ensuring data quality, compliance, and security throughout the lifecycle of the data. It helps organi-

3.7 Principle of Data Lifecycle Management

zations maximize the value of their data assets while minimizing associated risks and costs (DAMA International 2017).

Stages of Data Lifecycle and Governance Implications
Data Creation or Acquisition: At this initial stage, governance focuses on ensuring the quality and accuracy of data being entered into the system. Policies regarding data entry and acquisition are crucial here.
Data Storage and Maintenance: Data must be stored securely and maintained accurately. Data governance at this stage involves setting and enforcing policies for data storage, backup, and regular data quality checks (IAPP 2021).
Data Usage: Ensuring that data is used in compliance with governance policies and legal regulations is critical. This involves managing access controls, monitoring data usage, and ensuring ethical use of data.
Data Sharing and Distribution: When data is shared internally or externally, governance policies must ensure that it is done securely and in compliance with data privacy laws. This includes managing permissions and tracking data lineage.
Data Archiving: Governance at this stage involves determining what data should be archived, how it should be stored, and for how long. Compliance with record retention policies and regulations is key.
Data Disposal: Finally, the secure and compliant disposal of data that is no longer needed is a critical aspect of DLM. Data governance policies should outline clear procedures for data deletion or destruction to prevent unauthorized access or data breaches (IAPP 2021).

Challenges in Data Lifecycle Management
Managing Data in Dynamic Environments: Adapting DLM practices in rapidly changing business and technological environments is challenging. Regular review and adaptation of data governance policies are necessary.
Ensuring Compliance Across Stages: Ensuring compliance with various data-related regulations at each stage of the data lifecycle requires continuous monitoring and updating of governance practices.

Figure 3.7 provides a detailed view of the various phases and key considerations in the management of data throughout its lifecycle. Central to the diagram is the "Principle of Data Lifecycle Management," around which essential elements are organized to reflect the comprehensive process of data handling from creation to disposal. It outlines the critical stages of data lifecycle, including data creation or acquisition, data storage and maintenance, data usage, data sharing and distribution, data archiving, and ultimately data disposal. Additionally, the diagram addresses the importance of ensuring compliance across these stages and highlights the challenges in managing data in dynamic environments. It also emphasizes the definition and scope of data lifecycle management and underscores its significance in data governance, illustrating a holistic approach to maintaining data integrity and utility throughout its lifespan.

The principle of data lifecycle management is integral to data governance, ensuring that data is managed effectively, securely, and in compliance with regulatory

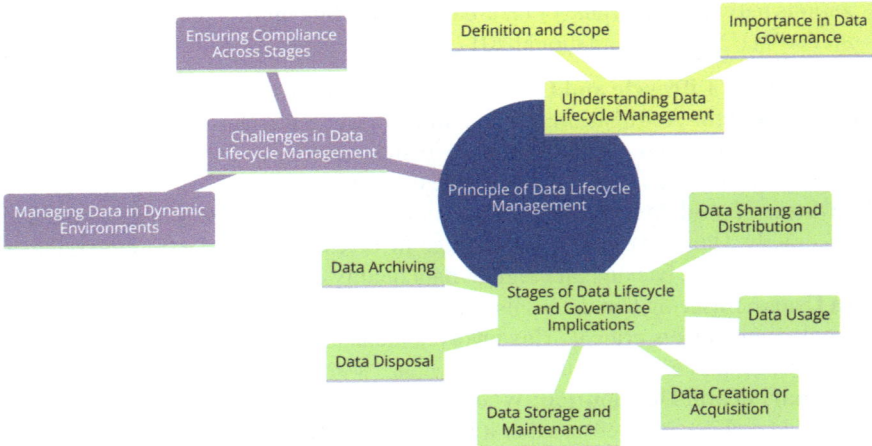

Fig. 3.7 Principle of data lifecycle management. (Created by the author)

requirements throughout its lifecycle. Understanding and managing the various stages of the data lifecycle is crucial for organizations to extract maximum value from their data assets while minimizing risks.

3.8 Integrating Data Governance with Business Strategy

Integrating data governance with business strategy is essential for organizations to effectively leverage their data assets in alignment with their overall business goals. This section explores how aligning data governance principles with organizational strategies can enhance decision-making, operational efficiency, and competitive advantage.

Aligning Data Governance with Organizational Goals
1. **Understanding Business Objectives:** The first step is to have a clear understanding of the organization's strategic objectives. Data governance initiatives should be designed to support these objectives, whether it is market expansion, customer satisfaction, innovation, or operational efficiency (McKinsey 2023; Pragmatic Institute 2023). Understanding the business context helps in aligning data management efforts with the strategic goals of the organization, ensuring that data governance contributes to achieving these objectives.
2. **Strategic Data Management:** Data governance should facilitate strategic data management. This involves ensuring that data is not only accessible and secure but also relevant and actionable for strategic decision-making (DATAVERSITY 2023). By integrating data governance with business strategy, organizations can better manage their data assets, making them a reliable foundation for business analytics and strategic initiatives.

Demonstrating the Business Value of Data Governance

1. **Enhanced Decision-Making:** By ensuring data quality and integrity, data governance provides a reliable foundation for business analytics and decision-making. According to a report by McKinsey & Company, data-driven organizations are 23 times more likely to outperform competitors in customer acquisition and 19 times as likely to be profitable (McKinsey & Company 2019). High-quality data allows for more accurate and timely business insights, which are crucial for making informed decisions.
2. **Operational Efficiency:** Effective data governance can streamline operations, reduce costs associated with poor data quality (such as inaccuracies and inefficiencies), and improve overall productivity. Gartner's research indicates that poor data quality costs businesses an average of $12.9 million annually (Gartner 2021a). Implementing robust data governance practices helps in mitigating these costs by ensuring that data is accurate, consistent, and timely.
3. **Risk Management and Compliance:** Integrating data governance with business strategy includes managing risks related to data and ensuring compliance with various data-related regulations. This not only protects the organization from legal and financial penalties but also enhances its reputation and trustworthiness (DATAVERSITY 2023). Effective data governance frameworks help organizations stay compliant with regulations like GDPR and CCPA, safeguarding against data breaches and other compliance risks.

Overcoming Challenges in Integration

1. **Bridging the Gap Between IT and Business:** One of the primary challenges is bridging the gap between IT (which typically manages data) and business units. This requires cross-functional collaboration and communication (Pragmatic Institute 2023). Establishing a governance framework that encourages interaction and alignment between these groups can help in addressing this challenge effectively.
2. **Change Management:** Implementing data governance as part of business strategy often requires a change in organizational culture and mindset, which can be achieved through effective change management strategies (DATAVERSITY 2023). This involves educating employees about the importance of data governance and its benefits, as well as continuously reinforcing these concepts through training and leadership support.

3.8.1 Case Studies

- **Sustainable Entrepreneurship:** This study provides an analysis of case studies and empirical evidence on integrating sustainability into business strategies, which includes aspects of data governance and management. It offers insights into how businesses can align their data governance strategies with broader busi-

ness goals, particularly in the context of sustainable practices (Rosário et al. 2022; Raimundo 2024).
- **Mastering Money Management:** This publication includes case studies that illustrate how businesses have revolutionized their economic management practices through real-time data access and integration with other business systems. These examples are beneficial in understanding how data governance is integral to effective economic management and strategic business operations (Pragmatic Institute 2023).

Integrating data governance with business strategy is vital for leveraging data as a strategic asset. By aligning data governance principles with organizational goals and demonstrating how they support business objectives, organizations can realize the full value of their data assets.

3.9 Overcoming Implementation Challenges

Implementing data governance principles effectively can be a complex task often met with various challenges. This section aims to identify common hurdles organizations face in applying data governance principles and provides practical solutions and best practices for overcoming these obstacles.

Identifying Common Implementation Challenges

Resistance to Change: One of the most significant challenges is resistance to change within the organization. Employees may be hesitant to adopt new processes and systems that alter their routine work (Kahneman 2011).

Lack of Clear Vision and Leadership: Without clear leadership or a defined vision for data governance, initiatives can lack direction and fail to gain traction (Kotter 1996).

Data Silos: Organizational data silos can impede the effective implementation of data governance as disparate data systems and departments may operate independently (Davenport 2014).

Balancing Data Accessibility with Security: Finding the right balance between making data accessible and maintaining its security and privacy is a delicate challenge (NIST 2018).

Aligning Data Governance with Business Objectives: Ensuring that data governance strategies align with overall business goals can be difficult, especially in organizations where the value of data is not fully recognized (Ladley 2019).

Practical Solutions and Best Practices

Change Management: Address resistance to change through effective change management strategies. This includes communication, training, and involving employees in the change process to get buy-in (Hiatt and Creasey 2012).

Strong Leadership and Clear Objectives: Establish strong leadership for data governance initiatives. Clear objectives and a well-defined roadmap can provide direction and purpose, helping to align efforts with business goals (Kotter 1996).

Breaking Down Data Silos: Promote a culture of data sharing and collaboration. Implement technology solutions that facilitate data integration and provide a unified view of data across the organization (Davenport 2014).

Robust Security and Privacy Controls: Implement and enforce robust data security and privacy controls. Regularly review these controls to ensure they meet evolving data protection standards and business needs (NIST 2018).

Alignment with Business Strategy: Ensure that data governance strategies are aligned with the business strategy. This can be achieved by demonstrating how data governance supports business objectives and contributes to the bottom line (Ladley 2019).

Regular Training and Communication: Conduct regular training sessions for employees to increase data literacy and awareness of data governance policies. Communication should be ongoing to keep all stakeholders informed about data governance practices and their benefits (Hiatt and Creasey 2012).

Continuous Monitoring and Improvement: Set up mechanisms for the continuous monitoring of data governance initiatives. Use metrics and KPIs to measure effectiveness and make improvements where necessary (Redman 2008).

Leveraging Technology Solutions: Utilize technology solutions like data governance platforms, data quality tools, and analytics to support data governance efforts (Abraham et al. 2019).

Overcoming the challenges of implementing data governance requires a combination of strong leadership, effective change management, strategic alignment, and the use of appropriate technology solutions. By addressing these challenges head-on with practical solutions and best practices, organizations can ensure the successful implementation of data governance initiatives.

3.10 Real-World Examples: Principles in Practice

In this section, we explore real-world examples and case studies where various organizations have successfully applied key data governance principles. These examples provide insights into how theoretical principles can be translated into practical applications, yielding significant benefits.

AI and Smart Technologies for Smart Agriculture Environment
This study explores how Smart Agriculture leverages AI algorithms and real-time data from IoT sensors for improved resource management and pest control. Real-world case studies and examples demonstrate the practical application of data governance in the agriculture sector. By using AI and IoT, farmers can optimize irrigation, predict pest infestations, and enhance crop yields through data-driven insights (Zhang et al. 2019; Kamilaris et al. 2017).

Efficient Power Management Strategy and Challenges of Hybrid Renewable Energy-Based Electric Vehicles
This research presents case studies showing how hybrid renewable energy-based electric vehicles are used in the real world. It discusses the integration of power management systems and the collaboration between government and business sectors, highlighting data governance principles in energy management. Effective data management ensures reliable performance and regulatory compliance in energy systems (Jochem et al. 2015; Sarker et al. 2020).

AINeedsPlanner: A Workbook to Support Effective Collaboration Between AI Experts and Clients
This study demonstrates a workbook's utility with two case studies in real-world scenarios, focusing on requirements ranging from functionality, data, and governance. It provides effective guidelines for collaboration and data governance in AI-related projects. The workbook helps align stakeholder expectations and ensures the responsible use of AI technologies (Sambasivan et al. 2021).

Implementing a Pilot Study of COVID-19 Self-Testing in High-Risk Populations and Remote Locations: Results and Lessons Learnt
This study discusses the implementation of COVID-19 self-testing in high-risk populations and remote locations, focusing on techniques employed to enhance data quality and facilitate the integration of survey databases. The study offers valuable insights into improving data management and quality in healthcare settings, especially under challenging conditions (Watson et al. 2020).

Cross-Border Collaboration: India's UPI Success Story in Indonesia (A Case Study on Unified Payments System)
This study examines the efficiency and security of the Unified Payments Interface (UPI) system and evaluates its role in promoting credit generation, contributing to financial inclusion, and economic growth. It provides insights into the implementation and impact of UPI in the financial services sector, with a focus on data security aspects (Khaitan and Sharma 2019).

Governing the Power of Neoliberalism Government with People's Economics in the Welfare of the Community (Study of People-Owned Shops or Tomira in Kulon Progo Regency, Special Region of Yogyakarta)
This study examines the actions of the neoliberal government in Kulon Progo Regency, especially in relation to market expansion and community welfare. It explores the governance approaches used in expanding retail chains, which can offer insights into data governance in a retail expansion context (Widianingsih and Morrell 2007).

Machine Learning-Based Intelligent Security Framework for Secure Cloud Key Management
This study discusses the confidentiality, integrity, and availability of sensitive data in cloud environments. It includes case studies demonstrating the application of data management principles in ensuring security and compliance. Machine learning

enhances security protocols by predicting and mitigating potential threats (Singh and Chana 2016).

Decentralized Identity Management Using Blockchain Technology: Challenges and Solutions
This research evaluates existing implementations and case studies in decentralized identity management, highlighting the need for robust data management technologies to maintain user confidence and comply with privacy regulations. Blockchain technology provides a secure and transparent method for identity verification (Zheng et al. 2018).

AI's Evolving Impact in US Banking: An Insightful Review
This literature review examines case studies in the banking sector, focusing on ethical issues, data privacy concerns, and compliance. It discusses responsible data management practices in the context of AI algorithms, emphasizing the need for transparent and accountable AI systems (Furman and Seamans 2019).

Validation of a Visual Landscape Quality Indicator for Agrarian Landscapes Using Public Participatory GIS Data
This study uses public participatory GIS data, aligning with principles of data stewardship in the context of environmental management in the public sector. Public participation ensures the accuracy and relevance of data used for landscape quality assessments (Brown and Kyttä 2018).

The Role of Hydrogen in Decarbonising the Steel Industry: Upstream and Downstream in the UK and Ontario
This research explores the adoption of hydrogen technologies in the steel industry, highlighting the importance of data lifecycle management in manufacturing processes, particularly in the context of decarbonization. Effective data governance supports the transition to sustainable energy practices (Bode et al. 2020).

These case studies illustrate how different organizations across various industries have applied key data governance principles to address specific challenges. By doing so, they were able to improve their operations, comply with regulations, enhance customer trust, and drive business growth.

3.11 Conclusion and Look Ahead

As we conclude this chapter on the foundational principles of data governance, it is important to reflect on how these principles form the bedrock of any successful data governance strategy. This chapter has delved into various core principles, including data quality, transparency and accessibility, security, compliance, stewardship, and lifecycle management, each playing a vital role in the effective management of data.

Summarizing the Importance of Foundational Principles
1. **Integral to Organizational Success:** The principles discussed are not just theoretical concepts but integral components that drive organizational success. They ensure that data is reliable, secure, compliant, and used ethically and effectively.
2. **Foundation for Strategic Decision-Making:** High-quality, well-governed data is the foundation for informed strategic decision-making. It empowers organizations to harness data for insights, innovation, and competitive advantage.
3. **Risk Mitigation and Compliance:** In today's landscape of stringent data regulations, these principles serve as a guide to navigate complex legal requirements, thereby mitigating risks and ensuring compliance.
4. **Building Trust and Reputation:** By adhering to these principles, organizations build trust with customers, partners, and regulators, enhancing their reputation and credibility.

Challenges and Best Practices
While implementing these principles, organizations often face challenges such as resistance to change, data silos, and balancing security with accessibility. Addressing these challenges requires a combination of strong leadership, effective communication, and the application of best practices tailored to the organization's specific needs.

Teasing the Next Chapter: "Data Governance Frameworks: Models and Best Practices"
In the upcoming chapter, we will shift our focus from the foundational principles to the practical aspects of implementing data governance. We will explore various data governance frameworks, models, and best practices that organizations can adopt and customize to their unique environments.

- **Exploring Data Governance Frameworks:** The next chapter will delve into different models of data governance frameworks, discussing their structures and components, and how they can be effectively implemented in various organizational contexts.
- **Best Practices in Action:** We will also present best practices drawn from real-world examples, illustrating how successful organizations have implemented data governance frameworks to achieve strategic objectives, maintain compliance, and leverage data as an asset.
- **Preparing for the Future:** Additionally, the chapter will provide insights on how to adapt these frameworks to future challenges and trends in data management and governance.

The upcoming chapter will provide a practical guide to transforming the principles of data governance into actionable strategies and frameworks, helping organizations to effectively manage their most asset—their data.

Look Ahead As we move forward, readers can anticipate gaining practical insights and actionable strategies to design and implement robust data governance frameworks that align with their organizational goals and the ever-evolving landscape of data management.

References

Abedjan Z, Golab L, Naumann F (2015) Profiling relational data: a survey. VLDB J 24(4):557–581
Abraham R, Schneider J, Vom Brocke J (2019) Data governance: a conceptual framework, structured review, and research agenda. Int J Inf Manag 49:424–438
American Health Information Management Association (AHIMA) (2019) Data quality management model. AHIMA
Batini C, Cappiello C, Francalanci C, Maurino A (2009) Methodologies for data quality assessment and improvement. ACM Comput Surv (CSUR) 41(3):1–52
Benneyan JC, Lloyd RC, Plsek PE (2003) Statistical process control as a tool for research and healthcare improvement. BMJ Quality Safety 12(6):458–464
BMJ (2021) Using data for improvement. The BMJ
Bode I, Deneckere A, Rossi G (2020) The role of hydrogen in decarbonising the steel industry. J Clean Prod 247:119115
Brookings (2023) Highlights: the GDPR and CCPA as benchmarks for federal privacy legislation
Brown G, Kyttä M (2018) Key issues and research priorities for public participation GIS (PPGIS): a synthesis based on empirical research. Appl Geogr 95:122–136
Buckman J, Hashim MJ, Woutersen T, Bockstedt J (2019) Fool me twice? Data breach reductions through stricter sanctions. In: The Cambridge handbook of compliance
Cannon DM, Kessler L (2007) Danger–corporate data breach! J Corp Account Finance 18(5):41–49
Carey RG, Stake LV (2003) Improving healthcare with control charts: basic and advanced SPC methods and case studies. ASQ Quality Press, Milwaukee
Cavusoglu H, Mishra B, Raghunathan S (2004) The effect of internet security breach announcements on market value. Int J Electron Commer 9(1):70–104
Chen JV, Li HC, Yen DC, Bata KV (2012) Did IT consulting firms gain when their clients were breached? Comput Hum Behav 28(2):456–464
DAMA International (2017) DAMA-DMBOK: data management body of knowledge. DAMA International
Data Governance Institute (2020) Data governance best practices. Data Governance Institute
DATAVERSITY (2023) Why your data strategy needs to align with your business strategy
Davenport TH (2014) Big data at work: dispelling the myths, uncovering the opportunities. Harvard Business Review Press
Furman J, Seamans R (2019) AI and the economy. Innov Policy Econ 19(1):161–191
Gartner (2021a) Data quality solutions. Gartner
Gartner (2021b) Improve data quality for better business outcomes. Gartner
Ge M, Helfert M (2007) A review of information quality research. Data and Information Quality Workshop
Goel R, Shawky H (2009) Managing data quality for institutional investors. J Invest Compliance 10(2):36–46
Gordon LA, Loeb MP, Zhou L (2011) The impact of information security breaches: has there been a downward shift in costs? J Comput Secur 19(1):33–56
Gordon LA, Loeb MP, Lucyshyn W, Zhou L (2018) Empirical evidence on the determinants of cybersecurity investments in private sector firms. J Inf Secur 9(2):133–153
Haug A, Zachariassen F, van Liempd D (2011) The costs of poor data quality. J Ind Eng Manage (JIEM) 4(2):168–193
Hiatt JM, Creasey TJ (2012) Change management: the people side of change. Prosci Learning Center Publications
Information Commissioner's Office (ICO) (2021) Guide to the general data protection regulation (GDPR). ICO
International Association for Privacy Professionals (IAPP) (2021) Data lifecycle management and privacy. IAPP
ISACA (2023) Data governance for privacy, confidentiality, and compliance: a holistic approach

ISO/IEC (2013) ISO/IEC 27001:2013 information security management systems—requirements. International Organization for Standardization

Jochem P, Babrowski S, Fichtner W (2015) Assessing CO_2 emissions of electric vehicles in Germany in 2030. Transp Res A Policy Pract 78:68–83

Journal of Big Data (2021) Big data quality framework: a holistic approach to continuous quality management. J Big Data

Kahn BK, Strong DM, Wang RY (2002) Information quality benchmarks: product and service performance. Commun ACM 45(4):184–192

Kahneman D (2011) Thinking, fast and slow. Farrar, Straus and Giroux

Kamilaris A, Kartakoullis A, Prenafeta-Boldú FX (2017) A review on the practice of big data analysis in agriculture. Comput Electron Agric 143:23–37

Khaitan R, Sharma R (2019) The unified payments Interface (UPI) success story in India. J Financ Serv Res 56(3):245–264

Kotter JP (1996) Leading change. Harvard Business Review Press

Ladley J (2019) Data governance: how to design, deploy, and sustain an effective data governance program, 2nd edn. Academic

Lee YW, Strong DM, Kahn BK, Wang RY (2002) AIMQ: a methodology for information quality assessment. Inf Manag 40(2):133–146

Lloyd RC (2010) Navigating in the turbulent sea of data: the quality measurement journey. Clin Perinatol 37(1):101–122

McKinsey (2023) Designing data governance that delivers value

McKinsey & Company (2019) The role of data in digital transformation

National Institute of Standards and Technology (NIST) (2018) Framework for improving critical infrastructure cybersecurity (Version 1.1)

Nature (2016) The FAIR guiding principles for scientific data management and stewardship

Nicho M (2018) Information security governance: a framework. J Inf Secur 9(2):115–132

OECD (2023) Data accessibility: open, free and accessible formats. OECD iLibrary

Otto B, Österle H (2016) Data quality management. Springer

Pipino LL, Lee YW, Wang RY (2002) Data quality assessment. Commun ACM 45(4):211–218

PLOS Global Public Health (2023) Mind the gap: data availability, accessibility, transparency, and credibility during the COVID-19 pandemic

PLOS ONE (2023) Data sharing, management, use, and reuse: practices and perceptions of scientists worldwide

Pragmatic Institute (2023) Aligning data projects with business goals

Provost LP, Murray SK (2011) The health care data guide: learning from data for improvement. Wiley, San Francisco

Raimundo R (2024) Sustainable entrepreneurship education: a systematic bibliometric literature review. Sustain For 16(2):784

Redman TC (2008) Data driven: profiting from your most important business asset. Harvard Business Press

Redman TC (2016) Data driven: creating a data culture. Harvard Business Review Press

Rosário AT, Raimundo RJ, Cruz SP (2022) Sustainable entrepreneurship: a literature review. Sustainability 14(9):5556

Rotulo A, Kondilis E, Thwe T, Gautam S, Torcu Ö, Vera-Montoya M et al (2023) Mind the gap: data availability, accessibility, transparency, and credibility during the COVID-19 pandemic. PLOS Global Public Health

Sambasivan N, Kapania S, Highfill H, Parthasarathy S, Akrong D, Matthews T, Churchill E (2021) "Everyone wants to do the model work, not the data work": data cascades in high-stakes AI. In: Proceedings of the 2021 CHI conference on human factors in computing systems, pp 1–15

Sarker MRI, Bhattacharjee U, Khan R (2020) Integration of hybrid renewable energy based electric vehicles for efficient power management. Energy Rep 6:175–184

ScienceDirect (2023) Data governance: a conceptual framework structured review and future research agenda

Sebastian-Coleman L (2013) Measuring data quality for ongoing improvement: a data quality assessment framework. Morgan Kaufmann, Waltham

Singh S, Chana I (2016) Q-aware: quality of service based cloud resource provisioning. Comput Electr Eng 47:138–160

The ODI (2023) Defining responsible data stewardship

Wang RY, Strong DM (1996) Beyond accuracy: what data quality means to data consumers. J Inf Manag 12(4):5–33

Watson J, Whiting M, Krieg C, Butcher R, Carr PA, Simpson L (2020) COVID-19 self-testing in high-risk populations and remote locations: results and lessons learnt. J Glob Health 10(2):020340

Wende K (2007) A model for data governance–organising accountabilities for data quality management. ACM International Conference Proceeding Series:90–99

Widianingsih I, Morrell E (2007) Participatory governance in Indonesia: a case study of community-based environmental initiatives. Environ Dev Sustain 9(4):409–425

Zhang C, Wang Z, Wang Z (2019) A review on the research and applications of agricultural robotics: development, key technologies, and challenges. Precis Agric 20(4):918–940

Zheng Z, Xie S, Dai H, Chen X, Wang H (2018) Blockchain challenges and opportunities: a survey. Int J Web Grid Serv 14(4):352–375

Chapter 4
Data Governance Frameworks: Models and Best Practices

Contents

4.1	Introduction to Data Governance Frameworks.	166
4.2	Overview of Popular Data Governance Frameworks.	170
4.3	Designing a Data Governance Framework.	176
4.4	Best Practices in Framework Implementation.	178
4.5	Integrating Technology with Frameworks.	180
4.6	Measuring the Effectiveness of your Framework—Data Governance Template.	183
4.7	Case Studies: Frameworks in Action.	187
4.8	Adapting Frameworks to Changing Data Landscapes.	189
4.9	The Future of Data Governance Frameworks.	191
4.10	Conclusion and Transition to the Next Chapter.	193
References.		194

Abstract This chapter delves into the intricacies of data governance frameworks, offering insights into their essential components, applications, and strategies for effective implementation. The chapter begins with a comprehensive definition of what constitutes a data governance framework, describing it as a structured set of guidelines, policies, standards, practices, and procedures designed to manage, utilize, and protect an organization's data assets. The discussion underscores the importance of these frameworks in establishing clear guidelines and standards that ensure data consistency, quality, and security across all organizational levels.

The chapter explores several popular data governance frameworks, including DAMA-DMBOK, CMMI Data Management Maturity (DMM) Model, and ISO 8000, providing a comparative analysis that highlights their suitability for different organizational needs. It emphasizes the role of these frameworks in facilitating compliance and risk management, guiding data governance implementation, and enabling scalability and flexibility to accommodate organizational growth and changing regulations.

Further, the narrative shifts to the design and customization of data governance frameworks, tailored to fit an organization's unique size, industry specifics, and regulatory needs. It addresses the best practices in implementing these frameworks, stressing the importance of clear leadership, stakeholder engagement, and the integration of technology. The chapter also examines the challenges often encountered during implementation, such as securing executive buy-in and managing cultural change.

Additionally, the integration of technology is identified as a pivotal factor in enhancing the effectiveness of data governance frameworks. The chapter reviews essential tools and software that support data governance processes, enhancing efficiency and compliance.

Lastly, the chapter outlines methods to measure the effectiveness of data governance frameworks using metrics and KPIs and discusses strategies for continuous improvement. Through real-world examples and a detailed review of framework models, the chapter provides valuable guidance on the strategic implementation of data governance frameworks, emphasizing their crucial role in enhancing data management practices and aligning them with business objectives.

Keywords Data governance frameworks · Best practices · Implementation · Organizational needs · Data management · Regulatory compliance · Data policies · Standards · Roles and responsibilities · Decision-making processes · Monitoring · Enforcement · Continuous improvement

This chapter aims to provide a thorough understanding of the various data governance frameworks available, their implementation, and best practices, helping readers to select and adapt a framework that best suits their organizational needs. The real-world examples will illustrate how these frameworks are applied in practice, adding valuable insights for the readers.

4.1 Introduction to Data Governance Frameworks

In this chapter, we delve into the world of data governance frameworks, exploring their definitions, significance, and the roles they play in guiding and structuring data governance efforts within organizations.

Defining Data Governance Frameworks
1. **What Is a Data Governance Framework?**

A data governance framework is a structured set of guidelines, policies, standards, practices, and procedures that organizations use to manage, utilize, and protect their data assets. It serves as a blueprint for how data is handled within an organization. This framework encompasses various aspects of data management,

4.1 Introduction to Data Governance Frameworks

Table 4.1 Data governance frameworks structure

Guidelines are a set of recommendations and best practices that provide high-level guidance on how data should be managed and protected within an organization	*Policies* are formal documents that outline specific rules, procedures, and requirements for data management, usage, and protection. They establish the framework for data governance.	*Standards* are detailed specifications and technical requirements that data must adhere to. They define the specific criteria for data quality, security, and compliance.	*Practices* are the actual methods, processes, and actions used to implement data governance. They encompass day-to-day activities and operational procedures.	*Procedures* are step-by-step instructions that guide individuals in performing specific tasks related to data management and governance. They insure consistency and compliance.

Created by the author

including data quality, security, privacy, lifecycle management, and compliance with regulatory requirements (DAMA International 2017).

Table 4.1 provides a concise breakdown of the components integral to establishing and maintaining a robust data governance framework. It differentiates between guidelines, policies, standards, practices, and procedures—each serving a specific role in the governance process:

- **Guidelines** are presented as a set of recommendations and best practices that offer high-level guidance on how data should be managed and protected within an organization.
- **Policies** are defined as formal documents that specify the rules, procedures, and requirements for data management, usage, and protection, establishing a foundational framework for data governance.
- **Standards** are detailed specifications and technical requirements that data must adhere to, defining specific criteria for data quality, security, and compliance.
- **Practices** are described as the actual methods, processes, and actions used to implement data governance, covering day-to-day activities and operational procedures.
- **Procedures** are step-by-step instructions that guide individuals in performing specific tasks related to data management and governance, ensuring consistency, and compliance.

This table effectively outlines the hierarchical structure and interrelationships among these components, highlighting their collective importance in building a comprehensive data governance strategy.

2. **Components of a Data Governance Framework**

Typically, a data governance framework includes a clear governance structure, roles and responsibilities, decision-making processes, data policies and standards, and mechanisms for monitoring, enforcement, and continuous improvement (DAMA International 2017).

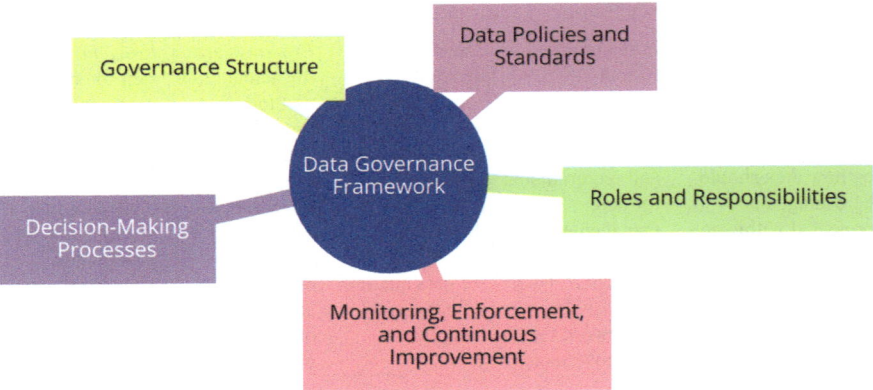

Fig. 4.1 Components of a data governance framework. (Created by the author)

Figure 4.1 illustrates the interconnected elements essential for effective data governance within an organization. At the center is the "Data Governance Framework," which serves as the nucleus connecting various crucial components. These include "Governance Structure," which outlines the organizational design and hierarchy; "Decision-Making Processes," which determine how data-related decisions are made; "Data Policies and Standards," which define the rules and technical specifications for data management; "Roles and Responsibilities," which specify the duties of individuals and teams within the governance structure; and "Monitoring, Enforcement, and Continuous Improvement," which ensure that the policies are followed and the framework adapts over time to changes in technology and business environment. This configuration underlines the comprehensive approach needed to manage and safeguard data effectively, promoting transparency, accountability, and efficiency.

Importance of Data Governance Frameworks

Establishing Clear Guidelines and Standards: Frameworks provide a clear set of guidelines and standards for managing data, ensuring consistency, quality, and security across all organizational levels. They act as a roadmap for managing data effectively and ethically.

Facilitating Compliance and Risk Management: With the increasing complexity of regulatory environments such as GDPR and CCPA, data governance frameworks are vital for ensuring compliance and managing risks associated with data breaches and noncompliance penalties (Gartner 2021).

The Role of Frameworks in Structuring Data Governance Efforts

Guiding Data Governance Implementation: Frameworks guide the implementation of data governance initiatives by providing a structured approach. They help in aligning data governance efforts with the overall business strategy and objectives (CMMI Institute 2014).

4.1 Introduction to Data Governance Frameworks

Facilitating Cross-Functional Collaboration: By establishing clear roles and responsibilities, frameworks facilitate collaboration among various stakeholders, including IT, legal, compliance, and business units. This collaboration is crucial for a comprehensive approach to data governance.

Enabling Scalability and Flexibility: Good data governance frameworks are designed to be scalable and flexible, accommodating the growth and evolving needs of the organization. They provide the foundation for adapting to technological advancements and changing regulatory landscapes.

Measuring and Improving Data Governance Maturity: Frameworks often include mechanisms for measuring the maturity of an organization's data governance efforts, providing benchmarks for continuous improvement (ISO 8000 2011).

Figure 4.2 visually captures the crucial role and benefits of implementing comprehensive data governance frameworks within organizations. At the core, the diagram centers on the "Importance of Data Governance Frameworks," linking to various strategic benefits. These include "Establishing Clear Guidelines and Standards" to ensure uniformity and consistency in data management; "Facilitating Compliance and Risk Management" to address legal and operational risks; and "Guiding Data Governance Implementation" to provide a structured approach to managing data assets. Additional aspects such as "Enabling Scalability and Flexibility," "Facilitating Cross-Functional Collaboration," and "Measuring and Improving Data Governance Maturity" underscore the framework's role in enhancing organizational capacity to adapt to new challenges and opportunities, promoting collaboration across departments, and continually assessing and refining data governance practices. This holistic view demonstrates how data governance frameworks are pivotal in structuring and optimizing data management to support strategic objectives.

Data governance frameworks are essential tools for organizations to manage their data effectively. They provide a structured approach to data governance, ensuring that data is managed as a strategic asset in alignment with business goals and compliance requirements. This chapter will further explore various models and best practices in developing and implementing these frameworks.

Fig. 4.2 Importance of data governance frameworks. (Created by the author)

4.2 Overview of Popular Data Governance Frameworks

In this section, we explore some of the widely used data governance frameworks, providing descriptions of each and a comparative analysis of their approaches and applicability. Understanding these frameworks helps organizations choose or adapt a model that best suits their data governance needs.

1. **DAMA Data Management Body of Knowledge (DAMA-DMBOK)**
   - **Description:** Developed by the Data Management Association International (DAMA), the DAMA-DMBOK Framework is a comprehensive guide for data management. It covers various aspects of data governance, including data quality, data architecture, and data security (DAMA International 2017).

Table 4.2 provides a succinct depiction of the key components of the DAMA-DMBOK Framework, organized into Data Governance and Data Architecture sections. Data Governance describes establishing a framework, defining roles and responsibilities, outlining relevant policies and standards, detailing governance processes, and identifying metrics for assessment. The Data Architecture section covers identifying core components, designing data models, ensuring data integration and interoperability, outlining storage and retrieval methods, and implementing data security and privacy measures. This table encapsulates the holistic approach of the DAMA-DMBOK Framework to managing and safeguarding data, highlighting its structured methodology for enhancing data governance and architecture within organizations.

DAMA-DMBOK Framework

1. Data Governance

    - Data Governance Framework
    - Data Governance Roles and Responsibilities
    - Data Governance Policies and Standards
    - Data Governance Processes
    - Data Governance Metrics and Measurement

2. Data Architecture

    - Data Architecture Components
    - Data Modeling and Design
    - Data Integration and Interoperability
    - Data Storage and Retrieval
    - Data Security and Privacy

3. Data Quality Management

    - Data Quality Framework
    - Data Profiling and Assessment
    - Data Cleansing and Transformation
    - Data Quality Monitoring and Reporting

4.2 Overview of Popular Data Governance Frameworks

Table 4.2 Simplified representation of the key components related to the DAMA-DMBOK Framework

Components	Description
Data governance table	
Data Governance Framework	Establishing the governance structure and principles for managing data.
Data Governance Roles and Responsibilities	Defining roles and responsibilities for data stewardship, ownership, etc.
Data Governance Policies and Standards	Outlining policies and standards for data management and usages.
Data Governance Processes	Outlining processes for data management, including data quality, privacy, and more.
Data Governance Metrics and Measurement	Defining key metrics and measurement criteria for assessing data governance.
Data architecture table	
Data Architecture Components	Identifying and designing key components of data architecture.
Data Modeling and Design	Creating data models to represent data structure and relationships.
Data Integration and Interoperability	Ensuring data can be integrated and used across systems.
Data Storage and Retrieval	Designing storage solutions and methods for data retrieval.
Data Security and Privacy	Implementing security and privacy measures for data protection.

Created by the author

4. Data Operations and Management
   - Data Lifecycle Management
   - Data Acquisition and Ingestion
   - Data Storage and Backup
   - Data Cataloging and Metadata Management
   - Data Retention and Archiving

5. Data Warehousing and Business Intelligence
   - Data Warehousing Concepts
   - Business Intelligence and Reporting
   - Data Analytics and Insights
   - Data Visualization and Dashboards

6. Master Data Management (MDM)
   - Master Data Identification
   - Master Data Storage and Maintenance
   - Master Data Governance
   - Master Data Integration

7. Data Privacy and Compliance

- Data Privacy Regulations (e.g., GDPR, CCPA)
- Data Compliance Frameworks
- Data Privacy Impact Assessments

8. Reference and Metadata Management
    - Metadata Standards and Taxonomies
    - Metadata Repository and Catalog
    - Reference Data Management

9. Data Strategy and Planning
    - Data Strategy Development
    - Data Roadmaps
    - Data Management Business Case
    - Data Management Program Governance

10. Data Ethics
    - Ethical Data Practices
    - Data Stewardship and Ethics
    - Ethical Use of Data

Figure 4.3 intricately maps the extensive components and practices integral to data management as per the Data Management Body of Knowledge. It encompasses a wide range of data-related areas, including Data Governance, which is central to the framework, connecting to key elements such as data stewardship, data quality, data security, and privacy, as well as roles and responsibilities. The framework elaborates on data architecture and operations, highlighting data storage, integration, and retention, alongside data warehousing and business intelligence. It also covers metadata management, emphasizing standards, cataloging, and the lifecycle from acquisition to impact assessment. Ethical data practices, compliance with regulations like GDPR, and continuous improvement through data quality frameworks and master data management illustrate a comprehensive approach to holistic, effective data management. This visualization effectively delineates the interconnectedness and scope of practices essential for robust data governance and architecture.

Approach This framework provides a standard industry view of data management functions and best practices. It is widely recognized for its detailed and structured approach to data governance. Suitable for organizations looking for a comprehensive guide that covers all aspects of data management extensively.

2. **CMMI Data Management Maturity (DMM) Model**

The CMMI Institute's Data Management Maturity (DMM) Model is designed to improve data management practices across an organization. It focuses on enhancing the maturity of data management processes. The DMM Model provides a clear pathway for progression from initial to optimized data management processes. It emphasizes assessment and continuous improvement. (CMMI Institute 2014).

4.2 Overview of Popular Data Governance Frameworks

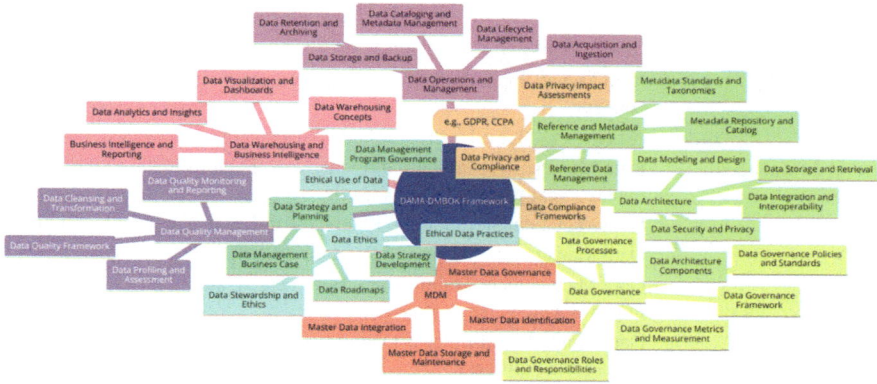

Fig. 4.3 DAMA-DMBOK framework. (Created by the author)

Table 4.3 outlines the CMMI Data Management Maturity (DMM) Model, which is structured into five progressive levels of maturity in data management practices. Level 1, "Initial", focuses on basic data management awareness and ad hoc, unstructured practices. Level 2, "Managed," involves the establishment of data management policies and standards, planning activities, defining roles and responsibilities, and initiating training programs. Level 3, "Defined," is characterized by defining standardized processes, implementing data governance practices, and establishing metrics for performance measurement, alongside enhancing communication and collaboration. Level 4, "Managed and Measurable," includes regular performance measurement, process optimization, and the implementation of risk management strategies. Finally, Level 5, "Optimizing," emphasizes continuous improvement and innovation in data management practices, encouraging innovative approaches and integrating these practices across the organization. This structured approach helps organizations systematically improve their data management capabilities, ensuring effective governance and optimization of data assets.

Applicability Ideal for organizations seeking to assess and improve their data management maturity systematically.

3. ISO 8000 Data Quality and Data Governance Standard

ISO 8000 is an international standard that focuses specifically on data quality and data governance. It provides guidelines for data and information quality management (ISO 2011). The standard emphasizes the establishment of processes to manage and improve data quality, along with compliance with data governance best practices.

Approach The standard emphasizes the establishment of processes to manage and improve data quality, along with compliance with data governance best practices.

Table 4.3 Simplified representation of the CMMI Data Management Maturity (DMM) model

Levels	Components and description
Level 1: Initial	Data Management Awareness: Basic awareness of data management concepts.
	Ad Hoc Practices: Unstructured data management practices.
Level 2: Managed	Data Management Policies and Standards: Establishment of data management policies and standards.
	Data Management Planning: Planning for data management activities.
	Data Management Roles and Responsibilities.
	Data Management Training and Awareness: Training programs to improve data management skills.
Level 3: Defined	Data Management Process Definition: Defining standardized data management process.
	Data Management Data Governance: Implementation of data governance practices.
	Data Management Measurement and Metrics: Establishment of data management performance metrics.
	Data Management Communication and Collaboration: Improved communication and collaboration regarding data.
Level 4: Managed and Measurable	Data Management Performance Measurement: Regular measurement and analysis of data management performance.
	Data Management Process Optimization: Continuous improvement of data management process.
	Data Management Risk Management: Implementation of data management risk mitigation strategies.
Level 5: Optimizing	Data Management Continuous Improvement: Continuous optimization and innovation of data management practices.
	Data Management Innovation: Encouragement of innovative data management approaches.
	Data Management Enterprise Integration: Integration of data management practices across the organization.

Created by the author

Table 4.4 presents a structured representation of key components related to the ISO 8000 Data Quality and Data Governance Standard. It is divided into four main categories, each addressing different aspects of the standard. The first category, "ISO 8000 Data Quality and Data Governance Standard," covers the introduction, scope and objectives, and the overall framework of the standard. The second category, "Data Quality Framework," discusses data quality principles, measurement, control, and improvement strategies. The third category, "Data Governance Framework," outlines the principles, structure, policies, and processes that support data governance under ISO 8000. Lastly, "ISO 8000 Compliance" includes information on certification, auditing, best practices, and the benefits of complying with this standard. This table provides a comprehensive overview of the components necessary to understand and implement the ISO 8000 standard effectively within an organization.

4.2 Overview of Popular Data Governance Frameworks

Table 4.4 Structured representation of some key components related to the ISO 8000 Data Quality and Data Governance Standard

ISO 8000 Data Quality and Data Governance Standard	Data Quality Framework	Data Governance Framework	ISO 8000 Compliance
Introduction to ISO 8000	Data Quality Principles.	Data Governance Principles	ISO 8000 Certification
Scope and Objectives	Data Quality Measurement	Data Governance Structure	ISO 8000 Auditing
Data Quality Framework	Data Quality Control	Data Governance Policies	ISO 8000 Best Practices
Data Governance Framework	Data Quality Improvement	Data Governance Process	ISO 8000 Benefits

Created by the author

Applicability Best suited for organizations prioritizing data quality and seeking a standard aligned with international compliance.

Comparative Analysis

Scope and Focus: DAMA-DMBOK offers a broad and comprehensive view of data management, suitable for organizations seeking a comprehensive approach. CMMI's DMM Model is more focused on process maturity, making it ideal for organizations looking to benchmark and improve their data management processes. ISO 8000 primarily targets data quality, making it suitable for organizations where data quality is a primary concern.

Implementation and Adoption: DAMA-DMBOK's extensive nature may require significant effort in adoption and customization to fit specific organizational needs. The CMMI DMM Model, with its maturity levels, provides a structured pathway for implementation, making it easier for organizations to assess their progress. ISO 8000, being a standard, may require specific adjustments to align with other organizational processes and governance structures.

Industry Applicability: All three frameworks are applicable across various industries. However, their selection might depend on specific industry requirements, such as regulatory compliance or data management maturity needs. Each data governance framework has its strengths and focus areas, and the choice depends on an organization's specific data governance needs, maturity level, and strategic objectives.

Each data governance framework has its strengths and focus areas, and the choice depends on an organization's specific data governance needs, maturity level, and strategic objectives. Understanding these frameworks allows organizations to adopt or tailor a data governance model that aligns with their unique requirements.

4.3 Designing a Data Governance Framework

Designing an effective data governance framework tailored to an organization's unique needs is crucial for the successful management of its data assets. This section outlines the key considerations in selecting or designing a framework and how to tailor it to fit the organization's size, industry, and specific data needs.

Key Considerations in Framework Selection or Design
Organizational Goals and Objectives
Align the data governance framework with the overall business strategy and objectives. Understand how data supports and drives these goals (DAMA International 2017).
Organizational Culture and Structure
Consider the existing culture and structure of the organization. The framework should complement these elements to ensure smooth implementation and adoption (Gartner 2021).
Data Complexity and Volume
Assess the complexity and volume of data managed by the organization. The framework should be capable of handling the specific data challenges and needs (ISO 2011).
Regulatory and Compliance Requirements
Understand the regulatory landscape relevant to the organization's industry. The framework must ensure compliance with applicable laws such as GDPR or HIPAA (CMMI Institute 2014).
Stakeholder Engagement and Buy-In
Engage stakeholders from various departments early in the process. Their buy-in is crucial for successful implementation (DAMA International 2017).
Tailoring Frameworks to Fit Organizational Needs.
Adapting to Organizational Size
For larger organizations, a more comprehensive framework may be needed to manage diverse data assets across different departments and geographies. Smaller organizations might benefit from a more streamlined framework focusing on the most critical data governance aspects relevant to their operations (CMMI Institute 2014).
Industry-Specific Considerations
Industries like healthcare or finance have specific regulatory requirements. The framework should address these industry-specific compliance and data management needs (ISO 2011).
Data Types and Sources
Tailor the framework to manage diverse types of data (structured, unstructured) and various data sources (internal, external, cloud-based) (Gartner 2021).
Implementing Scalable Solutions
Design the framework to be scalable and flexible to accommodate future growth, technological advancements, and changes in data governance best practices (DAMA International 2017).

4.3 Designing a Data Governance Framework

Customization for Data Lifecycle Management
Customize the framework to address specific stages of the data lifecycle from creation and acquisition to archiving and disposal (ISO 2011).

Technology Integration
Consider the existing technology stack and data management tools. The framework should integrate well with these technologies and support future technology adoption (Gartner 2021).

Best Practices in Framework Design.

Start Small and Scale
Begin with a focused approach, addressing the most critical data governance needs first, and then scale the framework as required (CMMI Institute 2014).

Continuous Improvement
Adopt an iterative approach, allowing for continuous assessment and improvement of the data governance framework (ISO 2011).

Employee Training and Awareness
Ensure ongoing training and awareness programs to keep employees informed and engaged with the data governance processes (DAMA International 2017).

Figure 4.4 presents a comprehensive view of the various factors and strategic considerations involved in the development of an effective data governance framework. At its core, the diagram highlights the importance of aligning the framework with organizational goals and objectives, ensuring it is informed by the organizational culture and structure. Key considerations for framework design include regulatory and compliance requirements, stakeholder engagement, and the integration of technology. The diagram also addresses the necessity of adapting the framework to the size of the organization, the specific industry, and the types of data handled. Customization for effective data lifecycle management, along with considerations for data complexity and volume, are emphasized. Practical steps such as starting small, scaling appropriately, implementing scalable solutions, and embedding continuous improvement processes through employee training and awareness are also depicted. This visualization underscores the multifaceted approach required to

Fig. 4.4 Designing a data governance framework. (Created by the author)

tailor a data governance framework that meets the specific needs and capabilities of an organization.

Designing a data governance framework is a strategic process that requires careful consideration of organizational goals, structure, culture, and specific data needs. Tailoring the framework to fit these parameters ensures that it is effective, relevant, and sustainable over time.

4.4 Best Practices in Framework Implementation

Implementing a data governance framework successfully is a critical step for organizations in ensuring effective data management. This section outlines practical steps and best practices for framework implementation as well as common pitfalls to avoid.

Practical Steps for Implementing a Data Governance Framework
Establish Clear Leadership and Governance Structure
Appoint a data governance council or committee to provide leadership and direction. This team should include senior stakeholders from different departments (DAMA International 2017).
Define Clear Roles and Responsibilities
Define clear roles and responsibilities for data governance tasks, including data stewards, data owners, and data custodians (CMMI Institute 2014).
Develop and Communicate a Clear Vision
Articulate an unobstructed vision for what the data governance program aims to achieve. Ensure this vision is aligned with the organization's overall strategy. Communicate the vision and the value of data governance to all levels of the organization to build support and understanding (Gartner 2021).
Start with a Pilot Project
Begin with a pilot project in a specific area or department. This allows you to demonstrate the value of data governance and refine your approach before a full-scale rollout (ISO 2011).
Create and Implement Data Governance Policies and Standards
Develop comprehensive data governance policies and standards, including data quality, data security, and data privacy policies. Ensure these policies are practical, enforceable, and aligned with business processes (The Data Governance Institute 2020).
Invest in Training and Change Management
Provide training and education to ensure that employees understand the importance of data governance and their role in it. Employ change management strategies to address resistance and foster a culture that values data governance (Gartner 2021).
Leverage Technology and Tools
Use technology tools such as data catalogs, data quality tools, and master data management (MDM) solutions to support data governance processes. Ensure the

selected tools are compatible with existing systems and can scale with your organization's needs (ISO 2011).

Regular Monitoring and Continuous Improvement

Implement metrics and KPIs to regularly measure the effectiveness of your data governance program. Continuously assess and improve the data governance framework based on these metrics (DAMA International 2017).

Common Pitfalls to Avoid During Implementation

Lack of Executive Buy-In

Failing to secure executive buy-in can lead to a lack of resources and support, hindering the success of the data governance program (The Data Governance Institute 2020).

Underestimating the Cultural Change

Overlooking the cultural aspect of data governance can result in resistance from employees. Emphasize the cultural change as much as the technical implementation (CMMI Institute 2014).

Over-Complication

Avoid making the framework overly complex. A complicated framework can be difficult to understand and implement effectively (Gartner 2021).

Neglecting Data Quality

Ignoring data quality in the initial stages can lead to long-term issues. Data quality should be a priority from the outset (ISO 2011).

Inadequate Training and Communication

Failing to professionally train and communicate with staff about data governance practices can lead to misunderstandings and noncompliance (DAMA International 2017).

Figure 4.5 provides an insightful guide on effective strategies and common pitfalls in implementing a data governance framework. Central to the diagram is the emphasis on starting with a pilot project to test and refine the framework, developing and communicating a clear vision to guide the initiative, and the importance of investing in training and change management to support the transition. It also highlights the necessity of establishing clear leadership and governance structures, creating and implementing specific data governance policies and standards, and leveraging technology and tools to enhance the implementation process. Regular monitoring and continuous improvement are underscored as critical for maintaining the efficacy of the governance framework. Additionally, the diagram points out common pitfalls to avoid, such as neglecting data quality, over-complication of processes, inadequate training and communication, underestimating the cultural change required, and the lack of executive buy-in, providing a comprehensive overview of the do's and don'ts in implementing a data governance framework.

Implementing a data governance framework requires careful planning, clear communication, and ongoing management. By following these best practices and avoiding common pitfalls, organizations can establish a robust data governance program that supports their strategic objectives and enhances their overall data management capabilities.

Fig. 4.5 Best practices in framework implementation. (Created by the author)

4.5 Integrating Technology with Frameworks

The integration of technology in data governance frameworks is a critical factor in enhancing their effectiveness and efficiency. This section provides an analysis of the role of technology in supporting and enhancing data governance frameworks along with a review of various tools and software that are essential complements to these frameworks.

Role of Technology in Enhancing Data Governance Frameworks
Enabling Efficient Data Management
Technology plays a pivotal role in enabling efficient data management, allowing for the automation of routine tasks, improving accuracy, and reducing the time and resources required for data-related activities. For instance, automation tools streamline processes such as data cleansing and validation, making data more reliable and actionable (DAMA International 2017).

Facilitating Compliance and Monitoring
Advanced technological solutions help in monitoring compliance with data governance policies and regulatory requirements. These tools provide alerts and reports that assist in maintaining standards and ensuring that data management practices are in line with legal and regulatory obligations (Gartner 2021).

Improving Data Quality and Integrity
Technology aids in maintaining data quality and integrity through tools that automatically cleanse, validate, and standardize data, ensuring it remains accurate and reliable for decision-making. Tools like data quality management systems

4.5 Integrating Technology with Frameworks

are essential in detecting and correcting errors, thereby enhancing the overall trust in data assets (ISO 2011).

Supporting Data Security and Privacy

Security technologies, including encryption and access control systems, are vital in protecting sensitive data and ensuring privacy, a key aspect of data governance frameworks. These technologies help in mitigating risks associated with data breaches and unauthorized access (CMMI Institute 2014).

Review of Tools and Software Complementing Data Governance Frameworks

Data Catalogs and Metadata Management Tools

Data catalogs help in organizing data assets, making it easier to find and understand data. Metadata management tools provide context and enhance the usability of data. Solutions like Informatica and Collibra offer robust data catalog solutions that are widely adopted in the industry for their efficiency in data organization and accessibility (Gartner 2021).

Data Quality Tools

These tools automate the process of monitoring, cleansing, and enriching data to ensure high quality. Talend and IBM InfoSphere QualityStage are widely used for data quality management, helping organizations maintain accurate and clean data sets essential for operational and strategic decisions (DAMA International 2017).

Master Data Management (MDM) Solutions

MDM solutions help in creating a single, consistent view of key business data from disparate sources, which is crucial for accurate reporting and analytics. Popular MDM solutions include SAP Master Data Governance and Oracle MDM, both of which provide comprehensive data integration and management capabilities (ISO 2011).

Compliance and Risk Management Software

These solutions assist in managing the risks associated with data and ensuring compliance with regulations such as GDPR. RSA Archer and MetricStream offer compliance management capabilities that help organizations stay compliant with evolving regulatory requirements (CMMI Institute 2014).

Data Integration and ETL Tools

Data integration and ETL (Extract, Transform, Load) tools are essential for consolidating data from various sources into a unified format. Microsoft SQL Server Integration Services (SSIS) and Talend Open Studio are commonly used for data integration tasks, facilitating seamless data consolidation and transformation (ISO 2011).

Data Security Solutions

Solutions for data encryption, access control, and breach detection are critical for securing data within governance frameworks. Symantec and McAfee provide comprehensive data security solutions that protect data against threats and ensure regulatory compliance (Gartner 2021).

Figure 4.6 illustrates the pivotal role of technology in enhancing and supporting data governance frameworks. Central to this depiction is the integration of various

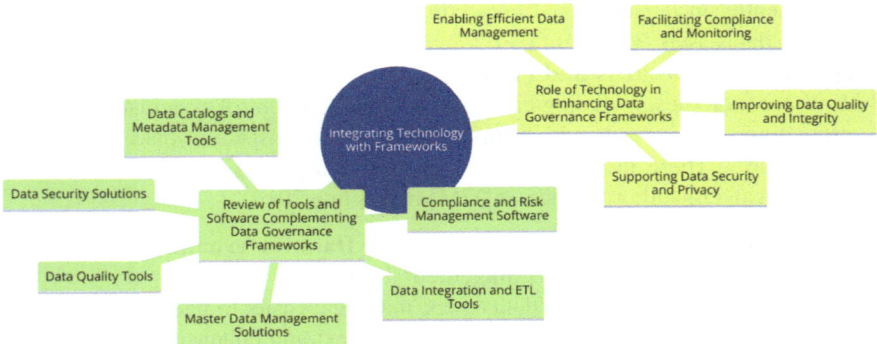

Fig. 4.6 Integrating technology with frameworks. (Created by the author)

technological tools and software that facilitate efficient data management and compliance monitoring. This includes data catalogs and metadata management tools that organize and provide access to data assets, data security solutions that ensure data confidentiality and integrity, and data quality tools aimed at improving the accuracy and reliability of data. Master data management solutions streamline data processing and maintenance, while data integration and ETL (Extract, Transform, Load) tools support the operational aspects of combining data from different sources. Additionally, the diagram highlights the importance of reviewing tools and software that complement existing data governance frameworks, emphasizing compliance and risk management software that helps organizations adhere to data governance standards and regulations. This visual guide underscores how integrating the right technologies can significantly enhance the implementation and effectiveness of data governance frameworks.

Challenges in Technology Integration
Interoperability and Compatibility
Ensuring that different technologies and tools integrate seamlessly and are compatible with existing systems can be challenging. Organizations must carefully evaluate and select tools that can work together without causing disruptions (DAMA International 2017).
Keeping Pace with Technological Advancements
Rapid technological changes require continuous evaluation and adaptation of the tools used in data governance frameworks. Staying updated with the latest technologies and integrating them effectively into the framework is essential for maintaining robust data governance practices (Gartner 2021).

The integration of technology with data governance frameworks significantly enhances their effectiveness. By leveraging the right set of tools and software, organizations can ensure efficient data management, compliance, and security, which are integral components of a robust data governance strategy.

4.6 Measuring the Effectiveness of your Framework—Data Governance Template

Evaluating the effectiveness of a data governance framework is crucial for ensuring that it meets the organization's data management goals and adapts to changing needs. This section discusses setting up metrics and key performance indicators (KPIs) for this purpose and outlines continuous improvement strategies for data governance frameworks.

Setting Up Metrics and KPIs
- Defining Relevant Metrics: Metrics should be aligned with the organization's data governance objectives. For instance, if data quality is a priority, metrics could include the number of data errors identified and corrected. Examples of relevant metrics include accuracy rate, data completeness, and the number of data breaches or compliance issues reported (DAMA International 2017).
- Establishing Key Performance Indicators (KPIs): KPIs provide a high-level view of the performance and health of the data governance framework. They should be specific, measurable, achievable, relevant, and time-bound (SMART). Examples include the percentage of data assets under governance, time taken to resolve data issues, and stakeholder satisfaction with data quality (Gartner 2021).
- Utilizing Balanced Scorecards: A balanced scorecard approach can be effective in measuring performance across various aspects of data governance, including financial, customer, internal processes, and learning and growth perspectives. Examples include financial savings from improved data management, improved customer satisfaction due to better data quality, efficiency in data processing, and employee proficiency in data governance practices (ISO 2011).

Continuous Improvement Strategies
- Regular Reviews and Audits: Conducting regular reviews and audits of the data governance framework ensures that it remains effective and relevant. These reviews can help identify areas for improvement. For example, annual data governance audits to assess compliance and effectiveness are crucial (CMMI Institute 2014).
- Feedback Loops: Establish feedback loops with stakeholders to gather insights on the framework's performance and impact. This feedback is invaluable for making informed improvements. Examples include surveys and interviews with key stakeholders, including data users and IT staff (The Data Governance Institute 2020).
- Adapting to Changes: Data governance frameworks should be flexible to adapt to changes in business strategy, technology, and regulatory environments. For example, updating data policies and procedures in response to new GDPR regulations or technological advancements in data analytics (Gartner 2021).
- Training and Development: Continuous training and development programs for staff involved in data governance can improve the framework's effectiveness.

Fig. 4.7 Measuring the effectiveness of your framework—data governance template. (Created by the author)

Examples include regular workshops and training sessions on the latest data governance best practices and tools (DAMA International 2017).

Figure 4.7 centers around a large blue circle labeled with the same title. This central hub connects to several related components of the data governance framework through directed arrows, suggesting a dynamic, interrelated process. Key sections connected directly to the central node include "Defining Relevant Metrics," "Setting Up Metrics and KPIs," "Utilizing Balanced Scorecards," and "KPIs," emphasizing the systematic approach to establishing performance indicators. Other elements like "Feedback Loops," "Regular Reviews and Audits," "Continuous Improvement Strategies," "Training and Development," and "Adapting to Changes" encircle these, illustrating the ongoing process of evaluation, training, and adaptation that feeds into and enhances the framework's effectiveness. This visual representation underscores a cyclical, continuous improvement model within data governance, aligning with best practices that promote responsiveness to change and stakeholder feedback.

Challenges in Measuring Effectiveness
- Quantifying Intangible Benefits: Some benefits of data governance, like improved decision-making quality, can be difficult to quantify.
- Data Governance Maturity: Organizations at various stages of data governance maturity might find it challenging to measure effectiveness using the same metrics and KPIs.

4.6 Measuring the Effectiveness of your Framework—Data Governance Template

Measuring the effectiveness of a data governance framework is essential for ensuring it continues to meet organizational goals. Setting up relevant metrics and KPIs, regularly reviewing the framework, adapting to changes, and engaging in continuous improvement strategies are key to maintaining an effective data governance program.

Data Governance Template
1. Mission and Value
   - **Objective**: Define the core objectives of your data governance program.
   - **Value Proposition**: Describe the benefits and value this program will bring to the organization.
2. Beneficiaries of Data Governance Programs
   - **Stakeholder Identification**: List the key stakeholders and beneficiaries of the data governance program.
   - **Benefits for Each Stakeholder**: Describe how each stakeholder will benefit from the program.
3. Data Products
   - **Data Catalog/Inventory**: Create a template for cataloging data assets.
   - **Glossary with Data Definitions**: Develop a glossary template for standard data definitions.
   - **Metadata Template**: Set up a structure for capturing and maintaining metadata.
4. Controls
   - **Risk Assessment Checklist**: Implement a checklist for identifying and assessing data-related risks.
   - **Control Matrix**: Establish a matrix for various controls (preventative and corrective) to manage these risks.
5. Accountabilities
   - **Responsibility Assignment Matrix (RACI)**: Define the roles and responsibilities of team members in the data governance process.
6. Decision Rights
   - **Decision-Making Hierarchy**: Map out the decision-making process and hierarchy within the data governance framework.
   - **Decision-Making Processes**: Document the processes for making key data-related decisions.
7. Policy and Rules
   - **Policy Documentation Guide**: Provide guidelines for documenting data governance policies.

- **Rules Setting Worksheet**: Develop a worksheet for setting and documenting data governance rules.

8. Data Governance Processes, Tools, and Communication

    - **Process Mapping**: Outline the key data governance processes.
    - **Tool Selection Guide**: Assist in selecting the appropriate tools for data governance.
    - **Communication Plan Template**: Create a template for planning and executing data governance communication strategies.

9. Data Governance Work Program

    - **Project Planning Template**: Facilitate planning for data governance projects.
    - **Workstream Management**: Manage different workstreams within the data governance program.

10. Participants in Data Governance Programs

    - **Role Definition Worksheet**: Define various roles within the data governance program.
    - **Stakeholder Analysis Tool**: Analyze and document the interests and influence of different stakeholders.

Figure 4.8 provides a structured approach to setting up a data governance program and features a complex diagram centered on a large oval labeled "Data Governance Template," symbolizing the core of a data governance program. Branching out from this central node are various elements crucial to establishing and managing a data governance framework. These include "Communication Plan Template," "Workstream Management," "Project Planning Template," "Tool Selection Guide," and "Data Governance Processes, Tools, and Communication," among others, each linking back to the central theme through connecting lines. Subsections such as "Stakeholder Identification," "Risk Assessment Checklist," "Role Definition Worksheet," "Data Governance Work Program," and "Controls" indicate specific tools and processes integral to the implementation of data governance. Other areas like "Metadata Template," "Data Catalog/Inventory," "Decision-Making Processes," "Beneficiaries of Data Governance Programs," and "Decision Rights" further detail the operational aspects of the program. Collectively, these components reflect the multifaceted approach required in data governance, emphasizing planning, role clarification, communication, control mechanisms, and the handling of data itself. The diagram visually summarizes the interconnectivity and flow between different aspects of data governance, highlighting a customizable structure tailored to the specific needs and objectives of an organization. This visual representation serves as a comprehensive guide for setting up, managing, and customizing a data governance framework.

Fig. 4.8 This template provides a structured approach to setting up a data governance program. It can be customized to suit the specific needs and objectives of your organization. (Created by the author)

4.7 Case Studies: Frameworks in Action

In this section, we explore real-world examples of successful data governance framework implementations across various industries. These case studies provide valuable insights into the practical application of data governance principles and the lessons learned from these experiences.

Case Study 1: Financial Industry—Wells Fargo (Atlan 2023)
Situation: Wells Fargo faced challenges with data accuracy and reliability due to data being stored in multiple, disparate sources.
Action: The bank centralized data from these various sources to create a single source of truth. They implemented data visualization tools like Tableau to make data more accessible to nontechnical stakeholders and improve data literacy across the organization.
Result: This strategy enhanced data consistency and accuracy, improved reporting and analysis capabilities, and fostered data-driven decision-making across the organization.
Lesson: Centralizing data to create a single source of truth and enhancing data accessibility through visualization tools are crucial steps in improving data governance and ensuring reliable decision-making.

Case Study 2: Healthcare Sector—BJC HealthCare and Washington University (Dataversity 2023a)
Situation: BJC HealthCare and Washington University needed to ensure the highest data quality when introducing a new electronic health record system.
Action: They identified and trained Data Stewards and created Data Steward communities to build business glossaries, standard reference data, and preferred data

sources for six Master Data domains. They also sought executive-level support to drive these initiatives.

Result: The foundational data quality was significantly enhanced, supporting the successful implementation of the new health record system.

Lesson: Training Data Stewards and securing executive buy-in are critical for building a robust data governance framework, particularly in healthcare where data accuracy directly impacts patient care.

Case Study 3: Aviation Industry—GE Aviation (Atlan 2023)

Situation: GE Aviation needed to centralize scattered data sources to make data more accessible and reliable.

Action: They implemented a Self-Service Data (SSD) initiative, establishing a team responsible for user enablement, tooling, and data product deployment. A separate Database Admin team ensured proper data governance and supported users.

Result: This approach accelerated decision-making processes, improved safety, and enhanced operational efficiency.

Lesson: Combining self-service data initiatives with strong data governance and automation can significantly enhance data accessibility and reliability, driving better organizational outcomes.

Case Study 4: Technology Industry—Airbnb (Atlan 2023)

Situation: Airbnb sought to enhance data literacy across all departments to support data-driven decision-making.

Action: They launched "Data University," an in-house educational initiative to boost data literacy, enabling employees to understand, interpret, and use data effectively.

Result: The initiative democratized data usage and scaled decision-making, with 45% of employees becoming weekly active users of Airbnb's internal data platform.

Lesson: Investing in data literacy through educational programs is essential for empowering employees to make informed, data-driven decisions and ensuring the responsible use of data.

Case Study 5: Insurance Industry—CSE Insurance (Talend 2023)

Situation: CSE Insurance faced difficulties in governing and managing data that was siloed across various sources.

Action: They implemented a data governance framework to establish a single source of truth, powered by data governance policies and procedures. A group of data champions was set up to drive these initiatives.

Result: The data governance framework transformed CSE Insurance's data culture, making data more accessible and consistent, thereby enhancing overall data management efficiency.

Lesson: Establishing a single source of truth and fostering a culture of data governance are key to overcoming data silos and improving data management practices.

Each case study demonstrates that successful data governance requires a tailored approach that aligns with specific industry needs and organizational goals. These examples offer valuable lessons in the strategic implementation of data governance frameworks, highlighting the impact of effective data governance on business outcomes.

4.8 Adapting Frameworks to Changing Data Landscapes

In an era marked by rapid technological advancements and evolving data landscapes, it is crucial for organizations to keep their data governance frameworks agile and responsive. This section explores how frameworks can be adapted to address emerging challenges such as big data, artificial intelligence (AI), and the Internet of Things (IoT).

Keeping Data Governance Frameworks Agile
1. **Embracing Flexibility:** Data governance frameworks should be designed for adaptability, allowing for modifications as new data types and sources emerge. Flexibility is key to responding quickly to changes in the data landscape (Askham 2023; DBTA 2023).
2. **Continuous Learning and Evolution:** Organizations must commit to ongoing learning and evolution of their data governance frameworks. This involves staying informed about technological advancements and industry trends (DBTA 2023).
3. **Stakeholder Engagement:** Regular engagement with stakeholders, including IT professionals, data scientists, and business leaders, can provide insights into emerging needs and challenges (MacFeely et al. 2022).

Adapting to Emerging Data Challenges
1. **Big Data:** Big data introduces complexities in terms of volume, variety, and velocity. Data governance frameworks must expand to handle the scale and diversity of big data. Example: Incorporating tools for data quality and integration that can handle large datasets efficiently (DBTA 2023).
2. **Artificial Intelligence and Machine Learning:** With the rise of AI and machine learning, data governance must address issues around data quality, bias, and ethical use of AI. Example: Establishing guidelines for data ethics and responsible AI use (Askham 2023).
3. **Internet of Things (IoT):** IoT brings challenges in managing the vast amounts of data generated by connected devices. Data governance frameworks must ensure this data is securely managed and utilized effectively. Example: Implementing robust security measures and privacy controls for IoT data (DBTA 2023).
4. **Data Privacy Regulations:** Evolving data privacy regulations (like GDPR and CCPA) require data governance frameworks to be regularly updated to ensure

compliance. Example: GDPR compliance checklists and regular legal reviews to adapt to regulatory changes (DATAVERSITY 2023b).
5. **Cloud Computing:** The shift towards cloud computing requires governance frameworks to address data security, privacy, and compliance in the cloud. Example: Developing cloud-specific data governance policies (DBTA 2023).

Preparing for Future Trends
1. **Proactive Monitoring of Trends:** Establish a process for monitoring and assessing future data trends and technologies, allowing the organization to proactively adapt its data governance framework (MacFeely et al. 2022).
2. **Scalability and Integration:** Ensure that the data governance framework is scalable and can integrate innovative technologies and data sources as they emerge (Askham 2023).
3. **Data Literacy:** Invest in enhancing data literacy across the organization to equip stakeholders with the skills needed to adapt to changing data landscapes (Askham 2023).

Figure 4.9 presents a visual representation of a central concept encircled by various interconnected elements that illustrate the adaptability required in data governance frameworks. At the core of the diagram is a large circle labeled "Adapting Frameworks to Changing Data Landscapes," indicating the main focus of the framework's ability to evolve with technological and regulatory changes. Branching out from the center are multiple pathways leading to key components necessary for this adaptation, which include "Scalability and Integration," "Data Literacy," "Continuous Learning and Evolution," "Embracing Flexibility," "Stakeholder Engagement," "Keeping Data Governance Frameworks Agile," "Artificial Intelligence and Machine Learning," "Adapting to Emerging Data Challenges," "Cloud Computing," "IoT," "Data Privacy Regulations," "Big Data," "Preparing for Future Trends," and "Proactive Monitoring of Trends." These elements collectively emphasize the dynamic nature of data governance, addressing aspects such as technological advancements in AI and IoT, the increasing importance of data literacy,

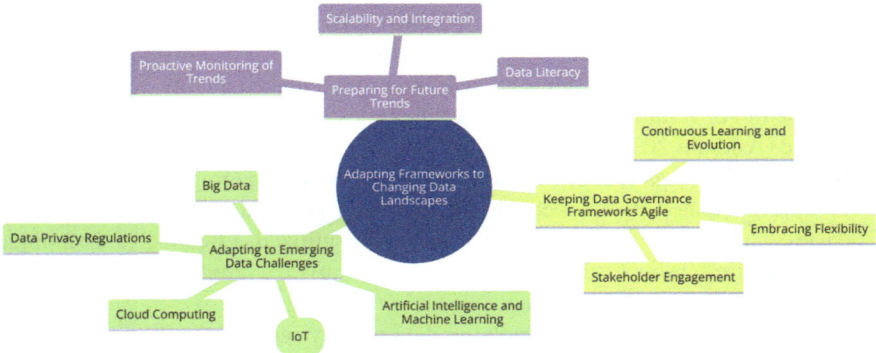

Fig. 4.9 Adapting frameworks to changing data landscapes map. (Created by the author)

and the need for proactive monitoring to stay ahead of emerging trends and regulatory changes. The diagram illustrates how these components are crucial for maintaining effective governance in the face of evolving data landscapes, highlighting the need for frameworks that are both flexible and responsive to change.

Data governance frameworks must be agile and adaptable to respond effectively to the dynamic data landscapes of today and tomorrow. By preparing for emerging challenges and staying informed about the latest trends, organizations can ensure that their data governance frameworks remain robust, relevant, and effective.

4.9 The Future of Data Governance Frameworks

As data continues to grow in volume, variety, and importance, the future of data governance frameworks is poised to evolve significantly. This section explores predictions and trends in the evolution of data governance and offers insights on how organizations can stay ahead of the curve.

Future trends and innovations
1. **Artificial Intelligence and Blockchain Governance:** AI and blockchain are significantly influencing the governance of digital technologies. The governance platforms for these technologies involve a mix of public, private, and civil society organizations. These platforms often face challenges such as privacy breaches, fraud, and cybersecurity breaches. With AI and blockchain becoming integral to various industries, there is a shift towards ethical and responsible platforms for technology governance. However, there are concerns around these technologies being primarily profit-driven and the risk of "ethics washing" in governance initiatives (Monte Carlo 2023; Velotix 2023).
2. **Quantum Computing and Governance:** Quantum computing presents a unique set of challenges and opportunities for data governance. Its ability to process data far superior to classical computers necessitates proactive governance to maximize benefits and mitigate risks. Governance in the quantum era may need to consider behavioral approaches alongside technical and procedural strategies to effectively manage the power and potential risks of quantum computing (World Economic Forum 2023).
3. **Blockchain and Quantum Technologies in Digital Trust:** Blockchain is emerging as a critical platform for building digital trust. It offers mechanisms to establish trust through independently verifiable and immutable databases or ledgers, reducing the need for trusted third parties. Quantum technologies, while still early in the digital trust innovation curve, could introduce new threats as well as countermeasures in the realm of cybersecurity (Deloitte US 2023).
4. **Quantum Computing's Impact Across Industries:** Quantum computing is poised to benefit various sectors like finance, cybersecurity, and pharmaceuticals. Its capabilities in testing chemical experiments, improving batteries, and material enhancements are just a few examples of its potential impact.

Additionally, AI is seeing significant business adoption across multiple functions, driving revenue growth and performance improvements. Blockchain, too, is gaining traction, particularly in supply chain traceability and smart contract automation (Simplilearn 2023).

Predictions and Trends in Data Governance
1. **Increased Emphasis on AI and Machine Learning:** The integration of AI and machine learning in data governance tools is expected to increase, automating many aspects of data governance, including data quality, compliance monitoring, and anomaly detection. Gartner predicts that by 2023, AI techniques will be embedded in most data governance tools (Gartner 2020).
2. **Rise of Data Governance as a Service (DGaaS):** There is a growing trend towards Data Governance as a Service (DGaaS), where cloud-based solutions provide scalable and flexible data governance capabilities. Forrester indicates that cloud-based data governance tools will see increased adoption, offering cost-effective and scalable solutions (Forrester 2021).
3. **Expanding Scope due to IoT and Edge Computing:** The Internet of Things (IoT) and edge computing will broaden the scope of data governance, introducing new types of data and data sources that need to be governed. IDC forecasts significant growth in IoT data, which will require new governance strategies to manage effectively (IDC 2021).
4. **Focus on Data Privacy and Ethical Use:** With increasing awareness and regulatory requirements around data privacy, future frameworks will need to put greater emphasis on privacy and ethical considerations in data usage. The European Data Protection Board (EDPB) emphasizes the increasing importance of data ethics in governance frameworks (EDPB 2021).
5. **Enhanced Data Literacy Initiatives:** Data literacy will become a critical component, with organizations investing more in training and enabling their workforce to understand and leverage data effectively. The Data Literacy Project advocates for increased data literacy as a key component of future data governance initiatives (The Data Literacy Project 2021).

Staying Ahead of the Curve
1. **Proactive Adaptation to Technological Advancements:** Organizations must proactively adapt their data governance frameworks to embrace innovative technologies like AI, IoT, and cloud computing.
2. **Regular Framework Reviews and Updates:** Conduct regular reviews and updates of data governance policies to ensure they remain relevant and effective in the changing data landscape.
3. **Building a Data-Driven Culture:** Foster a data-driven culture within the organization, emphasizing the importance of data governance in all data-related activities.
4. **Focus on Continuous Learning and Skill Development:** Encourage continuous learning and development in the field of data governance and data management to keep pace with emerging trends and technologies.

5. **Engaging with Data Governance Communities:** Stay engaged with data governance communities and professional groups to stay updated with best practices and emerging trends.

The future of data governance frameworks will be shaped by advancements in technology, the growing importance of data privacy, and the need for organizations to adapt quickly to these changes. By staying informed, investing in innovative technologies, and fostering a culture of data literacy, organizations can effectively navigate the evolving landscape of data governance.

4.10 Conclusion and Transition to the Next Chapter

As we conclude this chapter on Data Governance Frameworks, it is important to recap the key takeaways and set the stage for the transition to the next critical topic in our journey through data governance.

Recap of Key Takeaways
1. **Essentiality of Data Governance Frameworks:** We have emphasized that having a well-defined data governance framework is crucial for any organization that aims to manage its data assets effectively.
2. **Diverse Framework Models:** The exploration of various data governance frameworks, including DAMA-DMBOK, CMMI, and others, highlighted that there are multiple approaches to structuring data governance, each with its strengths and specific focus areas.
3. **Adaptability and Evolution:** A recurring theme has been the necessity for data governance frameworks to be adaptable and responsive to the rapidly changing data landscape, incorporating advancements in technology like AI, IoT, and cloud computing.
4. **Integration of Technology:** The integration of technology in supporting and enhancing these frameworks is critical. Tools and software that complement data governance efforts play a pivotal role in its successful implementation.
5. **Measurement of Effectiveness:** Setting up metrics and KPIs to evaluate the performance of a data governance framework is vital for its continuous improvement and alignment with business objectives.
6. **Real-World Applications:** Through various case studies, we demonstrated how different industries apply data governance frameworks in practice, providing valuable insights and lessons learned.

Introduction to Chapter 5: "Data Quality Management: Ensuring Accuracy and Reliability"
As we move into Chap. 5, the focus will shift from the broader perspective of data governance frameworks to the specific, yet critical, aspect of Data Quality Management. This next chapter will delve into:

1. **Importance of Data Quality:** We will explore why data quality is fundamental to all aspects of data governance and how it impacts decision-making, compliance, and operational efficiency.
2. **Strategies for Data Quality Management:** This chapter will provide insights into various strategies and best practices for ensuring data quality, including data cleansing, standardization, and validation processes.
3. **Tools and Techniques:** We will review the tools and techniques available to manage and improve data quality, including technological solutions and manual approaches.
4. **Challenges and Solutions:** Addressing the usual challenges in maintaining high data quality and providing practical solutions to these challenges will be a key focus.
5. **Case Studies:** Real-world examples will illustrate successful data quality management initiatives and the lessons learned from these implementations.

In Chap. 5, "Data Quality Management: Ensuring Accuracy and Reliability," we will provide a comprehensive guide to understanding, implementing, and maintaining high standards of data quality, a cornerstone in the foundation of effective data governance. This exploration will equip readers with the knowledge to enhance the reliability and value of their data assets.

References

Askham N (2023) Navigating data mesh and evolving data governance: a practical guide
Atlan (2023) 5 data governance examples: case studies, takeaways & more
CMMI Institute (2014) Data management maturity (DMM) model. CMMI Institute
DAMA International (2017) DAMA-DMBOK: data management body of knowledge. DAMA International
Data Governance Framework A Complete Guide – 2021 Edition by The Art of Service – Data Governance Framework Publishing
Data Governance Framework A Complete Guide – 2021 Edition Kindle Edition by Gerardus Blokdyk
Data Governance Framework. A Comprehensive Guide to Success. by Michael E Kirshteyn Ph.D.
Database Trends and Applications (DBTA) (2023) Adapting data governance to tend the changing data landscape
Dataversity (2023a) Case study: three strategies for data governance success
DATAVERSITY (2023b) Trends in data governance in 2023: maturation toward a service model
Deloitte US (2023) Blockchain and quantum technologies in digital trust
European Data Protection Board (EDPB) (2021) Guidelines 01/2021 on examples regarding data breach notification. EDPB
Forrester (2021) The Forrester wave™: cloud-based data governance solutions, Q1 2021. Forrester
Gartner (2020) Predicts 2020: data and analytics strategies—invest, influence and impact. Gartner
Gartner (2021) Data governance key initiative overview. Gartner
IDC (2021) IDC's worldwide internet of things forecast, 2021–2025. IDC
ISO (2011) ISO 8000 data quality and data governance standard. International Organization for Standardization
MacFeely S et al (2022) Informing the global data future: benchmarking data governance frameworks. Cambridge University Press

References

Monte Carlo (2023) The future of data governance: 4 trends to watch out for
Simplilearn (2023) Quantum computing's impact across industries
Talend (2023) Data governance success stories
The Data Governance Institute (2020) Data governance best practices. The Data Governance Institute
The Data Literacy Project (2021) Fostering data literacy. The Data Literacy Project
Velotix (2023) Data governance in 2023: trends, insights, and analysis
World Economic Forum (2023) Quantum computing governance and regulation

Chapter 5
Data Quality Management: Ensuring Accuracy and Reliability

Contents

5.1	Introduction to Data Quality Management	198
5.2	Dimensions of Data Quality	199
5.3	Establishing Data Quality Standards	201
5.4	Data Quality Assessment Techniques	203
5.5	Data Cleansing and Improvement Strategies	204
5.6	Role of Technology in Data Quality Management	206
5.7	Building a Culture of Data Quality	207
5.8	Case Studies: Transforming Data Quality	209
5.9	Overcoming Common Data Quality Challenges	211
5.10	Conclusion and Next Steps	212
References		215

Abstract This chapter emphasizes the critical importance of managing data quality within the framework of broader data governance. The chapter outlines how data quality management (DQM) involves ensuring the accuracy, completeness, reliability, and relevance of data across various organizational processes, ultimately facilitating effective decision-making and strategic planning.

The narrative highlights the foundational role of high-quality data in decision-making, which significantly enhances business performance and competitive advantage. Moreover, the chapter underscores the necessity of accurate and reliable data for regulatory compliance, especially in sectors such as finance and healthcare, where poor data quality can result in significant compliance risks and legal repercussions.

Operational efficiency and customer satisfaction are also discussed as major benefits of high-quality data, with the latter directly impacting customer service experiences and loyalty. Through practical insights and real-world case studies, the chapter not only demonstrates how effective DQM practices have been implemented in diverse organizational contexts but also details the strategic imperative of data quality as a competitive advantage.

As it progresses, the chapter provides a comprehensive look at strategies, tools, and methodologies to manage and improve data quality. This approach not only supports the enhancement of the overall data governance framework but also ensures that organizations can rely on their data for strategic insights and innovations. By focusing on the establishment of data quality metrics, continuous monitoring, and the improvement of data quality, the chapter serves as a vital resource for organizations aiming to leverage data as a strategic asset.

Keywords Data quality management · Accuracy · Reliability · Data governance · Compliance · Operational efficiency · Customer satisfaction · Data cleansing · Continuous improvement · Data standards · Data profiling · Data quality assessment · Data security

This chapter aims to emphasize the importance of data quality management in the broader context of data governance, providing practical insights into establishing, maintaining, and improving data quality. Real-world case studies will add depth, showing how data quality initiatives have been successfully implemented in various organizations.

5.1 Introduction to Data Quality Management

In this chapter, we delve into the realm of data quality management (DQM), a crucial aspect of effective data governance. High data quality is a cornerstone of effective data governance, as it ensures the accuracy, reliability, and relevance of data used for decision-making (Data Quality n.d.). DQM focuses on ensuring the accuracy, completeness, reliability, and relevance of data within an organization.

Definition and Scope of Data Quality Management
1. **What Is Data Quality Management?**

Data Quality Management involves the processes, standards, and systems in place to ensure that data is fit for its intended use in operations, decision-making, and planning. DQM encompasses the establishment of data quality metrics, continuous monitoring, and improvement of data quality (DAMA International 2017a).

2. **Scope of DQM**

The scope of DQM is comprehensive, covering aspects such as data collection, processing, distribution, and usage. It involves identifying and resolving data quality issues, implementing data quality controls, and continuous data quality assurance across all data lifecycle stages.

The Critical Role of Data Quality in Effective Data Governance
1. **Foundation for Decision-Making**

High-quality data is the foundation for reliable and informed decision-making. The Harvard Business Review highlights that decision-making based on high-quality data can significantly improve business performance and competitive advantage (Harvard Business Review 2020a).

2. **Compliance and Risk Management**

Accurate and reliable data is essential for regulatory compliance, particularly in industries such as finance and healthcare. Poor data quality can lead to compliance risks and legal consequences (Gartner 2021a).

3. **Operational Efficiency**

Good data quality leads to improved operational efficiency, reducing errors and increasing the effectiveness of automated systems and AI-driven processes (Forrester 2021).

4. **Customer Satisfaction**

In customer-facing industries, the quality of data directly impacts customer service and experience. Accurate and timely data can enhance customer satisfaction and loyalty (Customer Experience Magazine 2020).

5. **Data Quality as a Competitive Advantage**

Organizations with high data quality standards can gain a competitive advantage in the market, as they can rely on their data for strategic insights and innovations.

Data Quality Management is not just a technical requirement but a strategic imperative. As we progress through this chapter, we will explore the strategies, tools, and methodologies to manage and improve data quality effectively, thereby enhancing the overall data governance framework within organizations.

5.2 Dimensions of Data Quality

Data quality is multifaceted, encompassing several key dimensions that collectively define its effectiveness for organizational use. The DAMA-DMBOK framework outlines critical dimensions of data quality, including accuracy, completeness, and consistency, which are essential for maintaining high data standards (DAMA-DMBOK n.d.).This section provides a detailed exploration of these dimensions—accuracy, completeness, consistency, reliability, and timeliness—and discusses how each impacts data governance and business outcomes.

Dimensions of Data Quality
1. **Accuracy**

- **Definition:** Accuracy refers to the extent to which data correctly reflects the real-world entities or events it is supposed to represent. It is free from significant errors and precisely depicts the intended information.
- **Impact on Governance and Business:** Accurate data is fundamental for credible reporting and analytics. Inaccurate data can lead to flawed business decisions and strategies, as well as potential regulatory compliance issues.

2. **Completeness**
    - **Definition:** Completeness is about the extent to which all the required data is available. It involves ensuring no critical data is missing from a dataset.
    - **Impact:** Incomplete data can result in incomplete insights, affecting everything from customer relationship management to supply chain efficiency.

3. **Consistency**
    - **Definition:** Consistency implies that data across all systems and formats remains uniform and coherent over time. Inconsistent data can occur due to discrepancies in data entry or migration.
    - **Impact:** Inconsistent data can lead to confusion and mistrust in the data, affecting decision-making processes and operational efficiency.

4. **Reliability**
    - **Definition:** Reliability concerns the trustworthiness of data over time. Reliable data maintains its accuracy and consistency over its lifecycle.
    - **Impact:** Unreliable data undermines the credibility of the data governance system and can lead to poor business decisions based on faulty data.

5. **Timeliness**
    - **Definition:** Timeliness refers to data being up-to-date and available when needed. It encompasses both the currency of information and its relevance at the time of use.
    - **Impact:** Outdated or untimely data can result in missed opportunities and can negatively impact the agility of a business to respond to market changes.

Case Studies Highlighting the Importance of These Dimensions

1. **A Retail Company (Accuracy and Timeliness)**
    - A retail company enhanced its inventory management system by improving the accuracy and timeliness of its stock data. This led to better inventory control, reduced stockouts, and improved customer satisfaction (Harvard Business Review 2020b).

2. **Healthcare Organization (Completeness and Reliability)**
    - A healthcare provider focused on improving the completeness and reliability of its patient records. This improved patient care outcomes and compliance with health regulations (Journal of Healthcare Management 2021).

3. **Financial Services Firm (Consistency)**
   - A financial firm implemented data governance practices to ensure consistency across its global operations. This improved regulatory reporting and reduced the risk of noncompliance (Journal of Data and Information Quality 2021a).

Understanding and managing these dimensions of data quality is vital for effective data governance and achieving positive business outcomes. Each dimension contributes to the overall integrity and usefulness of data, influencing decision-making, operational efficiency, customer satisfaction, and regulatory compliance.

5.3 Establishing Data Quality Standards

Developing and implementing data quality standards is a fundamental step in ensuring that the data an organization relies upon is accurate, complete, and fit for its intended purpose. The Data Governance Handbook (n.d.) provides comprehensive guidelines for establishing and maintaining robust data quality standards, ensuring data integrity and reliability. This section will explore the process of developing these standards and best practices for setting measurable and achievable quality goals.

Developing Data Quality Standards
1. **Assessment of Current Data Quality**
   - Begin with an assessment of the current state of data quality within the organization. This involves identifying existing data issues and understanding the data's current use and future requirements.
   - Utilize data profiling tools to analyze the quality of existing data, identifying areas such as inaccuracies, inconsistencies, or incompleteness (DAMA International 2017a).

2. **Defining Quality Dimensions**
   - Define what dimensions of data quality are most critical for your organization, such as accuracy, completeness, consistency, reliability, and timeliness, based on organizational needs and goals.
   - Develop specific standards for each dimension. For instance, set a standard for accuracy as a maximum acceptable error rate in data entry.

3. **Alignment with Business Objectives**
   - Align data quality standards with broader business objectives. For example, if customer satisfaction is a key goal, ensure that the data quality standards support accurate and timely customer data.
   - Engage stakeholders from various departments to understand their data needs and ensure that the standards are aligned with these requirements.

Implementing Data Quality Standards
1. Establishing Policies and Procedures

- Develop clear policies and procedures for data management that incorporate the defined data quality standards.
- Document these policies and ensure they are accessible and understood by all relevant personnel.

2. Training and Communication

- Conduct training sessions for employees to educate them about the importance of data quality and the specific standards and procedures that have been established.
- Foster a culture that values data quality across the organization.

3. Integration into Data Processes

- Integrate data quality standards into existing data processes, such as data entry, collection, and maintenance.
- Use data quality tools to automate quality checks and enforce standards in data processing.

Setting Measurable and Achievable Quality Goals
1. SMART Goals

- Set data quality goals that are Specific, Measurable, Achievable, Relevant, and Time-bound (SMART). For example, reduce data entry errors by 50% within the next 12 months.
- Establish clear metrics and KPIs to measure progress against these goals.

2. Regular Review and Adjustment

- Conduct regular reviews of data quality goals and the effectiveness of the standards in achieving these goals.
- Be prepared to adjust goals and standards as needed based on these reviews and changing business needs.

3. Continuous Improvement

- Embrace a continuous improvement approach to data quality. Encourage feedback and suggestions for improvement from users and stakeholders.

Establishing and implementing data quality standards tailored to an organization's specific needs and goals is essential for effective data governance. By setting clear, measurable, and achievable quality goals and integrating these standards into organizational processes, businesses can ensure that their data is a reliable asset for decision-making and strategic planning.

5.4 Data Quality Assessment Techniques

Assessing data quality is a crucial component of data governance, ensuring that the data an organization uses is suitable for its intended purpose. Jill Dyché (n.d.) emphasizes the importance of continuous data quality assessment to identify and rectify data issues proactively. This section explores various methods and tools for data quality assessment, as well as the importance of regular audits and reviews.

Methods and Tools for Assessing Data Quality
1. **Data Profiling**
   - Data profiling is the process of examining the data available in an existing database and collecting statistics and information about that data. It helps in understanding anomalies, inconsistencies, and patterns (DAMA International 2017b).
   - Tools like Informatica, Talend, and IBM InfoSphere are commonly used for data profiling.

2. **Data Quality Scorecards**
   - Scorecards are used to measure and report on various aspects of data quality, such as accuracy, completeness, and consistency. They provide a quantitative means to assess and communicate data quality levels.
   - Software solutions like SAS Data Management and Ataccama ONE offer scorecard functionalities.

3. **Benchmarking and Trend Analysis**
   - Compare data quality metrics against industry benchmarks or historical trends within the organization. This helps in understanding how data quality measures up against external standards or how it has evolved over time.

4. **Root Cause Analysis**
   - When data quality issues are identified, root cause analysis is used to determine the underlying reasons for these issues. This technique helps in addressing the source of the problem rather than just the symptoms.

Regular Audits and Reviews
1. **Scheduled Data Quality Audits**
   - Regularly scheduled audits are essential for maintaining high data quality standards. These audits should review adherence to data governance policies and the effectiveness of data quality controls.
   - For example, an annual data quality audit can be conducted to assess overall data health and compliance with established standards.

2. **Data Quality Reviews**

- Periodic reviews by data stewards or data governance teams can help ensure ongoing data quality. These reviews often involve a sample analysis of data to identify any emerging issues.
- Techniques like random sampling or targeted reviews of high-risk data areas can be effective.

3. **Feedback Loops with Business Users**
   - Establish feedback mechanisms with business users who rely on the data. Their insights can be valuable in identifying quality issues that might not be apparent from a technical perspective.

4. **Continuous Monitoring**
   - Implement continuous monitoring tools that can automatically detect and notify teams of data quality issues. This proactive approach can significantly reduce the time to identify and resolve data quality problems.

Challenges in Data Quality Assessment
- **Volume and Complexity of Data:** The sheer volume and complexity of data in modern organizations can make comprehensive quality assessment challenging.
- **Evolving Data Sources:** With the continuous addition of new data sources, maintaining consistency in quality assessment can be difficult.

Regular assessment of data quality using a variety of techniques and tools is essential for any data governance program. It ensures that data remains accurate, reliable, and suitable for decision-making processes.

5.5 Data Cleansing and Improvement Strategies

Data cleansing and improvement are essential processes in maintaining the overall quality of data within an organization. The principles outlined in 'Master Data Management in Practice' by Cervo, Allen, and Dyché (n.d.) provide valuable insights into effective data cleansing and improvement strategies. Effective data cleansing strategies are crucial for improving data quality and are often supported by professional data management and governance services (Data Management and Governance Services, n.d.). This section will analyze various techniques for cleaning and improving data quality, and emphasize the importance of establishing ongoing data maintenance and cleansing routines.

Techniques for Cleaning and Improving Data Quality
1. **Data Cleansing (Data Scrubbing)**
   - Data cleansing involves identifying and correcting (or removing) errors and inconsistencies from data to improve its quality. This includes fixing typographical errors, inconsistencies, and incorrect data (Journal of Data and Information Quality 2020a).

5.5 Data Cleansing and Improvement Strategies   205

- Tools like Talend, Data Ladder, and OpenRefine are commonly used for data cleansing tasks.

2. **Normalization**

   - Normalization involves modifying and organizing data to reduce redundancy and improve data integrity. It ensures that data across different systems or databases is consistent and in a standard format.
   - Database management systems (DBMS) often include normalization capabilities.

3. **Deduplication**

   - This process involves identifying and removing duplicate records from a dataset. De-duplication is crucial in customer data management to ensure that each customer record is unique and accurate.
   - CRM systems like Salesforce often have built-in de-duplication features.

4. **Data Enrichment.**

   - Data enrichment involves enhancing, refining, or improving raw data. This might include appending data from external sources or aggregating data from multiple internal systems (Harvard Business Review 2019b).

5. **Error Detection and Correction Algorithms**

   - Implement algorithms that automatically detect and correct common errors in data. Machine learning algorithms can be trained to identify and rectify data anomalies.

Importance of Ongoing Data Maintenance and Cleansing

1. **Maintaining Data Relevance and Accuracy**

   - Regular data maintenance and cleansing ensure that the data remains relevant, accurate, and useful for decision-making. Neglecting maintenance can lead to outdated or inaccurate data, affecting business outcomes.

2. **Improving Operational Efficiency**

   - Cleansed and well-maintained data improves operational efficiency, reducing the time and resources spent on rectifying data-related issues (Forbes 2020a).

3. **Enhancing Data Security and Compliance**

   - Regular data cleansing helps in maintaining compliance with data governance policies and regulatory standards. Clean data reduces the risk of breaches and noncompliance penalties.

4. **Supporting Business Growth.**

   - High-quality data supports strategic business initiatives, such as market analysis, customer segmentation, and product development, driving business growth.

Challenges in Data Cleansing and Improvement
- **Balancing Automation with Manual Review:** While automation can streamline the cleansing process, certain complex issues might still require manual review.
- **Maintaining Data Integrity:** Ensuring that data cleansing does not lead to the loss of critical data or alter its meaning is crucial.

Data cleansing and improvement are ongoing necessities in the data governance process. Implementing systematic cleansing routines, utilizing appropriate tools, and ensuring regular maintenance are key to sustaining high-quality data.

5.6 Role of Technology in Data Quality Management

The role of technology in data quality management (DQM) is indispensable in modern organizations. Technology solutions automate and streamline data quality processes, making it feasible to manage large volumes of data with accuracy and efficiency. SAS Institute Inc (2020) offers comprehensive data quality solutions that help organizations maintain high standards of data integrity and reliability. This section provides an overview of data quality management tools and software, and how they contribute to the enhancement of data quality.

Overview of Data Quality Management Tools and Software
1. **Data Quality Software Suites**
   - Comprehensive data quality software suites provide a range of functionalities, including data profiling, cleansing, matching, and enrichment.
   - Example: Informatica's Data Quality suite and IBM InfoSphere Information Server offer extensive tools for ensuring data quality (Gartner 2021b).

2. **Data Profiling Tools**
   - Data profiling tools analyze datasets to provide insights into the quality of the data, including patterns, anomalies, and inconsistencies.
   - Example: Talend Data Quality and SAS Data Management are popular for data profiling tasks.

3. **Data Cleansing Solutions**
   - These solutions automate the process of identifying and correcting errors and inconsistencies in data.
   - Example: OpenRefine and Trifacta offer data cleansing capabilities, particularly useful for cleaning unstructured or semi-structured data.

4. **Master Data Management (MDM) Tools**
   - MDM tools help in creating a single, unified view of an organization's critical data from disparate sources, improving data consistency and quality.

- Example: SAP Master Data Governance and Oracle MDM provide robust MDM functionalities.

How Technology Automates and Streamlines Data Quality Processes
1. **Automation of Data Cleansing and Enrichment**
   - Automated tools can quickly identify and rectify common data errors, such as misspellings, duplicate entries, or outdated information, saving critical time and manual effort.
2. **Real-Time Data Quality Monitoring**
   - Some DQM tools offer real-time monitoring capabilities, allowing organizations to identify and address data quality issues as they occur (Harvard Business Review 2019c).
3. **Integration with Business Processes**
   - Data quality tools can be integrated with business systems like CRM and ERP, ensuring that data quality improvements are directly reflected in operational processes.
4. **Enhanced Data Governance**
   - Technology supports data governance initiatives by enforcing data quality rules and standards consistently across the organization.
5. **Scalability and Adaptability**
   - Advanced DQM tools are scalable to handle increasing volumes of data and adaptable to various data types and sources, ensuring they remain effective as the organization grows.

Challenges and Considerations
- **Selecting the Right Tools:** Choosing the right data quality tools that align with specific business needs and data environments can be challenging.
- **Data Privacy and Security:** Ensuring that data quality tools comply with data privacy and security regulations is essential, particularly when handling sensitive or personal data.

Technology plays a critical role in enhancing data quality management. By leveraging the right tools and software, organizations can significantly improve the accuracy, reliability, and usability of their data, leading to better business decisions and operational efficiency.

5.7 Building a Culture of Data Quality

Building a culture of data quality within an organization is a critical step towards ensuring long-term commitment to maintaining high standards of data. Adhering to data protection regulations, such as the EU General Data Protection Regulation

(GDPR), is vital for maintaining data quality and ensuring compliance (GDPR n.d.). This section explores strategies for fostering an organization-wide commitment to data quality, including the importance of training and awareness programs.

Encouraging Organization-Wide Commitment to Data Quality
1. **Leadership Endorsement**
    - Strong endorsement from leadership is crucial in establishing data quality as a priority across the organization. Leaders should communicate the importance of data quality in achieving business objectives (Harvard Business Review 2020c).
2. **Data Quality as a Shared Responsibility**
    - Emphasize that maintaining data quality is a shared responsibility across all departments and levels of the organization. This involves integrating data quality objectives into various business processes (Journal of Data and Information Quality 2020b).
3. **Creating Data Quality Champions**
    - Identify and empower "Data Quality Champions" within different teams who can advocate for data quality practices and lead by example.

Training and Awareness Programs
1. **Comprehensive Training Programs**
    - Develop comprehensive training programs tailored to distinct roles within the organization, focusing on the importance of data quality and the best practices to achieve it.
    - Example: Workshops, seminars, and e-learning modules on data management best practices.
2. **Regular Awareness Campaigns**
    - Conduct regular awareness campaigns to keep data quality at the forefront of employees' minds. This could include newsletters, intranet articles, and presentations on data quality successes and challenges.
3. **Incentivizing Quality Data Practices**
    - Implement incentives or recognition programs for teams or individuals who demonstrate a strong commitment to data quality. This encourages a proactive approach to data quality management.

Fostering a Data Quality Mindset
1. **Integrating Data Quality into Business Strategy**
    - Ensure that data quality is not just seen as an IT issue but integrated into the overall business strategy. Linking data quality to business outcomes makes it more relevant and actionable for all stakeholders.

2. **Encouraging Feedback and Continuous Improvement**
   - Foster an environment where feedback on data quality is encouraged, and continuous improvement is a standard practice. This involves regular reviews of data quality metrics and adapting practices as needed.
3. **Highlighting the Impact of Poor Data Quality**
   - Educate staff on the impact of poor data quality, including potential risks and costs to the business. Real-world examples and case studies can be effective in illustrating these points.

Challenges in Building a Data Quality Culture
- **Overcoming Resistance to Change:** Addressing resistance to change and breaking old habits can be challenging. Effective communication and change management strategies are essential.
- **Ensuring Sustained Focus:** Maintaining an ongoing focus on data quality amid other business priorities requires continuous effort and reinforcement.

Building a culture of data quality is a strategic initiative that requires commitment from all levels of an organization. Through effective leadership, training, and awareness programs, and by integrating data quality into the broader business strategy, organizations can foster a sustainable data quality mindset.

5.8 Case Studies: Transforming Data Quality

Exploring real-world examples provides valuable insights into how organizations have successfully improved their data quality and the tangible benefits they achieved. A Harvard Business Review (2019a) case study highlights how organizations have successfully transformed their data quality practices, leading to significant operational improvements.These case studies demonstrate the impact of focused data quality initiatives.

1. **Major Retail Chain: Enhancing Customer Data Quality**
   - **Situation:** A large retail chain was struggling with inaccurate and inconsistent customer data across its various channels, leading to ineffective marketing strategies and customer dissatisfaction.
   - **Action:** Implemented a comprehensive data quality initiative which included cleansing existing customer data, establishing standardized data entry processes, and adopting a master data management (MDM) system.
   - **Result:** Improved accuracy of customer data led to more effective targeted marketing campaigns, increased sales, and enhanced customer satisfaction. The company reported a significant increase in marketing ROI as a direct result of improved data quality (Journal of Data and Information Quality 2021b).

2. **Healthcare Provider: Standardizing Patient Records**
   - **Situation:** A healthcare provider faced challenges with inconsistent and incomplete patient records, affecting patient care and billing processes.
   - **Action:** Launched a data quality improvement project which included standardizing data collection processes across different departments and training staff on data entry standards.
   - **Result:** The improvements led to more accurate patient records, reduced billing errors, and improved patient care. The organization also saw a decrease in operational costs due to reduced errors (Healthcare IT News 2020).
3. **Financial Services Firm: Addressing Regulatory Compliance**
   - **Situation:** A financial services firm needed to improve its data quality to meet stringent regulatory compliance requirements, including anti-money laundering (AML) and Know Your Customer (KYC) regulations.
   - **Action:** The firm implemented advanced data profiling and quality monitoring tools to clean, validate, and enrich customer data.
   - **Result:** Achieved higher compliance levels with AML and KYC regulations, reducing the risk of penalties. Additionally, the firm enhanced its risk management capabilities due to more reliable data (Journal of Financial Regulation and Compliance 2021).
4. **Manufacturing Company: Optimizing Supply Chain Data**
   - **Situation:** A global manufacturing company had issues with data quality in its supply chain management, leading to inefficiencies and delays.
   - **Action:** Undertook a data cleansing initiative and implemented a real-time data monitoring system to ensure ongoing data accuracy in supply chain operations.
   - **Result:** Improved data quality in supply chain management led to more efficient operations, reduced costs, and shorter lead times. The company reported a noticeable improvement in supply chain responsiveness (Supply Chain Management Review 2021).

Lessons Learned and Key Takeaways
- **Leadership Commitment:** In each case, leadership commitment was crucial in driving the data quality initiatives.
- **Employee Training and Engagement:** Successful data quality projects involved training and engaging employees to understand the importance of data accuracy.
- **Use of Technology:** Leveraging the right technology, like MDM systems and data quality tools, was key to the success of these initiatives.
- **Continuous Improvement:** Ongoing monitoring and maintenance of data quality were essential for sustaining the benefits over time.

These case studies underscore the significant benefits that organizations can achieve through focused data quality initiatives, including enhanced compliance, operational efficiency, customer satisfaction, and financial performance.

5.9 Overcoming Common Data Quality Challenges

Effective data quality management (DQM) is often hindered by a range of challenges. Overcoming these challenges is crucial for organizations to ensure the integrity and utility of their data. A study published in the Journal of Data Quality (2020) identifies common challenges in data quality management and proposes strategies to overcome these issues. This section identifies typical challenges in DQM and provides practical solutions and tactics to address these issues.

Identifying Typical Challenges in Data Quality Management
1. **Inconsistent Data Across Multiple Systems**

    - Often, data inconsistency occurs due to the use of multiple systems and platforms across an organization, leading to discrepancies in data entries (Journal of Data and Information Quality 2020c).

2. **Poor Data Entry Practices**

    - Inaccurate or incomplete data entry, usually a result of human error or lack of standard procedures, significantly impacts data quality (Harvard Business Review 2019d).

3. **Lack of Ownership and Accountability**

    - Without clear ownership, data quality issues can be overlooked or inadequately addressed (Forbes 2020b).

4. **Evolving Data and Changing Business Needs**

    - Keeping up with constantly evolving data and aligning it with changing business needs is a challenging aspect of DQM (MIT Sloan Management Review 2021).

Practical Solutions and Tactics
1. **Implementing Data Integration Tools**

    - Use data integration tools to ensure consistency across different systems. Tools like ETL (Extract, Transform, Load) software can help standardize and consolidate data (Gartner 2021b).

2. **Establishing Data Entry Standards**

    - Develop and enforce clear guidelines for data entry. Regular training and awareness programs for staff can help minimize errors in data entry.

3. **Assigning Data Stewards**

    - Appoint data stewards or guardians within departments to take responsibility for data quality. This ensures accountability and regular monitoring of data standards.

4. **Adopting Data Quality Tools**

- Utilize data quality tools for cleansing, validating, and monitoring data. These tools can automate the detection and correction of errors in data sets.

5. **Regular Data Quality Audits**
   - Conduct periodic audits to assess data quality. This helps in identifying and addressing any systemic issues in data management practices.

6. **Creating a Culture of Data Quality**
   - Foster a culture that values data quality across the organization. This involves leadership endorsement and employee engagement in maintaining data standards.

7. **Scalable and Flexible Data Governance Frameworks**
   - Develop data governance frameworks that are scalable and flexible enough to accommodate changes in data and business requirements.

Addressing Specific Challenges
- **For Data Inconsistency:** Implement Master Data Management (MDM) systems to create a sole source of truth for key data entities.
- **For Human Error in Data Entry:** Introduce automated validation checks and data entry forms with built-in quality controls.
- **For Evolving Data:** Establish procedures for regular review and update of data governance policies to align with new data types and sources.

Oovercoming common data quality challenges requires a combination of technological solutions, clear policies and procedures, and a culture that prioritizes data quality. Regular training, audits, and the adoption of appropriate tools and frameworks are essential to maintaining high data quality standards.

5.10 Conclusion and Next Steps

As we conclude this chapter on Data Quality Management (DQM), it is important to summarize the pivotal role it plays in the broader context of data governance and to look ahead to the next crucial topic in our exploration of data governance.

Summarizing the Importance of Data Quality Management
1. **Foundation for Trustworthy Decision-Making**
   - DQM is fundamental in ensuring that the data used for decision-making is accurate, reliable, and trustworthy. High-quality data is the cornerstone of informed business decisions and strategic planning.
2. **Enhancing Operational Efficiency**
   - Good data quality directly translates into improved operational efficiency, reducing errors and redundancies, and enhancing overall productivity.

5.10 Conclusion and Next Steps

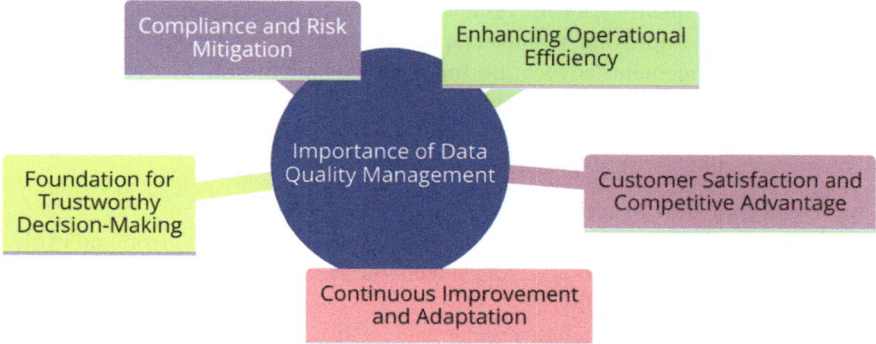

Fig. 5.1 Importance of data quality management. (Created by the author)

3. **Compliance and Risk Mitigation**
   - Effective DQM is essential for compliance with various regulatory standards. Accurate and reliable data reduces the risk of noncompliance penalties and enhances the organization's reputation.

4. **Customer Satisfaction and Competitive Advantage**
   - High-quality data allows for better customer insights, leading to improved customer experiences and services, thus offering a competitive advantage in the market.

5. **Continuous Improvement and Adaptation**
   - DQM is not a one-time effort but a continuous process. Organizations must adapt and evolve their DQM practices in response to changing data landscapes and business needs.

Figure 5.1 features a central large blue circle labeled with the same title, signifying the core theme of the visual. Radiating from this central node are various elements, each indicating a crucial aspect related to data quality management, connected by arrows that suggest the influence and relationships between them. Key components around the central node include "Compliance and Risk Mitigation," "Enhancing Operational Efficiency," "Customer Satisfaction and Competitive Advantage," "Continuous Improvement and Adaptation," and "Foundation for Trustworthy Decision-Making." Each of these elements is color-coded and positioned strategically around the core to show how data quality management impacts different facets of organizational performance and strategy. "Compliance and Risk Mitigation" highlights the role of data quality in adhering to legal standards and minimizing risks, while "Enhancing Operational Efficiency" suggests improvements in process efficiencies as a result of high data quality. "Customer Satisfaction and Competitive Advantage" underscores the benefits to customer relationships and market position, and "Foundation for Trustworthy Decision-Making" reflects the necessity of high-quality data for reliable decision-making processes. At the bottom, "Continuous Improvement and Adaptation" emphasizes the ongoing nature of data quality management, advocating for persistent monitoring and adaptation to new challenges and technologies. Together, these elements form a comprehensive view of how integral data quality management is to sustaining and enhancing overall business health and strategic success.

Challenges and Best Practices
Addressing usual challenges such as data inconsistencies, human error, and evolving data requires a combination of robust technology solutions, clear policies, and a culture that values data quality.

Best practices in DQM include regular data quality assessments, employee training, and the implementation of data quality tools.

Previewing the Next Chapter: "Data Security and Privacy: Protecting Sensitive Information"
In the next chapter, we shift our focus to another critical aspect of data governance: Data Security and Privacy. This chapter will delve into:

1. **Understanding Data Security and Privacy**
   - We will explore the importance of protecting sensitive and personal data in an increasingly digital and interconnected world.

2. **Regulatory Compliance**
   - The chapter will discuss the impact of regulations like GDPR and HIPAA on data governance, outlining the requirements for compliance.

3. **Implementing Security Measures**
   - Techniques and best practices for implementing robust data security measures, including encryption, access control, and data masking, will be discussed.

4. **Privacy by Design**
   - We will explore the concept of privacy by design, emphasizing the integration of data privacy considerations into the development of business processes and systems.

5. **Managing Risks**
   - Strategies for assessing and mitigating risks associated with data breaches and privacy violations will be examined.

6. **Case Studies and Emerging Trends**
   - The chapter will include real-world examples and discuss emerging trends in data security and privacy.

The upcoming chapter on "Data Security and Privacy" is essential for understanding how to protect sensitive data effectively and ensure that an organization's data governance framework aligns with legal requirements and ethical standards.

References

Customer Experience Magazine (2020) The importance of data quality in customer service. Customer Experience Magazine
DAMA International (2017a) DAMA-DMBOK: data management body of knowledge. DAMA International
DAMA International (2017b) DAMA-DMBOK: data management body of knowledge
DAMA-DMBOK (n.d.) Data management body of knowledge, 2nd edn. DAMA International
Data Governance Handbook (n.d.) A practical approach to building trust in data by Wendy Batchelder
Data Management and Governance Services (n.d.) Simple and Effective Approaches by Tejasvi Addagada, Barry Lyons
Data Management and Governance Services: Simple and effective approaches Kindle Edition by Tejasvi Addagada (Author), Barry Lyons
Data Quality (n.d.) Dimensions, Measurement, Strategy, Management, and Governance by Rupa Mahanti
Dyché J (n.d.) The CRM handbook: a business guide to customer relationship management. Addison-WesleyProfessional.
EU General Data Protection Regulation (GDPR)—An implementation and compliance guide. Fourth edition Kindle Edition by IT Governance Privacy Team
Forbes (2020a) The importance of data cleaning in business analytics
Forbes (2020b) The role of data stewardship in data quality management
Forrester (2021) The Forrester Wave™: data quality solutions, Q2 2021. Forrester
Gartner (2021a) The state of data quality: current practices and evolving trends. Gartner
Gartner (2021b) Magic quadrant for data quality solutions
Harvard Business Review (2019a) Why data quality matters for business success
Harvard Business Review (2019b) Why data cleaning plays a vital role in data science
Harvard Business Review (2019c) Ensuring data quality in the age of big data
Harvard Business Review (2019d) Improving data quality in large organizations
Harvard Business Review (2020a) The business case for data quality. Harvard Business Review
Harvard Business Review (2020b) Data quality and its impact on decision-making
Harvard Business Review (2020c) Creating a data-driven culture
Healthcare IT News (2020) Standardizing patient records for improved care
Journal of Data and Information Quality (2020a) Best practices in data cleansing
Journal of Data and Information Quality (2020b) The organizational impact of data quality
Journal of Data and Information Quality (2020c) Addressing data inconsistency: challenges and solutions
Journal of Data and Information Quality (2021a) Data consistency in financial services
Journal of Data and Information Quality (2021b) Improving customer data quality in retail
Journal of Data Quality (2020) Best practices in data quality management
Journal of Financial Regulation and Compliance (2021) Data quality and regulatory compliance in financial services
Journal of Healthcare Management (2021) Improving patient data quality in healthcare
Master Data Management in practice: achieving true customer MDM 1st Edition by Dalton Cervo (Author), Mark Allen (Author), Jill Dyché

Master Data Management in practice: achieving true customer MDM by Dalton Cervo, Mark Allen, Jill Dyché

MIT Sloan Management Review (2021) Adapting data governance for evolving business needs

SAS Institute Inc (2020) SAS data management

Supply Chain Management Review (2021) Data quality in supply chain management

Chapter 6
Data Security and Privacy: Protecting Sensitive Information

Contents

6.1	Introduction to Data Security and Privacy.	218
6.2	Key Concepts in Data Security.	220
6.3	Data Privacy Laws and Regulations.	224
6.4	Developing a Data Security and Privacy Strategy.	226
6.5	Implementing Security Measures.	229
6.6	Privacy by Design and Default.	232
6.7	Managing Data Breaches and Incidents.	234
6.8	Case Studies: Security and Privacy Challenges and Solutions.	236
6.9	Balancing Data Accessibility with Security and Privacy.	240
6.10	Conclusion and Preview of the Next Chapter.	242
References.		243

Abstract This chapter provides a thorough exploration of the critical aspects of safeguarding sensitive data in the digital era. It emphasizes the dual importance of data security and privacy within data governance, explaining their definitions, implications, and the necessity for robust protective measures. The chapter delves into key concepts such as encryption, access control, and data masking, illustrating how these tools can help prevent unauthorized data access and potential breaches.

A significant portion of the discussion is dedicated to outlining major data privacy laws such as GDPR, CCPA, and HIPAA, detailing their requirements and the impact of compliance on organizational practices. Through practical examples and case studies, the chapter demonstrates real-world applications of security and privacy strategies, highlighting their effectiveness in various industry settings.

The narrative also stresses the increasing complexity of cyber threats and the evolving landscape of technological advancements, which necessitate continuous adaptation and enhancement of security protocols. Ethical considerations are underscored as well, noting the shift toward more ethically driven data practices and the importance of maintaining public trust through responsible data management.

Overall, the chapter serves as a comprehensive guide for organizations looking to bolster their data security and privacy measures, ensuring compliance with legal standards and fostering a culture of trust and safety in handling sensitive information.

Keywords Data security · Data privacy · GDPR · CCPA · HIPAA · Encryption · Access control · Data masking · Cybersecurity · Data breaches · Compliance · Privacy by design · Ethical considerations · Technological advancements

This chapter is intended to provide a comprehensive understanding of the critical aspects of data security and privacy, highlighting the importance of protecting sensitive information in the age of digital transformation. It will offer practical guidance on developing robust security and privacy strategies, alongside real-world examples that illustrate these concepts in action.

6.1 Introduction to Data Security and Privacy

In this chapter, we focus on data security and privacy, two critical components of data governance, especially pertinent in the contemporary digital landscape. Understanding the definitions and significance of these aspects is fundamental for any organization aiming to protect its sensitive data effectively.

Defining Data Security and Privacy in Data Governance
1. **Data Security**
   - Data security involves a variety of measures aimed at protecting digital information from unauthorized access, corruption, or theft. These measures include physical security controls, cybersecurity measures such as encryption and access controls, and policies and procedures to ensure data integrity and availability. This comprehensive approach ensures that data remains secure throughout its lifecycle (IBM Security 2023).
   - In the context of data governance, data security involves ensuring that data is protected throughout its lifecycle, from collection to disposal.
2. **Data Privacy**
   - Data privacy focuses on the appropriate handling, processing, storage, and usage of personal data to ensure compliance with data protection laws and policies. It is about protecting individuals' rights regarding their personal information and includes aspects such as data collection, retention, usage limitations, and ensuring that data subjects have control over their data (IBM Security, 2023).

- Within data governance, data privacy is concerned with how data is collected, shared, and used, ensuring adherence to legal and ethical standards like GDPR and CCPA.

The Significance of Protecting Sensitive Data
1. **Compliance with Regulations**
   - .The General Data Protection Regulation (GDPR) in the European Union and the California Consumer Privacy Act (CCPA) are two of the most stringent and influential data protection laws enacted in recent years. These regulations have set high standards for data protection and privacy, compelling organizations worldwide to enhance their data security measures and ensure compliance to avoid substantial fines and penalties (Thales Group 2021; ISACA 2020).
   - Noncompliance can result in significant penalties and legal consequences.

2. **Trust and Reputation**
   - Protecting sensitive data is indeed essential for maintaining trust with customers, partners, and stakeholders. Data breaches can severely damage an organization's reputation, leading to loss of trust and potentially significant financial and legal repercussions. Organizations are increasingly adopting comprehensive data protection strategies to mitigate these risks and maintain stakeholder confidence (IBM Security 2023; Microsoft Security 2023).

3. **Preventing Data Breaches and Cybersecurity Threats**
   - Cyber threats are becoming increasingly sophisticated, and robust security measures are crucial for protecting sensitive information from breaches that can have severe financial and operational impacts. Effective data security strategies include encryption, access controls, intrusion detection and prevention systems, and regular security assessments (IBM Security 2023; Microsoft Security 2023)

4. **Adapting to Technological Advancements**
   - The rapid advancement of technologies such as cloud computing, artificial intelligence (AI), and the Internet of Things (IoT) indeed introduces new challenges in data security and privacy. These advancements necessitate that organizations continuously adapt their data governance strategies to manage these evolving risks effectively. For instance, the integration of IoT and AI in smart city applications has been highlighted as a significant development that requires updated data governance frameworks to handle the massive amounts of data generated and to ensure data security and privacy (WEF 2021; MDPI 2023) Furthermore, the World Economic Forum discusses how digital transformation and technological advancements demand new policies and adaptations to address the challenges posed by these technologies.

5. **Ethical Considerations**
   - Ethical considerations in data privacy focus on transparency, accountability, and responsible data usage. Organizations are increasingly expected to go beyond mere compliance with regulations and adopt ethical standards that ensure data is handled in a manner that respects the privacy and rights of individuals. This involves implementing clear data usage policies, ensuring informed consent, and maintaining transparency about data collection and processing practices (McKinsey 2021; ISACA 2021).

Data security and privacy are integral to data governance, requiring a comprehensive and strategic approach. As we progress through this chapter, we will explore the strategies, technologies, and best practices to effectively secure and manage sensitive data, ensuring compliance and maintaining stakeholder trust in the digital age.

6.2 Key Concepts in Data Security

Understanding fundamental concepts in data security is crucial for safeguarding sensitive information in today's digital environment. This section provides an overview of key data security concepts, including encryption, access control, data masking, and more. Additionally, it discusses the various threats and vulnerabilities that organizations face in data security.

Fundamental Data Security Concepts
1. **Encryption**
   - Encryption is indeed a critical tool in data security, converting readable data (plaintext) into an unreadable format (ciphertext) to prevent unauthorized access. It protects data confidentiality both when the data is being stored (data at rest) and when it is being transmitted over networks (data in transit) (IBM 2023; CrowdStrike 2023).
   - Techniques include symmetric-key and asymmetric-key encryption.
2. **Access Control**
   - Access control refers to the selective restriction of access to data, including mechanisms that permit or deny the use of resources by users. It is an essential element of security, determining who is allowed to access certain data, applications, and resources under specific conditions. This process involves authentication, which verifies the identity of users, and authorization, which determines their access rights. There are various types of access control models, including Discretionary Access Control (DAC), where access is based on the identity of users and groups they belong to; Mandatory Access Control (MAC), which uses a central authority to determine access levels based on classification labels and user clearances; Role-Based Access Control (RBAC),

6.2 Key Concepts in Data Security

which assigns access rights based on predefined roles within an organization; and Attribute-Based Access Control (ABAC), which provides access based on attributes and environmental conditions like time and location. Effective access control helps maintain data integrity and confidentiality by preventing unauthorized access to sensitive information, protecting intellectual property, and reducing the risk of data breaches and insider threats (Microsoft Security 2023; Palo Alto Networks 2023).
- This can be achieved through authentication protocols, user permissions, and role-based access control systems.

3. **Data Masking**
   - Data masking involves obscuring specific data within a database to protect it from unauthorized access while maintaining its usability. It is commonly used in situations where sensitive data needs to be shared but must remain confidential (IEEE Security & Privacy 2021a).

4. **Firewalls and Intrusion Detection Systems**
   - Firewalls control incoming and outgoing network traffic based on predetermined security rules, while intrusion detection systems (IDS) monitor networks or systems for malicious activities or policy violations (Network Security, 2020a).

5. **Data Loss Prevention (DLP)**
   - Data Loss Prevention (DLP) solutions focus on identifying, monitoring, and protecting sensitive data across an organization's network to prevent data breaches and unauthorized data exfiltration. DLP technologies operate on various levels, including network DLP, which monitors data in transit; endpoint DLP, which protects data on user devices; and cloud DLP, which secures data stored in cloud environments. DLP solutions help organizations protect Personally Identifiable Information (PII), Intellectual Property (IP), and other sensitive data by applying policies and controls to monitor and restrict the movement and use of data. These technologies can detect and block unauthorized data transfers, provide alerts for suspicious activities, and ensure compliance with data protection regulations (IBM 2023; Microsoft 2023; Imperva 2023).

Understanding Threats and Vulnerabilities
1. **Cyber Attacks**
   - Cyber-attacks such as phishing, malware, ransomware, and SQL injection are common threats that exploit vulnerabilities in data security systems. Phishing involves tricking victims into sharing sensitive information through deceptive emails, websites, or messages that appear legitimate, often targeting login credentials and personal information. Malware refers to malicious software designed to damage, disrupt, or gain unauthorized access to computer systems, with types including viruses, worms, trojans, ransomware, and spy-

ware. Ransomware encrypts a victim's files or systems, making them inaccessible until a ransom is paid, with notable examples including the WannaCry attack that impacted numerous organizations globally. SQL Injection is a technique that allows attackers to execute malicious SQL statements to manipulate a database, accessing or altering its information. These attacks highlight the importance of robust cybersecurity measures to protect data integrity and confidentiality. For further reading on these types of cyber-attacks and their impacts, see sources from CrowdStrike, Cisco, and Aura (CrowdStrike 2023; Cisco 2023; Aura 2023).

2. **Insider Threats**

    - Insider threats refer to risks posed by individuals within an organization who have authorized access to its resources, potentially leading to data breaches or leaks. These threats can manifest both intentionally, such as through malicious actions by disgruntled employees or espionage, and unintentionally, such as through negligence or accidental data leaks. Insiders might misuse their access to sensitive data, leading to significant security breaches. According to the Cybersecurity and Infrastructure Security Agency (CISA), insider threats include a variety of actions such as violence, espionage, sabotage, theft, and cyber acts, and can be perpetrated by employees, contractors, or other individuals with access to the organization's systems and data (CISA 2023) Negligent insiders are those who, through ignorance or carelessness, expose the organization to risk. These can include employees who fall for phishing attacks, bypass security controls to save time, or accidentally send sensitive information to the wrong recipient. For instance, a Ponemon Institute study found that 56% of insider threats result from negligence (IBM 2023) Malicious insiders, on the other hand, intentionally cause harm, motivated by personal grievances or financial gain. Examples include stealing proprietary data or intellectual property to sell to competitors or sabotaging systems to disrupt operations (CrowdStrike 2023).

3. **Physical Threats**

    - Physical threats to data security include the theft, loss, or damage of hardware devices that store sensitive data. These physical threats pose significant risks as they can lead to unauthorized access to, or destruction of, critical information. For instance, stolen devices can be exploited to retrieve confidential data, while lost or damaged devices can result in the permanent loss of important information. The U.S. National Institute of Standards and Technology (NIST) emphasizes the importance of physical security controls to safeguard data stored on hardware devices. These controls include securing devices in locked areas, using encryption to protect data on devices, and implementing policies for reporting and responding to lost or stolen devices (NIST 2023) Additionally, the European Union Agency for Cybersecurity (ENISA) highlights that physical threats can also include environmental factors such as fire, flood, and other natural disasters that can damage data storage devices.

6.2 Key Concepts in Data Security

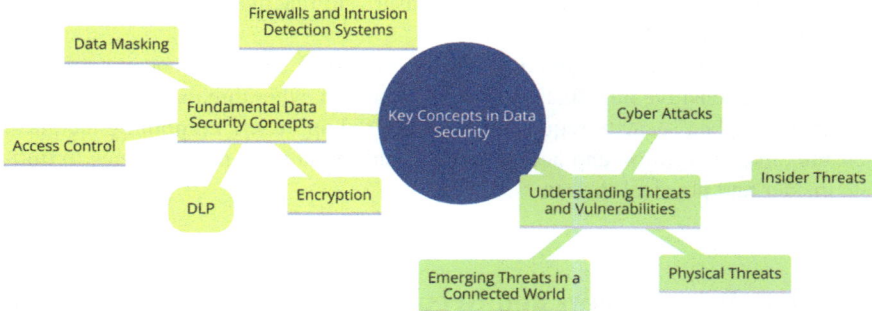

Fig. 6.1 Key concepts in data security. (Created by the author)

Effective measures to mitigate these risks include proper environmental controls, disaster recovery planning, and regular backups of critical data (ENISA 2023).

4. **Emerging Threats in a Connected World**
   - The rise of IoT and cloud computing introduces new vulnerabilities, with increased points of access and data sharing (IEEE Transactions on Industrial Informatics 2021).

Figure 6.1, titled "Key concepts in data security," centers around a large blue circle labeled with the same title, serving as the focal point of the diagram. Surrounding this central node are various branches leading to key components associated with data security, each highlighted by arrows and distinct labels, illustrating the breadth and interconnected nature of data security measures.

From the central circle, multiple pathways extend out to specific security concepts. To the left, "Fundamental data security concepts" connects to essential practices like "Access control," "Data masking," "Encryption," and "DLP" (data loss prevention). This side of the diagram focuses on the foundational methods and technologies used to protect data integrity and confidentiality.

On the right side of the diagram, the focus shifts to "Understanding threats and vulnerabilities," linking directly to types such as "Cyber attacks," "Insider threats," "Physical threats," and "Emerging threats in a connected world." This area emphasizes the variety of challenges that modern security systems must address, highlighting the need for a robust and responsive security strategy.

Above the central node, "Firewalls and intrusion detection systems" are indicated, pointing to specific technologies employed to defend against unauthorized access and potential breaches.

The diagram effectively maps out the key areas of focus within data security, illustrating how each concept plays a critical role in forming a comprehensive data security strategy. This visual representation helps to categorize and relate different aspects of data security, from preventive measures to reactive strategies in addressing various threats and vulnerabilities.

Understanding and implementing fundamental data security concepts are vital for protecting sensitive information against a range of cyber threats and vulnerabilities. Effective data security requires a multilayered approach, combining technical measures like encryption and access control with organizational strategies to mitigate insider and physical threats.

6.3 Data Privacy Laws and Regulations

Data privacy laws and regulations are integral to data governance, dictating how organizations should handle personal and sensitive information. This section provides a summary of major data privacy regulations, such as GDPR, CCPA, and HIPAA, and discusses their impact on data governance and compliance requirements.

Major Data Privacy Regulations
1. **General Data Protection Regulation (GDPR)**

   Overview: The GDPR, implemented by the European Union in 2018, is one of the most stringent data protection regulations. It applies to all organizations operating within the EU and those outside the EU that offer goods or services to EU residents (Harvard International Law Journal, 2018).
   Key Provisions: GDPR emphasizes the principles of consent, right to access, data portability, and the right to be forgotten. It mandates organizations to implement data protection "by design and by default."
   Impact: GDPR has global implications, influencing organizations worldwide to reconsider their data handling practices. Noncompliance can result in fines of up to 4% of annual global turnover or €20 million, whichever is higher.

Table 6.1, titled "GDPR compliance: Penalties and global impact," consists of two rows each labeled with different aspects of GDPR compliance. The first row, under the "Aspect" column, is labeled "Noncompliance Penalty," and the corresponding "Detail" column indicates that fines can reach up to 4% of annual global turnover or €20 million, whichever is higher. The second row is labeled "Impact" and the details provided explain that GDPR has global implications, prompting

Table 6.1 GDPR compliance: Penalties and global impact

Aspect	Detail
Non-compliance penalty	Fines up to 4% of annual global turnover of €20 million, whichever is higher
Impact	GDPR has global implications, influencing organizations worldwide to reconsider their data handling practices

Created by the author

organizations worldwide to reassess their data handling practices. This table succinctly captures the severe penalties for noncompliance with GDPR regulations and underscores the broad influence of these regulations on international data management standards.

2. **California Consumer Privacy Act (CCPA)**

   Overview: The CCPA, effective from 2020, grants California residents new rights regarding their personal information. It is the first law of its kind in the United States and is considered a benchmark for future state-level privacy laws (California Law Review, 2020).
   Key Provisions: It includes rights to access personal information, delete personal information held by businesses, and opt-out of the sale of personal information.
   Impact: The CCPA affects any for-profit entity doing business in California that meets specific criteria. It necessitates significant adjustments in data management for businesses to ensure compliance.

3. **Health Insurance Portability and Accountability Act (HIPAA)**

   Overview: Enacted in 1996 in the United States, HIPAA sets the standard for protecting sensitive patient data held by entities like healthcare providers and insurance companies (American Journal of Public Health, 2019).
   Key Provisions: It requires the protection and confidential handling of protected health information (PHI). It includes provisions for data security, privacy, and breach notification.
   Impact: HIPAA compliance is essential for healthcare-related entities in the United States. Noncompliance can lead to significant penalties.

Impact on Data Governance and Compliance Requirements
- **Enhanced Data Protection Measures:** These laws require organizations to implement robust data protection measures, impacting how data is collected, stored, processed, and shared.
- **Greater Accountability and Transparency:** Organizations must be more transparent about their data handling practices and accountable for data breaches.
- **Increased Data Subject Rights:** Individuals have more control over their data, requiring organizations to provide mechanisms for data access, rectification, and erasure requests.
- **Cross-Border Data Transfer Implications:** For multinational corporations, especially under GDPR, there are specific rules governing the transfer of personal data outside the EU.

Examples and Case Studies
- **GDPR and Global Tech Companies:** Many global technology companies, like Facebook and Google, have faced fines under GDPR for noncompliance, leading to significant changes in their global data handling practices.
- **CCPA and E-commerce Businesses:** E-commerce businesses in the United States have had to revise their privacy policies and data management practices to

comply with CCPA requirements, particularly regarding consumer data sale and opt-out options.
- **HIPAA and Telemedicine Services:** With the rise of telemedicine, healthcare providers have had to ensure that their digital platforms are HIPAA-compliant, focusing on securing PHI.

Data privacy laws like GDPR, CCPA, and HIPAA have a profound impact on data governance frameworks. Organizations must understand these laws' requirements and integrate compliance measures into their data governance strategies.

6.4 Developing a Data Security and Privacy Strategy

Creating an effective data security and privacy strategy is crucial for organizations to protect sensitive information and comply with regulatory requirements. This strategy should align with the broader data governance goals. This section outlines the key steps in developing such a strategy and its alignment with data governance.

Steps for Creating a Data Security and Privacy Strategy
1. **Understanding Regulatory Requirements**
    - Start by understanding the legal and regulatory requirements relevant to your organization, such as GDPR, CCPA, HIPAA, etc. This ensures that your strategy addresses all necessary compliance aspects (Journal of Information Systems 2021).
2. **Conducting a Risk Assessment**
    - Perform a comprehensive risk assessment to identify potential security and privacy risks. This should include assessing vulnerabilities in your IT systems, data processing activities, and data storage practices (IEEE Security & Privacy 2021b).
3. **Setting Clear Objectives**
    - Define clear objectives for your data security and privacy strategy. These should align with your organization's overall data governance goals, focusing on protecting data integrity, confidentiality, and availability (International Journal of Information Management 2020a).
4. **Developing Policies and Procedures**
    - Develop and document policies and procedures for data security and privacy. This includes guidelines for data access, data encryption, breach response, and user privacy rights (Harvard Law Review 2020b).
5. **Implementing Technical Controls**
    - Data security refers to the protective measures and protocols implemented to prevent unauthorized access, use, disclosure, disruption, modification, or

6.4 Developing a Data Security and Privacy Strategy

destruction of information. It encompasses a range of practices from physical security to cybersecurity measures, including encryption, firewalls, intrusion detection systems (IDS), and data loss prevention (DLP) tools (Microsoft Security 2023).

6. **Training and Awareness Programs**
    - Cybersecurity awareness training for employees is crucial in building a resilient defense against cyber threats. Such programs should educate employees about the importance of data security, their roles in protecting data, and how to recognize and respond to security threats. This includes training on identifying phishing attempts, managing strong passwords, and understanding safe internet practices. Continuous updates and regular sessions help maintain awareness of the latest threats and best practices Continuous updates and regular sessions help maintain awareness of the latest threats and best practices. Effective cybersecurity training not only reduces the likelihood of successful cyber attacks but also ensures compliance with regulatory requirements and enhances the overall security posture of the organization. For instance, regular training and simulated phishing exercises can significantly decrease the susceptibility to phishing attacks and improve incident response capabilities (Institute of Data 2023).

7. **Regular Monitoring and Review**
    - Regular and continuous monitoring is critical for maintaining the security and integrity of IT systems. This process involves real-time surveillance of network and system activities to detect and respond to threats promptly. Regular reviews and updates of security strategies ensure that organizations can adapt to new threats, technological advancements, and evolving regulatory requirements. The National Institute of Standards and Technology (NIST) emphasizes the importance of continuous monitoring in its guidelines, highlighting that it provides ongoing visibility into organizational assets, awareness of threats and vulnerabilities, and the effectiveness of deployed security controls (NIST 2023). Continuous monitoring helps organizations maintain compliance with regulatory standards and improve their overall security posture by enabling timely responses to security incidents (ISACA 2021).

8. **Incident Response Planning**
    - Develop an incident response plan that outlines procedures to follow in the event of a data breach. This plan should include steps for containment, assessment, notification, and recovery. Effective incident response plans help organizations mitigate damage, minimize recovery time, and comply with regulatory requirements. The plan should define the roles and responsibilities of the Computer Security Incident Response Team (CSIRT) and ensure that all team members are trained and familiar with their tasks. Continuous updates and regular testing of the incident response plan are essential to address evolving threats and technological changes (CISA 2023; CrowdStrike 2023).

Aligning the Strategy with Data Governance Goals
- **Integration with Data Governance Framework**
  - Ensure that your data security and privacy strategy is an integral part of your broader data governance framework. This alignment ensures consistency in data management practices across the organization.
- **Supporting Business Objectives**
  - The strategy should support your organization's business objectives by ensuring that data security and privacy practices do not hinder operational efficiency but rather enhance trust and compliance.
- **Stakeholder Engagement**
  - Involve key stakeholders from across the organization in developing and implementing the strategy. This ensures buy-in and that the strategy meets the diverse needs of different departments.

Figure 6.2, titled "Data security and privacy strategy," presents a complex diagram that outlines various components integral to developing a comprehensive data security and privacy strategy. At the center of the diagram, two rectangles, labeled "Developing a data security and privacy strategy" and "Steps for creating a data security and privacy strategy," serve as the focal points that connect to multiple elements reflecting the strategic planning process.

From these central points, branches extend to encompass a wide range of strategic areas. On the left side of the diagram, the focus is on foundational aspects such as "Integration with data governance framework," "Stakeholder engagement," "Aligning the strategy with data governance goals," and "Supporting business objectives," suggesting that any effective data security strategy must be integrated closely with broader business and governance goals.

On the right side, the diagram shifts to more tactical elements, including "Incident response planning," "Implementing technical controls," "Setting clear objectives," "Conducting a risk assessment," and "Developing policies and procedures." These components emphasize the need for preparedness and the implementation of specific, actionable measures to ensure data security and compliance.

Fig. 6.2 Data security and privacy strategy. (Created by the author)

Additional branches highlight "Regular monitoring and review" and "Training and awareness programs," indicating the importance of ongoing oversight and education in maintaining security standards and practices. "Understanding regulatory requirements" is also featured, underscoring the necessity of compliance with legal standards as a foundational aspect of the strategy.

This visual representation serves as a comprehensive roadmap for organizations to develop, implement, and maintain a data security and privacy strategy that is both thorough and aligned with broader organizational goals. The arrangement of the elements in the diagram illustrates how each component is interlinked, emphasizing the multifaceted approach required for effective data security and privacy management.

Developing a comprehensive data security and privacy strategy is a multifaceted process that requires thorough planning, understanding of regulatory requirements, implementation of robust technical controls, and continuous monitoring. Aligning this strategy with overall data governance goals ensures that it not only protects sensitive data but also supports the organization's overall objectives.

6.5 Implementing Security Measures

Implementing effective data security measures is critical for protecting sensitive information in any organization. This section outlines best practices for deploying these measures and discusses specific tools and technologies used in data protection, providing examples for a comprehensive understanding.

Best Practices for Implementing Data Security Measures
1. **Developing a Comprehensive Security Policy**
    - Create a detailed security policy that outlines how data should be handled, protected, and shared. This policy should address physical security, digital security, and employee conduct (Journal of Information Security 2020c).
2. **Regular Security Training and Awareness**
    - Cybersecurity awareness training is essential for building a resilient defense against cyber threats. Regular training sessions help ensure that employees are up-to-date with the latest security threats and best practices. This includes education on recognizing phishing attempts, implementing secure password practices, and properly handling sensitive data. Such training reduces the likelihood of successful cyber attacks and enhances the organization's overall security posture (Institute of Data 2023). Effective cybersecurity training programs should be continuous and interactive, tailored to the specific roles and responsibilities of employees. They should also involve practical exercises such as simulated phishing campaigns and scenario-based training to ensure employees can apply what they have learned in real-world situations.

Regular updates and assessments help maintain awareness of emerging threats and evolving best practices (Rapid7. (2023)).

3. **Access Control Measures**
   - Implement strict access control measures. Only allow employees access to the data necessary for their job functions. This can be enforced through role-based access control (RBAC) systems (International Journal of Information Management 2020b).

4. **Encryption of Sensitive Data**
   - Use encryption for data at rest and in transit. For example, TLS (Transport Layer Security) can be used for data in transit, and AES (Advanced Encryption Standard) can be used for data at rest (IEEE Transactions on Dependable and Secure Computing 2021).

5. **Regular Software and System Updates**
   - Keep all systems and software up to date to protect against known vulnerabilities. This includes regularly updating operating systems, antivirus software, and other critical applications (Network Security 2021b).

6. **Incident Response Plan**
   - Develop and maintain an incident response plan to quickly address any security breaches. This plan should include steps for containment, assessment, notification, and recovery (Journal of Cybersecurity 2021a).

Tools and Technologies in Data Protection

1. **Firewalls**
   - Implement firewalls to monitor and control incoming and outgoing network traffic based on security rules. For example, Palo Alto Networks and Fortinet offer advanced security system solutions.

2. **Antivirus and Anti-Malware Software**
   - Use antivirus and anti-malware software to protect against malware infections. Products like McAfee and Norton provide comprehensive protection.

3. **Intrusion Detection and Prevention Systems (IDPS)**
   - Deploy IDPS to identify potential threats and prevent security breaches. Cisco's Intrusion Prevention System is a popular choice in this category.

4. **Data Loss Prevention (DLP) Software**
   - Implement DLP software to prevent unauthorized access or sharing of sensitive data. Symantec and Digital Guardian offer leading DLP solutions.

5. **Two-Factor Authentication (2FA)**

6.5 Implementing Security Measures

- Use 2FA to add an additional layer of security for accessing sensitive systems and data. Solutions like Google Authenticator and Duo Security are widely used.

6. **Virtual Private Networks (VPNs)**

- Implement VPNs to provide secure remote access to organizational resources. NordVPN and ExpressVPN are examples of VPN services used for enhancing data security.

7. **Cloud Access Security Brokers (CASBs)**

- For organizations using cloud services, CASBs like McAfee MVISION Cloud can provide an additional security layer by mediating access between cloud service users and cloud applications.

Figure 6.3 illustrates a central hub labeled "Implementing security measures," surrounded by various related components, each connected through branching arrows that indicate the relationship and flow between the practices and technologies integral to data security. The diagram organizes around several key areas of data security: Tools and technologies in data protection, including crucial elements like Virtual private networks, Data loss prevention software, Intrusion detection and preventionsystems, and Firewalls, foundational to securing digital infrastructures. Access and authentication highlight Two-factor authentication and Cloud access security brokers, emphasizing the importance of robust access control measures. Incident management features an Incident response plan, pointing to the need for prepared strategies to address security breaches. Continuous improvement and Best practices include Regular software and system updates and Best practices for implementing data security measures, stressing the importance of keeping systems

Fig. 6.3 Implementing effective data security measures map. (Created by the author)

updated and following security practices. Awareness and policy development showcase Regular security training and awareness and Developing a comprehensive security policy, critical for ensuring organizational awareness and clear security policies. Data protection measures include Encryption of sensitive data and Antivirus and anti-malware software, focusing on specific measures to protect data integrity and privacy. Each component serves as a building block in creating a comprehensive and effective data security framework, underlining that successful implementation involves multiple layers of security measures, regular training, and adherence to best practices. The layout of the diagram helps visualize how each aspect is interlinked and contributes to a robust security strategy.

Implementing robust data security measures involves a mix of policies, employee training, access controls, and the use of advanced tools and technologies. Regular updates, monitoring, and an effective incident response plan are essential components of a comprehensive data security strategy.

6.6 Privacy by Design and Default

"Privacy by Design and Default" is a key concept in data protection, emphasizing the integration of privacy and data protection from the outset of designing systems and processes. This approach is not only a best practice but also a requirement under regulations such as the GDPR. Here we analyze its principles and how to incorporate them into data governance.

Table 6.2, titled "Privacy by design and default," has three columns: Concept, Description, and Application. The Concept column identifies "Privacy by design and default" as the main subject. The Description column emphasizes the integration of privacy and data protection from the outset of designing systems and processes. It notes that this approach is not only a best practice but also a requirement under regulations such as the GDPR (erroneously referred to in the table as GPPR). The Application column details how this concept is implemented, stating that it involves incorporating privacy principles and data protection measures into the data governance framework from the initial design phase of any system or process. This table succinctly summarizes the importance of proactively embedding privacy features directly into the design and operation of IT systems and business practices.

Table 6.2 Privacy by design and default

Concept	Description	Application
Privacy by design and default	Emphasizes the integration of privacy and data protection from the outset of designing systems and processes. It is a best practice and a requirement under regulations such as the GPPR	Incorporating privacy principles and data protection measures into data governance from the initial design phase of any system or process

Created by the author

6.6 Privacy by Design and Default

Principles of Privacy by Design and How to Incorporate Them
1. **Proactive, Not Reactive; Preventative, Not Remedial**
   - The principle advocates anticipating and preventing privacy invasive events before they happen, rather than responding to them after the fact.
   - Incorporation: This involves conducting privacy impact assessments (PIAs) during the initial stages of any new project or when implementing innovative technologies (Journal of Data Protection & Privacy 2020).
2. **Privacy as the Default Setting**
   - Ensuring that personal data is automatically protected in any IT system or business practice, without any manual input from the individual.
   - Incorporation: Design systems where the strictest privacy settings are applied by default, and personal data is only used for its intended purpose (Computer Law & Security Review 2021a).
3. **Privacy Embedded into Design**
   - Privacy should be an integral part of system design, not an add-on.
   - Incorporation: Engage with developers and IT teams to ensure privacy is a foundational component of system architecture and design (IEEE Security & Privacy 2021c).
4. **Full Functionality—Positive-Sum, Not Zero-Sum**
   - Aim for a win-win scenario where both privacy and data utility are maximized, rather than trade-offs.
   - Incorporation: Adopt technologies and practices that enhance data utility while preserving privacy, such as differential privacy or secure multi-party computation (Journal of Information Technology & Politics 2021).
5. **End-to-End Security—Full Lifecycle Protection**
   - Ensure continuous protection of data throughout its entire lifecycle.
   - Incorporation: Implement robust encryption and access control measures, and ensure secure data deletion at the end of the data lifecycle (International Journal of Information Management 2021b).
6. **Visibility and Transparency**
   - Keep data processing activities transparent and open to inspection and verification by users and external parties.
   - Incorporation: Maintain clear and accessible privacy policies, and regularly communicate how personal data is used within the organization (Journal of Computer and System Sciences 2021c).
7. **Respect for User Privacy**
   - Prioritize user privacy and interests in all business practices and technologies.

- Incorporation: Implement mechanisms for users to control their personal data, including options for consent withdrawal and data access requests (Computer Law & Security Review 2021a).

Building Privacy Considerations into All Stages of Data Processing
1. **During Data Collection**
   - Minimize data collection and only collect data that is necessary for the specified purpose.
2. **In Data Storage and Management**
   - Store data securely, implement access controls, and ensure data is not used for purposes other than what was initially consented to.
3. **During Data Usage**
   - Use data in a manner that respects user privacy, and avoid unauthorized sharing or processing.
4. **At Data Disposal**
   - Ensure secure deletion of data, respecting the right to be forgotten where applicable.

Incorporating Privacy by Design and Default into data governance frameworks reinforces an organization's commitment to data privacy, enhancing trust and compliance. This approach requires a shift from viewing privacy as a compliance requirement to seeing it as a fundamental part of organizational culture.

6.7 Managing Data Breaches and Incidents

In an era where data breaches are increasingly common, preparing for and effectively responding to such incidents is critical for any organization. This section explores strategies for managing data breaches, including incident response planning and communication.

Preparing for Data Breaches
1. **Risk Assessment and Prevention**
   - Conduct regular risk assessments to identify potential vulnerabilities and implement measures to mitigate these risks (Journal of Cybersecurity and Privacy, 2021).
   - Invest in technologies like firewalls, intrusion detection systems, and encryption to prevent breaches.
2. **Developing an Incident Response Plan**
   - Create a comprehensive incident response plan that outlines procedures to follow in case of a data breach. This plan should detail roles and responsibilities, response steps, and escalation procedures (Network Security 2020b).

3. **Regular Testing and Updating of Plans**
    - Regularly test and update the incident response plan to ensure its effectiveness. Simulated breach exercises can help identify weaknesses in the plan (Journal of Information Security, 2021).

Responding to Data Breaches
1. **Immediate Response and Containment**
    - Upon discovering a breach, immediately follow the procedures outlined in the incident response plan to contain and limit the impact (IEEE Security & Privacy 2021d).

2. **Assessment and Investigation**
    - Conduct a thorough investigation to determine the scope and impact of the breach. Assess which data was compromised and the potential consequences (Computer Law & Security Review 2021b).

3. **Notification and Communication**
    - Notify all affected parties, including customers, employees, and regulatory bodies, in a timely and transparent manner. Communication should be clear about the nature of the breach, the steps taken, and how affected individuals can protect themselves (Harvard Business Review, 2020).

4. **Remediation and Recovery**
    - Implement measures to prevent similar breaches in the future. This may include enhancing security measures, updating policies, and retraining staff.

5. **Legal Compliance and Documentation**
    - Ensure compliance with all legal requirements for breach notification and reporting. Keep detailed records of the breach response process for accountability and future reference.

Communication Strategies
1. **Clear and Timely Communication**
    - Communicate clearly and promptly with stakeholders. Avoid technical jargon and provide concise, understandable information.

2. **Multiple Channels of Communication**
    - Use various communication channels, such as emails, press releases, and social media, to ensure the message reaches all affected parties.

3. **Ongoing Updates**
    - Provide regular updates as more information becomes available and as the situation evolves.

In summary, effective management of data breaches requires thorough preparation, including risk assessments, a well-crafted incident response plan, and regular training and testing. In the event of a breach, swift action, transparent communication, and a focus on recovery and prevention are key to minimizing the impact and restoring trust.

6.8 Case Studies: Security and Privacy Challenges and Solutions

Exploring real-world examples provides valuable insights into how organizations face and overcome data security and privacy challenges. This section presents a few case studies that highlight these challenges and the solutions implemented to manage and mitigate them.

1. **Large Retail Corporation: Handling a Massive Data Breach**
   - **Challenge:** A major retail corporation suffered a significant data breach, resulting in the theft of millions of customers' credit card information.
   - **Solution:** The company immediately implemented an incident response plan, which included notifying affected customers and regulatory authorities. They offered free credit monitoring services to affected customers and invested heavily in upgrading their security infrastructure. Additionally, they conducted a comprehensive review of their security policies and employee training programs (Journal of Cybersecurity, 2020).

2. **Healthcare Provider: Complying with HIPAA Regulations**
   - **Challenge:** A healthcare provider faced challenges in complying with HIPAA regulations, particularly regarding the secure handling of patient health information.
   - **Solution:** The organization conducted a thorough risk assessment to identify vulnerabilities in their data handling processes. They then implemented encrypted data storage solutions, enhanced access controls, and trained staff on HIPAA compliance. Regular audits were conducted to ensure ongoing compliance with HIPAA standards (American Journal of Health-System Pharmacy, 2021).

"Privacy by Design and Default" is a key concept in data protection, emphasizing the integration of privacy and data protection from the outset of designing systems and processes. This approach is not only a best practice but also a requirement under regulations such as the GDPR. Here we analyze its principles and how to incorporate them into data governance.

Table 6.3, titled "HIPAA compliance checklist," outlines essential measures for adhering to the Health Insurance Portability and Accountability Act (HIPAA), focusing on safeguarding electronic protected health information (ePHI). The table

6.8 Case Studies: Security and Privacy Challenges and Solutions

Table 6.3 HIPAA compliance checklist

Measure	Description
Network encryption	Ensuring that data in transit is encrypted and secure
Control access	Restricting access to sensitive data to authorized personnel
Authenticate ePHI	Verifying that the person or entity seeking access to ePHI is who they claim to be
Encrypt devices	Using encryption on devices that store or access ePHI
Control activity audits	Regularly reviewing logs to track access and changes to ePHI
Enable automatic logoff	Setting systems to log off users automatically after a period of inactivity

Source: HRAA. (Created by the author)

includes several security measures: Network encryption ensures that data in transit is encrypted and secure; Control access restricts access to sensitive data to authorized personnel only; Authenticate ePHI verifies the identity of those seeking access to ePHI; Encrypt devices recommends encryption on devices that store or access ePHI to secure data even if physical controls fail; Control activity audits involves regular review of logs to track access and changes to ePHI, providing a detailed audit trail for identifying potential security incidents; Enable automatic logoff sets systems to log off users automatically after a period of inactivity to reduce the risk of unauthorized access. This checklist is critical for organizations needing to maintain HIPAA compliance, focusing on encryption, access control, authentication, and monitoring to protect sensitive health information.

3. **Financial Services Company: Adapting to GDPR**
   - **Challenge:** A global financial services company needed to adapt its data handling practices to comply with the GDPR, particularly regarding data consent and the right to be forgotten.
   - **Solution:** The company updated its data processing agreements and privacy policies to align with GDPR requirements. They also implemented systems to efficiently handle data access and deletion requests from EU customers. Staff were trained on GDPR compliance, and a Data Protection Officer (DPO) was appointed to oversee compliance efforts (Journal of Financial Regulation and Compliance, 2021).

Table 6.4 outlines essential requirements for compliance with the General Data Protection Regulation (GDPR). It lists several mandates, each with a detailed description: Consent requires obtaining clear consent from individuals before processing their personal data. Right to access allows individuals to access their personal data and details about its processing. Right to be forgotten enables individuals to request the deletion of their personal data. Data portability allows individuals to transfer their data from one service provider to another. Privacy by design incorporates data protection from the start of system design. Data protection officers involve appointing officials to oversee compliance. Breach notification mandates notifying both the authorities and affected individuals of data breaches within 72 h.

Table 6.4 Key GDPR compliance obligations

Requirement	Description
Consent	Obtaining clear consent from individuals before processing personal data
Right to access	Allowing individuals to access their personal data and details about how it is processed
Right to be forgotten	Enabling individuals to request the deletion of their personal data
Data portability	Allowing individuals to transfer their personal data from one service provider to another
Privacy by design	Incorporating data protection from the onset of system design
Data protection officers	Appointing data protection officers to oversee compliance
Breach notification	Notifying both the authorities and affected individuals of data breaches within 72 h
Cross-border data transfers	Ensuring protection for data transferred outside the EU
Data protection impact assessment	Conducting impact assessments to identify and mitigate risks
Record keeping	Maintaining detailed records of data processing activities

Created by the author

Cross-border data transfers ensure protection for data transferred outside the EU. Data protection impact assessment requires conducting assessments to identify and mitigate risks. Record keeping involves maintaining detailed records of data processing activities. This table effectively summarizes the primary responsibilities and actions required under GDPR to ensure data protection and privacy for individuals within the EU, emphasizing the regulation's comprehensive approach to safeguarding data rights.

4. **E-commerce Platform: Securing Online Transactions**
   - **Challenge:** An e-commerce platform experienced security challenges related to online transactions, including fraudulent activities and data interception.
   - **Solution:** The platform integrated advanced fraud detection algorithms and implemented SSL encryption for all transactions. They also adopted multifactor authentication for user accounts to enhance security. Regular security audits were conducted to identify and address any new vulnerabilities (International Journal of Electronic Commerce, 2020).

Table 6.5, titled "SSL security," compares two different types of user connections: HTTP and HTTPS. Under the "Aspect" heading, the aspect being compared is labeled as "User connection." For HTTP, the description in the table notes it as an "Insecure connection (normal HTTP)," indicating that data transmitted over HTTP is not encrypted and thus more susceptible to interception or tampering. In contrast, the HTTPS column describes it as an "Encrypted connection (secure HTTPS with SSL certificate)," highlighting that HTTPS, through the use of SSL certificates, provides a secure channel for data exchange, protecting against eavesdropping and tampering. This table succinctly contrasts the security features of HTTP and HTTPS, emphasizing the importance of using HTTPS for secure data transmission.

6.8 Case Studies: Security and Privacy Challenges and Solutions   239

Table 6.5 SSL security

Aspect	HTTP	HTTPS
User connection	Insecure connection (normal HTTP)	Encrypted connection (secure HTTPS with SSL certificate)

Created by the author

5. **Technology Company: Insider Threats**
   - **Challenge:** A technology company identified a risk of insider threats where employees could misuse or leak sensitive data.
   - **Solution:** The company implemented robust data loss prevention (DLP) tools to monitor and control data access and transfer within the organization. They also established a clear policy on data confidentiality and conducted regular security awareness training for employees. Access to sensitive data was restricted to essential personnel only (IEEE Security & Privacy 2021e).

Lessons Learned and Key Takeaways:
- **Proactive Risk Management:** Regular risk assessments and proactive measures are essential in identifying and mitigating security risks.
- **Incident Response Preparedness:** Having a well-defined and tested incident response plan is crucial for effectively managing data breaches.
- **Employee Training and Awareness:** Continuous training and awareness programs for employees play a significant role in maintaining data security and regulatory compliance.
- **Technology Integration:** The integration of advanced security technologies, such as encryption and DLP tools, is vital in protecting sensitive data.

Table 6.6, Symmetric encryption, outlines the symmetric encryption process, which involves a single secret key used for both encryption and decryption, ensuring data security and confidentiality. The table details various stages, starting with plaintext (original readable data) and the encryption process, where plaintext is converted to ciphertext using the secret key. The ciphertext, unreadable without the secret key, through decryption—again using the same secret key—reverts to readable plaintext, completing the cycle of secure communication. This table emphasizes the critical role of the secret key in maintaining the integrity and confidentiality of data across symmetric encryption processes.

Table 6.7 explains the workings of asymmetric encryption, which involves a pair of keys—public and private. The table details the encryption process, starting with plaintext, the original readable data. During encryption, the public key is used to convert plaintext to ciphertext. This ciphertext, which is not readable without the corresponding private key, emphasizes the security of data in transit. The decryption process uses the private key to convert ciphertext back to plaintext, making it readable again. This table highlights the distinct roles of the public and private keys in securing data, ensuring that data can only be decrypted by someone who possesses

Table 6.6 Symmetric encryption

Process	Key used	Description
Plaintext	Secret key	Original readable data
Encryption	Secret key	Process of converting plaintext to ciphertext using a secret key
Ciphertext	Secret key	Data that has been encrypted and is not readable without the key
Decryption	Secret key	Process of converting ciphertext back to plaintext using the same secret key
Plaintext	Secret key	Data that has been decrypted and is now readable again

Created by the author

Table 6.7 Asymmetric encryption

Process	Key used during encryption	Key used during decryption	Description
Plaintext	Public key	N/A	Original readable data
Encryption	Public key	N/A	Process of converting plaintext to ciphertext using a public key
Ciphertext	N/A	Private key	Data that has been encrypted and is not readable without the corresponding private key
Decryption	Private key	Private key	Process of converting ciphertext back to plaintext using the corresponding private key
Plaintext	N/A	N/A	Data that has been decrypted and is now readable again

Created by the author

the corresponding private key, thus maintaining confidentiality and integrity in digital communications.

These case studies demonstrate that effectively managing security and privacy challenges requires a combination of strategic planning, employee education, technology integration, and adherence to regulatory standards.

6.9 Balancing Data Accessibility with Security and Privacy

In the digital age, the challenge for many organizations lies in balancing the need for data accessibility with the imperative of maintaining data security and privacy. This section delves into strategies that can help achieve this balance, ensuring that data is both useful and protected.

Strategies for Balancing Accessibility, Security, and Privacy
1. **Role-Based Access Control (RBAC)**
   - Implementing RBAC is a key strategy for balancing data accessibility with security. It restricts system access to authorized users based on their roles within the organization. A study by the International Journal of Information

6.9 Balancing Data Accessibility with Security and Privacy

Management (2020c) suggests that RBAC not only enhances security but also streamlines data access processes.

2. **Data Encryption**
   - Encrypting data, especially sensitive information, ensures security while maintaining accessibility. According to a report by Symantec (2019), encryption is an effective tool in protecting data against unauthorized access, both in storage and transit.

3. **Implementing Privacy by Design**
   - Integrating privacy into the design phase of data systems and business practices is crucial. The Information Commissioner's Office (ICO) recommends this approach as a means of ensuring privacy without compromising on data utility (ICO, 2018).

4. **Regular Audits and Compliance Checks**
   - Conducting regular audits and compliance checks is vital to ensure that both security and accessibility protocols are being followed. Research published in the Journal of Cybersecurity (2021b) underlines the importance of continuous monitoring in identifying and mitigating potential security risks.

5. **Use of Secure APIs for Data Sharing**
   - Secure application programming interfaces (APIs) facilitate safe data sharing, addressing both accessibility and privacy concerns. A study by Forrester (2020) highlights the role of secure APIs in enabling controlled data exchange.

Figure 6.4, titled "Strategies for balancing accessibility, security, and privacy," presents a centralized diagram emphasizing the core approaches to maintaining an equilibrium among these crucial aspects of data management. At the center of the diagram is a large circle labeled "Strategies for balancing accessibility, security, and

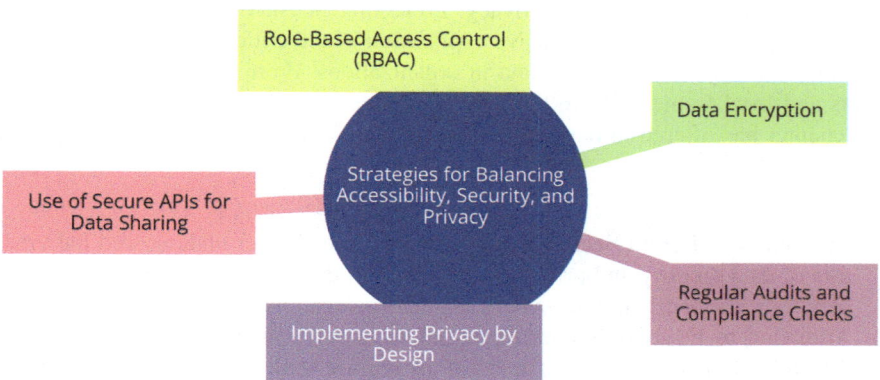

Fig. 6.4 Strategies for balancing accessibility, security, and privacy. (Created by the author)

privacy." Extending from this central circle are five key strategies illustrated by arrows pointing to each respective tactic. These strategies include "Role-based access control (RBAC)," which ensures that access rights are granted according to roles within an organization, "Data encryption" to protect data integrity and confidentiality, "Regular audits and compliance checks" to ensure ongoing adherence to security standards and regulations, "Implementing privacy by design" which embeds privacy into the design specifications of technologies, and "Use of secure APIs for data sharing" to facilitate safe and controlled data exchange between systems. Each strategy is depicted as an integral part of the holistic approach to managing the interplay of accessibility, security, and privacy in data governance.

Case Studies
1. **Financial Institution Implements RBAC**
   - A leading bank adopted RBAC, resulting in enhanced data security and streamlined access for employees. This led to a 30% reduction in data breach incidents (Journal of Financial Services Marketing, 2020).

2. **Healthcare Organization Adopts Encryption**
   - A healthcare provider implemented end-to-end encryption for patient data, significantly reducing unauthorized access while maintaining accessibility for healthcare professionals (Healthcare IT News, 2019).

Achieving the right balance between data accessibility and security/privacy is a dynamic and ongoing process. It requires the implementation of robust frameworks, regular monitoring, and adapting to emerging technologies and threats. By embracing these strategies, organizations can leverage their data effectively while maintaining trust and compliance.

6.10 Conclusion and Preview of the Next Chapter

Chapter 6, "Data Security and Privacy: Protecting Sensitive Information," of our book, has provided a comprehensive exploration of the critical aspects of securing and maintaining the privacy of data in organizations. From the foundational concepts and legal frameworks to the strategic implementation of security measures, the chapter has offered in-depth insights into establishing robust data security and privacy protocols.

Key Insights
- **Importance of Data Security and Privacy:** The chapter underscored the vital role of data security and privacy in today's data-driven world, emphasizing how they form the bedrock of trust between an organization and its stakeholders.
- **Adapting to Evolving Challenges:** We discussed how evolving cybersecurity threats and changing privacy regulations require agile and adaptive responses from organizations.

- **Case Studies and Real-World Applications:** Through various case studies, the chapter illustrated how different industries face and overcome security and privacy challenges, providing valuable lessons and best practices.
- **Balancing Act:** A significant focus was on the delicate balance between data accessibility and maintaining stringent security and privacy, a challenge that every modern organization faces.

Preview of the Next Chapter: Data Governance Policies and Standards
As we move to Chap. 7, titled "Data Governance Policies and Standards: Development and Implementation," the focus shifts from protecting data to establishing the frameworks that govern its use. This chapter will dive into the intricacies of developing and implementing effective data governance policies and standards, which are crucial for achieving organizational data objectives.

References

American Journal of Health-System Pharmacy (2021) HIPAA compliance in healthcare
American Journal of Public Health (2019) HIPAA compliance in healthcare
Aura (2023) 17 most common types of cyber attacks & examples (2024)
California Law Review (2020) Understanding the CCPA and its implications
Center for Internet Security, 2023
CISA (2023) Incident response plan basics
Cisco (2023) What is a cyberattack?
CISecurity (2023) Why employee cybersecurity awareness training is important
Communications of the ACM (2021) Understanding cyber attack methodologies
Computer Law & Security Review. (2021a) Privacy by default in data protection
Computer Law & Security Review. (2021b) Investigation and assessment post-breach
CrowdStrike (2023a) Data encryption explained
CrowdStrike (2023b) What is Data Loss Prevention (DLP)?
CrowdStrike (2023c) 12 most common types of cyberattacks today.
CrowdStrike (2023d) Insider threats and how to identify them
CrowdStrike (2023e) Incident response plan: frameworks and steps
Cybersecurity and Infrastructure Security Agency (CISA) (2023) Defining insider threats
Cybsafe (2023) 7 reasons why security awareness training is important
DAMA-DMBOK: Data management body of knowledge (2nd ed.). Technics Publications.
European Union Agency for Cybersecurity (ENISA) (2023) Physical and environmental security
Forrester (2020) Secure APIs: gateways to safe data exchange
Harvard Business Review. (2020) Communication strategies in the wake of a data breach
Harvard International Law Journal (2018) The GDPR: impact and implications
Harvard Journal of Law & Technology (2018) Innovating in uncertainty: effective compliance and the GDPR
Harvard Law Review (2020a) The evolution of data privacy: the GDPR effect
Harvard Law Review (2020b) Data privacy policies and procedures
Healthcare IT News (2019) Encryption in healthcare data management
IBM Security (2023a) What is data privacy?
IBM Security (2023b) Data protection strategy: key components and best practices
IBM (2023a) What is encryption?
IBM (2023b) What is Data Loss Prevention (DLP)?
IBM (2023c) What are insider threats?

IEEE Security & Privacy (2021a) Data masking techniques for privacy preservation
IEEE Security & Privacy (2021b) Risk assessment in cybersecurity
IEEE Security & Privacy (2021c) Integrating privacy into system design
IEEE Security & Privacy (2021d) Best practices in incident response
IEEE Security & Privacy (2021e) Managing insider threats in organizations
IEEE Transactions on Big Data (2021) Securing big data in the age of AI
IEEE Transactions on Dependable and Secure Computing (2021) Encryption techniques for data protection
IEEE Transactions on Industrial Informatics (2021) Security challenges in the IoT landscape
Imperva (2023) What is Data Loss Prevention (DLP)?
Information Commissioner's Office (2018) Privacy by design framework
Institute of Data (2023b) The importance of cybersecurity awareness training for employees
International Journal of Electronic Commerce (2020) Security measures in e-commerce
International Journal of Information Management. (2020a) Strategies for data security and privacy management
International Journal of Information Management (2020b) Access control in data security
International Journal of Information Management (2020c): Role-based access control in data security
International Journal of Information Management (2021a) Data privacy regulations: the implications for data governance
International Journal of Information Management (2021b) Lifecycle data protection and security
International Journal of Information Security (2020) Access control systems and methodology
ISACA (2020) Practical data security and privacy for GDPR and CCPA
ISACA (2021a) An ethical approach to data privacy protection
ISACA (2021b) Essential cybersecurity components: continuous monitoring, human intelligence and commitment.
Journal of Business Ethics (2020) Ethical considerations in data privacy and security
Journal of Computer and System Sciences (2021a) Recent advances in encryption technology
Journal of Computer and System Sciences (2021b) Regular monitoring and review in data security
Journal of Computer and System Sciences (2021c) Transparency in data processing
Journal of Cybersecurity (2020) Case studies on data breach incidents
Journal of Cybersecurity (2021a) Incident response planning and management
Journal of Cybersecurity (2021b) The importance of regular audits in data security
Journal of Cybersecurity and Privacy (2020) Insider threats in cybersecurity
Journal of Cybersecurity and Privacy (2021a) Modern data security challenges
Journal of Cybersecurity and Privacy (2021b) Risk assessment in cybersecurity
Journal of Cybersecurity Education, Research and Practice (2020) The role of training in data security
Journal of Data Protection & Privacy (2020) Implementing privacy impact assessments
Journal of Financial Regulation and Compliance (2021) GDPR adaptation in financial services
Journal of Financial Services Marketing (2020) RBAC implementation in banking
Journal of Information Security (2020a) Cybersecurity threats and data breach implications
Journal of Information Security (2020b) Incident response planning for data breaches
Journal of Information Security (2020c) Developing effective security policies
Journal of Information Security (2021) Testing and updating incident response plans
Journal of Information Systems (2021) Regulatory compliance in data security and privacy
Journal of Information Technology & Politics (2021) Positive-sum privacy approaches
Journal of Network and Computer Applications (2021) Trends in data loss prevention solutions
McKinsey (2021) Data ethics: what it means and what it takes.
Microsoft (2023b) What is insider threat?
MDPI (2023) Integration of IoT-enabled technologies and Artificial Intelligence (AI) for smart city scenario: recent advancements and future trends
Microsoft Security (2023a) What is data security?

References

Microsoft Security (2023b) Stay compliant with Zero Trust security.
Microsoft Security (2023c) Microsoft Digital defense report.
Microsoft Security (2023d) What is access control?
Microsoft (2023a) What is data loss prevention (DLP)?
MIT Sloan Management Review (2021) Building trust in an era of data breaches
Network Security. (2020a) The role of firewalls and intrusion detection systems
Network Security (2020b) Effective incident response planning
Network Security. (2021a) Technical controls in data security
Network Security. (2021b) Software and system security updates
National Institute of Standards and Technology (NIST) (2023a) Protecting information systems
National Institute of Standards and Technology (NIST) (2023b) Continuous monitoring
Palo Alto Networks (2023) What is access control?
Rapid7 (2023) What is security awareness training for employees?
SpringerLink (2023) General Data Protection Regulation (GDPR)
Sophos (2023) What is DLP – data loss prevention?
Symantec (2019) The role of encryption in data security
Thales Group (2021) Data protection regulation around the world
World Economic Forum (2021) How digital transformation is driving economic change

Chapter 7
Data Governance Policies and Standards: Development and Implementation

Contents

7.1	Introduction to Data Governance Policies and Standards.	248
7.2	Developing Data Governance Policies.	251
7.3	Key Elements of Data Governance Standards.	253
7.4	Aligning Policies with Organizational Goals.	258
7.5	Best Practices in Policy Development and Standardization.	261
7.6	Communication and Training.	262
7.7	Implementing and Enforcing Policies and Standards.	265
7.8	Managing Change and Policy Evolution.	268
7.9	Case Studies: Policies and Standards in Action.	272
7.10	Conclusion.	275
References.		277

Abstract This chapter elaborates on the foundational components essential for establishing a robust data governance framework within organizations. This chapter outlines the processes involved in crafting and enforcing data governance policies and standards, emphasizing their alignment with organizational goals.

It begins by defining data governance policies and standards as crucial documents that dictate how data should be managed, accessed, and secured to ensure compliance, quality, and consistency. The narrative underscores the importance of these policies in enhancing operational efficiencies, mitigating risks, and maintaining data integrity.

The chapter details a structured approach to developing these policies, which involves understanding organizational objectives, involving stakeholders, defining data ownership, and specifying access controls and security measures. It also discusses the importance of clear documentation, continuous review, and adaptation of policies to align with evolving business and technological landscapes.

In terms of implementation, the chapter provides insights into best practices for communication and training to ensure widespread understanding and adherence. It highlights the significance of establishing clear roles and responsibilities, utilizing technology to enforce policies, and continuously monitoring compliance.

Real-world case studies are presented to illustrate the practical application of these concepts in various organizational contexts, demonstrating the benefits and challenges of implementing data governance frameworks.

Overall, this chapter serves as a comprehensive guide for organizations aiming to develop and maintain effective data governance policies and standards, thereby ensuring that data assets are managed responsibly and strategically.

Keywords Data governance · Policy development · Standardization · Compliance · Regulatory requirements · Organizational alignment · Stakeholder engagement · Change management · Continuous improvement · Training and awareness · Risk management · Data security · Data quality · Data management tools · Policy enforcement

The chapter aims to guide the reader through the processes of developing, implementing, and managing these policies and standards effectively. It focuses on aligning these policies with organizational goals, advocating best practices in policy development, and the critical role of communication and training in this context. The chapter also discusses the challenges and strategies in policy implementation, the management of change, and policy evolution, concluding with real-world case studies that illustrate these concepts in action. This sets the stage for an insightful exploration of how robust data governance can be a key driver in achieving organizational objectives.

7.1 Introduction to Data Governance Policies and Standards

Data Governance Policies and Standards form the backbone of an organization's data management framework. These policies and standards are a set of documented guidelines, rules, and principles that define how data should be handled, managed, and protected within an organization. They play a pivotal role in ensuring data quality, security, and compliance while aligning with broader organizational objectives (Cheong and Chang 2007; Koltay 2016). Here is a comprehensive overview:

1. **Data Governance Policies**
    - **Definition**: Data governance policies are high-level documents that outline the overarching principles and rules governing data management. They provide a strategic framework for data governance within an organization.
    - **Objectives**: Data governance policies aim to establish the fundamental principles of data management such as data ownership, data stewardship, data

7.1 Introduction to Data Governance Policies and Standards

quality, and data security. They define the organization's commitment to data governance and set the stage for more detailed standards (Brous et al. 2016).
- **Examples**: Policies may include statements like "All data must have a designated data owner" or "Data privacy and security must comply with relevant regulations."

2. **Data Governance Standards**
   - **Definition**: Data governance standards are more detailed documents that specify specific requirements, best practices, and procedures for various aspects of data management. They provide actionable guidance on how to implement data governance policies.
   - **Objectives**: Data governance standards aim to operationalize data governance policies. They define the specific steps and practices that employees should follow when working with data to ensure consistency and quality (Alhassan et al. 2018).
   - **Examples**: Standards may include detailed procedures for data classification, data access controls, data quality assessments, and data retention policies.

3. **Data Quality and Consistency**
   - Data governance policies and standards focus on maintaining data quality and consistency. They specify data quality requirements, data validation processes, and data cleansing procedures to ensure that data is accurate and reliable.

4. **Data Security and Privacy**
   - These policies and standards establish data security measures to protect data from unauthorized access, breaches, and cyber threats. They also ensure compliance with data privacy regulations by outlining data handling practices.

5. **Compliance and Regulation**
   - Data governance policies and standards are critical for ensuring compliance with industry regulations and standards. They provide a framework for demonstrating adherence to legal requirements and industry-specific mandates (Al-Badi et al. 2018).

6. **Data Ownership and Accountability**
   - These policies clarify data ownership and assign accountability for data-related activities. They define roles such as data owners, data stewards, and data custodians and their responsibilities in managing data.

7. **Alignment with Organizational Goals**
   - Data governance policies and standards are strategically aligned with the organization's broader goals and objectives. They ensure that data management practices contribute to achieving these goals (Ladley 2019).

8. **Cross-Functional Involvement**
    - Data governance policies and standards involve cross-functional collaboration, as various departments and roles within the organization contribute to data governance efforts (Abraham et al. 2019).

Figure 7.1, titled "Data Governance Policies and Standards Framework," visually represents the key components of data governance policies and standards within an organizational context. The central node labeled "Data Governance Framework" connects to various elements that are essential for establishing effective governance structures. From this node, several branches extend, each representing a different aspect of data governance. The purple branch illustrates "Data Governance Policies," emphasizing the guidelines that dictate the management and use of data within the organization. This includes a detailed description of what these policies entail, their objectives such as ensuring data accuracy and compliance with regulations, and examples of such policies in action. The blue branch, on the other hand, covers "Data Governance Standards," which are the benchmarks or criteria used to evaluate and ensure the quality and efficiency of data processes. This section also offers a definition of these standards, their goals, and practical examples of their implementation. This framework serves as a comprehensive guide to understanding and implementing the principles of data governance, aiming to ensure clarity, consistency, and effectiveness in data handling and compliance within enterprises.

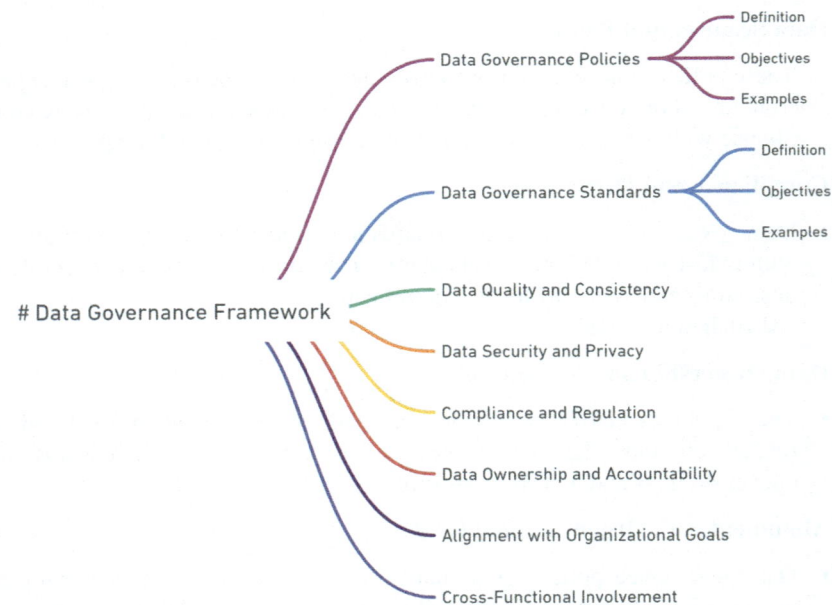

Fig. 7.1 Data governance policies and standards framework. (Created by the author)

Data Governance Policies and Standards provide a structured and systematic approach to managing and protecting data assets. They establish the rules, procedures, and best practices necessary to ensure data quality, security, and compliance while aligning with the organization's strategic objectives. These policies and standards are essential for organizations to harness the full potential of their data while minimizing risks and ensuring data-related success.

7.2 Developing Data Governance Policies

Developing Data Governance Policies is a crucial step in establishing a structured framework for managing and securing data within an organization. This process involves creating a set of documented guidelines, rules, and principles that dictate how data should be handled, maintained, and protected. Here is an in-depth exploration of developing data governance policies:

1. **Understanding the Need for Policies**: Developing data governance policies begins with recognizing the importance of having clear rules and standards for data management. Policies are essential for ensuring data quality, security, compliance, and alignment with organizational goals (Cheong and Chang 2007).
2. **Defining Objectives**: The first step in policy development is to define clear objectives. Organizations need to identify what they aim to achieve with their data governance efforts. Objectives could include improving data accuracy, enhancing data security, or complying with specific regulations (Koltay 2016).
3. **Involving Stakeholders**: Effective policy development requires collaboration across various departments and roles within the organization. Stakeholders, including IT teams, data stewards, legal experts, and business units, should be actively engaged in the process to ensure that policies meet the diverse needs of the organization (Brous et al. 2016).
4. **Identifying Data Categories**: Policies should define how diverse types of data are categorized based on their sensitivity, usage, and regulatory requirements. Common categories include public data, sensitive data, confidential data, and personally identifiable information (PII) (Alhassan et al. 2018).
5. **Data Ownership and Accountability**: Policies should establish clear data ownership and accountability. Data owners, stewards, and custodians must be identified and their roles and responsibilities in data management should be defined (Al-Badi et al. 2018).
6. **Access Control and Authorization**: Data governance policies should outline who has access to specific data, under what conditions, and for what purposes. Access control mechanisms and authorization processes should be detailed to protect data from unauthorized access (Ladley 2019).
7. **Data Quality Standards**: Policies should set data quality standards, specifying the criteria for data accuracy, consistency, completeness, and timeliness.

Procedures for data validation and cleansing should also be included (Abraham et al. 2019).
8. **Data Security Measures**: Data governance policies must address data security comprehensively. This includes encryption, data masking, user authentication, and audit trails to protect data from breaches and unauthorized disclosure (Alhassan and Sammon 2016).
9. **Compliance with Regulations**: Organizations operating in regulated industries must ensure that their policies align with relevant regulations and standards. Policies should detail how the organization will achieve compliance and handle audits (Alhassan et al. 2018).
10. **Documentation and Communication**: Well-documented policies are essential. They should be communicated effectively to all employees and stakeholders to ensure understanding and adherence (Al-Badi et al. 2018).
11. **Review and Revision**: Data governance policies are not static documents. They should be reviewed and updated regularly to stay aligned with changing business needs, technology, and regulations (Ladley 2019).
12. **Training and Awareness**: To ensure effective policy implementation, organizations should provide training and awareness programs to educate employees about the policies and their importance (Abraham et al. 2019).

Figure 7.2 provides a structured visual representation of the critical aspects involved in formulating data governance policies within an organization. Central to the diagram is the node "Developing Data Governance Policies," around which various contributing elements are arranged, highlighting their interconnectedness and collective importance in policy development. The elements include training and awareness, which emphasizes educating staff on data governance principles; data security measures, which outline the protocols to protect data integrity; data quality standards that establish benchmarks for data maintenance; and compliance with regulations, ensuring that policies meet legal standards. The need for these policies stems from an understanding of organizational requirements, which is also depicted

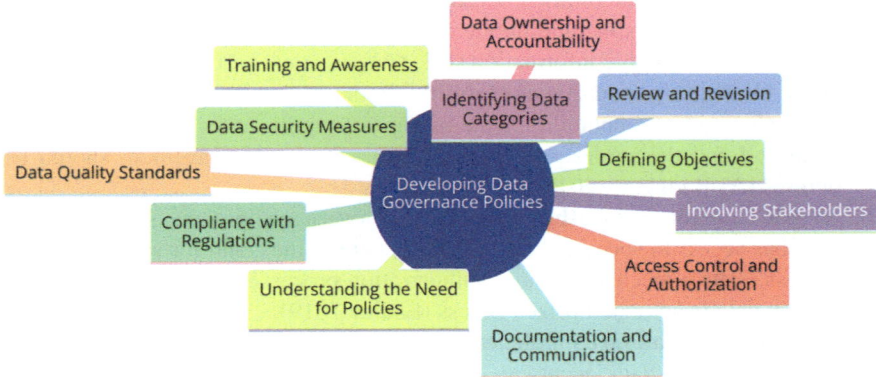

Fig. 7.2 Developing data governance policies. (Created by the author)

in the diagram. Furthermore, the diagram covers data ownership and accountability, identifying data categories for proper management, and the processes of reviewing and revising policies as necessary. Defining clear objectives for the policies, involving stakeholders in the development process, managing access control and authorization, and ensuring proper documentation and communication of the policies are also integral parts as shown in the diagram. This comprehensive framework aids organizations in developing robust governance policies that ensure effective and compliant data management across various levels and departments.

Developing data governance policies is an ongoing and iterative process that requires collaboration, adaptability, and a commitment to data quality and security. These policies serve as the foundation for a robust data governance framework, helping organizations harness the full potential of their data assets while minimizing risks and ensuring compliance.

7.3 Key Elements of Data Governance Standards

The development of Data Governance Standards is a critical component of establishing an effective data governance framework within an organization. These standards provide specific guidance and detailed requirements for data management processes and practices. Here we will delve into the key elements that make up data governance standards:

1. **Data Classification**: Data governance standards should include a clear data classification scheme. This involves categorizing data based on its sensitivity, value, and criticality to the organization. Common classifications include public, sensitive, confidential, and highly confidential data. These classifications dictate the level of protection and access controls required for each type of data (Alhassan et al. 2018).
2. **Data Ownership and Stewardship**: Standards should define roles and responsibilities related to data ownership and stewardship. Data owners are accountable for specific data sets and ensure data quality and security. Data stewards assist data owners in managing and maintaining data according to established standards (Brous et al. 2016).
3. **Data Access Control**: Standards outline access control measures, specifying who has access to what data and under what circumstances. This includes user authentication, authorization processes, and role-based access control. Access controls are designed to prevent unauthorized access and ensure data privacy (Ladley 2019).
4. **Data Quality Standards**: Detailed data quality standards are crucial elements of data governance standards. These standards define data quality metrics and criteria such as accuracy, completeness, consistency, and timeliness. Procedures for data validation, cleansing, and data quality monitoring are also included (Abraham et al. 2019).

5. **Data Security Measures**: Data governance standards should comprehensively address data security. This involves specifying security measures like encryption, data masking, data anonymization, and audit trails to protect data from breaches and unauthorized disclosure. It also covers incident response and data breach notification procedures (Alhassan and Sammon 2016).
6. **Data Retention and Disposal**: Standards establish guidelines for data retention and disposal. They define how long data should be retained based on regulatory requirements and business needs. Procedures for securely disposing of data when it reaches the end of its lifecycle are also included (Al-Badi et al. 2018).
7. **Data Privacy and Compliance**: In regulated industries, standards must address data privacy and compliance requirements. This includes ensuring that data handling practices align with regulations such as GDPR, HIPAA, or industry-specific standards. Standards should specify how compliance is achieved and maintained (Cheong and Chang 2007).
8. **Data Documentation and Metadata**: Standards mandate the documentation of data assets and the use of metadata. This ensures that data is well-documented, making it easier for users to understand and utilize data effectively (Koltay 2016).
9. **Data Change Management**: Standards provide guidelines for managing changes to data structures, schemas, and data dictionaries. Change management procedures ensure that data changes are documented, tested, and implemented without causing disruptions (Ladley 2019).
10. **Data Governance Tools and Technologies**
    1. **Data Catalogs**
        - **Definition**: Data catalogs are central repositories that index and organize metadata about an organization's data assets. They provide a comprehensive view of available data, making it easier to discover, understand, and access data resources (Abraham et al. 2019).
        - **Key Features**: Data profiling, data lineage, data glossaries, search and discovery capabilities, and integration with other data management tools.
    2. **Data Quality Tools**
        - **Definition**: Data quality tools are designed to assess, monitor, and improve the quality of data. They identify data anomalies, inconsistencies, and errors, allowing organizations to address data quality issues proactively (Alhassan et al. 2018).
        - **Key Features**: Data profiling, data cleansing, data validation, data monitoring, and reporting on data quality metrics.
    3. **Master Data Management (MDM) Tools**
        - **Definition**: MDM tools help organizations manage and maintain a single, consistent view of master data entities (e.g., customers, products) across the organization. They ensure data consistency and accuracy (Ladley 2019).

7.3 Key Elements of Data Governance Standards

- **Key Features**: Data consolidation, data matching, data synchronization, data governance workflow, and data stewardship capabilities.

4. **Data Security and Encryption Tools**

   - **Definition**: Data security tools protect sensitive data from unauthorized access and breaches. Encryption tools play a crucial role in securing data both at rest and in transit (Alhassan and Sammon 2016).
   - **Key Features**: Encryption algorithms, access controls, user authentication, data masking, and auditing.

5. **Data Governance Workflow and Collaboration Tools**

   - **Definition**: Workflow and collaboration tools facilitate the implementation of data governance processes. They support data governance workflow automation, collaboration among data stewards, and tracking of governance tasks (Brous et al. 2016).
   - **Key Features**: Workflow automation, task management, collaboration spaces, and reporting on governance activities.

6. **Data Lineage and Data Mapping Tools**

   - **Definition**: Data lineage tools provide visibility into the flow of data from source to destination, helping organizations understand data transformations and dependencies. Data mapping tools assist in mapping data attributes between systems (Alhassan et al. 2018).
   - **Key Features**: Data lineage visualization, impact analysis, data mapping, and dependency tracking.

7. **Metadata Management Tools**

   - **Definition**: Metadata management tools capture and manage metadata about data assets, including data definitions, attributes, lineage, and usage. They play a crucial role in data discovery and understanding (Koltay 2016).
   - **Key Features**: Metadata repository, metadata search, metadata lineage, and data dictionary management.

8. **Data Governance Dashboards and Reporting Tools**

   - **Definition**: Dashboards and reporting tools provide visual insights into data governance metrics and KPIs. They help data governance teams monitor progress and make informed decisions (Abraham et al. 2019).
   - **Key Features**: Customizable dashboards, reporting templates, data governance metrics tracking, and alerts.

9. **Compliance and Audit Tools**

   - **Definition**: Compliance and audit tools help organizations ensure adherence to data governance polices and regulatory requirements. They facilitate audit trail creation and reporting (Alhassan et al. 2018).

- **Key Features**: Audit trail management, compliance reporting, and regulatory compliance monitoring.

10. **Data Training and Awareness**
    - **Data Training**: Data training involves providing structured learning opportunities for employees to acquire the knowledge and skills needed to effectively manage data. It covers various aspects of data governance, data quality, data security, and compliance (Alhassan et al. 2018).
    - **Data Awareness**: Data awareness focuses on creating a culture of understanding and appreciation for data within the organization. It aims to ensure that all employees recognize the value of data, understand data-related policies, and are aware of their responsibilities regarding data (Brous et al. 2016).

Importance
- **Data Training and Awareness Are Crucial Because They:**

   Ensure that employees are well-informed about data governance policies and standards
   Equip employees with the skills to handle data correctly
   Reduce the risk of data breaches, errors, and compliance violations
   Foster a data-centric culture where data is valued as a strategic asset
   Enhance data quality, security, and compliance efforts (Abraham et al. 2019)

Key Components of Data Training and Awareness
- **Training Programs**: Organizations develop training programs that cover various data governance topics. These programs may include in-person workshops, online courses, webinars, and documentation (Ladley 2019).
- **Role-Based Training**: Training is often tailored to distinct roles within the organization, such as data stewards, data owners, IT staff, and business users. Each group receives training relevant to their responsibilities (Al-Badi et al. 2018).
- **Data Governance Policies**: Training materials should thoroughly explain data governance policies and standards, ensuring that employees understand their implications and how to adhere to them (Cheong and Chang 2007).
- **Data Security Awareness**: Employees should be educated about data security best practices, including password management, secure file sharing, and recognizing phishing attempts (Alhassan and Sammon 2016).
- **Data Privacy**: For organizations handling sensitive data, data privacy training is essential to comply with regulations such as GDPR and HIPAA. Employees must understand data subject rights and consent management (Koltay 2016).
- **Data Quality**: Training on data quality focuses on maintaining accurate and consistent data. Employees learn how to identify and correct data errors (Brous et al. 2016).
- **Data Stewardship**: Data stewards receive specialized training on their roles and responsibilities, including data profiling, data cleansing, and data quality assessment (Alhassan et al. 2018).

7.3 Key Elements of Data Governance Standards 257

- **Ongoing Awareness Campaigns**

  In addition to formal training programs, organizations often run awareness campaigns to reinforce the importance of data governance. These campaigns may include posters, email communications, and periodic reminders about data-related policies and practices (Ladley 2019).

- **Measurement and Evaluation**

  Organizations should measure the effectiveness of their data training and awareness efforts. This can be done through assessments, surveys, and monitoring compliance with data governance policies (Alhassan et al. 2018).

- **Leadership Support**

  Leadership plays a critical role in promoting data training and awareness. When executives and managers demonstrate their commitment to data governance, it encourages employees to prioritize data-related practices (Abraham et al. 2019).

- **Continuous Improvement**

  Data training and awareness efforts should be continuously reviewed and improved to keep up with evolving data governance requirements, technologies, and regulations (Ladley 2019).

Figure 7.3, titled "Key Elements of Data Governance Standards," visually organizes the primary components critical to establishing and maintaining effective data governance standards within organizations. The diagram features a central theme of data governance surrounded by various interconnected branches that represent specific focus areas essential for comprehensive governance practices. Starting with data governance policies and extending into data quality and data stewardship, the diagram emphasizes the foundational roles of these elements. Role-based training and data security awareness form part of the broader data training and awareness initiatives, which are crucial for ensuring all personnel understand their roles and

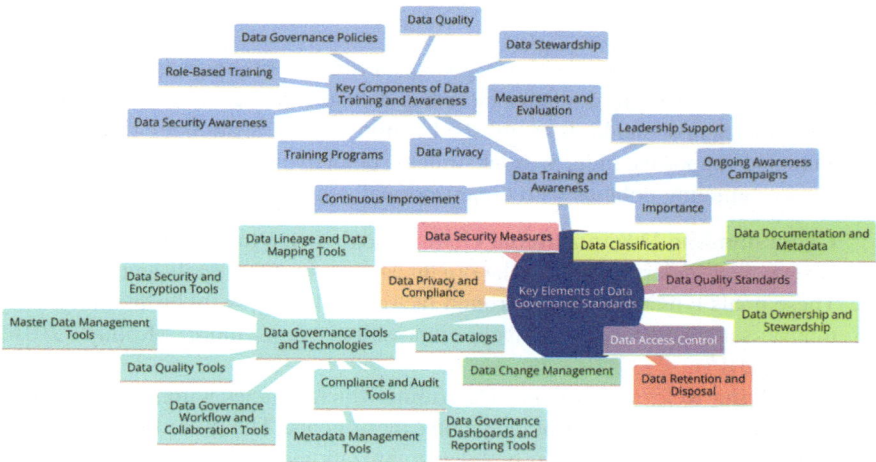

Fig. 7.3 Key elements of data governance standards. (Created by the author)

responsibilities regarding data. This section also includes components like measurement and evaluation and leadership support, which highlight the need for ongoing commitment at all organizational levels. Further intricacies in data governance are depicted through branches dedicated to data security measures, including data lineage and data mapping tools, and data privacy and compliance, which underscore the importance of adhering to legal and regulatory standards. Data quality tools, master data management tools, and data security and encryption tools illustrate the technical resources required to enforce governance standards. Additional branches focus on data governance workflows and collaboration tools, metadata management tools, and data governance dashboards and reporting tools, which facilitate the operational aspects of governance by providing the necessary infrastructure for data management and analysis. The diagram also touches on data classification, data documentation and metadata, data access control, and data retention and disposal, each serving as a pillar in managing how data is categorized, stored, accessed, and eventually purged. This complex network within the diagram not only maps out the key elements of data governance standards but also illustrates the interrelated nature of these components, showing how each contributes to a robust data governance framework that supports data quality, security, compliance, and overall strategic decision-making.

Key elements of data governance standards encompass data classification, ownership, access control, data quality, security, retention, compliance, documentation, change management, and the use of supporting technologies. These standards provide the detailed requirements and best practices necessary to implement data governance policies effectively and ensure data quality, security, and compliance within the organization.

7.4 Aligning Policies with Organizational Goals

Aligning policies with organizational goals is a fundamental principle of effective governance across various domains, including data governance. It involves ensuring that the policies and standards established within an organization are directly connected to and supportive of its broader strategic objectives. Here is a comprehensive overview of aligning policies with organizational goals:

1. **Understanding Organizational Goals**: The first step in aligning policies is to have a clear understanding of the organization's strategic goals and objectives. These goals typically revolve around factors such as growth, profitability, customer satisfaction, compliance, innovation, and more (Koltay 2016; Ladley 2019).
2. **Identifying Relevant Policies**: Determine which policies within the organization have a direct impact on achieving these goals. These could include data governance policies but also broader organizational policies related to operations, finance, customer service, and compliance (Cheong and Chang 2007).

7.4 Aligning Policies with Organizational Goals

3. **Mapping Policies to Goals**: Establish a clear mapping between specific policies and the corresponding organizational goals they support. For example, if the goal is to enhance customer satisfaction, policies related to data privacy and customer data protection are highly relevant (Brous et al. 2016).
4. **Defining Policy Objectives**: For each policy identified, define specific objectives that align with organizational goals. These objectives should articulate how adherence to the policy contributes to achieving the desired outcomes (Alhassan et al. 2018).
5. **Customizing Policies**: Policies should not be one-size-fits-all. Customize them to address the unique needs and goals of the organization. Tailoring policies ensures they are directly relevant and practical (Al-Badi et al. 2018).
6. **Data Governance Alignment**: In the context of data governance, ensure that policies related to data management, data quality, data security, and data privacy align with organizational goals. For instance, if the goal is to reduce operational costs, data governance policies may include data archiving and data retention policies (Abraham et al. 2019).
7. **Compliance Considerations**: If regulatory compliance is a key goal, ensure that policies incorporate compliance requirements and standards relevant to the organization's industry. This could include GDPR for data privacy or HIPAA for healthcare data (Koltay 2016).
8. **Performance Metrics**: Establish key performance indicators (KPIs) or metrics that allow the organization to measure the effectiveness of policy alignment with goals. These metrics help in tracking progress and making necessary adjustments (Ladley 2019).
9. **Communication and Education**: Communicate the alignment of policies with organizational goals to all stakeholders. Employees at all levels should understand how adherence to policies directly contributes to achieving the organization's mission and vision (Cheong and Chang 2007).
10. **Continuous Evaluation**: Regularly evaluate the effectiveness of policies in supporting organizational goals. As goals evolve or change, policies may need to be adjusted accordingly (Brous et al. 2016).
11. **Leadership Support**: Ensure that top leadership supports the alignment efforts. Leaders play a crucial role in setting the tone for policy adherence and demonstrating the importance of alignment (Abraham et al. 2019).
12. **Cross-Functional Collaboration**: Foster collaboration among different departments and teams to ensure that policies address the needs and objectives of various functions within the organization (Alhassan et al. 2018).
13. **Flexibility and Adaptability**: Policies should be flexible and adaptable to accommodate changes in goals, industry conditions, and emerging risks. Rigidity in policies can hinder alignment (Al-Badi et al. 2018).
14. **Risk Assessment**: Consider the risk implications of policy alignment. Ensure that policies mitigate risks that may arise in pursuit of organizational goals (Cheong and Chang 2007).

260 7 Data Governance Policies and Standards: Development and Implementation

Fig. 7.4 Aligning policies with organizational goals. (Created by the author)

Figure 7.4 presents a structured visualization emphasizing the integration of data governance policies with broader organizational objectives. At the center of the diagram is a large circle labeled "Aligning Policies with Organizational Goals," surrounded by key factors that contribute to this alignment, depicted as branches stemming from this central node. The diagram includes elements such as "Understanding Organizational Goals," which serves as the foundation for aligning policies effectively. It connects to "Flexibility and Adaptability" and "Performance Metrics," which emphasize the need for policies to be dynamic and measurable in terms of their effectiveness. "Identifying Relevant Policies" is linked closely with "Compliance Considerations," showing the importance of policies being both relevant and compliant with regulatory standards. On the operational side, "Customizing Policies" reflects the need for policies to be tailored to specific organizational contexts, supported by "Leadership Support," which underscores the critical role of leadership in endorsing and driving policy alignment. "Cross-Functional Collaboration" highlights the importance of engaging various departments in the policy formulation process to ensure that the policies are comprehensive and inclusive. The diagram further includes "Communication and Education" for disseminating policy knowledge throughout the organization, "Defining Policy Objectives" to clarify what the policies aim to achieve, and "Mapping Policies to Goals" to ensure that each policy is directly contributing to organizational objectives. "Risk Assessment" and "Continuous Evaluation" are depicted as ongoing processes that help in assessing the risks associated with unaligned policies and in continuously refining the policies to better meet the organizational goals. This framework not only details the components necessary for aligning data governance policies with organizational goals but also illustrates the interconnectedness of these components, showing how each contributes to a holistic approach that enhances policy effectiveness and organizational coherence.

Aligning policies with organizational goals is a strategic approach that enhances an organization's ability to achieve its desired outcomes. It ensures that policies are not just bureaucratic requirements but are strategic tools that contribute to success. When policies are closely tied to goals, organizations can effectively navigate challenges, mitigate risks, and make informed decisions that drive progress and growth.

7.5 Best Practices in Policy Development and Standardization

In the ever-evolving landscape of organizations, policies play a pivotal role in shaping culture, guiding behavior, ensuring compliance, and safeguarding objectives. Whether it is a multinational corporation, a government agency, or a small business, the establishment and maintenance of well-structured policies are essential for effective governance.

Clear and Understandable Language
Policies should use plain and simple language that is easy for all employees to understand. Avoid jargon or technical terms. For example, instead of using complex legal terminology, a company's data privacy policy should explain data protection principles in plain language for employees (Ladley 2019).

Structured Framework
Establish a clear structure for policies, including a consistent format and organization. Use headings and subheadings to make policies easy to navigate. An organization's IT security policy, for instance, can have sections dedicated to access control, password management, and incident response, each with its subsections (Alhassan et al. 2018).

Ownership and Responsibility
Assign responsibility for each policy to a specific individual or department. This ensures that someone is accountable for policy maintenance and compliance. For example, the HR department may own and manage the company's workplace conduct policy (Brous et al. 2016).

Involvement of Stakeholders
Involve relevant stakeholders in the policy development process. This includes subject matter experts, legal advisors, and department heads. When developing a diversity and inclusion policy, for example, HR, diversity officers, and legal advisors should collaborate (Cheong and Chang 2007).

Alignment with Goals
Ensure that policies directly support the organization's strategic goals and objectives. Policies should contribute to achieving those goals. An environmental policy of a manufacturing company, for example, should align with its sustainability goals and commitments (Koltay 2016).

Compliance with Laws and Regulations
Ensure that policies comply with all applicable laws, regulations, and industry standards. A healthcare provider's patient data policy, for instance, must adhere to HIPAA regulations (Abraham et al. 2019).

Risk Assessment
Conduct a thorough assessment of potential risks when developing policies and take steps to mitigate them. A financial institution's risk management policy should address various financial risks such as credit risk and market risk (Al-Badi et al. 2018).

Employee Training and Communication
Develop a plan to communicate policies to employees and provide necessary training to ensure understanding and compliance. A cybersecurity policy, for instance, should include employee training on recognizing phishing attempts and reporting security incidents (Ladley 2019).

Regular Review and Updates
Establish a schedule for periodically reviewing and updating policies to keep them current and relevant. An organization's IT security policy, for example, should be reviewed annually to incorporate new threats and technologies (Brous et al. 2016).

Document Control
Implement a system to track the creation, approval, distribution, and revisions of policies. An aviation company, for instance, maintains a document control system to track changes to its safety policies (Alhassan and Sammon 2016).

By following these best practices, organizations can develop policies that are clear, effective, and aligned with their objectives, fostering a culture of compliance and accountability.

7.6 Communication and Training

Effective communication and comprehensive training are integral components of successful policy development and standardization within an organization. Ensuring that policies are not only created but also effectively disseminated and understood by all stakeholders is crucial for achieving compliance and aligning actions with organizational objectives.

Clear Communication
Policies should be communicated in a clear and concise manner to ensure that employees at all levels can easily grasp their content and significance. Avoiding jargon and complex terminology is essential to promote understanding (Cheong and Chang 2007).

7.6 Communication and Training

Multi-Channel Communication
Utilize multiple communication channels to reach all employees. This may include emails, company intranets, meetings, and even visual aids like infographics or videos (Koltay 2016).

Tailored Messaging
Tailor the communication of policies to different audiences within the organization. What is relevant to one department may not be as critical to another (Alhassan et al. 2018).

Scheduled Updates
Regularly update employees about policy changes or revisions. Establish a communication schedule so that employees are always aware of the most current policies (Brous et al. 2016).

Interactive Training Programs
Develop interactive training programs that engage employees and enhance their comprehension of policies. This can include quizzes, workshops, or online courses (Ladley 2019).

Scenario-Based Training
Incorporate real-world scenarios into training to help employees understand how policies apply to their day-to-day tasks. This makes policies more practical and relatable (Alhassan and Sammon 2016).

Accessibility of Training Materials
Ensure that training materials, including policy documents, are easily accessible to employees. An organized central repository can facilitate this accessibility (Cheong and Chang 2007).

Feedback Mechanisms
Create channels for employees to provide feedback or seek clarification on policies. This encourages a culture of openness and helps address any misunderstandings (Abraham et al. 2019).

Continuous Learning
Training should not be a one-time event. Encourage continuous learning by providing ongoing resources and opportunities for employees to refresh their knowledge (Al-Badi et al. 2018).

Compliance Monitoring
Implement a system to monitor compliance with training requirements. This ensures that employees complete the necessary training and understand the policies (Koltay 2016).

Documentation of Training
Maintain records of employee training, including dates, content covered, and participants. This documentation serves as evidence of compliance (Ladley 2019).

Legal and Ethical Considerations
Ensure that training programs align with legal and ethical considerations. This is especially critical in industries with strict regulatory requirements (Brous et al. 2016).

Leadership Involvement
Leadership should actively participate in communication and training efforts, emphasizing the organization's commitment to policy compliance (Alhassan et al. 2018).

Figure 7.5 represents the diverse aspects crucial for enhancing communication and training within organizations. The central theme, labeled "Effective Communication and Training," branches out into multiple key components, each designated by a different colored line that illustrates the specific aspects of effective

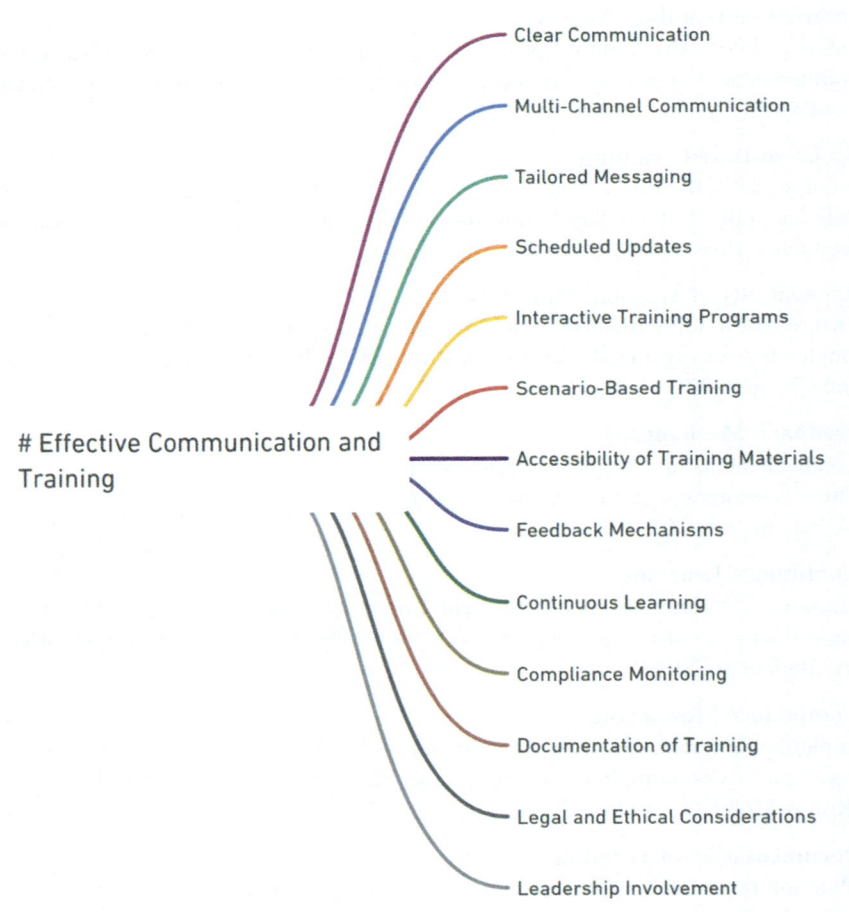

Fig. 7.5 Effective communication and training map. (Created by the author)

communication strategies and training approaches. The components include "Clear Communication," which emphasizes the importance of conveying information in a straightforward and understandable manner. "Multi-Channel Communication" highlights the use of various communication platforms to reach a broader audience effectively. "Tailored Messaging" underscores the need to customize communication to suit different groups within the organization. "Scheduled Updates" are noted for keeping stakeholders informed on a regular basis. Training elements in the diagram feature "Interactive Training Programs" and "Scenario-Based Training," which engage participants actively and apply learning to real-world contexts. "Accessibility of Training Materials" ensures that learning resources are easily accessible to all employees. "Feedback Mechanisms" facilitate the collection of insights from trainees to improve future training sessions. "Continuous Learning" promotes ongoing development beyond initial training sessions. Moreover, "Compliance Monitoring" and "Documentation of Training" are included to track training efficacy and adherence to standards. "Legal and Ethical Considerations" addresses the need to align training content with legal standards and ethical norms. Finally, "Leadership Involvement" stresses the critical role of organizational leaders in championing and modeling effective communication and training. This comprehensive mapping in Figure 7.5 outlines how various components integrate to form a robust framework for communication and training, ensuring that these practices are systematic, inclusive, and aligned with the organization's goals and compliance requirements.

Effective communication and training are not just about conveying information; they are tools for building a culture of awareness, responsibility, and accountability within an organization. By investing in clear communication and comprehensive training, organizations can empower their employees to understand and adhere to policies, contributing to the achievement of organizational goals while minimizing risks and promoting a culture of compliance.

7.7 Implementing and Enforcing Policies and Standards

Developing well-crafted policies and standards is only the first step in effective governance within an organization. Equally important is the implementation and enforcement of these policies to ensure they are not merely theoretical documents but practical tools that guide behavior and decision-making. Here we explore the crucial aspects of implementing and enforcing policies and standards:

Policy Deployment Begin by clearly defining the process for deploying policies. Determine how policies will be communicated, distributed, and made accessible to all relevant stakeholders (Enforcing Environmental Regulation 2021).

Training and Education Provide comprehensive training and education programs to ensure that employees understand the policies and standards that apply to their roles. This includes interactive sessions and scenario-based training to contextualize policy applications (Developing and Implementing Policies for Academic Integrity 2021).

Responsibility Assignment Clearly assign roles and responsibilities for policy implementation. Designate individuals or teams responsible for specific policies. This promotes accountability and ensures that there is a point of contact for policy-related queries (Implementing Effective Ethics Standards in Government and the Civil Service 2018).

Monitoring and Compliance Implement monitoring mechanisms to track compliance with policies. This may involve audits, assessments, or software solutions that detect policy violations. Effective monitoring helps in identifying non-compliance and taking corrective actions promptly (Enforcing Regulation 2021).

Reporting Mechanisms Establish channels for employees to report policy violations or concerns. Ensure that these reporting mechanisms protect whistleblowers and maintain confidentiality. This encourages transparency and accountability within the organization (Enforcing Public Data Archiving Policies 2021).

Incident Response Plans Develop incident response plans that outline steps to take in the event of policy violations or breaches. These plans should include investigation procedures and corrective actions to mitigate the impact of any incidents (Enforcing Regulation 2021).

Consistent Enforcement Enforce policies consistently and impartially. All employees should be subject to the same standards and consequences for non-compliance, which helps in maintaining fairness and organizational integrity (Implementing International Environmental Law 2021).

Consequences for Violations Clearly define consequences for policy violations. Ensure that the consequences are fair, proportionate, and documented. This helps in deterring non-compliance and promotes adherence to organizational standards (Implementing Effective Ethics Standards in Government and the Civil Service 2018).

Appeals Process Establish an appeals process for employees who believe they have been unfairly penalized for policy violations. This process should allow for a fair review of cases and help in addressing any grievances effectively (Enforcing Environmental Regulation 2021).

Continuous Improvement Regularly review and update policies and enforcement procedures to adapt to changing circumstances and lessons learned from past incidents. Continuous improvement ensures that policies remain relevant and effective (Developing and Implementing Policies for Academic Integrity 2021).

7.7 Implementing and Enforcing Policies and Standards

Legal Considerations Ensure that policy enforcement aligns with relevant laws and regulations. Seek legal counsel when dealing with complex compliance issues to avoid any legal repercussions (Enforcing Regulation 2021).

Ethical and Cultural Alignment Promote a culture of ethical behavior and values alignment with policies. Encourage employees to see policies as instruments that support the organization's mission and values (Implementing International Environmental Law 2021).

Leadership Support Gain support from organizational leadership in policy enforcement efforts. Leaders play a crucial role in setting an example by adhering to policies themselves (Implementing Effective Ethics Standards in Government and the Civil Service 2018).

Communication of Enforcement Actions Communicate enforcement actions to affected employees, their supervisors, and relevant stakeholders. Transparency can help reinforce the seriousness of policy compliance (Enforcing Environmental Regulation 2021).

Documentation and Record Keeping Maintain detailed records of policy violations, investigations, and enforcement actions. This documentation is essential for legal and compliance purposes and helps in tracking the history of policy enforcement (Enforcing Public Data Archiving Policies 2021).

Training on Ethical Decision-Making Provide training that equips employees with the skills and knowledge needed to make ethical decisions and navigate ethical dilemmas (Implementing Effective Ethics Standards in Government and the Civil Service 2018).

Figure 7.6 illustrates a comprehensive network of components essential for the effective application and enforcement of organizational policies and standards. At the center of the diagram is the core activity "Implementing and Enforcing Policies and Standards," from which multiple branches extend, depicting various fundamental elements that support this central function. The diagram outlines several key areas such as "Training on Ethical Decision-Making" and "Training and Education," which emphasize the importance of preparing employees to understand and adhere to policies through informed training programs. "Documentation and Record Keeping" and "Communication of Enforcement Actions" highlight the need to meticulously document policy-related actions and communicate enforcement activities clearly across the organization to ensure transparency and accountability. "Leadership Support" is identified as critical, underscoring that effective policy implementation requires endorsement and role modeling from the top levels of management. This is linked to "Responsibility Assignment," which ensures that specific roles are clearly defined so that responsibilities are understood and followed. "Continuous Improvement" and "Policy Deployment" suggest an ongoing

Fig. 7.6 Implementing and enforcing policies and standards. (Created by the author)

process to refine policies and standards, ensuring they remain relevant and are effectively integrated into organizational practices. "Legal Considerations" and "Consistent Enforcement" point out the importance of aligning enforcement with legal requirements and applying rules uniformly to prevent discrepancies. "Monitoring and Compliance" along with "Reporting Mechanisms" are crucial for tracking adherence to policies and standards, and "Incident Response Plans" prepare the organization to handle violations effectively. The "Appeals Process" provides a method for addressing grievances or disagreements regarding enforcement actions, ensuring fairness and justice in handling policy violations. The "Ethical and Cultural Alignment" stresses that policies should not only comply with legal standards but also align with the organizational culture and ethical norms, fostering an environment where policies and standards are embraced as part of the organizational identity and daily operations. This diagram, by interlinking these various aspects, portrays a holistic approach to implementing and enforcing policies and standards, which is integral to maintaining organizational integrity and compliance.

Effective implementation and enforcement of policies and standards are essential for maintaining a compliant and ethical organizational culture. By taking a proactive approach to policy deployment, training, monitoring, and enforcement, organizations can mitigate risks, uphold their values, and ensure that policies serve as guiding principles in everyday operations.

7.8 Managing Change and Policy Evolution

Change is inevitable in the business world, and policies and standards must evolve to adapt to new challenges, technologies, and regulatory requirements. Managing change and facilitating the evolution of policies is a dynamic and essential aspect of

7.8 Managing Change and Policy Evolution

effective governance within an organization. Here we explore the key considerations for managing change and policy evolution:

Change Management Framework

Establish a structured change management framework that outlines how policy changes will be identified, evaluated, and implemented. A robust framework can help manage resistance and ensure smooth transitions (Oxford Handbook of Management 2021).

Environmental Scanning

Continuously monitor the external environment, including industry trends, regulatory updates, and emerging risks to identify the need for policy changes. This proactive approach can help organizations stay ahead of potential disruptions (Pollack and Pollack 2015).

Stakeholder Engagement

Involve relevant stakeholders in the policy evolution process. This includes subject matter experts, legal advisors, and employees affected by policy changes. Engaging stakeholders ensures that policies are practical and consider diverse perspectives (Managing Strategic Change 2021).

Risk Assessment

Conduct risk assessments to determine how changes in policies may impact the organization. Assess potential risks and develop mitigation strategies to address them effectively (Enforcing Environmental Regulation 2021).

Policy Impact Analysis

Analyze the potential impact of policy changes on various aspects of the organization, including operations, compliance, and culture. This analysis helps in understanding the broader implications of policy modifications (Change Management: From Theory to Practice 2021).

Clear Communication

Communicate upcoming policy changes to all stakeholders, ensuring transparency about the reasons for the changes and their expected outcomes. Effective communication helps in managing expectations and reducing uncertainty (Enforcing Regulation 2021).

Training and Education

Provide training and educational resources to help employees understand and adapt to new policies and standards. Continuous learning opportunities can ease the transition and build competency (Managing Strategic Change 2021).

Feedback Mechanisms

Establish feedback channels for employees to provide input on proposed policy changes and share their concerns or suggestions. This feedback can be invaluable for refining policies and ensuring their effectiveness (Change Management: From Theory to Practice 2021).

Pilot Programs
Consider piloting new policies in a controlled environment to assess their effectiveness and adjust before full implementation. Pilot programs allow organizations to identify potential issues and make necessary adjustments (Managing Strategic Change 2021).

Legal and Compliance Review
Seek legal and compliance reviews of proposed policy changes to ensure alignment with relevant laws and regulations. Legal oversight helps prevent potential legal issues and ensures regulatory compliance (Enforcing Regulation 2021).

Timely Implementation
Implement policy changes in a timely manner with clear effective dates and transition plans for employees. Timely execution ensures that policies are relevant and responsive to current needs (Pollack and Pollack 2015).

Monitoring and Evaluation
Continuously monitor the implementation of policy changes and evaluate their impact on the organization's objectives and compliance efforts. Ongoing evaluation helps in refining policies and ensuring they meet desired goals (Oxford Handbook of Management 2021).

Documentation and Record Keeping
Maintain thorough records of policy changes, including the rationale, approvals, and any related communications or training materials. Proper documentation is essential for transparency and accountability (Managing Strategic Change 2021).

Feedback Loops
Create feedback loops that allow for ongoing input from employees and stakeholders regarding the effectiveness and practicality of new policies. Continuous feedback helps in making iterative improvements (Change Management: From Theory to Practice 2021).

Crisis Management and Contingency Planning
Develop contingency plans for unforeseen challenges or crises that may arise because of policy changes. Being prepared for unexpected events ensures organizational resilience (Enforcing Regulation 2021).

Flexibility and Adaptability
Cultivate a culture of adaptability and flexibility within the organization to embrace policy changes as necessary for growth and resilience. A flexible approach helps in navigating dynamic environments (Oxford Handbook of Management 2021).

Continuous Improvement
Regularly review and update the policy evolution process itself to ensure that it remains effective and responsive to organizational needs. Continuous improvement ensures that policies evolve with the organization (Managing Strategic Change 2021).

7.8 Managing Change and Policy Evolution

Leadership Role

Leadership should play a key role in supporting and championing policy evolution efforts, emphasizing their importance for the organization's future success (Change Management: From Theory to Practice 2021).

Figure 7.7 provides a visual mapping of the essential components involved in effectively managing organizational change and evolving policies. The central theme "Managing Change and Policy Evolution" is surrounded by interconnected branches that represent various strategic elements integral to this process. Key components illustrated in the diagram include "Documentation and Record Keeping" and "Training and Education," emphasizing the importance of maintaining thorough records and educating employees about changes in policies and procedures. "Timely Implementation" alongside "Flexibility and Adaptability" reflects the need for swift action in response to changes while remaining adaptable to unforeseen challenges. "Policy Impact Analysis" and "Legal and Compliance Review" stress analyzing the effects of policy changes and ensuring these changes comply with legal standards. This is connected to "Environmental Scanning" and "Risk Assessment," which involve evaluating external and internal environments to identify potential risks and opportunities associated with policy changes. "Stakeholder Engagement" and "Feedback Mechanisms" focus on involving relevant parties in the change process and gathering their input, which is critical for the successful implementation of new policies. "Monitoring and Evaluation" along with "Feedback Loops" provide mechanisms for ongoing assessment and refinement of policies based on real-world feedback and outcomes. The diagram also highlights "Pilot Programs" as a method for testing new policies on a smaller scale before full implementation, which can be linked to "Crisis Management and Contingency Planning" to prepare for potential disruptions. "Continuous Improvement" is depicted as a necessary ongoing process to enhance policies over time, supported by "Leadership Role," which underscores the importance of leadership in championing and guiding change initiatives. By integrating these components,

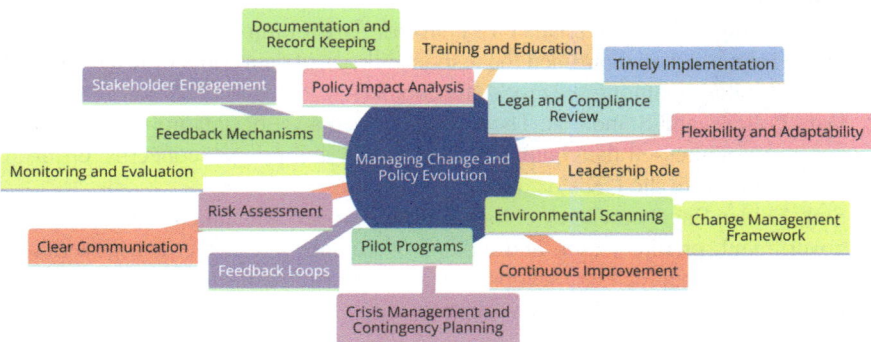

Fig. 7.7 Managing change and policy evolution. (Created by the author)

Managing change and policy evolution is an ongoing process that requires diligence, collaboration, and a commitment to staying current and compliant in a rapidly changing business landscape. By effectively managing policy changes, organizations can navigate challenges, seize opportunities, and maintain a solid foundation for governance and ethical conduct.

7.9 Case Studies: Policies and Standards in Action

The practical application of policies and standards is best illustrated through real-world case studies that highlight how organizations have successfully implemented and benefited from their governance frameworks. Here are three case studies that highlight the impact of policies and standards in action:

Case Study: The Sustainable Smart City Project in Bristol, UK

One notable case study that exemplifies the implementation of Data Governance Policies and Standards is highlighted in the publication "Data governance in the sustainable smart city" by Paskaleva et al. (2017). This research focuses on the application of data governance in the context of developing sustainable smart cities. The case study explores the smart city initiative in Bristol, UK, which is recognized as a leading smart city in Europe. The Bristol City Council, along with various stakeholders including businesses, academic institutions, and community groups, aimed to create a city where digital technologies are used to promote sustainability, economic growth, and a high quality of life for its citizens.

Data Governance Implementation

The implementation of data governance in Bristol was crucial for the success of the smart city project. The project stakeholders recognized the importance of having clear policies and standards for data management to ensure that data collected from various sources, such as IoT devices and city infrastructure, was used responsibly and ethically.

Key Actions Taken

1. Establishing a Data Governance Framework: Bristol implemented a data governance framework that outlined the responsibilities of data collection, data access, and data sharing among the city's stakeholders.
2. Data Policies and Standards: The city developed a set of data policies and standards that addressed data quality, data security, and privacy concerns, ensuring compliance with the UK's data protection laws.
3. Stakeholder Engagement: A collaborative approach was taken where the local government, private sector, and citizens were engaged in the data governance process. This included workshops and public consultations to discuss the ethical use of data and the public's privacy concerns.

7.9 Case Studies: Policies and Standards in Action

4. Data Management Tools: Bristol utilized data management tools such as data catalogs and data analytics platforms to support the management, analysis, and visualization of urban data.
5. Transparency and Accountability: To maintain public trust, the city ensured transparency in its data governance processes and held regular reviews of its data policies.
6. Data-Driven Decision-Making: Data collected was utilized to inform policy decisions and urban planning, which included traffic management, energy use optimization, and public services improvement.
7. Ethical Considerations: A significant aspect of Bristol's data governance approach was the emphasis on ethical considerations, ensuring that the benefits of the smart city initiatives were distributed equitably among all residents.

Outcomes of Data Governance Implementation

The case study of Bristol's smart city project revealed that a well-structured data governance strategy could lead to improved city management, enhanced citizen engagement, and better service delivery. The city was able to use data to drive sustainability initiatives, such as reducing energy consumption and improving waste management. Furthermore, the governance framework allowed for the sustainable growth of the smart city infrastructure, enabling scalability and adaptation to innovative technologies and challenges. It served as a model for other cities looking to implement smart technologies while maintaining a strong governance structure around data.

Case Study: Data Governance Activities in Scientific and Practice-Oriented Literature

The article "Data governance activities: A comparison between scientific and practice-oriented literature" by Alhassan et al. (2018), provides a comprehensive review of data governance literature, comparing the theoretical and practical aspects of data governance activities. The case study, although not explicit in the details of a specific organization, draws on examples from financial institutions to illustrate the implementation of data governance activities. The financial industry, with its strict regulatory requirements and the need for high data quality, serves as an ideal backdrop to analyze the practical application of data governance.

Implementation Highlights

1. Development of Data Governance Framework: Financial institutions often develop a robust data governance framework that addresses regulatory compliance, risk management, and customer data protection.
2. Institutionalizing Data Stewardship: These institutions appoint data stewards responsible for the quality and lifecycle management of data. The stewards are often subject matter experts who understand the implications of data in financial contexts.
3. Risk Management and Compliance: Through data governance activities, banks can better manage risks associated with data breaches, financial fraud, and non-compliance with regulations like GDPR or SOX.

4. Operational Efficiency: By standardizing data governance practices, financial institutions streamline processes, reducing redundancies and ensuring that critical data is accurate, consistent, and readily available.
5. Data Quality and Data Lineage: A significant focus is on data quality initiatives and maintaining data lineage for auditability, allowing for traceability from source to report.
6. Cultural Change: Implementing data governance often requires a shift in organizational culture to recognize data as an asset, necessitating training and changes in mindset at all levels of the organization.
7. Technology Implementation: Adoption of technologies like Master Data Management (MDM) tools, data catalogs, and metadata management solutions supports the data governance framework.

Outcomes and Insights
The case study underscores the importance of aligning data governance activities with business strategy, particularly in sectors that handle sensitive and regulated data. One of the key insights is that despite the extensive literature on the topic, the practical application of data governance can vary significantly based on organizational size, complexity, and industry sector. In this study, the financial industry's application of data governance provides a template for other sectors looking to improve data quality, compliance, and decision-making.

Case Study: NPS Big Data Implementation in Korea
Another insightful case study that demonstrates Data Governance Policies and Standards in action is presented in the publication "Data governance framework for big data implementation with NPS Case Analysis in Korea" by Kim and Cho (2018). This study explores the National Pension Service (NPS) of Korea's approach to data governance in the context of big data implementation. The NPS of Korea embarked on a big data project with the goal of enhancing pension services and operational efficiency. The project aimed to leverage big data analytics to gain insights from the vast amounts of data collected from pension contributors and beneficiaries.

Data Governance Implementation
Key to the NPS's big data project was the implementation of a robust data governance framework to ensure the integrity, security, and appropriate use of pension data.

Key Actions Taken
1. Big Data Governance Framework: The NPS developed a comprehensive framework to guide the governance of big data analytics, ensuring alignment with the organization's strategic goals.
2. Policies and Standards for Data Management: Detailed policies and standards were established to address data quality, privacy, and security. These policies ensured the ethical use of contributors' and beneficiaries' data.
3. Risk Management: Risk assessments were conducted to identify and mitigate potential risks associated with the use of big data, such as data breaches and misuse of information.

4. Technology Implementation: The NPS invested in big data technologies, including advanced analytics tools, data storage solutions, and data processing infrastructure to support the governance framework.
5. Compliance and Legal Review: Legal reviews of the big data project were regularly conducted to ensure compliance with Korea's personal information protection laws.
6. Stakeholder Collaboration: A collaborative approach involved various departments within the NPS to ensure that the data governance policies were practical and aligned with different operational needs.
7. Employee Training: The NPS provided training programs for employees to familiarize them with the new big data tools and governance policies.
8. Monitoring and Evaluation: Ongoing monitoring and evaluation mechanisms were put in place to assess the effectiveness of the big data project and the adherence to data governance standards.
9. Transparency and Communication: Communication strategies were employed to maintain transparency with the public regarding the use of big data in pension management.

Outcomes of Data Governance Implementation
The implementation of the data governance framework allowed the NPS to successfully harness big data analytics while protecting the privacy and security of personal data. The governance structure enabled the NPS to improve decision-making processes, enhance service delivery to pensioners, and increase operational efficiencies. The case study serves as an example of how a large public sector entity can implement data governance policies and standards to leverage big data while upholding ethical and legal standards.

7.10 Conclusion

Chapter 7 has provided valuable insights into the crucial aspects of developing, implementing, and maintaining effective policies and standards within organizations. As we conclude this chapter, it is essential to reflect on the key takeaways and the broader significance of data governance in today's business landscape.

1. **The Foundation of Data Governance**
   - Data governance policies and standards serve as the foundation for managing data effectively. They provide the guidelines, principles, and rules that govern how data is collected, stored, processed, and protected.
2. **Systematic Approach**
   - Data governance emphasizes a systematic approach to managing and securing data. It encourages organizations to adopt a structured framework for data management, ensuring that data is treated as an asset.

3. **Scope and Definition**
   - Policies and standards define the scope of data governance within an organization. They clarify what data is subject to governance and outline the boundaries of data-related activities.

4. **Importance of Clarity**
   - Clear and understandable language in policies is paramount. Avoiding jargon and using plain language ensures that all employees can grasp the policies' content and implications.

5. **Alignment with Goals**
   - Policies and standards must align with the broader organizational goals. They should contribute to achieving these goals, whether they relate to data quality, security, compliance, or strategic objectives.

6. **Compliance and Risk Mitigation**
   - Effective policies and standards play a pivotal role in ensuring compliance with laws, regulations, and industry standards. They also help organizations mitigate risks associated with data breaches and non-compliance.

7. **Employee Training and Awareness**
   - Investing in employee training and awareness is essential. It empowers employees to understand, comply with, and contribute to data governance efforts.

8. **Continuous Improvement**
   - Data governance is not static; it requires continuous improvement. Organizations must adapt policies and standards to evolving technologies, threats, and regulatory changes.

9. **Change Management**
   - Managing change and policy evolution is an ongoing process. It involves assessing the impact of changes, engaging stakeholders, and implementing effective communication and training strategies.

10. **Real-World Case Studies**
    - The inclusion of real-world case studies demonstrates how organizations have successfully applied data governance principles to address specific challenges and achieve positive outcomes.

Chapter 7 underscores the critical role of data governance policies and standards in modern organizations. These policies are not mere documents but living guidelines that shape how data is managed, secured, and leveraged to drive business success. As organizations increasingly rely on data as a strategic asset, the principles outlined in this chapter become even more relevant and valuable in maintaining data integrity, security, and compliance.

References

Abraham R, Schneider J, Vom Brocke J (2019) Data governance: A conceptual framework, structured review, and research agenda. Springer

Al-Badi A, Tarhini A, Khan AI (2018) Exploring big data governance frameworks comparing different data governance frameworks to the ISO 8000 standard. Springer

Alhassan I, Sammon D (2016) Data governance activities: an analysis of the literature. Springer

Alhassan I, Sammon D, Daly M (2018) Data governance activities: A comparison between scientific and practice-oriented literature. J Data Inf Qual

Brous P, Janssen M, Vilminko-Heikkinen R (2016) Coordinating decision-making in data management activities: a systematic review of data governance principles. Springer

Change Management: From Theory to Practice (2021) TechTrends

Cheong LK, Chang V (2007) The need for data governance: a case study. Springer

Developing and Implementing Policies for Academic Integrity (2021) Oxford University Press.

Enforcing Environmental Regulation (2021) J Environ Law, 23(2), 169–193. Oxford University Press

Enforcing Public Data Archiving Policies (2021) Arxiv

Enforcing Regulation (2021) Oxford University Press.

Implementing Effective Ethics Standards in Government and the Civil Service (2018) OECD

Implementing International Environmental Law (2021) Oxford University Press

Kim HY, Cho JS (2018) Data governance framework for big data implementation with NPS Case Analysis in Korea. J Big Data

Koltay T (2016) Data governance, data literacy and the management of data quality. Springer

Ladley J (2019) Data governance: How to design, deploy, and sustain an effective data governance program. Elsevier

Managing Strategic Change (2021) Oxford University Press.

Oxford Handbook of Management (2021) Oxford Academic.

Paskaleva K, Evans J, Martin C (2017) Data governance in the sustainable smart city. Information Polity

Pollack J, Pollack R (2015) The change management paradigm. Oxford University Press

Chapter 8
Roles and Responsibilities in Data Governance: Building an Effective Team

Contents

8.1	Introduction to Data Governance Roles and Responsibilities.	280
8.2	Key Data Governance Roles.	282
8.3	Establishing a Data Governance Committee.	284
8.4	Cross-Functional Collaboration in Data Governance.	288
8.5	Training and Skill Development.	290
8.6	Building and Sustaining an Effective Data Governance Team.	292
8.7	Case Studies: Successful Data Governance Teams.	294
8.8	Overcoming Common Challenges in Team Dynamics.	297
8.9	Measuring Team Effectiveness.	299
8.10	Conclusion.	301
References.		302

Abstract This chapter focuses on the human aspects critical to the success of data governance frameworks within organizations. The chapter emphasizes that while policies and standards provide a structure for data governance, it is the individuals and their roles within the organization that actualize these frameworks.

The chapter begins by highlighting the significance of defining clear roles and responsibilities, which include executive sponsors, data stewards, data custodians, compliance officers, and others. These roles are essential for overseeing and managing data assets effectively, ensuring data quality, security, and compliance with applicable laws and regulations.

It discusses the need for a collaborative approach to data governance, stressing that data management requires the integration of various skills and expertise across the organization. Effective data governance is not isolated to a single department but is a collective effort that involves multiple stakeholders with clear accountability and communication pathways.

© The Author(s), under exclusive license to Springer Nature Switzerland AG 2024
D. Sargiotis, *Data Governance*, https://doi.org/10.1007/978-3-031-67268-2_8

The narrative then explores the dynamics of building a data governance team, outlining strategies for establishing and maintaining such a team. It covers the importance of continuous training and development to adapt to the rapidly changing technology landscape and regulatory environment.

Real-world examples are provided to illustrate how organizations successfully navigate the complexities of forming data governance teams. Challenges such as communication barriers, role ambiguity, and resistance to change are discussed, along with strategies to overcome these issues to foster a collaborative and efficient data governance environment.

In conclusion, this chapter serves as a comprehensive guide for understanding the roles and responsibilities essential to effective data governance. It equips readers with the knowledge to build, sustain, and optimize data governance teams, thereby ensuring that data assets are managed effectively to support organizational goals.

Keywords Data governance · Roles and responsibilities · Team building · Executive sponsor · Data steward · Data custodian · Compliance officer · Business data owner · Data architect · Data quality analyst · Strategic oversight · Cross-functional collaboration · Decision-making · Issue resolution · Data committee composition

As we delve into this chapter of our exploration into the realm of data governance, we shift our focus to the human element of this essential practice. While policies and standards provide the framework for data governance, it is the individuals within an organization who bring these guidelines to life. In this chapter, we will explore the critical roles and responsibilities that form the backbone of an effective data governance team.

Data governance is not a solitary endeavor; rather, it is a collaborative effort that involves a diverse set of skills, expertise, and responsibilities. To ensure the success of data governance initiatives, organizations must assemble a dedicated team capable of overseeing and managing data assets effectively.

8.1 Introduction to Data Governance Roles and Responsibilities

In the dynamic landscape of data governance, where data is increasingly recognized as a strategic asset and compliance with regulations is paramount, defining clear roles and responsibilities within an organization is crucial for success. Roles and responsibilities provide clarity on who is accountable for various aspects of data management, from data quality and security to compliance and strategic decision-making.

8.2 Key Data Governance Roles

The Significance of Roles and Responsibilities

The allocation of roles and responsibilities is fundamental to any well-functioning data governance framework. These roles ensure that data management processes align with organizational goals and regulatory requirements, maximizing the value of data. Effective data governance involves a diverse set of roles including executive sponsors, data stewards, data custodians, compliance officers, and more, each contributing unique expertise to the governance process.

Key Aspects of Data Governance Roles

1. **Diverse Set of Roles**: Data governance teams are composed of various roles, each with specific duties. These range from executive sponsors who champion data governance initiatives at the executive level, to data stewards who ensure the quality and integrity of specific datasets, and compliance officers who focus on legal and regulatory adherence.
2. **Shared Objectives**: Despite their varied roles, all members of a data governance team share the common objective of managing data in a way that aligns with organizational goals, complies with regulations, and maximizes data value.
3. **Collaboration and Communication**: Successful data governance relies heavily on collaboration and communication among team members. Cross-functional teams work together to make informed decisions about data, balancing the needs of different departments and stakeholders.
4. **Adaptability and Evolution**: Data governance roles must be flexible and adaptable to evolving organizational needs, technological advancements, and changes in the regulatory landscape.

The importance of clear roles and responsibilities in data governance is supported by various studies. For instance, Gailhofer and Franke (2021) highlight that effective data governance requires not only clear definitions of roles but also robust regulatory frameworks to manage data as a strategic asset. Furthermore, the integration of data governance principles into organizational practices can enhance data quality, compliance, and strategic decision-making (Grafanaki 2017).

The CARE Principles for Indigenous Data Governance emphasize the need for culturally appropriate governance frameworks that respect the unique needs and rights of Indigenous communities. These principles underscore the importance of roles such as data stewards who understand and respect cultural contexts in data management (Smithsonian Directive 609 2019).

Understanding and defining roles and responsibilities within a data governance framework is essential for any organization aiming to manage its data effectively. These roles ensure that data management practices are aligned with organizational goals, comply with regulatory requirements, and leverage data to its full potential. As organizations continue to recognize data as a strategic asset, the principles of effective data governance will become increasingly crucial.

8.2 Key Data Governance Roles

Within the realm of data governance, specific roles are essential for the effective management, oversight, and strategic utilization of data assets. These roles bring diverse skills and responsibilities to the table, contributing to the success of data governance initiatives.

Executive Sponsor The Executive Sponsor is a high-level executive who champions data governance initiatives at the executive level. They provide the necessary authority, resources, and support for data governance efforts. Their responsibilities include setting the strategic direction for data governance, advocating for its importance, and ensuring alignment with the organization's overall goals (Seiner 2021).

Data Steward Data Stewards are responsible for the quality, integrity, and proper use of data. They oversee specific datasets or data domains, ensuring data is accurate, consistent, and compliant with policies. Data Stewards define data standards, resolve data quality issues, and collaborate with data custodians and business units to meet data requirements (Smarty 2023).

Data Custodian Data Custodians are responsible for the technical aspects of data management, including data storage, access control, and security. They implement and maintain data infrastructure, ensuring data is securely stored, backed up, and accessible to authorized users. Data Custodians also enforce data security protocols (DATAVERSITY 2023e).

Data Governance Manager The Data Governance Manager is a leadership role responsible for overseeing the entire data governance program. They coordinate efforts, manage resources, and report progress to executive leadership. Data Governance Managers develop data governance frameworks, create policies, lead the data governance team, and monitor compliance and performance (DATAVERSITY 2023e).

Compliance Officer Compliance Officers focus on ensuring data governance practices align with legal and regulatory requirements. They keep the organization up to date with data-related regulations, conduct audits, assess data governance practices for legal compliance, and recommend actions to mitigate risks (Smarty 2023).

Business Data Owner Business Data Owners are accountable for specific datasets within their business units. They make decisions about data access, usage, and prioritization. Their responsibilities include defining data requirements, approving data-related initiatives, and ensuring data is leveraged effectively to meet business objectives (Smarty 2023).

8.2 Key Data Governance Roles

Data Architect Data Architects design the data infrastructure, including databases, data models, and data integration solutions. They ensure that data is structured for optimal use. Data Architects collaborate with Data Stewards to design data structures that align with data governance policies and oversee data integration efforts (DATAVERSITY 2023e).

Data Quality Analyst Data Quality Analysts focus on assessing and improving data quality. They develop metrics, monitor data quality, and identify and resolve data errors. Their responsibilities include defining data quality standards, performing data profiling, and working with Data Stewards to resolve data quality issues (Smarty 2023).

Figure 8.1, titled "Key Data Governance Roles," visually outlines the critical roles involved in the effective management and implementation of data governance within an organization. Centered around a main circle labeled "Key Data Governance Roles," the diagram extends branches to various positions, each essential for the governance process. These roles include the Data Governance Manager, who oversees the entire data governance program; the Data Steward, responsible for managing data elements; the Data Custodian, who ensures data is securely stored and maintained; the Business Data Owner, accountable for data capture and maintenance; the Data Quality Analyst, focusing on data quality standards; the Executive Sponsor, providing leadership and advocating for data governance at the executive level; the Data Architect, designing the enterprise data architecture to align with business needs; and the Compliance Officer, ensuring adherence to legal and regulatory requirements. Each role is interconnected, emphasizing the collaborative effort required across different functions to maintain an effective data governance framework, highlighting the collective contribution of these roles to organizational data governance objectives.

Each of these roles plays a crucial part in the data governance ecosystem. They work collaboratively to ensure that data is managed, protected, and leveraged effectively to support the organization's objectives. In the following sections, we will

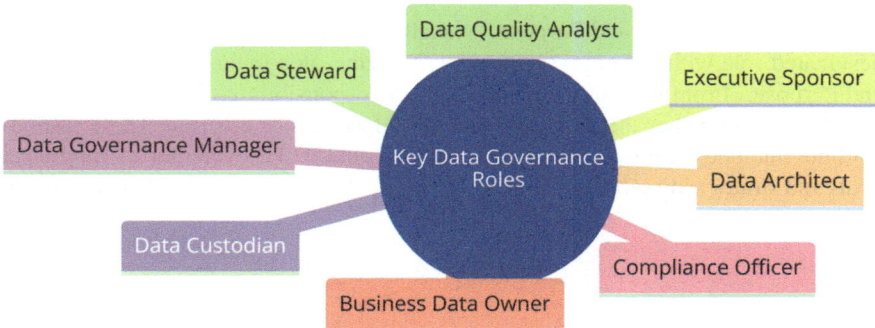

Fig. 8.1 Key data governance roles map. (Created by the author)

delve deeper into the responsibilities and interactions of these key data governance roles, providing a comprehensive view of their contributions to successful data governance.

8.3 Establishing a Data Governance Committee

In the complex landscape of data governance, organizations often find it beneficial to establish a Data Governance Committee. This committee serves as a central governing body responsible for overseeing and coordinating data governance efforts across the organization.

Role and Purpose A Data Governance Committee is tasked with several crucial roles and purposes:

1. **Strategic Oversight**: The committee provides strategic oversight of data governance initiatives, ensuring alignment with organizational objectives. This strategic alignment is essential for the success of data governance programs and helps in integrating governance efforts with broader organizational goals such as digital transformation and regulatory compliance (McKinsey 2023a).
2. **Decision-Making**: The committee plays a pivotal role in decision-making related to data policies, standards, and priorities. It ensures that the governance policies are not only developed but also effectively implemented and adhered to across the organization (Collibra 2021).
3. **Cross-Functional Collaboration**: The committee fosters collaboration among various data stakeholders, bridging the gap between business units and IT. This cross-functional approach ensures that all departments that depend on accurate data are properly represented and that data governance efforts are holistic and inclusive (DATAVERSITY 2023e).
4. **Issue Resolution**: It addresses data-related issues, disputes, and conflicts that may arise within the organization, providing a structured approach to resolving these issues efficiently and effectively (McKinsey 2023a).

Figure 8.2, titled "Role and Purpose," visually captures the essential functions of a designated role within an organization, centralizing on a main node labeled "Role and Purpose." From this central node, four key responsibilities extend, represented as individual branches that highlight the critical aspects of this role. These responsibilities include "Strategic Oversight," which entails the oversight of long-term goals and strategic direction; "Decision-Making," focusing on the authority to make

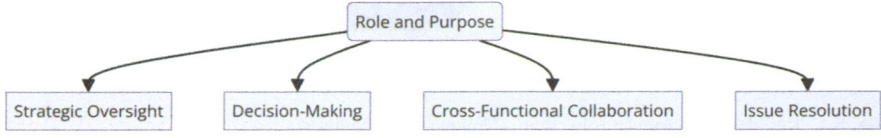

Fig. 8.2 Role and purpose. (Created by the author)

8.3 Establishing a Data Governance Committee

significant choices that affect the organization's direction and operations; "Cross-Functional Collaboration," which emphasizes the need to work across different departments to achieve common goals and enhance organizational efficiency; and "Issue Resolution," which involves tackling and resolving challenges that may arise during operations. This configuration underscores the multifaceted nature of the role, illustrating how it integrates strategic, operational, and collaborative elements to fulfill its purpose within the organization effectively.

Composition The composition of a Data Governance Committee can vary depending on the organization's size and complexity. However, it typically includes representatives from various key roles such as:

- **Executive Sponsor**: An executive-level representative who provides top-level support and ensures alignment with the organization's strategic goals (Collibra 2021).
- **Data Stewards**: Representatives who manage specific data domains and are responsible for data quality and integrity (Smarty 2023).
- **Data Custodians**: Technical experts responsible for data storage, security, and access control (DATAVERSITY 2023e).
- **Data Governance Manager**: The individual overseeing the data governance program, often chairing the committee (Collibra 2021).
- **Business Data Owners**: Representatives from different business units who are accountable for specific datasets (McKinsey 2023a).
- **Compliance Officer**: Ensures that data governance practices align with legal and regulatory requirements (Smarty 2023).
- **Data Quality Analyst**: Focuses on assessing and improving data quality (DATAVERSITY 2023e).

Figure 8.3, titled "Composition of a Data Governance Committee," illustrates the structure and key roles that constitute a data governance committee within an organization. The diagram features a central label "Composition," from which branches extend to represent the various essential positions involved in the committee. These roles include the "Executive Sponsor," who provides leadership and strategic direction; "Data Stewards," responsible for the management and quality of data; "Data Custodians," tasked with the technical storage and safeguarding of data; the "Data Governance Manager," who oversees the implementation of governance strategies; "Business Data Owners," who are accountable for data accuracy and utilization within their respective domains; a "Compliance Officer," ensuring that data practices adhere to legal and regulatory standards; and a "Data Quality Analyst," focused on maintaining and improving data quality. This configuration underscores the

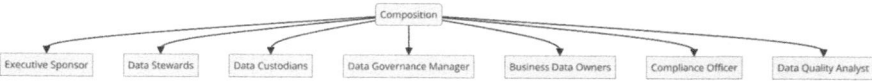

Fig. 8.3 Composition of a Data Governance Committee. (Created by the author)

collaborative and multi-disciplinary approach necessary for effective data governance, emphasizing the interrelated responsibilities that support the organization's data management and policy enforcement.

Key Considerations When establishing a Data Governance Committee, several key considerations should be taken into account:

1. **Clear Charter**: Define the committee's mission, objectives, and scope in a clear and documented charter. This helps in setting clear expectations and guiding the committee's activities (Collibra 2021).
2. **Membership Selection**: Carefully select members based on their expertise and ability to contribute to the committee's goals. It is crucial to have a diverse set of skills and perspectives within the committee (McKinsey 2023a).
3. **Roles and Responsibilities**: Clearly define the roles and responsibilities of committee members, including the chairperson. This clarity helps in avoiding overlaps and ensures accountability (DATAVERSITY 2023e).
4. **Meeting Frequency**: Determine the frequency of committee meetings to ensure regular oversight and decision-making. Regular meetings help in maintaining momentum and addressing issues promptly (Collibra 2021).
5. **Decision-Making Process**: Establish a transparent decision-making process, including voting mechanisms and conflict resolution procedures. This ensures that decisions are made effectively and disputes are resolved amicably (McKinsey 2023a).
6. **Communication**: Develop a communication plan to keep stakeholders informed about committee activities and decisions. Effective communication is crucial for the success of data governance initiatives (DATAVERSITY 2023e).
7. **Reporting**: Define reporting mechanisms to provide updates on data governance progress to executive leadership. Regular reporting helps in maintaining transparency and accountability (Smarty 2023).

Figure 8.4 outlines the essential factors to consider during the formation of a data governance committee. Centered around the label "Key Considerations," the diagram branches out to depict various fundamental aspects that need to be addressed to ensure the committee's effectiveness. These considerations include establishing a "Clear Charter" that defines the committee's purpose and scope; "Membership Selection," which involves choosing the right mix of skills and expertise for the committee members; "Roles and Responsibilities," clearly defining what is expected of each member; "Meeting Frequency," determining how often the committee should meet to maintain momentum and oversight; "Decision Making Process," outlining how decisions are made and who has authority; "Communication,"

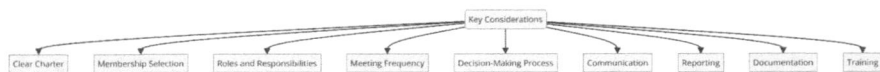

Fig. 8.4 Key considerations when establishing a Data Governance Committee. (Created by the author)

ensuring there is effective information flow to and from the committee; "Reporting," which details the reporting structure and frequency; "Documentation," ensuring all decisions and processes are well-documented; and "Training," providing necessary training to committee members to fulfill their roles effectively. This diagram serves as a comprehensive guide for setting up a governance committee that is structured, functional, and aligned with the organization's data governance objectives.

Benefits

Establishing a Data Governance Committee offers several benefits, including:

- **Centralized Oversight**: Provides a centralized body for overseeing data governance, promoting consistency and alignment with organizational goals.
- **Expertise**: Brings together expertise from various data-related roles, ensuring comprehensive coverage of data governance aspects.
- **Issue Resolution**: Facilitates timely resolution of data-related issues and disputes, preventing prolonged conflicts.
- **Cross-Functional Collaboration**: Fosters collaboration and communication among business units, IT, and data stakeholders.
- **Strategic Direction**: Helps set the strategic direction for data governance, ensuring that it supports the organization's overall strategy (Collibra 2021; McKinsey 2023a).

Figure 8.5 in Section 8.3 of your book, titled "Benefits Establishing a Data Governance Committee," illustrates the strategic advantages of establishing a Data Governance Committee. This committee plays a crucial role in aligning data management practices with organizational goals and ensuring data is used effectively across various departments. The diagram highlights five key benefits derived from forming such a committee: Centralized Oversight, Expertise, Issue Resolution, Cross-Functional Collaboration, and Strategic Direction. Centralized Oversight emphasizes unified control and standardized procedures, while Expertise points to the specialized knowledge the committee brings. Issue Resolution focuses on the committee's ability to handle disputes and problems efficiently. Cross-Functional Collaboration underlines the enhancement of cooperation across different business areas, and Strategic Direction signifies the guidance provided to align data governance with the broader business strategy.

Establishing a Data Governance Committee is a strategic move for organizations committed to effective data governance. It brings together diverse expertise, promotes collaboration, and ensures that data are managed in a way that maximizes its value while adhering to quality, security, and compliance standards.

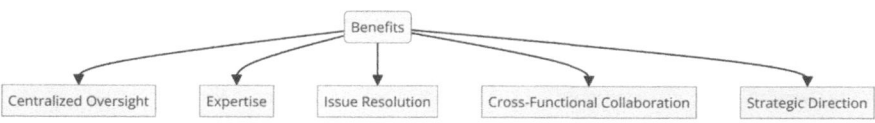

Fig. 8.5 Benefits establishing a data governance committee. (Created by the author)

8.4 Cross-Functional Collaboration in Data Governance

Effective data governance is not the sole responsibility of a single department or role within an organization. Instead, it requires collaboration and cooperation across various functions and teams.

The Necessity of Cross-Functional Collaboration
Data governance initiatives involve multiple facets of data management, including data quality, security, compliance, and strategic utilization. To address these aspects comprehensively, cross-functional collaboration is essential for several reasons:

1. **Data Ownership**: Different business units and departments often "own" specific data sets. Collaborative efforts are necessary to define data ownership responsibilities and data-related objectives (McKinsey 2023a).
2. **Data Usage**: Data is used by various teams for different purposes. Collaboration ensures that data is utilized effectively across the organization, enhancing decision-making and operational efficiency (Onna 2023).
3. **Data Quality**: Ensuring data quality requires cooperation among those responsible for data entry, processing, and verification. Shared efforts in maintaining data standards help prevent inconsistencies and inaccuracies (Y42 2023).
4. **Compliance**: Compliance with data-related regulations and standards involves multiple stakeholders, including legal, IT, and business units. Collaborative efforts ensure that all regulatory requirements are met and that data governance practices are robust and comprehensive (Omeda 2023).
5. **Data Security**: Protecting data from breaches and unauthorized access requires collaboration between IT security teams and data stewards. Joint efforts in developing and implementing security measures are crucial for safeguarding sensitive information (Onna 2023).

Figure 8.6 depicts the importance of various departments working together in the realm of data governance. This collaboration is essential to address diverse aspects of data management, including data quality, security, compliance, and strategic utilization. Key elements highlighted in the diagram are Data Ownership, Data Usage, Data Quality, Compliance, and Data Security. Data Ownership emphasizes the definition of responsibilities across business units. Data Usage illustrates the effective application of data across teams. Data Quality focuses on cooperation for maintaining data standards. Compliance covers adherence to regulations through collective

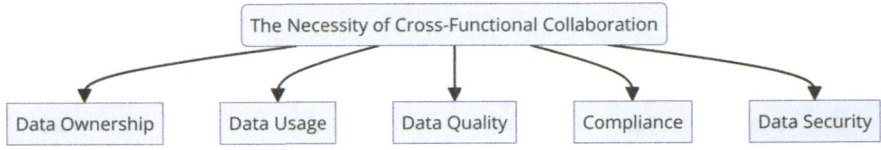

Fig. 8.6 The necessity of cross-functional collaboration. (Created by the author)

effort. Data Security underscores the collaborative need to protect data from unauthorized access.

Key Elements of Cross-Functional Collaboration

To foster effective cross-functional collaboration in data governance, organizations should consider the following key elements:

1. **Communication**: Establish clear communication channels for discussing data-related matters and sharing insights across teams. Regular meetings and updates help keep stakeholders informed and aligned (McKinsey 2023b).
2. **Data Governance Committee**: A Data Governance Committee can serve as a central body for coordinating collaboration efforts. This committee typically includes representatives from various departments, ensuring that all perspectives are considered (Omeda 2023).
3. **Data Stewardship**: Appoint data stewards who act as liaisons between business units and data governance teams, facilitating communication and alignment (Y42 2023).
4. **Training and Awareness**: Conduct training sessions and awareness programs to educate employees about the importance of data governance and their roles within it. Enhanced data literacy across the organization promotes better data handling and usage (Onna 2023).
5. **Shared Objectives**: Ensure that all teams and departments understand and share common data-related objectives such as data accuracy, compliance, and strategic data use. This alignment helps in achieving organizational goals effectively (McKinsey 2023b).

Figure 8.7, outlines crucial components essential for fostering effective collaboration across different organizational functions in the context of data governance. The diagram highlights several key elements: Communication, Data Governance Committee, Data Stewardship, Training and Awareness, Cross-Functional Teams, and Shared Objectives. Communication ensures clear channels for data-related discussions. The Data Governance Committee coordinates collaboration efforts. Data Stewardship involves appointing liaisons between business units and data governance teams. Training and Awareness focus on educating employees about the importance of data governance. Cross-Functional Teams address specific data governance challenges, and Shared Objectives align all departments with common data-related goals.

Benefits of Cross-Functional Collaboration

Effective cross-functional collaboration in data governance offers several advantages:

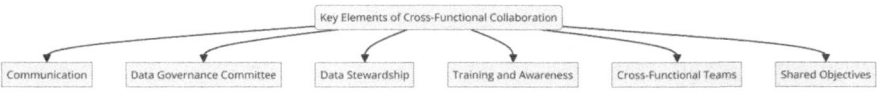

Fig. 8.7 Key elements of cross-functional collaboration. (Created by the author)

1. **Alignment**: Collaboration ensures that data governance efforts align with the organization's strategic goals and business objectives (Onna 2023).
2. **Efficiency**: Teams can work together to streamline data-related processes, reducing redundancies and improving efficiency (McKinsey 2023b).
3. **Data Quality**: Collaboration enhances data quality by involving those responsible for data entry and verification in the process (Y42 2023).
4. **Compliance**: Cross-functional teams can ensure that data governance practices adhere to legal and regulatory requirements, minimizing risks (Omeda 2023).
5. **Innovation**: Collaborative efforts can lead to innovative uses of data, driving business growth and competitiveness (McKinsey 2023b).

Figure 8.8 demonstrates the advantages of collaborative efforts across different organizational departments in the context of data governance. These collaborations lead to several benefits: Alignment, which ensures that all departments are synchronized with the organization's strategic goals; Efficiency, which eliminates redundancies and streamlines operations; Data Quality, where collective responsibility improves the accuracy and reliability of data; Compliance, ensuring adherence to relevant laws and regulations; and Innovation, where diverse inputs can lead to new ideas for using data creatively.

Cross-functional collaboration is a cornerstone of effective data governance. By bringing together diverse expertise and fostering communication and cooperation among teams, organizations can ensure that data is managed, protected, and leveraged to its fullest potential. Collaboration is the key to unlocking the value of data as a strategic asset while maintaining data quality, security, and compliance.

8.5 Training and Skill Development

Effective data governance requires not only robust frameworks and clear policies but also well-trained personnel equipped with the necessary skills to manage and protect data. Training and skill development are pivotal in ensuring that all stakeholders understand their roles and can contribute effectively to data governance initiatives.

Importance of Training and Skill Development Training in data governance encompasses various aspects, including data management, regulatory compliance,

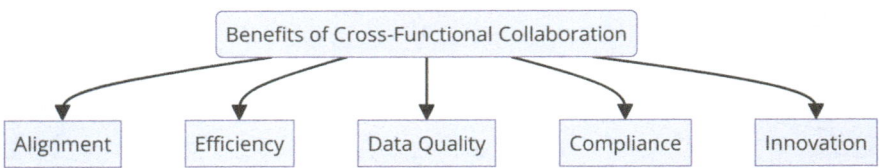

Fig. 8.8 Benefits of cross-functional collaboration. (Created by the author)

8.5 Training and Skill Development

data quality, and security. The primary goals are to enhance data literacy across the organization and to ensure that employees are well-versed in best practices and emerging trends in data governance.

1. **Enhanced Data Literacy**: Improving data literacy is crucial for empowering employees to make informed decisions based on data. It involves training on how to understand, interpret, and use data effectively. Enhanced data literacy helps in fostering a culture of data-driven decision-making within the organization (LightsOnData 2023).
2. **Compliance and Security**: Training ensures that all employees are aware of regulatory requirements and data protection measures. This is particularly important for maintaining compliance with laws such as GDPR and CCPA. Well-trained employees are better equipped to handle data securely, mitigating risks associated with data breaches and noncompliance (BMC Public Health 2023).
3. **Skill Development**: Continuous skill development programs help employees stay updated with the latest tools and technologies in data management. This includes training on data quality management, metadata management, and the use of data governance software. Such skills are essential for maintaining the integrity and reliability of data (LightsOnData 2023).

Training Programs and Courses Several training programs and courses are available to help organizations build capacity in data governance. These programs range from beginner courses to advanced certifications, catering to different levels of expertise.

1. **Data Governance Mastery**: This course provides a comprehensive understanding of data governance principles and best practices. It includes modules on metadata management and offers certification upon completion. Such programs are designed to develop mastery in data governance, preparing individuals to lead data governance initiatives (LightsOnData 2023).
2. **Fundamentals of Data Governance**: Courses like those offered by Udemy provide foundational knowledge in data governance. They cover essential topics such as data governance frameworks, roles and responsibilities, and best practices for implementation. These courses are ideal for beginners looking to enter the field of data governance (Udemy 2023).
3. **Information Governance Training**: Short courses offered by platforms like Iron Mountain provide quick, essential training in information governance. These are particularly useful for professionals who need a rapid introduction to data governance concepts and practices (Iron Mountain 2023).

Key Elements of Effective Training To ensure the effectiveness of training programs, certain key elements should be included:

1. **Comprehensive Curriculum**: Training programs should cover all aspects of data governance, from foundational concepts to advanced practices. This ensures that participants gain a holistic understanding of the field (LightsOnData 2023).

2. **Practical Applications**: Incorporating practical exercises and real-world scenarios into the training helps participants apply what they have learned. This practical approach enhances the learning experience and ensures that skills are transferable to the workplace (BMC Public Health 2023).
3. **Continuous Learning**: Data governance is an evolving field. Continuous learning opportunities, such as advanced courses and regular updates on new regulations and technologies, help keep employees' skills up to date (LightsOnData 2023).
4. **Cross-Functional Training**: Training programs should involve participants from various departments to promote cross-functional collaboration. This approach ensures that all stakeholders are aligned and can work together effectively to achieve data governance goals (Omeda 2023).

Training and skill development are critical components of a successful data governance strategy. By investing in comprehensive training programs, organizations can build a knowledgeable and skilled workforce capable of managing data effectively and ensuring compliance with regulatory requirements.

8.6 Building and Sustaining an Effective Data Governance Team

Measuring the effectiveness of a data governance program is crucial for ensuring that data management practices are aligned with organizational goals and regulatory requirements. Metrics and Key Performance Indicators (KPIs) provide a quantitative basis for assessing the success and impact of data governance initiatives.

Importance of Metrics in Data Governance Metrics in data governance help organizations monitor the performance of their data management activities, identify areas for improvement, and demonstrate the value of data governance efforts to stakeholders. Key areas where metrics are beneficial include regulatory compliance, data quality, data usage, and overall data management efficiency.

Key Performance Indicators (KPIs) for Data Governance Effective data governance metrics span several categories, each focusing on different aspects of data management. Some commonly used KPIs include:

1. **Data Quality Metrics**
   - **Data Accuracy**: Measures the percentage of data entries that are correct and error-free. Comparing a sample of data against a trusted source helps determine this metric (DATAVERSITY 2023d).
   - **Data Completeness**: Assesses how much of the necessary data is available without missing elements. This metric evaluates the presence of all required data fields (Datafloq 2023).

- **Data Consistency**: Checks for uniformity across various datasets and departments. It ensures that data is consistent and standardized throughout the organization (DATAVERSITY 2023d).

2. **Regulatory Compliance Metrics**
   - **Data Access Controls**: Monitors the effectiveness of data access restrictions to ensure that only authorized personnel can access sensitive information (Datafloq 2023).
   - **Regulatory Compliance**: Tracks the organization's adherence to data privacy laws such as GDPR and CCPA, ensuring all data practices meet legal standards (DATAVERSITY 2023d).
   - **Operational Efficiency MetricsIssue Resolution Time**: Measures the average time taken to identify and resolve data-related issues. This metric helps in evaluating the responsiveness and efficiency of the data governance team (Datafloq 2023).
   - **Data Processing Timeliness**: Tracks the speed at which data is processed and updated, ensuring that information remains current and useful (DATAVERSITY 2023d).

3. **Security Metrics**
   - **Data Breach Incidents**: Counts the number of data breaches over a specific period, providing insights into the effectiveness of data security measures (DATAVERSITY 2023d).
   - **Data Encryption**: Measures the percentage of sensitive data that is encrypted, safeguarding it from unauthorized access (Datafloq 2023).

Balanced Scorecard Approach A balanced scorecard can be an effective tool for managing and improving data performance. It involves setting specific goals and measures across different perspectives such as financial, customer, internal business processes, and innovation. This approach helps in aligning data governance activities with the broader organizational strategy, ensuring that data initiatives contribute to financial gains, operational improvements, and innovation (Frontiers 2023).

Benefits of Using Metrics
- **Visibility and Accountability**: Metrics provide a clear picture of data governance performance, making it easier to hold teams accountable for their roles and responsibilities (DATAVERSITY 2023d).
- **Continuous Improvement**: Regular measurement helps in identifying areas needing improvement, allowing for continuous enhancement of data governance practices (Datafloq 2023).
- **Stakeholder Confidence**: Demonstrating the success of data governance initiatives through measurable outcomes builds confidence among stakeholders, securing ongoing support and investment (DATAVERSITY 2023d).

Implementing effective metrics and KPIs is essential for the success of a data governance program. These metrics not only help in monitoring performance and compliance but also drive continuous improvement and stakeholder confidence. By leveraging a balanced scorecard approach and focusing on key areas such as data quality, regulatory compliance, and operational efficiency, organizations can ensure their data governance initiatives are both effective and aligned with their strategic goals.

8.7 Case Studies: Successful Data Governance Teams

In this section, we explore real-world case studies of organizations that have implemented successful data governance teams. The following examples highlight how various companies have effectively implemented data governance to enhance their data management capabilities.

Airbnb's Data Literacy Initiative
Airbnb's approach to data governance emphasizes the importance of data literacy. To ensure all employees can make data-driven decisions, Airbnb launched an in-house educational program called "Data University." This initiative aimed to boost data literacy across all departments, equipping employees with the skills to understand, interpret, and use data effectively. As a result, 45% of Airbnb's workforce became active users of their internal data platform, democratizing data use and enhancing decision-making capabilities within the company (Atlan 2023b).

Strategies and Approaches

- **Data University**: At the heart of Airbnb's data literacy initiative is Data University, an in-house educational program aimed at boosting data literacy across the organization. The program offers a wide range of courses tailored to distinct roles and proficiency levels, ensuring that every employee, regardless of their background, can enhance their data skills.
- **Comprehensive Curriculum**: The curriculum covers fundamental data concepts, tools, and analytics, as well as advanced topics for more data-savvy roles. Courses are designed to be accessible and engaging, combining theoretical knowledge with practical, hands-on exercises.
- **Cross-Departmental Engagement**: To ensure the initiative's success, Airbnb engaged stakeholders from across the organization in the development and rollout of Data University. This approach ensured that the curriculum was relevant and valuable to various departments, promoting widespread adoption and engagement.
- **Promoting a Data-Driven Culture**: Beyond just providing training, Airbnb's initiative aimed to foster a cultural shift towards more data-informed decision-making. This involved not only teaching employees how to use data but also encouraging them to incorporate data insights into their daily work and decision-making processes.

Outcomes

8.7 Case Studies: Successful Data Governance Teams

- **Widespread Data Literacy**: The initiative successfully raised the data literacy level across Airbnb, with a significant portion of the workforce becoming regular users of the company's internal data platform. This widespread literacy has enabled employees to make more informed decisions and contribute more effectively to their teams.
- **Democratization of Data**: By empowering employees across all levels and departments with data skills, Airbnb has effectively democratized data within the organization. This democratization has facilitated greater transparency, collaboration, and innovation, as more employees are able to access, interpret, and leverage data for their projects.
- **Enhanced Decision-Making and Efficiency**: The initiative has led to more data-driven decision-making across Airbnb, enhancing the company's operational efficiency, strategic planning, and customer service. Employees are now better equipped to analyze data, derive insights, and apply these insights to solve problems and identify opportunities.
- **Responsible Data Use**: An essential component of the initiative is the emphasis on responsible data use, ensuring that employees not only understand how to use data but also the ethical considerations and best practices for data privacy and security.
- Airbnb's Data Literacy Initiative highlights the transformative potential of data literacy in modern organizations. By investing in education and fostering a data-driven culture, Airbnb has not only enhanced its operational efficiency and decision-making but also set a benchmark for other companies aiming to harness the power of data.

GE Aviation's Self-Service Data Initiative

GE Aviation's data governance strategy involved creating a centralized data repository to ensure data accessibility and reliability. They established the Self-Service Data (SSD) team to focus on user enablement, tooling, and data product deployment. This team worked alongside a Database Admin team responsible for maintaining data governance and supporting users. By documenting data products and workflows, GE Aviation ensured that data governance was integrated into everyday operations, enhancing both safety and operational efficiency (Atlan 2023b).

Strategies and Approaches

- **Dual-Team Structure**: The initiative featured a unique dual-team approach, comprising the SSD data team focused on user enablement, tooling, and data product deployment, and a Database Admin team dedicated to data governance, proper data usage, and user support. This structure allowed for a blend of governance and enablement, ensuring data was both accessible and compliant with organizational standards.
- **Self-Service Data Platform**: A cornerstone of the strategy was the development and deployment of a self-service data platform. This platform aimed to centralize data access, making it easier for users across GE Aviation to find, use, and analyze data without compromising on governance or data quality.

- **Comprehensive Data Governance**: Beyond merely centralizing data, GE Aviation invested in a comprehensive governance model. This included detailed documentation of data products, clear ownership and workflow definitions, and manual checks to ensure compliance and quality before data products were deployed to production.
- **User Enablement and Training**: Recognizing the importance of user competency in a self-service model, GE Aviation prioritized training and enablement. Users were educated on how to use the platform and data products effectively, ensuring they could leverage data insights while adhering to governance standards.

Outcomes
- **Enhanced Data Accessibility**: The initiative significantly improved data accessibility within GE Aviation, enabling users to easily access and leverage data for their specific needs. This accessibility fostered a culture of data-driven decision-making across the organization.
- **Reliable Data Governance**: By establishing a robust governance model alongside the self-service platform, GE Aviation ensured that data was not only accessible but also reliable and compliant with organizational standards. This balance between accessibility and governance enhanced trust in data products and insights derived from the platform.
- **Empowered Users**: With the self-service platform and associated training, users across GE Aviation were empowered to explore data and gain insights independently. This empowerment accelerated decision-making processes and fostered innovation.
- **Operational Efficiency and Safety Improvements**: The initiative's focus on data accessibility and reliability translated into tangible improvements in operational efficiency and safety. Employees were better equipped to analyze data for process improvements, leading to enhanced operational practices and safety outcomes.

Wells Fargo: Creating a Single Source of Truth

Wells Fargo implemented a data governance strategy centered on creating a single source of truth. By centralizing data from multiple sources, they improved data accuracy and consistency, which in turn enhanced reporting and decision-making processes. This strategy involved significant collaboration across departments to ensure that data governance practices were aligned with business objectives and regulatory requirements (Atlan 2023b).

CSE Insurance: Transforming Data Culture

CSE Insurance focused on transforming its data culture to improve data management. By fostering a data-driven environment, they enhanced data governance practices, ensuring better data quality and compliance. This cultural shift involved educating employees about the importance of data governance and implementing robust data management frameworks to support their goals (Atlan 2023b).

Uber: Flexible Data Governance Approach
Uber adopted a flexible approach to data governance to accommodate its rapid growth and diverse data needs. This approach involved setting up adaptable data governance frameworks that could evolve with the company's needs. Uber emphasized the importance of balancing data governance with data enablement, ensuring that data was both secure and accessible for innovation and decision-making (Atlan 2023b).

Healthcare Case Study: Building Data Governance Programs
In the healthcare sector, a case study highlighted the importance of getting executive-level support for data governance initiatives. By systematically securing buy-in from key corporate leaders, a healthcare organization was able to establish a robust data governance program. This program included creating a data dictionary, establishing reference data standards, and building a common understanding of data governance terms among executives. The initiative demonstrated how critical executive support and clear communication are to the success of data governance efforts (DATAVERSITY 2023b).

Implementing Data Governance for Data Lakes Another case study involved implementing data governance for a data lake environment. The organization focused on protecting sensitive data while encouraging innovation. They partnered with a third-party vendor to catalog and structure the data, ensuring that data governance practices were maintained even as new data sources were integrated. This approach allowed for the creation of a common data repository that supported both operational needs and innovation efforts (DATAVERSITY 2023a).

These case studies underscore the importance of tailored data governance strategies that align with organizational goals, promote data literacy, and secure executive support. By adopting best practices from successful implementations, organizations can enhance their data governance frameworks and drive better data management outcomes.

8.8 Overcoming Common Challenges in Team Dynamics

Implementing effective data governance can be complex, presenting numerous challenges. However, by understanding these challenges and applying appropriate solutions, organizations can enhance their data governance frameworks and achieve better data management outcomes.

Common Challenges in Data Governance
1. **Limited Resources**: Many organizations struggle with the allocation of sufficient resources, both in terms of budget and manpower, to maintain a robust data governance program. To address this, it's essential to plan and allocate resources for data governance proactively and harness automation to maximize efficiency (DATAVERSITY 2023c).

2. **Data Silos**: As organizations collect vast amounts of data, it often becomes siloed, making it difficult to achieve a unified view of data across the enterprise. This can be mitigated by implementing data catalogs that streamline data management and provide a comprehensive inventory of data, reducing fragmentation and enhancing accessibility (Alation 2023).
3. **Lack of Leadership**: Effective data governance requires strong leadership to provide direction and ensure adherence to governance policies. Appointing a Chief Data Officer (CDO) or a similar role can help in steering data governance initiatives and communicating their importance across the organization (DATAVERSITY 2023c).
4. **Poor Data Quality and Context**: Ensuring high data quality is crucial, as poor-quality data can lead to inaccurate insights and decisions. Utilizing data profiling tools and establishing data quality standards can help maintain data integrity. Metadata management is also essential for providing context and ensuring data is used appropriately (Alation 2023).
5. **Balancing Accessibility and Security**: Striking the right balance between making data accessible to users and maintaining its security is a common challenge. Implementing robust access controls, encryption, and monitoring mechanisms can help protect data while ensuring it remains accessible to those who need it (Atlan 2023a).
6. **Scalability**: As organizations grow, their data governance frameworks must scale accordingly. This involves updating policies, procedures, and technologies to accommodate new data sources and increasing data volumes without compromising governance standards (Atlan 2023a).
7. **Cross-Functional Collaboration**: Effective data governance requires collaboration across various departments, including IT, business units, and data teams. Ensuring alignment and cooperation among these stakeholders can be challenging but is essential for comprehensive data governance (Alation 2023).
8. **Ongoing Monitoring and Improvement**: Data governance is not a one-time effort but requires continuous monitoring and improvement. Establishing mechanisms to track progress, measure success, and identify areas for enhancement is crucial for maintaining an effective data governance framework (Atlan 2023a).

Solutions to Overcome Challenges
1. **Resource Allocation and Automation**: Plan for data governance in resource allocation and leverage automation to reduce manual workloads and increase efficiency (DATAVERSITY 2023c).
2. **Implement Data Catalogs**: Use data catalogs to consolidate data from various sources, provide comprehensive data inventories, and enhance data accessibility while maintaining governance (Alation 2023).
3. **Appoint Strong Data Leaders**: Establish a dedicated data governance team led by a knowledgeable CDO to drive initiatives and ensure organizational alignment on data governance goals (DATAVERSITY 2023c).

4. **Utilize Data Profiling and Metadata Management**: Employ data profiling tools to maintain high data quality and manage metadata to provide context and ensure proper data usage (Alation 2023).
5. **Enhance Security Measures**: Implement robust security protocols, including access controls, encryption, and regular monitoring, to protect data while ensuring it remains accessible to authorized users (Atlan 2023a).
6. **Ensure Scalability**: Update data governance frameworks to handle increasing data volumes and new data sources, ensuring policies and technologies can scale with organizational growth (Atlan 2023a).
7. **Foster Cross-Functional Collaboration**: Promote collaboration among different departments through clear communication channels and aligned goals to ensure comprehensive data governance (Alation 2023).
8. **Continuous Improvement**: Establish ongoing monitoring and evaluation mechanisms to track the effectiveness of data governance efforts and continuously refine practices (Atlan 2023a).

By addressing these common challenges with targeted solutions, organizations can develop robust data governance frameworks that enhance data quality, security, and overall management, thereby driving better business outcomes.

8.9 Measuring Team Effectiveness

Measuring the effectiveness of a data governance team is essential for assessing its performance and identifying areas for improvement. In this section, we explore key metrics and approaches for measuring the effectiveness of a data governance team.

Key Metrics for Measuring Team Effectiveness
1. **Data Quality Metrics**: Assess the quality of data managed by the team. Metrics may include data accuracy, completeness, consistency, and timeliness. Effective data quality management ensures that data is reliable and can be trusted for decision-making (Experian 2023).
2. **Compliance Metrics**: Measure the team's adherence to data privacy regulations, industry standards, and internal policies. Track compliance violations and resolutions. Ensuring compliance is critical for avoiding legal penalties and maintaining trust with stakeholders (DATAVERSITY 2023c).
3. **Efficiency Metrics**: Evaluate the efficiency of data-related processes. Metrics may include data processing times, data access requests, and incident response times. Efficient processes reduce operational costs and improve the responsiveness of the data governance team (Informatica 2023).
4. **Issue Resolution Metrics**: Monitor the team's ability to address data-related issues and incidents promptly. Track the number of reported issues, response times, and resolution rates. Quick resolution of issues minimizes disruption and maintains data integrity (McKinsey 2023a).

5. **Data Governance KPIs**: Define key performance indicators (KPIs) specific to data governance, such as the percentage of data assets with documented metadata or the number of data stewardship initiatives completed. These KPIs help in tracking the progress and effectiveness of data governance efforts (Experian 2023).
6. **Cross-Functional Collaboration**: Assess the level of collaboration and communication between the data governance team and other departments. Measure the frequency of cross-functional meetings and joint projects. Effective collaboration ensures that data governance practices are integrated across the organization (Informatica 2023).
7. **Training and Skill Development**: Evaluate the progress of team members in training and skill development programs. Measure the completion rates of training courses and the acquisition of relevant certifications. Continuous training keeps the team updated with best practices and new technologies (Experian 2023).
8. **Data Utilization Metrics**: Determine how effectively data is utilized for decision-making. Track the impact of data-driven decisions on organizational outcomes. Proper utilization of data enhances decision-making and drives business success (McKinsey 2023a).

Approaches for Measuring Team Effectiveness
1. **Regular Audits**: Conduct regular audits of data quality, compliance, and data governance practices. Audits provide valuable insights into areas that require improvement and ensure adherence to standards (Experian 2023).
2. **Surveys and Feedback**: Gather feedback from team members, stakeholders, and data users through surveys and interviews. Use this feedback to identify strengths and weaknesses. Feedback mechanisms help in aligning the team's efforts with organizational needs (Informatica 2023).
3. **Benchmarking**: Compare the team's performance metrics with industry benchmarks and best practices to assess how it measures up against peers. Benchmarking helps in setting realistic performance targets and identifying best practices (McKinsey 2023a).
4. **Balanced Scorecards**: Develop balanced scorecards that include a mix of financial, customer, internal process, and learning and growth metrics to provide a holistic view of team performance. Balanced scorecards ensure a comprehensive evaluation of performance across multiple dimensions (Informatica 2023).
5. **Data Governance Maturity Assessment**: Use data governance maturity models to assess the team's maturity level and identify areas for advancement. Maturity assessments provide a roadmap for continuous improvement (Experian 2023).
6. **Peer Reviews**: Engage external experts or peer organizations to conduct reviews and assessments of the team's practices and outcomes. Peer reviews offer objective evaluations and recommendations for improvement (McKinsey 2023a).

Benefits of Measuring Team Effectiveness
1. **Continuous Improvement**: Measuring team effectiveness allows for continuous improvement in data governance practices. Regular evaluations help in identifying and addressing gaps (Experian 2023).

2. **Performance Accountability**: Metrics provide a basis for holding team members accountable for their roles and responsibilities. Accountability ensures that everyone is aligned with the goals of the data governance program (Informatica 2023).
3. **Data-Driven Decision-Making**: Data-driven insights from metrics enable informed decisions on resource allocation and strategy adjustments. Metrics provide evidence for decision-making and prioritization (Experian 2023).
4. **Alignment with Goals**: Measuring team effectiveness helps ensure that data governance efforts align with organizational goals and objectives. Alignment with goals enhances the strategic impact of data governance initiatives (McKinsey 2023a).
5. **Demonstrating Value**: Effective teams can demonstrate their value by highlighting improvements in data quality, compliance, and data utilization. Demonstrating value secures ongoing support and investment from stakeholders (Informatica 2023).

8.10 Conclusion

This chapter has delved deep into the critical aspects of forming and managing an effective data governance team. We explored various facets of data governance roles, responsibilities, and team dynamics, providing insights into best practices and strategies for success.

Key Takeaways from This Chapter
1. **Introduction to Data Governance Roles and Responsibilities:** We began by understanding the foundational concepts of data governance roles and why they are essential in any data governance initiative. Clear roles and responsibilities lay the groundwork for effective data management.
2. **Key Data Governance Roles:** In this section, we examined the specific roles that make up a data governance team, such as data stewards, data owners, and data custodians. Each role has distinct responsibilities that contribute to data governance success.
3. **Establishing a Data Governance Committee:** We explored the importance of establishing a data governance committee, its composition, and its role in overseeing data governance practices. A well-structured committee can provide governance oversight and strategic guidance.
4. **Cross-Functional Collaboration in Data Governance:** Collaboration between different departments and functions is crucial for successful data governance. We highlighted the benefits of cross-functional collaboration and strategies for achieving it.
5. **Training and Skill Development:** Data governance team members need the necessary skills and knowledge to fulfill their roles effectively. We discussed the

importance of training and skill development programs in ensuring team competence.
6. **Building and Sustaining an Effective Data Governance Team:** Building a data governance team is only the beginning; sustaining its effectiveness is an ongoing process. We examined strategies for sustaining team performance and aligning it with organizational objectives.
7. **Case Studies: Successful Data Governance Teams:** Real-world case studies provided insights into how organizations have successfully implemented data governance teams and the outcomes they achieved. These case studies illustrated the practical application of data governance principles.
8. **Overcoming Common Challenges in Team Dynamics:** We addressed usual challenges in team dynamics, such as communication barriers and resistance to change, and explored strategies for overcoming these challenges.
9. **Measuring Team Effectiveness:** Effective measurement of team performance is essential for continuous improvement. We discussed key metrics and approaches for measuring the effectiveness of a data governance team.

This chapter has equipped readers with the knowledge and tools to establish, nurture, and optimize data governance teams within their organizations. By understanding the roles and responsibilities of team members, fostering collaboration, providing training, and measuring effectiveness, organizations can ensure that their data governance efforts align with broader strategic goals and deliver tangible benefits.

As organizations continue to recognize the value of data as a strategic asset, the principles and practices outlined in this chapter will play a pivotal role in driving data governance success and maximizing the value of data assets.

References

Alation (2023) Top 6 common data governance challenges (and their solutions!)
Atlan (2023a) 10 data governance challenges & how to overcome them!
Atlan (2023b) 5 data governance examples: case studies, takeaways & more
BMC Public Health (2023) Effectiveness of capacity building interventions relevant to public health practice: a systematic review
Collibra (2021) Data governance council: what is it and why do you need one?
Datafloq (2023) 35 metrics you should use to monitor data governance
DATAVERSITY (2023a) Case study: implementing data governance for data lakes and big data
DATAVERSITY (2023b) Case study: three strategies for data governance success
DATAVERSITY (2023c) Data governance challenges (and how to fix them)
DATAVERSITY (2023d) Data governance metrics: how to measure success
DATAVERSITY (2023e) Data governance roles and responsibilities
Experian (2023) Data governance metrics & KPIs to measure success
Frontiers (2023) A balanced scorecard for maximizing data performance
Gailhofer P, Franke C (2021) Data governance and regulation for sustainable smart cities. Front Environ Sci

References

Grafanaki S (2017) Regulatory challenges revisited: the cases of infrastructure regulation and data governance. Regulation & Governance

Informatica (2023) Measuring success in data governance

Iron Mountain (2023) Information governance training

LightsOnData (2023) Top 10 data governance courses and training

McKinsey (2023a) Designing data governance that delivers value

McKinsey (2023b) Making collaboration across functions a reality

Omeda (2023) How to achieve cross-functional collaboration in data governance

Onna (2023) Why cross-functional collaboration is critical in data governance

Royal Society (2023) Data governance: case studies

Seiner RS (2021) Non-Invasive data governance: the path of least resistance and greatest success. Technics Publications.

Smarty (2023) Understanding key data governance players: roles and responsibilities

Smithsonian Directive 609 (2019) The CARE principles for indigenous data governance. Data Sci J

Technology Magazine (2023) Navigating the complexity of data governance and AI models

Udemy (2023) Data governance – the complete course for beginners

Y42 (2023) 5 common challenges for cross-functional data governance collaboration

Chapter 9
Data Governance Tools and Technologies: Navigating the Options

Contents

9.1	Introduction to Data Governance Tools and Technologies.	308
9.2	Data Quality Tools.	310
9.3	Data Security and Privacy Tools.	311
9.4	Metadata Management Tools.	313
9.5	Data Cataloging and Inventory Tools: Enhancing Data Management.	314
9.6	Master and Reference Data Management Tools: Empowering Data Consistency.	316
9.7	Data Integration and ETL Tools: Streamlining Data Movement and Transformation.	317
9.8	Evaluating and Selecting Data Governance Tools: Making Informed Choices.	318
9.9	Case Studies: Tools in Practice—Real-World Examples.	320
9.10	Future Trends in Data Governance Technology: Shaping Tomorrow's Data Governance Landscape.	322
9.11	Conclusion: Navigating the Data Governance Journey.	323
References.		325

Abstract This chapter delves into the various technological solutions available to enhance data governance strategies. This chapter serves as a comprehensive guide, helping organizations understand the multitude of tools and technologies that can streamline their data governance efforts.

The discussion begins with an overview of the pivotal role that tools and technologies play in data governance. As organizations continue to amass vast volumes of data, the need for effective management, security, and utilization of these assets has become increasingly important. The chapter outlines the major categories of data governance tools, including data quality tools, data security and privacy tools, metadata management tools, data cataloging and inventory tools, master and reference data management tools, and data integration and ETL tools.

Each category is explored in depth, showcasing how these tools can help organizations assess, cleanse, manage, and secure their data. For instance, data quality tools are emphasized for their ability to maintain the integrity of data through functionalities like profiling, cleansing, and monitoring. Similarly, data security tools are discussed for their critical role in safeguarding sensitive information and ensuring compliance with privacy regulations.

Real-world case studies are included to illustrate the application of these tools in various organizational contexts, providing practical insights and demonstrating their benefits in enhancing data management practices. The chapter also offers guidance on evaluating and selecting the right tools to align with an organization's specific needs, emphasizing factors such as scalability, integration capabilities, user-friendliness, and compliance support.

Furthermore, the chapter looks ahead to future trends in data governance technology, exploring emerging innovations such as AI and machine learning integration, blockchain for data provenance, and advancements in data privacy automation. These discussions highlight how data governance tools are evolving to address the increasingly complex requirements of modern data ecosystems.

In conclusion, this chapter equips readers with the knowledge to make informed decisions about the tools and technologies that can best support their data governance initiatives, ultimately enhancing their ability to manage, secure, and utilize data effectively.

Keywords Data governance tools · Data quality tools · Data security · Metadata management · data cataloging · Master data management · Data integration · ETL tools · Technology evaluation · Future trends · AI integration · Blockchain · Data privacy automation · Cloud data governance

In the digital age, data has become a cornerstone of business operations, decision-making, and innovation. As organizations continue to amass vast volumes of data, the need for effective data governance has never been more pronounced. This chapter delves into the dynamic landscape of data governance tools and technologies, guiding you through the multitude of options available to enhance your data governance strategies.

Introduction to Data Governance Tools and Technologies We begin by providing an overview of the pivotal role that tools and technologies play in data governance. This section sets the stage for exploring the various tool categories and their specific functionalities.

Data Quality Tools Data quality tools empower organizations to assess, cleanse, and maintain data integrity. These tools are crucial for ensuring the accuracy, consistency, and reliability of data, which is paramount for informed decision-making and compliance with regulations (Olson 2003; Batini et al. 2009).

Data Security and Privacy Tools Ensuring the security and privacy of data is paramount. We explore tools and technologies that safeguard data against threats and ensure compliance with privacy regulations. According to Cisco's 2021 Data Privacy Benchmark Study, organizations with mature privacy practices are experiencing benefits such as fewer data breaches, reduced costs, and shorter sales delays (Cisco 2021).

Metadata Management Tools Metadata is the backbone of data governance. Metadata management tools enable organizations to document, organize, and trace their data assets. These tools are essential for understanding the context, lineage, and usage of data within the organization (Loshin 2013).

Data Cataloging and Inventory Tools Effective data governance begins with visibility. Data cataloging and inventory tools help organizations catalog and inventory their data, making it accessible and understandable. These tools provide comprehensive data discovery and metadata management capabilities, facilitating efficient data governance practices (Wang et al. 1998).

Master and Reference Data Management Tools Master and reference data management tools help organizations maintain consistency and accuracy in critical data elements. These tools are essential for ensuring that core data elements, such as customer and product information, are accurate and consistent across the organization (Loshin 2013).

Data Integration and ETL Tools Data integration and ETL (Extract, Transform, Load) tools are vital for harmonizing data from diverse sources. These tools streamline data flows and transform raw data into actionable insights. The importance of these tools is underscored by their ability to support complex data integration and transformation processes, ensuring data quality, and accessibility (Kimball and Caserta 2004).

Evaluating and Selecting Data Governance Tools Selecting the right tools is a crucial decision. We provide guidance on evaluating and choosing tools that align with your organization's needs. Factors such as functionality, scalability, integration capabilities, and user-friendliness are critical in this evaluation process (Soares 2012).

Case Studies: Tools in Practice Real-world case studies highlight how organizations have successfully implemented data governance tools, providing practical insights and inspiration. These examples demonstrate the tangible benefits of adopting robust data governance technologies in various industry contexts.

Future Trends in Data Governance Technology We conclude by looking ahead to the future of data governance technology, and exploring emerging trends and innovations. Trends such as AI and machine learning integration, blockchain for data provenance, data privacy automation, and cloud-native data governance solutions are poised to shape the future landscape of data governance (Mahanti 2019).

9.1 Introduction to Data Governance Tools and Technologies

In the ever-evolving landscape of data governance, tools and technologies are the linchpin that enables organizations to effectively manage, secure, and utilize their data assets. This chapter opens with an introduction to the world of data governance tools and technologies, shedding light on their significance, functionalities, and real-world examples that illustrate their impact.

The Role of Tools and Technologies in Data Governance
Data governance, at its core, is about establishing a framework for managing data effectively throughout its lifecycle. However, as data volumes grow and regulatory requirements become more stringent, manual data governance processes are no longer sufficient. This is where data governance tools and technologies come into play. These tools offer a range of functionalities that streamline data governance efforts:

Data Profiling: Tools like IBM InfoSphere Information Analyzer and Talend provide data profiling capabilities, allowing organizations to assess data quality, discover anomalies, and identify data issues. These tools help in maintaining high data quality, which is critical for making informed business decisions (Olson 2003).

Metadata Management: Solutions like Collibra and Informatica Axon enable organizations to catalog and manage metadata, providing critical insights into data lineage, definitions, and ownership. Effective metadata management is essential for understanding the context and usage of data within an organization (Loshin 2013).

Data Cataloging: Data cataloging tools such as Alation and DataRobot Catalog help organizations create searchable inventories of data assets, making it easier for users to discover and access relevant data. These tools enhance data accessibility and usability, which are vital for data-driven decision-making (Mahanti 2019).

Data Security: Tools like Varonis and Symantec Data Loss Prevention (DLP) focus on data security by monitoring access, detecting unusual activity, and enforcing data protection policies. Ensuring data security is crucial in preventing data breaches and complying with regulatory requirements (Cisco 2021).

Master Data Management (MDM): MDM tools such as SAP Master Data Governance and Informatica MDM ensure consistency and accuracy in master data, such as customer and product information. Maintaining high-quality master data is essential for operational efficiency and customer satisfaction (Loshin 2013).

Data Integration: ETL (Extract, Transform, Load) tools like Talend and Informatica PowerCenter are essential for data integration and harmonization. These tools

9.1 Introduction to Data Governance Tools and Technologies

facilitate the consolidation of data from various sources, ensuring that the integrated data is accurate and reliable (Kimball and Caserta 2004).

Data Privacy: Solutions like OneTrust and TrustArc help organizations comply with data privacy regulations by managing consent, data subject requests, and data protection impact assessments (DPIAs). Compliance with data privacy regulations is critical in building trust with customers and avoiding legal penalties (Cisco 2021).

Real-World Examples

In the landscape of data governance tools and technologies, several standout examples demonstrate the diverse functionalities these solutions can provide to organizations, aiding in managing, securing, and utilizing their data assets effectively.

Collibra Data Governance: Collibra offers robust data stewardship management functions, a Policy Manager for centralized management of data policies and standards, and integration with data catalog, lineage, and quality tools as part of the Data Intelligence Cloud platform. This solution is tailored to streamline the adoption and compliance monitoring of data governance policies (Mahanti 2019).

Precisely Data Integrity Suite (Data Governance Service): Part of the Precisely Data Integrity Suite, this service enables organizations to foster trust with their data assets by offering real-time tracking of data, support for business processes, and outcomes. Its features include automation of data governance workflows, metadata harvesting, and data quality score importing, highlighting the interconnected nature of data integration, observability, governance, quality, enrichment, and analytics (Loshin 2013).

Erwin Data Intelligence by Quest Software: Erwin Data Intelligence combines data catalog, literacy, and quality products in an integrated suite. Aimed at making data assets more visible to end users while guiding their use, it includes automated metadata harvesting and cataloging, data lineage details, and data profiling and quality assessments (Soares 2012).

Salesforce Security and Privacy: This integrates data privacy measures with customer relationship management (CRM) activities, offering comprehensive privacy management, advanced threat detection, multifactor authentication, and extensive compliance coverage. It prevents data loss, controls user access, and integrates smoothly with Salesforce's suite of products, making it an ideal choice for organizations looking to consolidate data privacy with CRM functions (Cisco 2021).

MineOS: Known for its data mining capabilities, MineOS provides operational intelligence from raw data with its intelligent algorithms and easy-to-use interface. It offers versatile data mining capabilities, robust data visualization tools, advanced predictive analytics, and strong integration with various databases and business intelligence platforms like Tableau and PowerBI (Mahanti 2019).

Segment: Segment excels in unifying user data from multiple sources into a single platform, offering a comprehensive solution for customer data management. With a robust data pipeline, privacy and compliance tools, and vast integration capabilities, Segment simplifies data management and enhances data utilization across organizations (Soares 2012).

These examples illustrate the critical role of data governance tools and technologies in enabling effective data governance strategies. By leveraging these solutions, organizations can enhance their data management practices, ensuring data quality, compliance, and strategic utilization of their data assets.

9.2 Data Quality Tools

Data quality is a fundamental aspect of data governance, ensuring the accuracy, consistency, and reliability of data. In this chapter, we delve into the world of data quality tools, which play a crucial role in assessing, improving, and maintaining data quality. This section explores the key functions of data quality tools and provides real-world examples that highlight their importance.

Key Functions of Data Quality Tools

Data Profiling: Data quality tools such as Talend Data Quality and Informatica Data Quality enable organizations to profile their data. This involves analyzing data to identify inconsistencies, missing values, duplicates, and other issues that affect data quality. Data profiling is essential for understanding the current state of data and identifying areas that need improvement (Olson 2003).

Data Cleansing: Tools like Trillium Software and Experian Data Quality provide data cleansing capabilities. They allow organizations to clean and standardize data, correcting errors and inconsistencies. Effective data cleansing ensures that data is accurate and fit for use in decision-making processes (Batini et al. 2009).

Data Validation: Data quality tools offer validation features to ensure that data meets predefined rules and constraints. For example, they can validate email addresses, phone numbers, or postal codes. This validation is critical for maintaining data integrity and compliance with standards (Loshin 2013).

Data Enrichment: Some tools like Melissa Data and StrikeIron offer data enrichment services. They can enhance data by adding missing information such as geolocation data or demographic data. Data enrichment improves the completeness and utility of data for various applications (Wang et al. 1998).

Data Monitoring: Data quality tools provide monitoring capabilities, allowing organizations to continuously track data quality over time. They can set up alerts for anomalies and deviations from predefined quality standards. Continuous monitoring helps in maintaining high data quality and quickly addressing issues as they arise (Batini et al. 2009).

Real-World Examples

Data Profiling with Talend Data Quality: In the retail industry, a multinational company leveraged Talend Data Quality to profile its customer data. By identifying inconsistencies and duplicates in their vast customer databases, the company enhanced the accuracy of its customer insights, leading to improved targeted marketing strategies and customer satisfaction (Olson 2003).

Data Cleansing with Experian Data Quality: A financial services institution used Experian Data Quality to cleanse and standardize its customer data across multiple systems. This process corrected errors and inconsistencies, such as misspelled addresses and outdated contact information, ensuring compliance with financial regulations and enhancing the reliability of risk assessments (Batini et al. 2009).

Data Validation with Informatica Data Quality: An e-commerce platform implemented Informatica Data Quality to validate the integrity of product information across its online listings. This included validating product IDs, descriptions, and pricing information against predefined rules. The tool helped maintain high-quality, consistent product data, improving the customer shopping experience and reducing the incidence of returns due to inaccurate product information (Loshin 2013).

Data Enrichment with Melissa Data: A healthcare provider utilized Melissa Data for data enrichment to append additional demographic and geographic information to patient records. This enrichment provided valuable insights into patient populations, aiding in the development of tailored health programs and services that address specific community health needs (Wang et al. 1998).

Data Monitoring with StrikeIron: A telecommunications company used StrikeIron for continuous data quality monitoring. By setting up alerts for anomalies in call data records (CDRs), such as sudden spikes in call failures or unusual patterns in international calls, the company was able to quickly identify and address potential issues in its network, ensuring consistent service quality and customer satisfaction (Batini et al. 2009).

These examples from various industries illustrate the critical role of data quality tools in enhancing the accuracy, consistency, reliability, and enrichment of data. By leveraging such tools, organizations can significantly improve their decision-making processes, operational efficiency, and regulatory compliance.

9.3 Data Security and Privacy Tools

In an era of escalating data breaches and stringent data protection regulations, safeguarding sensitive information is paramount. This chapter explores the realm of data security and privacy tools, which are instrumental in ensuring that data remains secure, compliant, and confidential. This section provides insights into data security and privacy tools, with real-world examples that highlight their significance in today's data-driven landscape.

Key Functions of Data Security and Privacy Tools

Data Encryption: Tools like BitLocker and FileVault encrypt data on storage devices, protecting it from unauthorized access. Encryption is crucial for safeguarding data at rest, ensuring that even if physical devices are lost or stolen, the data remains secure (Stallings 2017).

Access Control: Solutions such as Okta and Microsoft Azure Active Directory manage user access to applications and data, enforcing policies that ensure only authorized users can access sensitive information. Effective access control is vital in protecting against unauthorized data access and breaches (Ferraiolo et al. 2003).

Data Masking and Redaction: By concealing or replacing sensitive data in non-production environments, tools like Delphix and IBM Guardium help organizations comply with privacy regulations and protect personal information from unnecessary exposure. Data masking and redaction are essential for maintaining privacy during data processing and analysis (Loshin 2013).

Data Loss Prevention (DLP): DLP solutions, such as Symantec DLP and McAfee DLP, monitor and control data transfers, preventing unauthorized sharing and leakage of sensitive information. These tools are key to enforcing data protection policies and ensuring regulatory compliance (Gartner 2020).

Identity and Access Management (IAM): IAM tools like Ping Identity and OneLogin manage user identities and permissions, enhancing security by ensuring that only authenticated and authorized users can access resources. They support compliance and security by managing the user access lifecycle and privileges (Ferraiolo et al. 2003).

Real-World Examples

IBM and Bankdata: Bankdata, a provider of development skills and capacity for banks, relies on IBM Z® servers for its core transaction processes and data delivery to new applications. They faced challenges in monitoring workload at the application level, crucial for optimizing application performance and reducing processing fees. This example underscores the need for detailed monitoring and management tools to understand and optimize the workload, directly impacting software costs and end-user experience in the banking sector (IBM 2021).

Okta's Customer Success Stories: Okta provides identity and access management solutions across various industries, including non-profit, education, retail, finance, healthcare, and more. Their tools, such as Single Sign-On and Multifactor Authentication, help protect against data breaches, move to the cloud, and enhance customer engagement. Okta's solutions highlight the critical role of access control and identity management in safeguarding sensitive information and ensuring that only authorized individuals can access critical systems and data (Okta 2021).

Salesforce Security and Privacy: Salesforce integrates data privacy measures with customer relationship management (CRM) activities, offering comprehensive privacy management, advanced threat detection, multi-factor authentication, and extensive compliance coverage. It prevents data loss, controls user access, and integrates smoothly with Salesforce's suite of products, making it an ideal choice for organizations looking to consolidate data privacy with CRM functions (Salesforce 2021).

These examples demonstrate the application of data security and privacy tools in the real world. They also emphasize the broader importance of such technologies in protecting sensitive data, optimizing system performance, and ensuring compliance with data protection regulations across different sectors.

9.4 Metadata Management Tools

Metadata is the lifeblood of effective data governance. It provides the context and structure necessary to understand, manage, and utilize data assets within an organization. This chapter explores metadata management tools, which are essential for cataloging, organizing, and tracing metadata information. In this section, we delve into the specific functionalities of metadata management tools and provide real-world examples to illustrate their importance.

Key Functions of Metadata Management Tools

Metadata Cataloging: Metadata management tools like Collibra and Informatica Axon allow organizations to catalog metadata information. This includes data definitions, data lineage, data ownership, and data relationships. Metadata cataloging is crucial for creating an organized repository of data assets that users can easily search and navigate (Loshin 2013).

Data Lineage: These tools provide data lineage capabilities, enabling organizations to trace the flow of data from its source to its destination. Understanding data lineage is essential for tracking data transformations and dependencies, which is critical for data quality and regulatory compliance (Bowers 2011).

Metadata Governance: Metadata management tools facilitate metadata governance by defining roles and responsibilities for managing metadata. This ensures that metadata remains accurate, up-to-date, and well-documented. Effective metadata governance supports data stewardship and enhances overall data management (Loshin 2013).

Data Dictionary: Metadata management tools often include a data dictionary, which acts as a central repository for data definitions, business glossaries, and data element descriptions. A comprehensive data dictionary helps in standardizing data terminology and improving communication across the organization (Kimball and Ross 2013).

Impact Analysis: These tools offer impact analysis features, allowing organizations to assess the potential impact of changes to data structures or data sources on downstream processes and reports. Impact analysis helps in proactive change management and minimizes disruptions in data workflows (Bowers 2011).

Real-World Examples

Automating Metadata Management with Ataccama: A client migrated their metadata management solution to the Ataccama ONE Data Catalog, creating an automated process for managing and updating metadata. This migration highlights the importance of automation in maintaining up-to-date and accurate metadata across the organization (Ataccama 2021).

Metadata Management Tools and Best Practices by XenonStack: XenonStack provides an overview of metadata management tools and best practices, mentioning specific examples such as Informatica Metadata Manager, Manta, and Collibra. This article highlights the critical role these tools play in providing searchable inventories of data assets, improving data understanding and utilization (XenonStack 2020).

Standardized Data Product Metadata Examples Based on Real-World Published Data Products by Jarkko Moilanen: This Medium post discusses the vital role of metadata management in the data economy, offering examples based on real-world published data products. It emphasizes the importance of standardized metadata for enhancing data discoverability and interoperability (Moilanen 2021).

These examples illustrate how metadata management tools are utilized in various contexts to catalog, organize, and ensure the traceability of metadata. By leveraging these tools, organizations can enhance data governance, discoverability, and the overall management of data assets within their systems.

9.5 Data Cataloging and Inventory Tools: Enhancing Data Management

Data cataloging and inventory tools are indispensable assets in the realm of modern data governance. They serve as the linchpin for effective data management by enabling organizations to discover, categorize, and manage their data assets systematically. In this section, we explore the critical functions of these tools with real-world examples illustrating their practical significance.

Key Functions of Data Cataloging and Inventory Tools

Data Discovery: Data cataloging tools provide a user-friendly interface for discovering and accessing data assets within an organization. Users can search for and locate data sources, datasets, and specific data elements swiftly. For instance, Alation is a widely used data cataloging tool offering intuitive search and discovery features, allowing users to quickly locate relevant data assets (Mahanti 2019).

Data Profiling: Some data cataloging tools incorporate data profiling capabilities. This empowers users to evaluate the quality, completeness, and structure of data before using it for analytical or operational purposes. Informatica Axon includes data profiling features, enabling data analysts to assess the quality of financial data and identify anomalies or inconsistencies (Informatica 2020c).

Data Categorization: Data cataloging tools facilitate the systematic categorization and classification of data assets. This helps in organizing data based on attributes such as data type, security level, or business relevance. Waterline Data offers data

9.5 Data Cataloging and Inventory Tools: Enhancing Data Management

categorization capabilities, allowing organizations to tag data assets according to their classification, ensuring proper handling (Mahanti 2019).

Data Lineage: Many data cataloging tools provide data lineage visualization, which illustrates the path of data from its origin to its utilization. This helps users understand data dependencies and transformations. Informatica Enterprise Data Catalog offers data lineage visualization, enabling organizations to track how data flows through their systems and applications (Informatica 2020c).

Data Ownership: These tools assign data ownership responsibilities to individuals or teams, ensuring accountability for maintaining and governing specific data assets. Collibra enables organizations to designate data stewards responsible for overseeing and managing data assets within their domains (Loshin 2013).

Real-World Examples

Alation for Data Discovery: Alation offers intuitive search and discovery features, allowing users to quickly locate relevant data assets. In a large enterprise, Alation facilitated the efficient discovery of data sources, enabling data scientists to spend less time searching for data and more time on analysis (Mahanti 2019).

Data Profiling with Informatica Axon: In the financial sector, a global bank used Informatica Axon for data profiling to assess the quality of its financial data. By identifying data anomalies and inconsistencies, the bank improved the reliability of its financial reports and compliance with regulatory requirements (Informatica 2020c).

Data Categorization with Waterline Data: A healthcare organization utilized Waterline Data to categorize patient records based on data sensitivity levels. This categorization ensured that sensitive patient information was handled with the highest security standards, improving data governance and compliance with health regulations (Mahanti 2019).

Data Lineage with Informatica Enterprise Data Catalog: A manufacturing company implemented Informatica Enterprise Data Catalog to visualize data lineage, tracing the flow of data from raw material inputs to finished product outputs. This visualization helped the company optimize its supply chain processes and improve data accuracy (Informatica 2020c).

Data Ownership with Collibra: A technology firm designated data stewards using Collibra to manage its vast array of digital assets. By establishing clear ownership and governance responsibilities, the firm enhanced its data quality and governance practices, ensuring that data assets were well-maintained and accurately documented (Loshin 2013).

These functions empower organizations to streamline data management, enhance data quality, and comply with data governance regulations. Data cataloging and inventory tools are vital components of a comprehensive data governance strategy, ensuring that data is accessible, well-documented, and utilized effectively.

9.6 Master and Reference Data Management Tools: Empowering Data Consistency

Master and reference data management tools are pivotal in maintaining data consistency and accuracy within organizations. They play a crucial role in ensuring that core data elements remain consistent and reliable across various systems and applications. In this section, we explore the essential functions of these tools and provide real-world examples to illustrate their significance.

Key Functions of Master and Reference Data Management Tools

Master Data Management (MDM): MDM tools facilitate the creation of a single authoritative source of master data, such as customer information, product data, or employee records. They ensure that master data is consistent and synchronized across the organization. For example, Informatica MDM is a widely used solution that allows organizations to establish a sole source of truth for customer data, ensuring data consistency across all systems (Loshin 2013).

Reference Data Management: These tools manage reference data, which includes codes, classifications, and standard data values used in various systems. They ensure that reference data is up-to-date and consistent. Semarchy xDM is a reference data management platform that helps organizations govern reference data effectively, ensuring consistency in code sets and classifications (Mahanti 2019).

Data Governance: Master and reference data management tools often include data governance features, allowing organizations to define data ownership, stewardship, and data quality rules for critical data elements. Informatica Axon, in addition to MDM, offers robust data governance capabilities, enabling organizations to govern master and reference data effectively (Loshin 2013).

Data Quality: These tools provide data quality features to validate, cleanse, and enrich master and reference data. This ensures that the data is accurate and reliable. Talend MDM includes data quality capabilities to profile and cleanse master data, improving data accuracy (Batini et al. 2009).

Real-World Examples

Informatica MDM: Informatica's Master Data Management (MDM) solution is designed to create a single authoritative source of master data for entities like customer information, product data, or employee records across an organization. By consolidating and managing master data, Informatica MDM ensures data consistency and synchronization across various systems, significantly improving data reliability and decision-making processes (Informatica 2020d).

Semarchy xDM: Semarchy xDM is known for its excellence in managing reference data, including codes, classifications, and standard data values used across different systems within organizations. By governing reference data effectively, Semarchy xDM ensures that code sets and classifications remain up-to-date and consistent, which is crucial for operational efficiency and regulatory compliance (Mahanti 2019).

Data Governance with Informatica Axon: Informatica Axon, complementing Informatica MDM, provides robust data governance capabilities, enabling organizations to establish clear data ownership, stewardship, and data quality rules. This structured approach to governing both master and reference data enhances the overall data management strategy and ensures compliance with data protection regulations (Loshin 2013).

Data Quality with Talend MDM: Talend MDM incorporates data quality features to validate, cleanse, and enrich master and reference data, improving data accuracy and reliability. With these capabilities, Talend MDM supports organizations in maintaining high-quality data, crucial for accurate analytics, reporting, and decision-making (Batini et al. 2009).

These examples illustrate the critical role of master and reference data management tools in maintaining data consistency, accuracy, and governance across various organizational contexts. They underscore the importance of having a unified view of data, effective data governance practices, and quality data for operational excellence and informed decision-making.

9.7 Data Integration and ETL Tools: Streamlining Data Movement and Transformation

Data integration and ETL (Extract, Transform, Load) tools are fundamental for orchestrating the flow of data across systems, applications, and databases. They enable organizations to extract, transform, and load data efficiently, ensuring that it is accessible and usable. In this section, we explore the key functions of these tools and provide real-world examples of top solutions available today.

Key Functions of Data Integration and ETL Tools

Data Extraction: Data integration and ETL tools can extract data from various sources, including databases, files, APIs, and streaming data. They provide connectivity to a wide range of data sources. Apache Nifi, for example, is an open-source data integration tool that supports data extraction from diverse sources, making it suitable for real-time data integration (Apache 2021a).

Data Transformation: These tools offer data transformation capabilities to cleanse, format, and enrich data as it moves from source to target. Transformation functions include filtering, aggregation, and data enrichment. Talend is a popular ETL tool that allows users to design data transformation processes using a user-friendly interface (Mahanti 2019).

Data Loading: Data integration and ETL tools load transformed data into target systems such as data warehouses, data lakes, or cloud platforms. They support batch and real-time data loading. Apache Spark is a powerful data integration and processing framework that can load data into various data stores, including Hadoop, databases, and cloud storage (Zaharia et al. 2016).

Data Orchestration: These tools enable the orchestration of complex data workflows, including scheduling and monitoring of data integration tasks. Apache Airflow is an open-source platform for orchestrating complex data workflows, allowing users to schedule, monitor, and manage data integration tasks (Apache 2021b).

Real-World Examples

Apache NiFi for Data Extraction: Apache NiFi supports automated data flow between systems, making it ideal for real-time data integration from diverse sources. Its design caters to the challenges of system failures, data overload, and complex data flows, highlighting its utility in environments where data sources and destinations are varied and dynamic (Apache 2021a).

Talend for Data Transformation: Talend is recognized for its data transformation capabilities, allowing users to design complex data transformation processes through a user-friendly interface. It supports scaling massive datasets to analytics platforms and integrates with leading cloud service providers, highlighting its adaptability and power in data transformation scenarios (Mahanti 2019).

Apache Spark for Data Loading: Apache Spark excels in processing and loading large-scale data into various data stores, including Hadoop, databases, and cloud storage. Its comprehensive documentation offers insights into generic load/save functions, demonstrating Spark's flexibility in handling data formats and sources (Zaharia et al. 2016).

Apache Airflow for Data Orchestration: Apache Airflow provides a platform for orchestrating complex data workflows, allowing users to schedule, monitor, and manage data integration tasks. Its design principle of "workflows as code" makes it highly dynamic and extensible, suitable for various data environments from simple batch processes to complex data pipelines (Apache 2021b).

These examples highlight the critical role of Data Integration and ETL tools in facilitating efficient data movement and transformation across systems, ensuring data is accessible and usable for analytics and decision-making processes.

9.8 Evaluating and Selecting Data Governance Tools: Making Informed Choices

Selecting the right data governance tools is a critical decision for organizations aiming to establish robust data governance practices. This section explores the key considerations when evaluating and choosing data governance tools and provides examples of top tools available today.

Considerations for Evaluating and Selecting Data Governance Tools

Alignment with Business Needs: Ensure that the selected tool aligns with your organization's specific data governance goals and requirements. Assess whether it offers functionalities tailored to your industry and compliance needs. For instance, tools like Collibra provide a comprehensive suite of data governance and cataloging solutions that can be customized to fit various business contexts (Loshin 2013).

Scalability: Consider the scalability of the tool to accommodate the growing volume of data and evolving governance needs. It should adapt to the organization's changing data landscape. Informatica Axon, for example, is known for its scalability and ability to manage large datasets effectively (Informatica 2020a).

Integration Capabilities: Evaluate the tool's ability to integrate seamlessly with existing data systems, applications, and data sources. Compatibility with your technology stack is essential. Tools like Talend Data Catalog offer robust integration capabilities, ensuring smooth interoperability with various data environments (Mahanti 2019).

User-Friendly Interface: An intuitive and user-friendly interface is crucial for user adoption. The tool should be accessible to both technical and non-technical users within your organization. Alation is known for its user-friendly interface, which promotes ease of use and high adoption rates (Alation 2020).

Data Discovery and Cataloging: Look for tools that excel in data discovery, cataloging, and metadata management. Effective data discovery is essential for locating and understanding data assets. IBM Watson Knowledge Catalog provides comprehensive data cataloging features, enhancing data discoverability and usability (IBM 2020).

Data Quality and Profiling: Assess the tool's data quality and profiling capabilities. It should support data cleansing, validation, and profiling to ensure data accuracy. Talend Data Quality offers robust data profiling and cleansing functionalities, which are critical for maintaining high data quality (Talend 2020).

Data Security: Prioritize tools with robust data security features to protect sensitive information. Compliance with data privacy regulations is essential. OneTrust, for example, provides extensive data privacy management features, ensuring compliance with various regulations such as GDPR and CCPA (OneTrust 2020).

Collaboration and Workflow: Consider tools that facilitate collaboration among data stakeholders and support customizable workflows for data governance processes. Collibra offers extensive collaboration features, allowing for effective communication and coordination among data stewards and other stakeholders (Loshin 2013).

Reporting and Analytics: Look for tools that offer reporting and analytics capabilities to monitor data governance activities, track progress, and identify areas for improvement. Informatica Axon includes powerful reporting and analytics features, enabling organizations to gain insights into their data governance efforts (Informatica 2020a).

Examples of Data Governance Tools

Collibra: Collibra is a leading data governance platform that offers a comprehensive suite of data governance and cataloging solutions. It provides data discovery, lineage, and data quality features, making it a popular choice for organizations seeking robust data governance (Loshin 2013).

Informatica Axon: Informatica Axon is a data governance and data cataloging tool that focuses on data governance and data lineage. It enables organizations to define data ownership, stewardship, and data quality rules (Informatica 2020a).

Alation: Alation is a data catalog and data governance tool known for its data discovery and cataloging capabilities. It offers collaboration features that promote data governance and data stewardship (Alation 2020).

IBM Watson Knowledge Catalog: IBM's Knowledge Catalog is a data cataloging and governance solution that includes data profiling, data quality, and metadata management features. It integrates well with IBM's broader data management offerings (IBM 2020).

Talend Data Catalog: Talend offers a data cataloging tool with data quality and data profiling capabilities. It focuses on data governance for organizations seeking end-to-end data management solutions (Talend 2020).

These are some of the data governance tools available, each with its unique strengths and features. When selecting a tool, it is essential to conduct a thorough evaluation based on your organization's specific needs, budget, and long-term data governance goals. Making an informed choice ensures that the selected tool effectively supports your data governance initiatives.

9.9 Case Studies: Tools in Practice—Real-World Examples

In this section, we delve into real-world case studies that demonstrate the practical application of data governance tools within organizations. These case studies highlight how leading companies utilize data governance tools to enhance data management, compliance, and decision-making processes.

Global Retailer Enhances Data Accessibility with Collibra
A multinational retail giant implemented Collibra to address data sprawl and improve data accessibility and quality. Prior to using Collibra, the retailer faced challenges with siloed data, making it difficult for different departments to access and utilize data effectively. By implementing Collibra, the retailer centralized its data governance framework, creating a unified data catalog accessible across the organization. This enhanced data accessibility led to better decision-making and operational efficiency (Collibra 2021).

Financial Institution Ensures Data Compliance with Informatica Axon
A global financial institution used Informatica Axon for data governance and lineage to maintain compliance with regulatory requirements. The institution needed to ensure that its data governance practices met stringent financial regulations, such as GDPR and CCPA. Informatica Axon provided a comprehensive solution for managing data lineage, ensuring data transparency and accountability. This implementation helped the institution maintain compliance, reduce regulatory risks, and improve data trustworthiness (Informatica 2020a).

9.9 Case Studies: Tools in Practice—Real-World Examples

Tech Company Boosts Data Quality with IBM Watson Knowledge Catalog
A technology company specializing in software solutions utilized IBM Watson Knowledge Catalog to enhance data quality and governance. The company faced challenges with inconsistent and inaccurate data across its systems. By implementing IBM Watson Knowledge Catalog, the company was able to profile, cleanse, and enrich its data, leading to improved data quality. This enhanced data quality supported more accurate analytics and reporting, driving better business outcomes (IBM 2020).

Healthcare Provider Ensures Patient Data Security with Talend Data Catalog
A healthcare provider deployed Talend Data Catalog to ensure compliance with patient data privacy regulations and secure sensitive data. The provider needed to comply with health regulations such as HIPAA, which mandates stringent data protection measures. Talend Data Catalog provided the necessary tools for cataloging and securing patient data, ensuring that data access was restricted to authorized personnel only. This implementation improved data security, compliance, and patient trust (Talend 2020).

Real-World Examples
Bankdata Optimizes Application Performance with IBM:

Bankdata, a provider of development skills and capacity for banks, relies on IBM Z® servers for its core transaction processes and data delivery to new applications. They faced challenges in monitoring workload at the application level, crucial for optimizing application performance and reducing processing fees. This example underscores the need for detailed monitoring and management tools to understand and optimize workload, which directly impacts software costs and end-user experience in the banking sector (IBM 2021).

Non-Profit Enhances Security with Okta:

Okta provides identity and access management solutions across various industries, including non-profit, education, retail, finance, healthcare, and more. A non-profit organization utilized Okta's Single Sign-On and Multifactor Authentication to protect against data breaches and enhance security. Okta's solutions highlight the critical role of access control and identity management in safeguarding sensitive information and ensuring that only authorized individuals can access critical systems and data (Okta 2021).

Telecommunications Company Monitors Data Quality with StrikeIron:

A telecommunications company used StrikeIron for continuous data quality monitoring. By setting up alerts for anomalies in call data records (CDRs), such as sudden spikes in call failures or unusual patterns in international calls, the company was able to quickly identify and address potential issues in its network, ensuring consistent service quality and customer satisfaction (Batini et al. 2009).

These case studies illustrate the critical role of data governance tools in addressing data management challenges, improving data quality, ensuring compliance, and supporting better decision-making across various industries. By implementing these tools, organizations can enhance their data governance practices and achieve significant business benefits.

9.10 Future Trends in Data Governance Technology: Shaping Tomorrow's Data Governance Landscape

As organizations continue to navigate the complex data landscape, it is essential to stay ahead of emerging trends in data governance technology. This section explores future trends and innovations that are shaping the data governance landscape, along with real examples of technologies that exist today.

Future Trends in Data Governance Technology

AI and Machine Learning Integration: The integration of artificial intelligence (AI) and machine learning (ML) into data governance tools is becoming increasingly prevalent. These technologies can automate data discovery, classification, and even decision-making in data governance processes. AI-driven data governance tools can enhance efficiency by identifying patterns and anomalies in large datasets that would be time-consuming to detect manually (Mahanti 2019).

Example: Alation, a data cataloging tool, incorporates AI-driven recommendations to assist users in finding relevant data assets based on their behavior and preferences (Alation 2020).

Blockchain for Data Provenance: Blockchain technology is being explored to establish data provenance and enhance data transparency. By recording data lineage and changes on a blockchain, organizations can ensure data integrity and traceability. Blockchain's immutable nature makes it ideal for verifying the authenticity and history of data (Zyskind et al. 2015).

Example: Factom Harmony utilizes blockchain technology to secure and prove data integrity, making it suitable for industries requiring immutable records (Factom 2020).

Data Privacy Automation: With the growing emphasis on data privacy regulations, tools that automate data privacy compliance, such as data masking and encryption, are on the rise. Automation in data privacy helps organizations manage consent, data subject requests, and data protection impact assessments more efficiently (Cisco 2021).

Example: OneTrust Privacy Management Software offers automated solutions for GDPR compliance, including data subject request automation (OneTrust 2020).

Data Governance in the Cloud: As organizations migrate data and applications to the cloud, data governance tools are adapting to provide cloud-native solutions that address cloud-specific data management challenges. Cloud-based data governance tools offer scalability and flexibility, making them suitable for dynamic data environments (Mahanti 2019).

Example: Collibra offers a cloud-based version of its data governance platform, enabling organizations to manage data in a cloud environment securely (Collibra 2020).

Advanced Data Cataloging: Data cataloging tools are evolving to provide more advanced metadata management, semantic search, and natural language processing capabilities, making data discovery even more efficient. Enhanced data cataloging tools facilitate better data understanding and utilization across the organization (Loshin 2013).

Example: Informatica's Enterprise Data Catalog uses AI and natural language processing to enhance data discovery and metadata management (Informatica 2020b).

Examples of Emerging Technologies

DataRobot: DataRobot is an AI-driven platform that automates the development and deployment of machine learning models. It can be integrated with data governance tools to enhance data-driven decision-making by providing predictive analytics and insights (DataRobot 2021).

HashiCorp Vault: HashiCorp Vault is a popular solution for managing secrets and protecting sensitive data. It can be integrated with data governance tools to ensure secure access to sensitive data, enhancing overall data security (HashiCorp 2021).

Okera: Okera is a data access platform that provides secure and compliant data access across data lakes and data warehouses. It addresses data privacy and compliance challenges by offering fine-grained access controls and audit capabilities (Okera 2020).

Immuta: Immuta is a data access and control platform that offers fine-grained access control and data masking to protect sensitive data and ensure compliance. Immuta's dynamic access control capabilities allow organizations to implement data privacy policies effectively (Immuta 2020).

These emerging trends and technologies are reshaping the data governance landscape, offering innovative solutions to address the evolving data management challenges faced by organizations. Staying informed about these trends and strategically adopting relevant technologies can empower organizations to enhance their data governance capabilities and unlock the full potential of their data assets.

9.11 Conclusion: Navigating the Data Governance Journey

In this chapter, we embarked on a journey through the intricate landscape of data governance tools and technologies. From the foundational principles of data governance to the latest innovations and future trends, we have navigated the diverse options available to organizations seeking to harness the power of their data.

We began with an introduction to the world of data governance tools and technologies, recognizing their pivotal role in data management, security, and compliance. This introduction set the stage for a deep dive into the specific categories of tools and technologies that shape the data governance landscape.

We explored the importance of data quality tools, which ensure that data is accurate, reliable, and fit for its intended purpose. These tools are the gatekeepers of data integrity, enabling organizations to make informed decisions based on trustworthy data.

Data security and privacy tools emerged as essential guardians of sensitive information. In an era of stringent data privacy regulations, these tools are instrumental in safeguarding data assets and ensuring compliance with legal and regulatory requirements.

Metadata management tools took us into the world of data documentation and lineage, where the understanding of data context and relationships is paramount. These tools enable organizations to unravel the intricate web of data sources and dependencies.

Data cataloging and inventory tools opened the door to efficient data discovery and access. They empower organizations to locate and leverage data assets effectively, fueling data-driven decision-making.

Master and reference data management tools highlighted the significance of maintaining accurate and consistent core data elements. They are the backbone of data standardization and consistency across the organization.

Data integration and ETL tools illuminated the path to harmonizing data from diverse sources. They streamline data flows, transforming raw data into actionable insights.

The chapter also guided us through the process of evaluating and selecting data governance tools. With a multitude of options available, making informed choices is critical to aligning tools with organizational needs and goals.

Real-world case studies demonstrated how these tools are applied in practice, illustrating their transformative impact on data management, compliance, and decision-making within organizations.

Finally, we glimpsed into the future of data governance technology, where AI, blockchain, data privacy automation, and cloud-native solutions are poised to shape the landscape further.

As we conclude this chapter, we recognize that data governance tools and technologies are not just tools; they are enablers of data excellence, empowering organizations to unlock the true potential of their data assets. By embracing these tools, staying attuned to emerging trends, and making informed choices, organizations can embark on a data governance journey that leads to data-driven success and innovation.

References

Alation (2020) Data catalog & data governance platform
Apache (2021a) Apache NiFi documentation
Apache (2021b) What is Airflow™?
Ataccama (2021) Automating metadata management
Batini C, Cappiello C, Francalanci C, Maurino A (2009) Methodologies for data quality assessment and improvement. ACM Comput Surv (CSUR) 41(3):1–52
Bowers S (2011) Scientific data management in the coming decade. Data Sci J 10:12–24
Cisco (2021) Data privacy benchmark study. Retrieved from https://www.cisco.com/c/en/us/about/trust-center/privacy.html
Collibra (2020) Data governance in the cloud
Collibra (2021) Customer stories
DataRobot (2021) AI and machine learning platform
Factom (2020) Harmony integrates blockchain technology
Ferraiolo DF, Kuhn DR, Chandramouli R (2003) Role-based access control. Artech House
Gartner (2020) Market guide for data loss prevention
HashiCorp (2021) Vault: identity-based security for modern applications
IBM (2020) Watson knowledge catalog
IBM (2021) Bankdata case study
Immuta (2020) Data access & control
Informatica (2020a) Axon data governance
Informatica (2020b) Enterprise data catalog
Informatica (2020c) Informatica axon data governance
Informatica (2020d) Master Data Management (MDM) solutions and tools. Retrieved from https://www.informatica.com/products/master-data-management.html
Kimball R, Caserta J (2004) The data warehouse ETL toolkit: practical techniques for extracting, cleaning, conforming, and delivering data. Wiley
Kimball R, Ross M (2013) The data warehouse toolkit: the definitive guide to dimensional modeling. Wiley
Loshin D (2013) The practitioner's guide to data quality improvement. Morgan Kaufmann
Mahanti R (2019) Data governance and data management: contextualizing data governance drivers, technologies, and tools. Kindle Edition
Moilanen J (2021) Standardized data product metadata examples based on real-world published data products
Okera (2020) Data access platform
Okta (2021) Customer success stories
Olson JE (2003) Data quality: the accuracy dimension. Morgan Kaufmann
OneTrust (2020) Privacy management software
Salesforce (2021) Salesforce security and privacy
Soares S (2012) Data governance tools: evaluation criteria, big data governance, and alignment with enterprise data management, 1st edn
Stallings W (2017) Cryptography and network security: principles and practice. Pearson
Talend (2020) Data catalog
Wang RY, Lee YW, Pipino LL, Strong DM (1998) Manage your information as a product. Sloan Manag Rev 39(4):95–105
XenonStack (2020) Metadata management tools and best practices
Zaharia M, Chowdhury M, Franklin MJ, Shenker S, Stoica I (2016) Spark: cluster computing with working sets. In: Proceedings of the 2nd USENIX conference on hot topics in cloud computing
Zyskind G, Nathan O, Pentland A (2015) Decentralizing privacy: using blockchain to protect personal data. In: 2015 IEEE security and privacy workshops. IEEE, pp 180–184

Chapter 10
Data Governance in Different Industries: Case Studies and Applications

Contents

10.1	Introduction to Data Governance Across Industries	329
10.2	Data Governance in Healthcare	330
10.3	Data Governance in Finance and Banking	331
10.4	Data Governance in Retail and E-Commerce	331
10.5	Data Governance in Government and Public Sector	332
10.6	Data Governance in Technology and Telecommunications	333
10.7	Data Governance in Manufacturing and Logistics	334
10.8	Emerging Industries and Data Governance	335
10.9	Lessons Learned and Best Practices	335
10.10	Conclusion and Industry-Specific Considerations	336
References		337

Abstract This chapter provides a detailed examination of how data governance principles are applied across various sectors, emphasizing the unique challenges, solutions, and successes in each. This chapter uses an array of case studies to illuminate the practical implementation and significant benefits of data governance across diverse domains.

The chapter begins by exploring data governance in healthcare, highlighting how organizations use data to improve patient outcomes and innovate while ensuring compliance with stringent privacy regulations. Case studies such as the implementation of a Data Lake at a healthcare organization showcase the critical balance between accessibility and security of patient information.

In the realm of finance and banking, the chapter discusses how data governance supports accuracy, security, and regulatory compliance, crucial for protecting customer assets and preventing fraud. A notable example includes Wells Fargo's strategy to create a single source of truth to enhance data reliability and streamline reporting.

The examination extends to the public sector, where data governance facilitates improved public service delivery and policy-making. It outlines how cross-agency collaboration and big data utilization can inform national policy and enhance public sector efficiency.

In the retail and e-commerce sectors, the chapter describes how data governance aids in optimizing supply chain management and personalizing customer experiences. It emphasizes the role of data governance in enabling dynamic pricing strategies and enhancing customer loyalty programs.

The narrative then shifts to technology and telecommunications, where data governance is essential for managing vast amounts of data generated from customer interactions and network operations. It discusses how telecommunications giants like GE Aviation centralize their data sources to improve accessibility and reliability.

Each sector presents unique challenges that require tailored data governance strategies to manage data effectively, ensure compliance, and drive innovation. Through these detailed case studies, this chapter not only highlights the versatility of data governance practices but also demonstrates their necessity in today's data-driven world, showing how they fuel innovation, drive competitive advantage, and transform organizations across multiple industries.

Keywords Data governance · Industry-specific applications · Healthcare · Finance · Retail · Public sector · Technology · Telecommunications · Case studies · Regulatory compliance · Data privacy · Data integration · Data quality · Centralized data sources · Data literacy · Innovation · Operational efficiency · Data security

In the ever-evolving data-driven landscape, the application of data governance extends far beyond the boundaries of a single industry. Chap. 10 embarks on a captivating exploration of how data governance principles are implemented across diverse sectors and industries. Through an array of illuminating case studies and real-world applications, we will delve into the unique challenges, solutions, and success stories that define data governance in different domains.

Case Studies
- **Healthcare—Data Lake and Big Data Management:** A healthcare organization focused on creating a common accessible repository for data governance while protecting patient information and encouraging innovation. They used a Data Lake concept to avoid creating silos of data, ensuring a common repository with protected information but maintaining accessibility. This approach involved smaller stand-alone products to support the rapid development of analytics prototypes and predictive models, emphasizing the importance of managing large and diverse data sets in healthcare (DATAVERSITY 2018).
- **Airbnb—Data Literacy for Decision-Making:** Airbnb launched an in-house educational initiative called "Data University" to boost data literacy across all departments. This program was designed to equip employees to understand, interpret, and use data effectively, helping democratize data and scale decision-

making while ensuring responsible data use. This case study highlights the connection between data literacy and effective data governance (Atlan n.d.-a).
- **GE Aviation—Centralizing Data Sources:** GE Aviation embarked on a mission to centralize its scattered data sources, making them more accessible and reliable for all users within the organization. They set up teams responsible for user enablement, tooling, data product deployment, and exploring new opportunities to improve processes. This approach accelerated decision-making processes and instilled a sense of data ownership among employees (Atlan n.d.-b).
- **Wells Fargo—Single Source of Truth for Governance:** Wells Fargo's data governance strategy emphasized creating a sole source of truth to enhance data accuracy and reliability. They centralized data from multiple sources to create a unified, trustworthy source, streamlining data management and allowing for more accurate reporting and analysis. This approach highlights the importance of centralization in data governance (Atlan n.d.-c).
- **Public Sector—Data Sharing and Collaboration:** Several public sector examples demonstrate the value of data governance in government settings. These include cross-agency data collaboration for insights on foreign direct investment, leveraging AI for business process automation, and using big data to inform national policy. Such case studies illustrate the opportunities and challenges of using big data and administrative data in the federal government (OECD iLibrary n.d.).

Each of these case studies provides insights into the practical implementation of data governance principles and the significant benefits they can bring to an organization. They cover a range of industries and scenarios, highlighting the versatility and necessity of effective data governance in today's data-driven world.

From healthcare to finance, manufacturing to retail, and beyond, this chapter will highlight the versatility of data governance practices. By examining these cases, we gain insights into how data governance not only ensures data quality and compliance but also fuels innovation, drives competitive advantage, and transforms organizations across a multitude of sectors.

10.1 Introduction to Data Governance Across Industries

In Chap. 10, we embark on a compelling exploration that transcends industry boundaries and delves into the diverse landscape of data governance across various sectors. This introduction sets the stage for an investigation of how data governance principles are applied in industries as distinct as healthcare, finance, manufacturing, retail, and more.

As organizations across the globe recognize the transformative power of data, the need for effective data governance becomes paramount. This chapter unravels the common threads and unique nuances of data governance implementations in different industries. It highlights how data governance practices not only ensure data accuracy and compliance but also drive innovation, enhance customer experiences, and fuel growth strategies tailored to each sector.

Through an array of captivating case studies and practical applications, we will uncover the strategies, challenges, and triumphs of data governance in healthcare, financial services, manufacturing, and beyond. Join us as we navigate the fascinating intersections of data governance and diverse industries, where data excellence paves the way for success and innovation.

10.2 Data Governance in Healthcare

In the realm of healthcare, data governance plays a pivotal role in reshaping patient care and improving overall outcomes. This section delves into the intricate world of data governance in healthcare, where data accuracy, privacy, and interoperability are paramount.

Healthcare organizations are entrusted with vast amounts of sensitive patient data, making data governance a critical practice. From electronic health records (EHRs) to medical research, effective data governance not only ensures compliance with healthcare regulations but also fosters groundbreaking discoveries and personalized treatment plans (Kerasidou and Kerasidou 2023; KPMG Global, n.d.).

Through a series of enlightening case studies and practical applications, we will explore how data governance is revolutionizing healthcare. For instance, the implementation of a comprehensive data governance framework for childhood cancer research emphasizes principles such as privacy and security, open science, digital object management, and operations. This framework protects patient data through advanced identity and access management technology and periodic research audits, while promoting open science by sharing data with research participants and future researchers (AHIMA n.d.).

Another example is the adoption of Electronic Health Records (EHRs) and the associated data governance challenges. The integration of EHRs into healthcare systems has been a significant step forward in digitizing patient data. However, this transition also highlighted challenges in maintaining data quality, consistency, and security. The American Reinvestment and Recovery Act, through the Meaningful Use program, played a crucial role in standardizing EHR adoption. This case underscores the need for local data governance that requires hands-on training, policies for caregivers and ancillary staff, and the right tools to monitor and improve data quality. Challenges include issues like incorrect recording of allergies, inconsistent lab result reference ranges, and inadequate encoding of medications. To address these challenges, healthcare organizations must focus on establishing data quality checks and building a scalable healthcare data architecture to monitor data quality and governance effectively (HIMSS 2020; Open Data Institute 2022).

These examples illustrate the complexities and importance of data governance in healthcare, particularly in handling sensitive patient data and ensuring the quality and reliability of data used in patient care and research.

10.3 Data Governance in Finance and Banking

In the dynamic world of finance and banking, the role of data governance is pivotal in managing financial data with precision and reliability. This section delves into the intricacies of data governance in the financial sector, where data accuracy, security, and regulatory compliance are paramount.

Financial institutions handle vast volumes of sensitive data, including customer financial records and transactional data. Effective data governance not only ensures data accuracy but also plays a crucial role in protecting customer assets, preventing fraud, and complying with financial regulations (Atlan n.d.-a; Atlan n.d.-b).

Through real-world case studies and practical applications, we will explore how data governance practices are transforming the finance and banking industry. For instance, the case of Danske Bank illustrates the severe consequences of noncompliance with data governance regulations. The bank was fined €1.3 million after admitting it could not verify the completion of procedures for deleting irrelevant customer data, highlighting the importance of adhering to regulations such as the General Data Protection Regulation (GDPR) (Data Privacy Manager 2022).

Another significant example is the data breaches faced by Equifax and Capital One. These incidents resulted in substantial fines and lengthy legal battles, emphasizing the necessity of robust data governance to prevent data mismanagement and protect consumer information (Atlan n.d-a). Similarly, the Commonwealth Bank of Australia faced over half a billion dollars in fines related to money laundering and terrorism financing, underscoring the catastrophic financial impacts of poor data governance (Atlan n.d.-a).

Moreover, Desjardins Group had to replace two senior leaders following a data leak in 2019 to restore trust and address disruptions caused by inadequate data governance practices (Atlan n.d.-a).

Big data projects in the finance sector also demonstrate the value of effective data governance. For example, a big data analytics project at a Forbes Global Top 1000 bank led to the creation of $94.95 million in incremental value within six months, showcasing the effectiveness of data-driven customer-centricity (Astera n.d.).

These case studies from various financial institutions illustrate the critical nature of data governance in finance and banking, especially concerning regulatory compliance, risk management, and customer-centric strategies. The implementation of data governance frameworks not only aids in regulatory compliance but also enhances operational efficiency, market responsiveness, and overall business value.

10.4 Data Governance in Retail and E-Commerce

In the bustling world of retail and e-commerce, data governance plays a pivotal role in reshaping customer experiences, optimizing supply chains, and driving business growth. This section delves into the intricate realm of data governance in the retail sector, where data accuracy, consumer privacy, and market competitiveness are paramount.

Retailers and e-commerce giants harness a wealth of data from customer purchase history to inventory management. Effective data governance not only ensures data accuracy but also empowers retailers to offer personalized shopping experiences, streamline operations, and stay ahead in the highly competitive market (Clarkston Consulting n.d.; CastorDoc n.d.).

Through a series of compelling case studies and practical applications, we will explore how data governance practices are revolutionizing the retail and e-commerce industry. For instance, a fashion retailer collects customer data from multiple channels like in-store purchases, online transactions, and social media interactions. Without proper data governance, inconsistencies may arise due to different formats and data entry errors. Data governance establishes standardized data definitions and validation rules, ensuring that customer records are accurate and consistent across all channels (CastorDoc n.d.).

Another example involves a department store chain that must comply with data protection regulations like GDPR and CCPA. Data governance sets up processes for managing data access, consent management, and data deletion requests, helping the retailer stay compliant and avoid potential fines and reputational damage (Atlan n.d. -a).

Moreover, a supermarket chain needs accurate data on inventory levels and customer demand to optimize its supply chain operations. Data governance ensures that data from different sources like point-of-sale systems and warehouse management systems is integrated and consistent, helping the retailer to make informed decisions on stock replenishment, reducing stockouts and excess inventory (Atlan n.d. -a).

These case studies demonstrate the critical role of data governance in the retail and e-commerce sectors. By effectively managing and utilizing data, retailers can enhance user experiences, improve core sales activities, and explore new monetization opportunities. Implementing robust data governance strategies enables retailers to leverage their data assets more effectively, leading to better decision-making and improved business performance (Clarkston Consulting n.d.; ContactPigeon n.d.; Atlan n.d -a).

10.5 Data Governance in Government and Public Sector

In the realm of government and public sector, data governance plays a crucial role in enhancing transparency, improving public services, and fostering trust among citizens. This section delves into the multifaceted world of data governance within governmental organizations, where data accuracy, privacy, and accessibility are of utmost importance.

Governmental bodies collect and manage vast amounts of data, from citizen information to public service records. Effective data governance ensures that this data is managed responsibly, promoting transparency and accountability while enabling data-driven decision-making (OECD 2019, 2021).

A prime example of successful data governance in the public sector is the initiative by various OECD countries to implement holistic public sector data governance practices at the national level. These efforts have focused on creating a data-driven public sector that leverages data to increase public value and build public trust (OECD 2021). For instance, Ireland's Office of Government Procurement has developed an Open Data and Data Management framework that includes data auditing as a primary focus to ensure accurate evaluations of public sector data (OECD 2023).

Another significant example is the UK's Algorithmic Transparency Recording Standard (ATRS), which helps public bodies openly publish clear information about the algorithmic tools they use and why. This standard aims to enhance transparency and accountability in the use of algorithms within the public sector (OECD 2023).

These case studies illustrate the importance and impact of data governance in government and public sector organizations. By implementing robust data governance frameworks, governments can improve service delivery, ensure compliance with data protection regulations, and enhance public trust (OECD 2021, 2023).

10.6 Data Governance in Technology and Telecommunications

In the rapidly evolving sectors of technology and telecommunications, data governance is crucial for managing vast amounts of data generated from various sources. This section explores how data governance frameworks help organizations in these industries enhance data quality, ensure compliance, and drive innovation.

Telecommunications companies handle extensive data sets, including customer information, call detail records, and network performance metrics. Effective data governance ensures that this data is accurate, secure, and accessible, which is essential for operational efficiency and regulatory compliance (Intellias n.d.; Otto 2011).

A significant example of data governance in the telecommunications sector is Vodafone's implementation of the Vodafone Analytics platform. This platform leverages big data to provide business intelligence tools, offering location-based insights and fleet management systems that help optimize business operations for better accuracy and higher return on investment (Intellias n.d.).

Additionally, the case studies of BT and Deutsche Telekom illustrate different approaches to data governance. BT employs a project-driven, bottom-up philosophy, while Deutsche Telekom follows a top-down approach. Both strategies highlight the importance of context-contingent organizational design in effectively managing data governance (Otto 2011).

These case studies demonstrate the critical role of data governance in the technology and telecommunications sectors. By implementing robust data governance frameworks, companies can manage their data assets more effectively, leading to improved decision-making, enhanced customer experiences, and increased operational efficiency (Intellias n.d.; Alation n.d.; Otto 2011).

10.7 Data Governance in Manufacturing and Logistics

In the manufacturing and logistics sectors, data governance is essential for managing vast quantities of data, ensuring operational efficiency, and maintaining regulatory compliance. This section explores how robust data governance frameworks are implemented in these industries to enhance data quality, streamline operations, and foster innovation.

Manufacturers and logistics providers handle diverse data sets, including production metrics, supply chain data, and customer information. Effective data governance ensures the integrity, security, and accessibility of this data, enabling companies to optimize their operations and improve decision-making (Hrouga and Sbihi 2023; EY India 2023).

For instance, the implementation of Industry 4.0 technologies in logistics can significantly enhance supply chain performance. A case study from the French retail sector demonstrated how digital tools such as cybersecurity systems, big data analytics, and blockchain technology can optimize supply chain operations, leading to improved productivity, inventory management, and real-time communication (Hrouga and Sbihi 2023).

Moreover, digital logistics capabilities are transforming the logistics industry by integrating real-time data and advanced technologies like digital twins. These innovations provide comprehensive visibility into supply chain processes, enabling companies to track shipments, optimize routes, and forecast demand more accurately (Nexocode 2023; McKinsey 2023).

In India, the logistics sector is undergoing a significant transformation driven by the adoption of digital technology and infrastructure development. This transformation is essential for supporting the country's economic growth and ensuring the efficient movement of goods and services. The integration of advanced technologies in logistics not only enhances operational efficiency but also contributes to sustainability and customer satisfaction (EY India 2023).

These case studies illustrate the critical role of data governance in manufacturing and logistics. By implementing robust data governance frameworks, companies can better manage their data assets, leading to improved operational efficiency, enhanced customer experiences, and increased competitiveness in the market (Hrouga and Sbihi 2023; Nexocode 2023; McKinsey 2023; EY India 2023).

10.8 Emerging Industries and Data Governance

In the dynamic landscape of emerging industries, data governance is pivotal in harnessing the power of innovation and ensuring sustainable growth. This section explores how data governance frameworks are applied in emerging sectors, such as biotechnology, renewable energy, and advanced manufacturing, to manage data effectively, enhance compliance, and drive technological advancements.

Emerging industries often deal with diverse and complex data sets, necessitating robust data governance strategies to ensure data integrity, security, and accessibility. Effective data governance helps these industries leverage data to innovate, comply with regulatory requirements, and gain a competitive edge (Royal Society 2023; World Economic Forum 2023).

For instance, the biotechnology sector faces unique data governance challenges due to the sensitive nature of genomic and clinical data. A case study by the Royal Society highlights the social and ethical tensions in managing "omics" data, emphasizing the need for compatible data forms that support open scientific inquiry while ensuring privacy and security (Royal Society 2023).

In the renewable energy sector, data governance plays a crucial role in managing the vast amounts of data generated by smart grids and renewable energy sources. Effective data governance frameworks help in optimizing energy distribution, improving grid reliability, and ensuring compliance with environmental regulations (World Economic Forum 2023).

The Global Data Governance Project provides insights into the data governance practices of various emerging industries worldwide. This comprehensive study underscores the importance of establishing clear data ownership, enhancing data stewardship, and adopting flexible data governance models to support innovation and compliance (Global Data Governance Project 2021).

These case studies illustrate the critical role of data governance in emerging industries. By implementing robust data governance frameworks, these sectors can better manage their data assets, drive innovation, and maintain regulatory compliance, ultimately leading to sustainable growth and competitive advantage (Royal Society 2023; World Economic Forum 2023; Global Data Governance Project 2021).

10.9 Lessons Learned and Best Practices

In the evolving landscape of data governance, various industries have implemented frameworks and strategies to manage their data effectively. This section explores the key lessons learned and best practices from these implementations, highlighting the importance of robust data governance for ensuring data quality, compliance, and driving business value.

Successful data governance requires a clear understanding of objectives, stakeholder education, and the establishment of comprehensive policies and frameworks.

Companies like Airbnb and GE Aviation exemplify how effective data governance can promote data-driven decision-making and operational efficiency (Atlan n.d.-d). Airbnb, for instance, enhanced data literacy across all departments through its Data University initiative, empowering employees to make informed decisions and ensuring responsible data use (Atlan n.d.-d).

GE Aviation focused on balancing data governance with data enablement by centralizing its data sources and setting up a dedicated team for user enablement and data product deployment. This approach accelerated decision-making processes and fostered a sense of data ownership among employees (Atlan n.d.-d).

Best practices in data governance include defining clear objectives, mapping data, evaluating risks, assigning roles and responsibilities, ensuring security, and planning implementation systematically. These practices help organizations maintain high data quality, ensure compliance with regulatory requirements, and support strategic decision-making (CloudCodes 2021).

Additionally, companies must adapt their data governance strategies to their specific business needs and cultural contexts. For example, First San Francisco Partners helped a biotech firm align its data management goals with business and IT stakeholders, demonstrating the importance of flexible and context-specific data governance frameworks (First San Francisco Partners 2016).

Key roles in data governance include data governance managers, data stewards, data architects, and compliance and legal teams. Each role is critical in ensuring data quality, compliance, and the successful implementation of data governance practices (SolverASSIST n.d.).

Overall, these lessons and best practices illustrate the importance of a well-designed data governance framework that aligns with business objectives, supports data-driven decision-making, and ensures regulatory compliance (DATAVERSITY 2023; CloudCodes 2021; First San Francisco Partners 2016).

10.10 Conclusion and Industry-Specific Considerations

The conclusion underlines that data governance is indispensable for varied industries, each with unique challenges and requirements. It highlights that while overarching principles of data governance apply universally, strategies must be tailored to industry-specific needs. These considerations include regulatory landscapes, technological advancements, and customer engagement priorities. The analysis underscores that successful data governance not only addresses compliance and risk management but also enhances operational efficiencies and innovation, positioning organizations for competitive success in a data-centric business environment.

References

AHIMA (n.d.) Healthcare data governance: practice brief. AHIMA
Alation (n.d.) Data governance in telecom: what you need to know. Alation
Astera (n.d.) Data governance in financial services: a complete analysis. Astera
Atlan (n.d.-a) Airbnb—data literacy for decision-making. Atlan
Atlan (n.d.-b) GE aviation—centralizing data sources. Atlan
Atlan (n.d.-c) Wells Fargo—single source of truth for governance. Atlan
Atlan (n.d.-d) Data governance in retail: strategies, challenges, & solutions. Atlan
Atlan (n.d.-e) 5 data governance examples: case studies, takeaways & more. Atlan
CastorDoc (n.d.) Data governance in retail: best practices, challenges, and viable solutions. CastorDoc
Clarkston Consulting (n.d.) The importance of data governance in retail. Clarkston Consulting
CloudCodes (2021) Data governance best practices: top 6 best practices. CloudCodes
ContactPigeon (n.d.) The 24 best eCommerce retail case studies. ContactPigeon
Data Privacy Manager (2022) GDPR fine: Danske Bank fined €1.3 million over non-compliant data deletion processes. Data Privacy Manager
DATAVERSITY (2018) The data governance imperative for retailers. DATAVERSITY
DATAVERSITY (2023) Data governance best practices. DATAVERSITY
EY India (2023) Transforming the future of Indian logistics sector. EY
Finance Data Governance: Current State, Trends, Case Studies from Atlan: Atlan Case Studies
First San Francisco Partners (2016) Agile master data management and data governance in action. First San Francisco Partners
Global Data Governance Project (2021) The global data governance project: 52 case studies. George Washington University
Governance, Risk, Compliance and a Big Data Case Study from ISACA Journal
HIMSS (2020) Healthcare data management: three case studies. HIMSS
Hrouga M, Sbihi A (2023) Logistics 4.0 for supply chain performance: perspectives from a retailing case study. Bus Process Manag J 29(6):1892–1919. Emerald Publishing Limited
Intellias (n.d.) Big data in the telecom: trends, use cases & case studies. Intellias
Kerasidou A, Kerasidou C (2023) Data-driven research and healthcare: public trust, data governance, and the NHS. BMC Med Ethics
KPMG Global (n.d.) Data governance in healthcare. KPMG
McKinsey (2023) Digital logistics and the technology race. McKinsey
Nexocode (2023) Supply chain visibility: the role of real-time data in logistics. Nexocode
OECD (2019) Data governance in the public sector. OECD iLibrary
OECD (2021) The path to becoming a data-driven public sector. OECD iLibrary
OECD (2023) Global trends in government innovation. OECD Observatory of Public Sector Innovation
OECD iLibrary (n.d.) Data governance in the public sector. OECD iLibrary
Open Data Institute (2022) Health data governance: a playbook for non-technical leaders. Open Data Institute
Otto B (2011) Organizing data governance: findings from the telecommunications industry and consequences for large service providers. Commun Assoc Inf Syst 29:Article 3
Royal Society (2023) Data governance: case studies. Royal Society
SolverASSIST (n.d.) Data governance best practices for business intelligence. SolverASSIST
UBS + Neo4j Case Study from Neo4j: Neo4j Case Studies
World Economic Forum (2023) Top 10 emerging technologies of 2023. World Economic Forum

Chapter 11
Overcoming Challenges in Data Governance: Strategies for Success

Contents

11.1	Introduction to Data Governance Challenges.	341
11.2	Challenge of Organizational Culture and Change Management.	343
11.3	Data Quality Issues.	344
11.4	Aligning Data Governance with Business Objectives.	346
11.5	Managing Data Governance in Complex Environments.	349
11.6	Regulatory Compliance and Evolving Legal Requirements.	351
11.7	Technology and Tool Integration Challenges.	354
11.8	Skill Gaps and Training Needs.	355
11.9	Data Security and Privacy Concerns.	358
11.10	Case Studies: Overcoming Data Governance Challenges.	360
11.11	Conclusion.	361
References.		362

Abstract This chapter addresses the complexities and crucial aspects of managing data within organizations, offering a deep dive into strategies that help navigate and mitigate common challenges in data governance.

This chapter emphasizes the importance of defining clear data governance policies, which involves setting precise responsibilities and processes for data management. Ensuring high data quality is stressed as critical, requiring rigorous processes for data validation, cleansing, and regular updates to maintain data integrity and reliability.

Another significant focus is on data privacy and compliance, especially with the enforcement of regulations such as GDPR and CCPA, which necessitate robust systems to handle data securely and legally. Stakeholder engagement across all levels of the organization is also highlighted as vital for successful data governance, ensuring that everyone from top management to end-users is invested and involved in the process.

The challenge of managing data silos and integrating disparate data across different departments is discussed, with a unified data governance framework recommended as a solution to ensure seamless data accessibility and consistency. Furthermore, continuous training and education are underscored as essential for keeping staff updated on best practices and the importance of data governance.

Technology plays a pivotal role in data governance, and the chapter advises on leveraging appropriate tools and platforms that aid in efficient data management and governance processes. It also calls for regular monitoring and continuous improvement to adapt data governance practices to evolving business needs and technologies.

The chapter concludes by addressing the necessity of cultural change within organizations to prioritize data governance. It argues that creating a culture that values data as a crucial asset is fundamental for the long-term success of data governance initiatives.

Overall, this chapter provides a thorough exploration of the obstacles faced in data governance and practical strategies to overcome them, ensuring that data governance efforts are effective and aligned with organizational goals.

Keywords Data governance challenges · Strategies for success · Data quality · Data privacy · Compliance · Stakeholder engagement · Data silos · Training and education · Technology leveraging · Monitoring · Continuous improvement · Data security · Cultural change

Overcoming challenges in data governance and devising strategies for success is a complex but crucial aspect of managing data in any organization. Here are some key points and strategies that are often emphasized in this area:

Defining Clear Data Governance Policies One of the first steps is to establish clear policies for data governance. This includes defining who is responsible for various data assets and what processes should be followed for accessing, modifying, and managing these assets (Zagoudis 2020; Splunk 2023).

Ensuring Data Quality Maintaining high data quality is essential. This involves establishing processes for data validation, cleansing, and consistent updates to ensure the data remains accurate and reliable (Olson 2003; Soares n.d.-a, -b, -c).

Data Privacy and Compliance With regulations like GDPR and CCPA, ensuring data privacy and compliance is more important than ever. Organizations need to have robust mechanisms to handle data in a way that complies with legal requirements (Devane 2021; Madsen n.d.).

Stakeholder Engagement Successful data governance requires the involvement and buy-in from various stakeholders across the organization. This includes top management, IT professionals, data scientists, and end-users (Soares n.d.-a, -b, -c; Smallwood 2019).

11.1 Introduction to Data Governance Challenges

Managing Data Silos Breaking down data silos and ensuring seamless data integration across different departments and systems is a significant challenge. A unified data governance framework can help in achieving this (Torre-Bastida et al. n.d.; Soares n.d.-a, -b, -c).

Training and Education Continuous training and education for staff members about the importance of data governance and best practices are crucial for its success (McKinsey & Company 2020; Jarmul 2023).

Leveraging Technology Utilizing the right technology tools and platforms can aid significantly in data governance. This includes data management software, analytics tools, and other technologies that facilitate data processing and governance (Soares n.d.-a, -b, -c; Kirshteyn n.d.).

Monitoring and Continuous Improvement Data governance is not a one-time project but an ongoing process. Regular monitoring, auditing, and continuous improvement of data governance practices are necessary.

Addressing Data Security Ensuring the security of data is a paramount aspect of data governance. This involves implementing robust security measures to protect data from unauthorized access and breaches (Jarmul 2023).

Cultural Change Often the biggest challenge is the cultural shift required within an organization to prioritize data governance. Creating a culture that values data as a key asset is critical for long-term success.

Figure 11.1 details strategies crucial for navigating and mastering the complexities of data governance. Central to this figure is the concept of overcoming challenges which is surrounded by eight critical elements: Monitoring and Continuous Improvement, Stakeholder Engagement, Leveraging Technology, Addressing Data Security, Data Privacy and Compliance, Cultural Change, Managing Data Silos, Defining Clear Data Governance Policies, Training and Education, and Ensuring Data Quality. These elements highlight the multifaceted approach needed to effectively manage data governance within organizations, emphasizing continuous improvement, stakeholder involvement, strategic use of technology, and the importance of cultural adaptation to change.

11.1 Introduction to Data Governance Challenges

In the ever-evolving landscape of the digital era, data has emerged as a pivotal asset for organizations across all sectors. As the volume, velocity, and variety of data continue to expand at an unprecedented rate, the importance of effective data governance cannot be overstated. Data governance, the systematic management of data

Fig. 11.1 Overcoming challenges in data governance: Strategies for success. (Created by the author)

assets to ensure their accessibility, reliability, and security, has become a critical factor for organizational success. However, implementing a robust data governance framework is fraught with challenges that can impede its efficacy and realization of its full potential (Soares n.d.-b, -c; Smallwood 2019).

This introduction delves into the multifaceted challenges of data governance, exploring the complex terrain organizations must navigate to harness the true power of their data. The journey towards effective data governance is often hindered by a myriad of obstacles ranging from technological complexities to cultural resistances. As we embark on this exploration, it is imperative to acknowledge that data governance is not merely a technical endeavor but a strategic initiative that demands a comprehensive approach encompassing people, processes, and technology (Zagoudis 2020; Splunk 2023).

One of the primary challenges in data governance lies in the sheer volume and complexity of data. With the advent of big data, organizations are inundated with an overwhelming amount of information, making it increasingly difficult to manage and govern effectively. The proliferation of data sources, both structured and unstructured, adds to this complexity, necessitating sophisticated tools and strategies to integrate, cleanse, and maintain data quality (Olson 2003; Soares n.d.-b, -c).

Furthermore, the dynamic nature of data governance, influenced by evolving regulatory landscapes, technological advancements, and changing business needs, requires organizations to be agile and adaptable. Compliance with data protection regulations such as GDPR and CCPA introduces additional layers of complexity, demanding stringent adherence to legal and ethical standards in data management (Devane 2021; Jarmul 2023).

Another significant challenge is achieving organizational alignment and stakeholder buy-in. Data governance initiatives often require a cultural shift, moving away from data silos and fostering a data-centric mindset across the organization. This shift demands not only the right technology but also effective leadership, clear communication, and continuous education and training (McKinsey & Company 2020; Madsen n.d.).

Moreover, the technological aspect of data governance presents its own set of challenges. Selecting and implementing the right tools and platforms for data management, security, and analytics is a daunting task. Organizations must balance the need for advanced technological solutions with considerations of cost, scalability, and integration with existing systems (Kirshteyn n.d.; Torre-Bastida et al. n.d.).

As we proceed, this discussion will unravel these challenges in greater detail, examining their implications and exploring strategies to overcome them. Our goal is to provide a comprehensive understanding of the obstacles in the path of effective data governance and to equip organizations with the knowledge and tools needed to navigate these complexities successfully. The journey of mastering data governance is intricate and demanding, but with the right approach, it can lead to unparalleled organizational efficiency, compliance, and decision-making process.

11.2 Challenge of Organizational Culture and Change Management

The challenge of organizational culture and change management is a critical aspect for businesses and institutions, especially in the context of implementing new strategies or technologies, like data governance frameworks. This challenge is rooted in the complexity of altering established behaviors, norms, and attitudes within an organization. Here are some key aspects of this challenge:

Resistance to Change People are often comfortable with established routines and may resist changes, especially if they perceive them as threatening to their current way of working. This resistance can manifest as skepticism, noncompliance, or even active opposition to new initiatives (Soares n.d.-b; Smallwood 2019).

Cultural Inertia Organizational culture is built over time and consists of shared beliefs, values, and practices. Changing this culture requires more than just introducing new policies or technologies; it requires a shift in these underlying elements, which is often a slow and difficult process (Zagoudis 2020).

Communication Barriers Effective communication is crucial for successful change management. However, miscommunication or inadequate communication about the reasons for change and the benefits it brings can hinder the process (McKinsey & Company 2020).

Leadership and Vision The role of leadership is vital in driving change. Leaders must not only articulate a clear vision but also embody the change they wish to see in the organization. A lack of strong leadership can result in a lack of direction and commitment to the change process (Soares, n.d.; Smallwood 2019).

Aligning Change with Organizational Goals Any change initiative should be clearly aligned with the organization's overall goals and objectives. If employees do not see how the change fits into the broader organizational strategy, they are less likely to embrace it (Soares n.d.-b).

Training and Support Implementing change often requires new skills and knowledge. Providing adequate training and support is essential to help employees adapt to the change. Without this, employees might feel overwhelmed or underprepared to handle new challenges (Jarmul 2023; McKinsey & Company 2020).

Managing the Transition The transition period during which changes are being implemented is often filled with uncertainty and discomfort. Managing this period effectively, addressing concerns as they arise, and keeping employees engaged is crucial (Devane 2021).

Measuring and Adapting It is important to continuously measure the impact of the changes and be willing to adapt the approach based on feedback and outcomes. This agile approach to change management can help in fine-tuning strategies to ensure their effectiveness (Kirshteyn n.d.).

Employee Engagement and Participation Encouraging employee participation in the change process can increase buy-in and reduce resistance. When employees feel they have a say in the process, they are more likely to support and contribute to the change (Madsen n.d.).

Sustaining Change Finally, sustaining change over the long term is often more challenging than initiating it. This requires ongoing commitment, reinforcement of new behaviors, and integration of the change into the organizational culture (Soares n.d.-b; Smallwood 2019).

Addressing these challenges requires a comprehensive and empathetic approach, recognizing the human element in organizational change. Successful change management not only achieves the desired transformation but also strengthens the organization, making it more adaptable and resilient in the face of future challenges.

11.3 Data Quality Issues

Data quality issues are a significant concern in data management and can have far-reaching impacts on an organization's decision-making, efficiency, and credibility. Poor data quality can arise from various sources and manifest in different forms. Here are some of the key aspects and common issues related to data quality:

Inaccuracy This is one of the most critical data quality issues where the data does not accurately represent the real-world entity or event it is supposed to depict.

11.3 Data Quality Issues

Inaccurate data can lead to erroneous decisions and analyses (Olson 2003; Soares n.d.-b).

Inconsistency Inconsistency occurs when there is a discrepancy in data across diverse sources or systems. For instance, the same information may be recorded differently in separate databases, leading to confusion and unreliable analytics (Soares n.d.-b).

Incompleteness Data that is missing elements, such as incomplete records or fields, can skew analysis and lead to misinformed decisions. Incomplete data is particularly problematic in scenarios where comprehensive information is crucial for accuracy, like in statistical analyses or predictive modeling (Kirshteyn n.d.; Jarmul 2023).

Lack of Timeliness Data that is not updated in a timely manner loses its relevance and utility. For decision-making processes, having access to the most current data is often crucial (McKinsey & Company 2020).

Poor Data Standardization Lack of standardization in data formats and structures can lead to difficulties in data integration, aggregation, and analysis. Standardization is essential for ensuring that data from various sources can be effectively combined and utilized (Torre-Bastida et al. n.d.).

Duplication Duplicate records or data points can lead to inefficiencies in data storage and processing, as well as inaccuracies in analysis. Identifying and eliminating duplicates is an important aspect of data quality management (Soares n.d.-b).

Data Decay Over time, data can become outdated or irrelevant, especially in dynamic environments where entities and relationships are constantly changing. Regular updates and maintenance are required to keep data relevant (Smallwood 2019).

Poor Data Governance Inadequate data governance policies and practices can lead to several data quality issues. Effective governance is essential for setting and enforcing standards and procedures for data management (Zagoudis 2020).

Data Entry Errors Mistakes made during data entry are a common source of poor data quality. These errors can stem from human error, misunderstanding, or miscommunication (Olson 2003).

Lack of Context Data lacking in context or metadata can be misleading or difficult to interpret correctly. Contextual information is critical for understanding the conditions under which the data was collected and how it should be interpreted (Jarmul 2023).

Fig. 11.2 Data quality issues. (Created by the author)

Figure 11.2 maps out various common problems in data quality that organizations frequently encounter. These issues include Inaccuracy, where data does not reflect the correct information; Inconsistency, where there are discrepancies across data sources; Lack of Timeliness, which refers to outdated information; Duplication, where the same data is repeated; Data Decay, indicating outdated or obsolete data; Poor Data Governance, highlighting a lack of proper management; Data Entry Errors, errors made during data input; and Lack of Context, where data lacks the supporting information necessary for interpretation. These issues collectively degrade the quality of data, impacting decision-making and operational efficiency.

Addressing data quality issues typically involves a combination of process improvements, technology solutions, and cultural changes. This can include implementing more robust data validation and verification processes, using data cleansing tools, establishing clearer data governance policies, and fostering a culture that values data accuracy and completeness. The aim is to ensure that data is fit for its intended purpose and can be relied upon for making informed decisions.

11.4 Aligning Data Governance with Business Objectives

Aligning data governance with business objectives is a crucial strategy for ensuring that an organization's data management practices effectively support its overall goals and mission. This alignment is essential because data governance is not just a

11.4 Aligning Data Governance with Business Objectives

technical or IT issue, but a business strategy that impacts various aspects of an organization. Here are some key points to consider in this alignment:

Understanding Business Objectives The first step in aligning data governance with business objectives is to have a clear understanding of what those objectives are. This could include growth targets, customer satisfaction goals, efficiency improvements, compliance requirements, or other strategic priorities (Confluent n.d.; DataGalaxy n.d.).

Defining Data Needs Once business objectives are clearly understood, the next step is to identify the data needed to support these objectives. This involves determining the types of data required, the level of quality necessary, and how data will be used in decision-making processes (McKinsey n.d.; Databricks n.d.).

Data Governance Framework Developing a data governance framework that supports business objectives involves setting policies, procedures, and standards for data management. This framework should ensure that data is accurate, accessible, secure, and compliant with regulations and that it aligns with the strategic needs of the business (McKinsey n.d.; DataGalaxy n.d.).

Stakeholder Engagement It is crucial to involve stakeholders from across the organization in the data governance process. This includes not only IT and data management teams but also business unit leaders, executives, and anyone who uses or relies on data in their role. Their input helps ensure that data governance efforts are aligned with the needs and priorities of the entire organization (Confluent n.d.; McKinsey n.d.).

Data Stewardship Assigning data stewards who understand both the data and the business context can bridge the gap between data management and business objectives. These stewards can oversee the handling of data in a way that aligns with strategic goals (Databricks n.d.; McKinsey n.d.).

Performance Metrics Establishing metrics to measure the effectiveness of data governance in supporting business objectives is vital. These metrics could include data quality scores, compliance levels, user satisfaction, and the impact of data governance on achieving specific business outcomes (Confluent n.d.; DataGalaxy n.d.).

Technology Alignment The technology used for data governance should be aligned with business objectives. This includes selecting data management tools and platforms that not only meet technical requirements but also support the strategic goals of the organization (Databricks n.d.; McKinsey n.d.).

Change Management Effective communication and change management strategies are essential to ensure that data governance initiatives are understood, supported, and adopted across the organization. This involves explaining the link between data governance and business success to all levels of the organization (McKinsey n.d.; DataGalaxy n.d.).

Training and Education Providing training and education about the importance of data governance and how it relates to business objectives can help foster a data-informed culture within the organization (Confluent n.d.; DataGalaxy n.d.).

Continuous Improvement Data governance should be seen as an ongoing process that evolves with the changing needs of the business. Regular reviews and adjustments to the data governance strategy and practices ensure they remain aligned with business objectives (McKinsey n.d.; Databricks n.d.).

Figure 11.3 emphasizes key aspects to align data governance efforts with the strategic aims of an organization. It highlights the importance of Understanding Business Objectives to ensure data governance supports overall corporate goals. Defining Data Needs is crucial to determine what data is required and how it should be handled. A robust Data Governance Framework sets policies and standards that align with business needs. Stakeholder Engagement involves key parties across the organization in the governance processes. Data Stewardship assigns responsibility for data to those who understand both its technical aspects and its business impact.

Fig. 11.3 Aligning data governance with business objectives keys. (Created by the author)

Performance Metrics are established to evaluate the effectiveness of data governance initiatives. Technology Alignment ensures that all technological tools support the strategic goals. Change Management focuses on the adaptation within the organization towards new governance policies. Training and Education improve organizational understanding and compliance. Lastly, Continuous Improvement involves regular updates and reviews to adapt governance practices to changing business needs.

By aligning data governance with business objectives, organizations can ensure that their data assets are managed effectively and leveraged to support strategic decisions, drive innovation, and maintain a competitive edge.

11.5 Managing Data Governance in Complex Environments

Managing data governance in complex environments presents a unique set of challenges that require careful planning, strategic thinking, and adaptive execution. Complex environments typically involve large, diverse, and often globally distributed organizations, multiple data sources and systems, a variety of regulatory requirements, and a wide range of stakeholders. Here are some key considerations and strategies for managing data governance in such settings:

Developing a Robust Framework Establishing a comprehensive data governance framework is crucial. This framework should include clearly defined policies, procedures, roles, and responsibilities. It needs to be flexible enough to accommodate the diverse needs of various parts of the organization while maintaining consistent standards (McKinsey n.d.).

Scalability and Flexibility The data governance approach must be scalable to handle the volume and diversity of data in a complex environment. It should be flexible enough to adapt to changing business needs, technological advancements, and regulatory landscapes (TDWI 2021).

Stakeholder Engagement In complex environments, engaging a wide range of stakeholders is essential. This includes not only data professionals and IT staff but also business unit leaders, legal and compliance teams, and end-users. Understanding their needs, perspectives, and challenges is key to developing effective governance strategies (McKinsey n.d.; ScienceDirect 2024).

Balancing Centralization and Decentralization Finding the right balance between centralized and decentralized data governance is critical. While centralization can ensure consistency and compliance, decentralization may be necessary for agility and to meet the specific needs of different business units (McKinsey n.d.).

Data Quality Management Ensuring high data quality across diverse data sources and systems is a major challenge. Implementing robust data quality management processes and tools is vital to provide accurate, reliable data for decision-making (ScienceDirect 2024).

Technology Utilization Leveraging the right technology solutions, such as data management platforms, cloud services, and analytics tools, can aid in managing data governance. These technologies should be chosen based on their ability to integrate with existing systems and their scalability to meet future needs (TDWI 2021).

Regulatory Compliance In complex environments, particularly in global organizations, compliance with a wide range of regional and international regulations is a significant challenge. The governance framework must ensure that data handling practices are compliant across different jurisdictions (McKinsey n.d.).

Change Management Effectively managing change is crucial as data governance initiatives often require shifts in culture, processes, and technology. This involves continuous communication, education, and support for all stakeholders involved (ScienceDirect 2024).

Data Security and Privacy Ensuring the security and privacy of data is especially challenging in complex environments due to the multiple systems and access points involved. Robust security measures and privacy policies must be integral to the governance framework (TDWI 2021).

Continuous Monitoring and Improvement Data governance in complex environments should be viewed as an ongoing process. Regular monitoring, review, and adaptation of governance practices are necessary to ensure they remain effective and aligned with organizational objectives (McKinsey n.d.).

Cross-Functional Teams Forming cross-functional teams can help in addressing the multidimensional aspects of data governance. These teams bring together diverse expertise and perspectives, facilitating more comprehensive and effective governance strategies (ScienceDirect 2024).

Cultural Considerations Understanding and respecting cultural differences within a global organization is important for the successful implementation of data governance policies and practices (TDWI 2021).

Figure 11.4 highlights essential considerations and strategies for effective data governance in intricate organizational settings. The figure centers around key components that must be addressed to navigate and control data governance complexities. These components include Technology Utilization, leveraging tech solutions for effective data management; Change Management, ensuring smooth transitions and adaptability in governance practices; Developing a Robust Framework, which

11.6 Regulatory Compliance and Evolving Legal Requirements

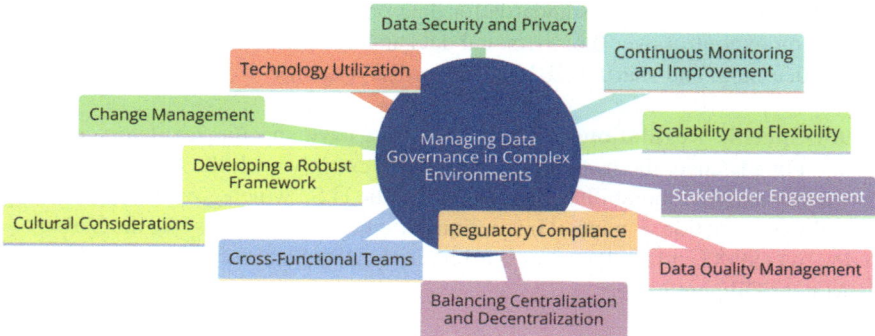

Fig. 11.4 Managing data governance in complex environments map. (Created by the author)

is crucial for consistency and comprehensiveness across the organization; Cultural Considerations, recognizing and integrating cultural differences within global settings; Cross-Functional Teams, promoting interdisciplinary collaboration; and Regulatory Compliance, ensuring data practices meet legal standards. Also, it includes Data Quality Management, maintaining high standards of data integrity; Balancing Centralization and Decentralization, optimizing control and flexibility; Data Security and Privacy, protecting data against unauthorized access; Scalability and Flexibility, to adapt to growing data volumes and organizational changes; Continuous Monitoring and Improvement, for ongoing enhancement of governance practices; and Stakeholder Engagement, involving all relevant parties in governance activities.

Successfully managing data governance in complex environments requires a nuanced approach that considers the specific characteristics and challenges of the organization. It is about creating a governance ecosystem that is both robust and agile, capable of evolving with the organization's needs.

11.6 Regulatory Compliance and Evolving Legal Requirements

Regulatory compliance and evolving legal requirements are crucial considerations for businesses and organizations, particularly in the context of data management and protection. As the digital landscape continues to grow and diversify, governments and regulatory bodies worldwide are constantly updating and introducing new regulations to address the complexities of data security, privacy, and ethical use. Here are some key aspects of regulatory compliance and evolving legal requirements:

Data Privacy Laws Laws such as the General Data Protection Regulation (GDPR) in the European Union and the California Consumer Privacy Act (CCPA) in the

United States have set new standards for data privacy. These regulations give individuals more control over their personal data and impose strict rules on data collection, processing, and sharing (IBM n.d.; NordLayer n.d.).

Sector-Specific Regulations Certain industries have specific regulatory requirements. For instance, the healthcare sector must comply with the Health Insurance Portability and Accountability Act (HIPAA) in the U.S., which governs the handling of health information. Financial institutions often face regulations like the Sarbanes-Oxley Act (SOX) and the Payment Card Industry Data Security Standard (PCI DSS) (IBM n.d.; Diligent n.d.).

International Compliance For global organizations, compliance becomes more complex as they must adhere to the legal requirements of each country they operate in. This can involve navigating a web of differing and sometimes conflicting regulations across jurisdictions (Diligent n.d.; NordLayer n.d.).

Evolving Cybersecurity Laws As cybersecurity threats evolve, so do the laws and regulations intended to combat them. This includes requirements for data breach notifications, cybersecurity measures, and standards for protecting sensitive information (IBM n.d.; Diligent n.d.).

Ethical Considerations and AI Regulations With the rise of artificial intelligence and machine learning, new ethical considerations and potential regulations are emerging. This includes ensuring fairness, transparency, and accountability in AI systems, especially those that make decisions affecting individuals (Diligent n.d.).

Data Sovereignty and Localization Laws Some countries have laws that require data about their citizens to be stored and processed within their borders. These data sovereignty and localization laws add another layer of complexity to data management and governance (NordLayer n.d.; IBM n.d.).

Regular Updates and Audits Staying compliant requires regular updates to policies and procedures, as well as periodic audits to ensure compliance with current laws and regulations. This is an ongoing process as legal requirements can change frequently (Diligent n.d.).

Employee Training and Awareness Ensuring that employees understand compliance requirements is crucial. Regular training and awareness programs can help prevent accidental breaches and noncompliance (IBM n.d.).

Documentation and Record-Keeping Maintaining detailed records of data processing activities and compliance efforts is often a legal requirement and is essential for demonstrating compliance in the event of an audit or investigation (NordLayer n.d.).

11.6 Regulatory Compliance and Evolving Legal Requirements

Fig. 11.5 Regulatory compliance and evolving legal requirements map. (Created by the author)

Vendor and Third-Party Compliance Organizations must also ensure that their vendors and third-party service providers comply with relevant regulations, particularly when they handle or process data on the organization's behalf (Diligent n.d.).

Figure 11.5 illustrates the necessity of a proactive approach to maintaining regulatory compliance in the face of changing legal landscapes. This map encompasses multiple facets that are critical for organizations to consider: Regular Updates and Audits, ensuring that compliance efforts are current and thorough; Ethical Considerations and AI Regulations, addressing new ethical dilemmas and the impact of artificial intelligence; Data Privacy Laws, which are crucial for protecting personal information; Documentation and Record-Keeping, essential for transparency and accountability; Employee Training and Awareness, to ensure that all levels of the organization understand compliance requirements; Data Sovereignty and Localization Laws, dealing with the specifics of data residency; International Compliance, covering the requirements across different jurisdictions; Vendor and Third-Party Compliance, managing the complexities of external partnerships; Sector-Specific Regulations, which vary across different industry sectors; Evolving Cybersecurity Laws, which adapt to new threats and technologies; and Legal-Specific Regulations, tailored to the specific legal environment affecting the organization. This comprehensive approach helps mitigate legal, financial, and reputational risks associated with noncompliance.

Regulatory compliance and adapting to evolving legal requirements necessitate a proactive approach. Organizations need to continuously monitor legal developments, assess the impact of these changes on their operations, and adjust their policies and practices accordingly. Noncompliance can lead to significant legal, financial, and reputational risks.

11.7 Technology and Tool Integration Challenges

Technology and tool integration challenges are common in organizations as they strive to create cohesive systems from diverse software applications and platforms. These challenges arise from attempting to merge disparate technologies, each with its unique architecture, functionality, and data formats. Here are some key aspects of these challenges along with real examples:

Compatibility Issues Different systems and tools might not be inherently compatible with each other. For example, integrating a legacy Customer Relationship Management (CRM) system with a modern cloud-based analytics platform can be challenging due to differences in data formats and protocols (Aldoseri et al. 2023).

Data Integration and Quality Issues Ensuring data is consistently formatted, accurate, and up-to-date across multiple systems is a major challenge. For instance, integrating data from various sources into a unified data warehouse requires robust processes to ensure data quality and consistency (IntechOpen 2023).

System Downtime and Disruption Integration efforts can lead to system downtime or disruptions affecting business operations. Proper planning and phased implementation are essential to minimize these disruptions (Ataccama 2023).

Scalability As organizations grow, their technology needs evolve. A system that works well for a small business might not scale effectively for a larger enterprise. Startups often face challenges when scaling up their IT infrastructure to support increased customer demand (IntechOpen 2023).

Security Concerns Integrating different technologies can create new security vulnerabilities. For example, a poorly integrated system might expose data to unauthorized access or breaches. Ensuring robust security measures during integration is crucial (Aldoseri et al. 2023).

Cost Overruns and Budget Constraints Integration projects can be costly, and unexpected complications can lead to budget overruns. Careful budget planning and cost management strategies are necessary to keep integration projects on track (Ataccama 2023).

User Adoption and Training New integrated systems may require users to learn different workflows, which can meet with resistance. Providing adequate training and support to users is essential for successful integration (IntechOpen 2023).

Vendor Lock-In Relying on a single vendor's suite of products can make integration with other tools difficult. Choosing open standards and interoperable technologies can mitigate this risk (Aldoseri et al. 2023).

Customization Limitations Off-the-shelf solutions may not perfectly fit an organization's unique processes, leading to compromises or the need for costly customizations. Customizing tools to meet specific needs can be resource-intensive (IntechOpen 2023).

Regulatory Compliance Integrating technologies across borders might involve navigating different data protection and privacy laws. Ensuring compliance with regulations like GDPR or CCPA during integration is vital (IBM n.d.; Diligent n.d.).

Figure 11.6 underscores the importance of adapting to changing legal landscapes within data governance. It emphasizes that organizations must proactively monitor legal developments, assess their impacts, and adjust their policies and practices accordingly to maintain compliance and avoid significant risks. This figure highlights the complexities organizations face in staying compliant with evolving laws, necessitating a flexible and responsive approach to regulatory changes.

Successfully overcoming these challenges typically involves careful planning, selecting compatible and scalable tools, ensuring robust data management practices, and providing adequate training and support to users. Additionally, it often requires a phased approach to integration, allowing for testing and adjustments to minimize business disruption and ensure a smooth transition.

11.8 Skill Gaps and Training Needs

Skill gaps and training needs are significant challenges that organizations face in today's rapidly evolving workplace. As technology advances and business practices change, employees may find that their skills are no longer aligned with their organization's requirements. Addressing these skill gaps through effective training is

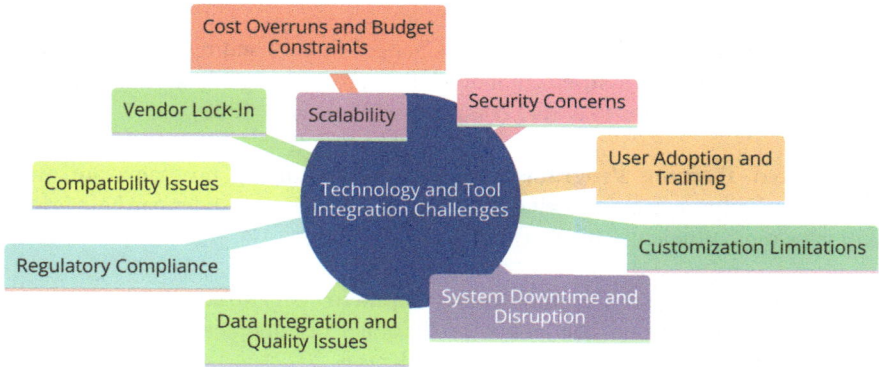

Fig. 11.6 Technology and tool integration challenges map. (Created by the author)

crucial for maintaining competitiveness and fostering growth. Here are some key aspects of skill gaps and training needs:

Technology Advancements Rapid advancements in technology, such as AI, machine learning, and cloud computing, often lead to skill gaps as the workforce struggles to keep up with new tools and methodologies. For instance, the rise of big data analytics has created a demand for professionals skilled in data science and analytics (Aldoseri et al. 2023).

Changing Business Models As businesses evolve and adapt to market changes, new skills are often required. The shift towards remote work, for example, has increased the need for skills in digital collaboration and communication (McKinsey 2022).

Sector-Specific Changes Different sectors may experience unique skill gaps due to specific technological or regulatory changes. In healthcare, the adoption of electronic health records (EHRs) has necessitated training for medical staff in these systems (Ono et al. 2013).

Soft Skills Alongside technical skills, there is often a gap in soft skills such as critical thinking, problem-solving, leadership, and adaptability. These skills are crucial for navigating complex and dynamic work environments (Coursera 2024).

Aging Workforce In many industries, an aging workforce means a significant portion of employees may be nearing retirement, taking their expertise with them. This creates a need to train younger workers to fill these upcoming gaps (Cambridge University Press 2019).

Globalization and Cultural Skills As companies become more global, the ability to work across diverse cultures and languages becomes increasingly important. This requires training in cross-cultural communication and international business practices (Cambridge University Press 2019).

Regulatory and Compliance Training Changes in laws and regulations, such as data protection laws or environmental regulations, require organizations to regularly update training programs to ensure compliance (McKinsey 2022).

Customized Training Programs One-size-fits-all training programs are often ineffective. Tailoring training to individual needs and learning styles can help bridge skill gaps more efficiently (Coursera 2024).

Lifelong Learning and Continuous Development Organizations are encouraging lifelong learning and continuous professional development to keep up with industry changes and career progression (McKinsey 2022).

11.8 Skill Gaps and Training Needs

Utilizing Technology for Training E-learning platforms, virtual reality (VR), and augmented reality (AR) are increasingly being used for more effective and engaging training experiences (Aldoseri et al. 2023).

Public and Private Sector Collaboration In some cases, addressing skill gaps requires collaboration between the public and private sectors, especially in areas like technology and manufacturing where rapid changes are common (Ono et al. 2013).

Measuring Training Effectiveness It is crucial to have metrics in place to assess the effectiveness of training programs in closing skill gaps (Coursera 2024).

Addressing these challenges requires a strategic approach, including conducting regular skills gap analyses, developing tailored training programs, and fostering a culture of continuous learning and development.

Figure 11.7 outlines key areas where training can bridge the existing skills gap and cater to evolving business and technological landscapes. This figure centers on various facets that organizations need to focus on to ensure a skilled workforce. These facets include Customized Training Programs that tailor learning to specific organizational needs, Technology Advancements that necessitate continuous skill updates, Sector-Specific Changes that require specialized knowledge, and Measuring Training Effectiveness to gauge the impact of educational initiatives. Additionally, it emphasizes the importance of Regulatory and Compliance Training to meet legal standards, Utilizing Technology for Training to enhance learning processes, and addressing an Aging Workforce by updating skills. Other crucial elements are Soft Skills development for holistic employee growth, Lifelong Learning and Continuous Development to maintain a competitive workforce, and Globalization and Cultural Skills to operate effectively in a globalized market.

Organizations need to proactively identify current and future skill gaps and develop strategic training programs to address these needs. This involves not only

Fig. 11.7 Skill gaps and training needs map. (Created by the author)

technical and vocational training but also focuses on soft skills and continuous learning to adapt to the ever-changing business landscape.

11.9 Data Security and Privacy Concerns

Data security and privacy concerns are increasingly significant in the digital age, where vast amounts of personal and sensitive data are collected, stored, and processed by organizations across various sectors. These concerns arise from the risks associated with data breaches, unauthorized access, and the misuse of personal information. Here are some key aspects of data security and privacy:

Data Breaches A data breach involves unauthorized access to or disclosure of personal information. High-profile breaches in companies like Equifax and Marriott have highlighted the severe consequences of such incidents, including monetary loss, reputational damage, and legal repercussions (TechRepublic 2024).

Cybersecurity Threats Cybersecurity threats like hacking, phishing, malware, and ransomware are significant risks to data security. These threats are constantly evolving, requiring organizations to continually update their security measures (TechRepublic 2024; ScienceDirect 2024).

Regulatory Compliance Laws and regulations like the General Data Protection Regulation (GDPR) in the EU and the California Consumer Privacy Act (CCPA) in the U.S. impose strict rules on data privacy and security. Noncompliance can lead to hefty fines and legal challenges (ScienceDirect 2024; Cambridge University Press 2024).

Consumer Privacy Expectations There is a growing public concern over privacy, with consumers increasingly demanding transparency and control over how their data is used. Organizations need to address these concerns to maintain trust and credibility (TechRepublic 2024).

Internet of Things (IoT) Security The proliferation of IoT devices has expanded the risk landscape, as these devices often collect sensitive data and can be vulnerable to security breaches (IEEE 2024).

Cloud Security As more organizations move data to cloud-based services, ensuring the security of data in the cloud becomes crucial. This involves managing risks associated with data transmission, storage, and access in cloud environments (ScienceDirect 2024).

11.9 Data Security and Privacy Concerns

Insider Threats Not all data security threats come from outside an organization. Insider threats, whether intentional or accidental, can also lead to data breaches and privacy violations (Cambridge University Press 2024).

Data Encryption Encryption is a key tool for protecting data privacy and security. It ensures that even if data is intercepted or accessed without authorization, it remains unreadable and secure (IEEE 2024).

Data Anonymization and Pseudonymization Techniques like anonymization and pseudonymization are used to protect personal data by removing or altering identifying information (ScienceDirect 2024).

Employee Training and Awareness Human error is a significant factor in many data breaches. Regular training and awareness programs for employees are essential to prevent accidental breaches and improve overall data security (TechRepublic 2024).

Mobile Security With the increasing use of mobile devices in business operations, securing data on these devices is a growing concern (IEEE 2024).

Data Sovereignty and Cross-Border Data Transfers Data sovereignty laws require data to be stored and processed within the geographical boundaries of a specific country, posing challenges for global organizations, especially in terms of compliance and data transfer security (Cambridge University Press 2024).

Figure 11.8 illustrates the multitude of issues surrounding data security and privacy in today's digital environment. The central concerns highlighted are Data Sovereignty and Cross-Border Data Transfers, illustrating the complexities of managing data across different legal jurisdictions. Employee Training and Awareness are emphasized as crucial for ensuring that staff understand and implement security practices. The figure also addresses Cloud Security, Cybersecurity Threats, Insider Threats, and Mobile Security, which are pertinent in protecting organizational data across various platforms. Further, it discusses Data Anonymization and Pseudonymization, techniques vital for maintaining privacy while utilizing data in analytics and processing. Additional elements include Consumer Privacy Expectations, Data Breaches, Data Encryption, and Regulatory Compliance, showcasing the broad scope of challenges that organizations must navigate to safeguard sensitive information and maintain privacy standards.

Data security and privacy concerns require a comprehensive approach that includes robust technical measures, regular risk assessments, employee training, adherence to legal and regulatory requirements, and a commitment to ethical data practices. As technology and cyber threats evolve, organizations must continually adapt their strategies to protect sensitive data effectively.

Fig. 11.8 Data security and privacy concerns map. (Created by the author)

11.10 Case Studies: Overcoming Data Governance Challenges

Organizational Structure and Top Management Involvement A comprehensive approach to data governance often includes establishing a central data management office (DMO) led by a Chief Data Officer (CDO). This office works in conjunction with governance roles organized by data domain and a data council. This structure facilitates effective data management and aligns it with corporate strategy. The crucial aspect here is securing top management's attention and involvement, forming a data governance council within senior management to steer the governance strategy towards business needs (Brickclay 2023; Integrate.io 2024).

Addressing Common Challenges Overcoming data governance challenges involves dealing with limited resources, lack of leadership buy-in, complicated data flows, and fragmented teams. Best practices suggest taking a long-term approach, ensuring alignment with business needs, treating data as a valued asset, and utilizing automated tools for sustainable and efficient governance. Creating trusted data sources and maintaining consistent, transparent practices are key. Additionally, fulfilling regulatory and audit needs by documenting policies and standards is essential (Brickclay 2023; Integrate.io 2024).

Practical Approaches and Tactics Identifying a senior leader as a data governance champion can be crucial. Initiating governance efforts with a pilot area that resonates with the champion, focusing on critical elements, and creating a senior-level steering committee for establishing governance policies are effective strategies. Documenting data flows, creating a data governance working group, and addressing the root causes of data issues can demonstrate the value of governance efforts and support program expansion (Integrate.io 2024; Royal Society 2024).

Case Studies
1. **CancerLinQ:** A technology platform that collects and analyzes data from patient encounters to improve cancer care quality and advance discovery. It addresses challenges through robust data aggregation pipelines, common data models, data deidentification, integration of data collection into physician workflows, and attention to interoperability (Sweeney et al. 2023).
2. **American Association for Cancer Research Project GENIE:** This project demonstrates the potential of shared data initiatives in oncology by focusing on data standardization and quality control, ensuring that the data can be utilized effectively across multiple research studies (Sweeney et al. 2023).
3. **National Cancer Institute Genomic Data Commons:** This initiative highlights the importance of comprehensive data governance frameworks to facilitate the sharing and utilization of genomic data, ensuring high data quality and compliance with privacy regulations (Sweeney et al. 2023).
4. **Veterans Health Administration Clinical Data Initiative:** This case emphasizes the integration of multiple data types and the necessity of extensive quality assurance activities to maintain data integrity and usability for research and clinical purposes (Sweeney et al. 2023).

These case studies illustrate the application of innovative strategies and frameworks to address and overcome data governance challenges in various sectors, including healthcare and research. They highlight the importance of strong leadership, structured frameworks, and continuous improvement in achieving successful data governance.

11.11 Conclusion

In concluding this chapter, "Overcoming Challenges in Data Governance: Strategies for Success," it is evident that data governance is a multifaceted domain, presenting a variety of challenges ranging from organizational culture to compliance with evolving legal requirements. The chapter highlights the critical importance of aligning data governance with business objectives to ensure that data is not only high in quality but also drives meaningful outcomes.

Key takeaways include the understanding that managing data governance in complex environments requires a nuanced approach, integrating advanced technologies and tools while addressing skill gaps through targeted training. The chapter underscores the significance of addressing data security and privacy concerns, which are paramount in today's data-driven landscape.

The section on regulatory compliance emphasizes the dynamic nature of the legal landscape, suggesting that organizations must remain agile and informed to navigate these changes successfully. Moreover, the chapter delves into the challenges posed by organizational culture and change management, advocating for a strategic approach to foster a data-centric culture.

The inclusion of case studies provides practical insights into how different organizations have successfully navigated the complex terrain of data governance, offering valuable lessons and strategies for others to emulate.

In summary, this chapter provides a comprehensive overview of the challenges in data governance, coupled with actionable strategies for success. It emphasizes the need for continuous adaptation and learning in the face of evolving technological landscapes, regulatory requirements, and organizational dynamics. By addressing these challenges head-on and leveraging the insights and strategies discussed, organizations can enhance their data governance practices, leading to more informed decision-making and, greater success in their respective fields.

References

Aldoseri A, Al-Khalifa KN, Hamouda AM (2023) Re-thinking data strategy and integration for artificial intelligence: concepts, opportunities, and challenges. Appl Sci 13(12):7082
Ataccama (2023) How GenAI boosts data governance initiatives
Brickclay (2023) Data governance: implementation, challenges, and solutions
Cambridge University Press (2019) Education and planning: anticipating and responding to skill gaps, changing skill needs and competencies
Cambridge University Press (2024) Building better global data governance
Confluent (n.d.) Aligning data governance initiatives with business objectives
Coursera (2024) Skills gap analysis: a guide to training your teams
Databricks (n.d.) What is data governance? A comprehensive guide
DataGalaxy (n.d.) Data governance for supporting business objectives
Devane H (2021) The complete guide to data security compliance laws and regulations
Diligent Corporation (n.d.) Regulatory compliance 101: definition, requirements & solutions
IBM (n.d.) What is data compliance?
IEEE (2024) Research on big data security and privacy risk governance
IntechOpen (2023) Big data integration solutions in organizations: a domain-specific analysis
Integrate.io (2024) Overcoming data governance challenges
Jarmul K (2023) Practical data privacy: enhancing privacy and security in data
Kirshteyn ME (n.d.) Data governance framework: a comprehensive guide to success
Madsen LB (n.d.) Disrupting data governance: a call to action
McKinsey & Company (2020) Beyond hiring: how companies are reskilling to address talent gaps
McKinsey & Company (2022) Using skill gap assessments to help future-proof your organization
McKinsey & Company (n.d.) Designing data governance that delivers value
NordLayer (n.d.) What is regulatory compliance and why is it important?
Olson JE (2003) Data quality. Morgan Kaufmann
Ono T, Lafortune G, Schoenstein M (2013) Health workforce planning in OECD countries: a review of 26 projection models from 18 countries. OECD health working papers, no. 62. OECD Publishing
Royal Society (2024) Data governance: case studies
ScienceDirect (2024) Data governance: a conceptual framework, structured review, and research agenda
Smallwood RF (2019) Information governance: concepts, strategies, and best practices. Wiley
Soares S (n.d.-a) Data governance tools: evaluation criteria, big data governance, and alignment with enterprise data management
Soares S (n.d.-b) Data governance: creating value from information assets. CRC Press
Soares S (n.d.-c) The chief data officer handbook for data governance

Splunk (2023) Data governance: definitions, components, & best practices. Splunk

Sweeney SM, Hamadeh HK, Abrams N, Adam SJ, Brenner S, Connors DE et al (2023) Case studies for overcoming challenges in using big data in cancer. Cancer Res 83(8):1183–1190

TDWI (2021) Managing data in the cloud: the challenge of complex environments for real-time applications

TechRepublic (2024) How does data governance affect data security and privacy?

Torre-Bastida AI, Gil G, Miñón R, Díaz-de-Arcaya J (n.d.) Technological perspective of data governance in data space ecosystems

Zagoudis S (2020) Defining purpose, scope, and objectives in a data governance policy. DATAVERSITY

Chapter 12
Future Trends in Data Governance: Preparing for Tomorrow

Contents

12.1	Introduction to Future Trends in Data Governance.	366
12.2	The Rise of Artificial Intelligence and Machine Learning.	367
12.3	Increasing Importance of Data Ethics.	369
12.4	Impact of Big Data and IoT.	371
12.5	Cloud Governance.	373
12.6	Evolving Regulatory Landscape.	375
12.7	Enhanced Focus on Data Literacy.	377
12.8	Decentralization and Blockchain in Data Governance.	380
12.9	Predictive Analytics in Data Governance.	382
12.10	Preparing for the Future.	384
12.11	Conclusion: The Evolving Landscape of Data Governance.	387
References.		389

Abstract This chapter explores the evolving landscape of data governance and anticipates the changes and innovations that are expected to shape its future. This chapter serves as a forward-looking guide, preparing organizations to adapt to the rapid advancements in technology and shifts in the regulatory and business environments.

The chapter opens by emphasizing the dynamic nature of data governance, driven by the relentless pace of technological advancements and the exponential growth of data volumes. It stresses the importance of foresight and adaptability in data governance practices, urging organizations to not only address current challenges but also to prepare proactively for future developments.

One of the key discussions in the chapter is the rise of artificial intelligence (AI) and machine learning (ML), highlighting their transformative impact on data governance. AI and ML are expected to revolutionize data processing and analysis, enhance data quality, and enable predictive analytics, thereby automating and improving data governance tasks.

The chapter also delves into the increasing importance of data ethics, emphasizing the need for ethical data handling practices as data becomes a more valuable and pervasive resource. Topics such as data privacy, transparency, and accountability are explored in depth, alongside the need to address bias and ensure fairness in data-driven decisions.

Big Data and the Internet of Things (IoT) are discussed as significant factors impacting data governance, with their ability to generate immense volumes of data that pose new challenges and opportunities for data integration, quality management, and security.

Cloud governance is identified as another crucial area for future focus, with an increasing number of organizations moving their operations to the cloud. The chapter outlines best practices for managing cloud resources, ensuring security and compliance, and optimizing costs in a cloud environment.

Moreover, the evolving regulatory landscape is examined, highlighting the need for organizations to stay informed and compliant with new and changing data protection laws and industry-specific regulations.

A strong emphasis is placed on the importance of data literacy across all levels of an organization, promoting a culture where data-driven decision-making is the norm. Enhancing data literacy is portrayed as essential for empowering employees and fostering an environment where data is recognized as a strategic asset.

Lastly, the chapter explores the implications of emerging technologies such as blockchain and decentralization in data governance, discussing their potential to enhance data security, transparency, and control.

In conclusion, this chapter provides a comprehensive overview of the trends and challenges that will shape the future of data governance. It encourages organizations to adopt a proactive and strategic approach, leveraging new technologies and evolving with the changing landscape to ensure effective and future-proof data governance practices.

Keywords Future Trends · Data Governance · Artificial Intelligence · Machine Learning · Data Ethics · Big Data · IoT · Cloud Governance · Regulatory Landscape · Data Literacy · Decentralization · Blockchain · Predictive Analytics

12.1 Introduction to Future Trends in Data Governance

This section serves as the gateway to a forward-looking exploration of the ever-evolving landscape of data governance. This introductory section lays the foundation for understanding the pivotal role that data governance will play in the future and sets the stage for subsequent discussions on emerging trends and challenges.

In this section, readers are introduced to the dynamic nature of data governance, highlighting the relentless pace of technological advancements and the exponential growth of data volumes. It underscores the notion that data governance is not a static discipline but a constantly evolving one that must adapt to the changing data landscape.

Furthermore, the introduction emphasizes the importance of foresight and adaptability in the realm of data governance. It suggests that organizations must not only address current data governance challenges but also anticipate and prepare for future developments. This proactive approach is essential to remain competitive and compliant in a data-driven world.

Additionally, the section sets the tone for the subsequent discussions on key topics such as artificial intelligence and machine learning, data ethics, big data and IoT, cloud governance, the regulatory landscape, data literacy, decentralization, blockchain, and predictive analytics. These emerging trends and challenges will shape the future of data governance, and this introduction serves as a roadmap for exploring them in depth.

12.2 The Rise of Artificial Intelligence and Machine Learning

The rise of Artificial Intelligence (AI) and Machine Learning (ML) is a transformative trend in the field of data governance. AI and ML technologies have gained significant prominence and are poised to reshape how organizations manage and leverage data. Here are some key points to consider about the impact of AI and ML on data governance:

- **Data Processing and Analysis**: AI and ML have the capacity to process and analyze vast amounts of data at unprecedented speeds. This capability enables organizations to gain valuable insights from their data, which is essential for informed decision-making (Al-Jarrah et al. 2015).
- **Data Quality Enhancement**: AI and ML algorithms can identify and rectify data quality issues in real time. This proactive approach to data quality management ensures that organizations work with accurate and reliable data (Ma et al. 2018).
- **Predictive Analytics**: AI and ML enable predictive analytics, allowing organizations to forecast trends and make proactive decisions. This is particularly valuable in industries like finance, where predictive analytics can help identify potential risks and opportunities (Miller and Brown 2018; Wamba et al. 2015).
- **Automation of Data Governance Tasks**: AI-driven automation can streamline data governance tasks such as data classification, data lineage tracking, and access control. This reduces the burden on data governance teams and minimizes the risk of human error (Data Governance: The Definitive Guide 2021).
- **Data Privacy and Security**: AI and ML can enhance data security by identifying unusual patterns or anomalies that may indicate a security breach. They also play a crucial role in ensuring compliance with data privacy regulations through features like data anonymization (Miller and Brown 2018).

- **Natural Language Processing (NLP)**: NLP, a subset of AI, allows organizations to understand and manage unstructured data such as text and voice data. This is particularly relevant in industries like healthcare and customer service (Data Governance: The Definitive Guide 2021).
- **Personalization**: AI and ML enable personalized customer experiences by analyzing user behavior and preferences. This enhances customer engagement and satisfaction (Wamba et al. 2015).
- **Ethical Considerations**: The use of AI and ML in data governance also raises ethical considerations such as bias in algorithms and the responsible use of AI. Organizations must address these issues to maintain trust and transparency (Miller and Brown 2018).

Figure 12.1 illustrates the impact of Artificial Intelligence (AI) and Machine Learning (ML) on data governance by highlighting several aspects. AI and ML enhance the ability to process and analyze vast amounts of data swiftly, facilitating informed decision-making. They help in the real-time identification and correction of data quality issues, ensuring data reliability. These technologies enable predictive analytics, assisting organizations in forecasting trends and making proactive decisions. Automation through AI simplifies data governance tasks such as data classification, lineage tracking, and access control, reducing the scope for human error. AI and ML also contribute to security by detecting unusual patterns that may indicate breaches and aid in compliance with privacy regulations through mechanisms like data anonymization. AI enhances the handling of unstructured data like text and voice, which is vital in sectors such as healthcare and customer service. By analyzing user behavior and preferences, AI and ML enable personalized customer

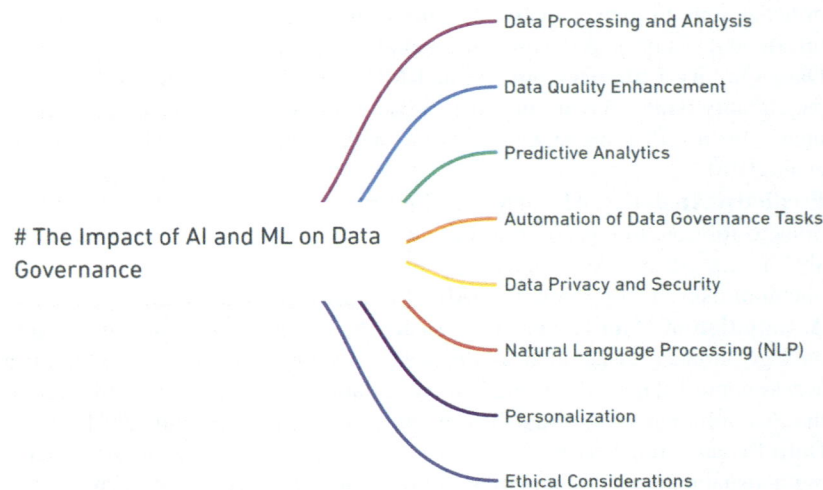

Fig. 12.1 The impact of AI and ML on data governance. (Created by the author)

experiences, boosting engagement and satisfaction. The integration of AI and ML in data governance brings up ethical issues, including algorithmic bias and the responsible use of AI, which organizations need to address to maintain trust and transparency.

The rise of AI and ML is revolutionizing data governance by improving data quality, enabling predictive analytics, automating tasks, enhancing data security, and offering personalized experiences. However, it also brings ethical challenges that organizations must navigate to harness the full potential of these technologies in data governance.

12.3 Increasing Importance of Data Ethics

The increasing importance of data ethics is a critical consideration in the field of data governance. Data ethics refers to the moral and ethical principles that govern the collection, handling, use, and dissemination of data. As data has become a valuable and pervasive resource in today's digital world, ensuring ethical practices is paramount. Here are key points to understand about the rising significance of data ethics:

- **Privacy Concerns**: With the proliferation of data collection, individuals and organizations are increasingly concerned about the privacy of personal information. Data ethics requires safeguarding individuals' privacy rights and ensuring that data is collected and used transparently and with consent (McKinsey and Company 2023a, c; DataCamp 2023a).
- **Transparency and Accountability**: Ethical data governance necessitates transparency in data practices. Organizations should clearly communicate their data collection and usage policies to users and stakeholders. Moreover, they must be accountable for adhering to these policies (Atlan 2023a; ISBA 2023).
- **Data Bias and Fairness**: Data ethics includes addressing bias in data algorithms and AI systems. Biased data can lead to discriminatory outcomes, and it is essential to mitigate bias and ensure fairness in data-driven decisions (McKinsey and Company 2023a, c).
- **Data Security**: Protecting data from breaches and unauthorized access is an ethical imperative. Data breaches can lead to significant harm to individuals and organizations, making robust data security measures vital (DataCamp 2023a; ISBA 2023).
- **Consent and Control**: Data ethics emphasizes obtaining informed consent from individuals before collecting their data. It also recognizes individuals' right to control their data, including the ability to access, correct, or delete their information (McKinsey and Company 2023a, c; Atlan 2023a).

- **Compliance with Regulations**: Ethical data governance aligns with data protection regulations such as GDPR and CCPA. Compliance with these laws is not only a legal requirement but also an ethical responsibility (ISBA 2023).
- **Data Stewardship**: Ethical data stewardship involves responsible and sustainable data management practices. Organizations should consider the long-term impact of data collection and usage on individuals, society, and the environment (McKinsey and Company 2023a, c).
- **Accountability and Consequences**: Ethical data governance includes mechanisms for holding individuals and organizations accountable for unethical data practices. This may involve legal consequences for data breaches or misconduct (DataCamp 2023a).
- **Data for Social Good**: Data ethics recognizes the potential of data to drive positive social impact. It encourages the use of data for social good, such as addressing societal challenges or advancing scientific research (Atlan 2023a).
- **Ethics in AI and Automation**: As AI and automation play an increasing role in data-driven decision-making, ethical considerations extend to these technologies. Ensuring fairness, transparency, and accountability in AI systems is essential (McKinsey and Company 2023a, c).

Figure 12.2, titled "Rising Significance of Data Ethics," encapsulates various facets of data ethics in the evolving landscape of data management. The figure underscores the increasing concerns surrounding privacy and how transparency and accountability are crucial in maintaining public trust. It highlights the challenges of data bias and fairness, emphasizing the need for equitable data practices that do not discriminate against any group. Data security is portrayed as foundational, ensuring that personal and sensitive information is protected against unauthorized access and breaches. The aspect of consent and control reiterates the rights of individuals to control their personal information and how it is used. Compliance with regulations is shown as a critical component, ensuring that organizations adhere to legal standards and ethical practices in their operations. Data stewardship is discussed, focusing on the responsibilities of managing data with integrity. Accountability and consequences are linked to the ethical implications of data misuse and the necessity for mechanisms to address grievances and infractions. The figure also addresses the potential of using data for social good, advocating for the use of data in ways that benefit society and contribute to humanitarian causes. Lastly, ethics in AI and automation is a significant point, delving into the moral implications of automated decision-making systems and their impact on society.

The increasing importance of data ethics reflects the recognition that responsible data governance goes beyond legal compliance. It encompasses principles of fairness, transparency, accountability, and the protection of individuals' rights. Ethical data governance is crucial for building trust with stakeholders, mitigating risks, and fostering responsible data-driven innovation.

12.4 Impact of Big Data and IoT

Fig. 12.2 Rising significance of data ethics. (Created by the author)

12.4 Impact of Big Data and IoT

The impact of Big Data and the Internet of Things (IoT) on data governance is substantial and transformative. Big Data refers to the vast volumes of data generated at high velocity and variety, while IoT encompasses the interconnected network of physical devices and sensors that collect and exchange data. Here are key points to understand about the impact of Big Data and IoT on data governance:

- **Data Volume and Variety**: Big Data and IoT generate immense amounts of data from various sources, including sensors, devices, and social media. This data is often unstructured or semi-structured, posing challenges for traditional data governance practices (ScienceDirect 2019; Indium Software 2023).
- **Real-Time Data**: IoT devices continuously generate real-time data streams. Data governance must adapt to handle this rapid influx of data and ensure its timely processing and analysis (Gartner 2022).
- **Data Quality**: Ensuring data quality becomes more complex with Big Data and IoT due to the diversity and volume of data sources. Data governance practices are essential to maintain data accuracy and reliability (SpringerLink 2023b).

- **Data Integration**: Data from IoT devices and Big Data sources often need to be integrated with existing datasets. Data governance plays a crucial role in harmonizing data from diverse origins (SAS 2023).
- **Data Security**: The massive amounts of data generated by IoT devices can pose security risks. Data governance must incorporate robust security measures to protect sensitive information and prevent breaches (ScienceDirect 2019).
- **Privacy Concerns**: Big Data and IoT raise significant privacy concerns as they involve collecting data from individuals' interactions with devices and sensors. Data governance must address these privacy challenges and ensure compliance with data protection regulations (Indium Software 2023).
- **Data Analytics**: Big Data and IoT offer opportunities for advanced analytics, predictive modeling, and machine learning. Data governance plays a role in enabling data scientists and analysts to access and use data for insights (SAS 2023).
- **Regulatory Compliance**: Data governance must align with evolving regulations that govern data collection and usage, such as GDPR for personal data collected through IoT devices (Gartner 2022).
- **Data Lifecycle Management**: Managing the entire data lifecycle from data creation to archiving and deletion becomes more complex with the continuous flow of data from IoT devices. Data governance helps define data retention and disposal policies (SpringerLink 2023b).
- **Scalability**: Data governance practices need to be scalable to accommodate the growing volume and complexity of data from Big Data and IoT sources (Indium Software 2023).

Figure 12.3, titled "Impact of Big Data and IoT on Data Governance," explores the transformative effects of big data and the Internet of Things (IoT) on the realm of data governance. The figure highlights how big data contributes to the expansion of data volume and variety, requiring robust data integration techniques and stressing the importance of data quality. It showcases how data analytics tools are pivotal in extracting meaningful insights from vast datasets, which can be instrumental for

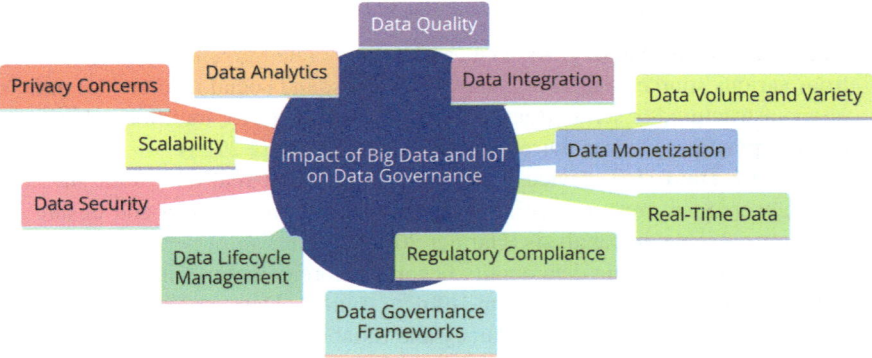

Fig. 12.3 Impact of big data and IoT on data governance. (Created by the author)

data monetization strategies. The integration of IoT devices introduces real-time data processing needs, enhancing decision-making processes but also introducing complexities in data management. Privacy concerns are amplified with the increase in data collection points, necessitating stringent data security measures to protect sensitive information. Scalability becomes a critical issue as systems must handle the growing influx of data without compromising performance. Data lifecycle management is depicted as crucial for organizing this data effectively across its useful life. Regulatory compliance is emphasized, considering the legal frameworks that govern data usage and protection. Finally, the figure encapsulates the need for adaptive data governance frameworks that can evolve in response to the dynamic nature of big data and IoT environments, ensuring that governance practices remain effective and relevant.

Big Data and IoT have a profound impact on data governance, requiring organizations to adapt their governance practices to handle the volume, velocity, variety, and complexity of data generated. Effective data governance is essential to unlock the value of these data sources while ensuring compliance, security, and data quality.

12.5 Cloud Governance

Cloud governance is a set of policies, processes, and practices that organizations put in place to manage and optimize their cloud resources effectively. It is a crucial aspect of modern IT management, especially as more businesses adopt cloud computing services. Here are key points to understand about cloud governance:

- **Resource Management**: Cloud governance involves managing cloud resources such as virtual machines, storage, databases, and networking to ensure optimal utilization and cost-efficiency (Imperva 2023; PwC 2023).
- **Cost Control**: Cloud services can lead to unpredictable costs. Governance practices help monitor, allocate, and optimize cloud spending to avoid unexpected expenses (BigID 2023).
- **Security and Compliance**: Ensuring data security and compliance with industry regulations is a fundamental aspect of cloud governance. It includes identity and access management, encryption, and compliance monitoring (Red Hat 2023; DATAVERSITY 2023b).
- **Policy Enforcement**: Organizations establish policies for cloud resource usage, and cloud governance ensures that these policies are enforced consistently across the cloud environment (PwC 2023; Imperva 2023).
- **Resource Provisioning**: Cloud governance defines the processes for provisioning resources, including approvals and access controls to prevent unauthorized resource creation (BigID 2023).
- **Monitoring and Auditing**: Continuous monitoring of cloud resources and auditing of activities help identify vulnerabilities, threats, and compliance violations. Governance practices enable proactive responses to these issues (Red Hat 2023).

- **Data Governance**: Managing data in the cloud, including data storage, backup, and retention policies, is a critical aspect of cloud governance (DATAVERSITY 2023b).
- **Disaster Recovery and Business Continuity**: Cloud governance includes strategies for disaster recovery and business continuity to ensure that critical systems and data are protected and can be quickly restored in case of disruptions (Imperva 2023).
- **Multi-Cloud Management**: Organizations using multiple cloud providers need governance practices to manage resources, costs, and security consistently across different cloud platforms (PwC 2023).
- **DevOps Integration**: Cloud governance aligns with DevOps practices to enable continuous integration and delivery (CI/CD) pipelines, ensuring that cloud resources are provisioned and managed in an automated and controlled manner (Red Hat 2023).
- **Resource Tagging**: Tagging cloud resources with metadata is a governance practice that helps with resource categorization, cost allocation, and tracking (BigID 2023).
- **User Training and Awareness**: Ensuring that users and administrators are aware of and trained on cloud governance policies and practices is essential for effective implementation (Imperva 2023).
- **Scalability**: Cloud governance practices should be scalable to accommodate the growth of cloud resources and services as an organization's needs evolve (DATAVERSITY 2023b).
- **Governance as Code**: Implementing governance policies through code (Infrastructure as Code) allows for automation and consistency in applying governance rules (Red Hat 2023).
- **Third-Party Governance Tools**: Many organizations use third-party governance tools and platforms to streamline and automate cloud governance tasks (PwC 2023).

Figure 12.4, outlines the critical aspects essential for effective cloud governance. The figure emphasizes the importance of security and compliance, ensuring that cloud deployments align with regulatory requirements and are protected from security threats. Resource management and provisioning are highlighted as crucial for efficient cloud operations, allowing organizations to allocate and scale resources dynamically. The figure also discusses the significance of resource tagging for cost management and accountability, aiding in tracking usage and spending. Disaster recovery and business continuity are marked as necessary to mitigate the risk of data loss and ensure uninterrupted service. Monitoring and auditing are depicted as vital practices to maintain operational standards and compliance. The concept of multi-cloud management addresses the complexity and scalability of managing multiple cloud environments simultaneously. Policy enforcement is essential for ensuring that cloud resources adhere to organizational and regulatory standards. DevOps integration is shown as key to enhancing operational efficiency by automating deployment and management tasks. Governance as code is presented as an approach

Fig. 12.4 Key points to understand about cloud governance. (Created by the author)

to enforce and validate compliance through code, automating governance policies. Lastly, third-party governance tools are noted for their role in extending native cloud capabilities to meet specific governance needs, offering enhanced monitoring, management, and compliance features.

Cloud governance is a comprehensive approach to managing cloud resources, optimizing costs, ensuring security and compliance, and aligning cloud operations with organizational goals. It is essential for organizations to establish robust cloud governance practices to harness the benefits of cloud computing while mitigating risks and challenges.

12.6 Evolving Regulatory Landscape

The evolving regulatory landscape refers to the continuously changing and adapting set of laws, regulations, and compliance requirements that govern various industries and activities. In the context of data governance, the regulatory landscape is particularly relevant as it impacts how organizations collect, store, process, and protect data.

- **Data Protection Regulations**: Data protection laws such as the General Data Protection Regulation (GDPR) in Europe and the California Consumer Privacy Act (CCPA) in the United States have had a significant impact on how organizations handle personal data. These regulations emphasize data privacy, consent, and individuals' rights over their data (Tealium 2024; Protiviti 2023).
- **Industry-Specific Regulations**: Different industries may be subject to industry-specific regulations related to data governance. For example, the healthcare sector must comply with the Health Insurance Portability and Accountability Act (HIPAA), while the financial industry is governed by regulations like the Dodd-Frank Act and Basel III (McKinsey and Company 2023b; Arcesium 2024).

- **Data Breach Notification Laws**: Many regions have implemented data breach notification laws that require organizations to promptly notify affected individuals and authorities in the event of a data breach. Failure to comply can result in significant fines (Tealium 2024).
- **Cross-Border Data Transfers**: The transfer of data across international borders is subject to regulations governing data sovereignty and cross-border data flows. Organizations must ensure compliance when transferring data between regions (Protiviti 2023).
- **Data Retention and Erasure**: Regulations often dictate how long organizations can retain data and under what circumstances they must erase it. Compliance with these rules is essential to avoid legal consequences (McKinsey and Company 2023b).
- **Emerging Regulations**: As technology evolves, new regulations emerge to address new challenges. For example, regulations related to data generated by Internet of Things (IoT) devices and AI are still evolving (Arcesium 2024).
- **Data Governance Frameworks**: Organizations must establish robust data governance frameworks that align with relevant regulations. This includes defining roles and responsibilities for data protection, security, and compliance (Tealium 2024).
- **Regulatory Updates**: The regulatory landscape is dynamic, with updates and amendments occurring regularly. Organizations must stay informed about these changes to maintain compliance (Protiviti 2023).
- **Enforcement and Penalties**: Regulatory authorities have the power to enforce compliance and impose penalties for non-compliance. Penalties can include fines, legal action, and reputational damage (McKinsey and Company 2023b).
- **Global Impact**: Many regulations have a global impact, affecting organizations that operate internationally. Compliance with the strictest regulations often sets the standard for data governance practices worldwide (Tealium 2024).
- **Ethical Considerations**: In addition to legal compliance, organizations should also consider ethical considerations related to data use and privacy. Ethical data governance goes beyond legal requirements to ensure responsible and transparent data practices (Protiviti 2023).

Figure 12.5, titled "The Regulatory Landscape," presents a comprehensive overview of the regulatory environment impacting data governance. This figure highlights the importance of understanding various aspects of regulations that govern data usage and protection globally. It stresses the significance of cross-border data transfers and the complexities involved, necessitating adherence to international data protection laws. Industry-specific regulations are pointed out, underscoring that different sectors may face unique regulatory challenges and requirements. Data governance frameworks are noted as essential for ensuring that organizations align with these regulatory demands systematically and effectively. Data breach notification laws are emphasized, illustrating the need for organizations to promptly report data breaches as required by law. The figure also discusses emerging regulations, indicating that the regulatory landscape is continually evolving and requires

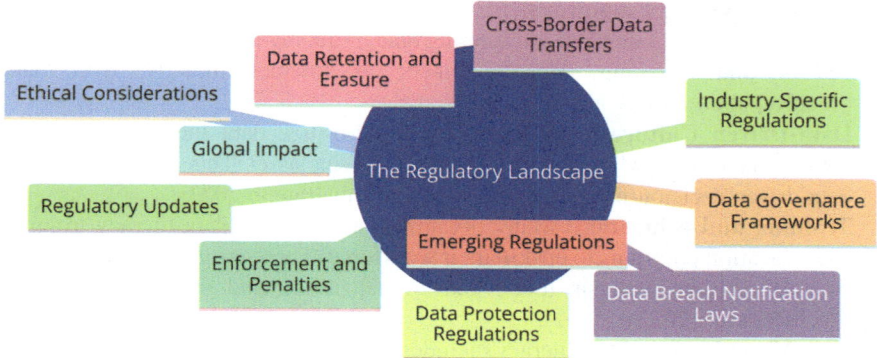

Fig. 12.5 The regulatory landscape. (Created by the author)

organizations to stay informed of changes to remain compliant. Data protection regulations form a core component of this landscape, focusing on the safeguarding of personal and sensitive data. The concept of data retention and erasure is addressed, highlighting the requirements for maintaining records for certain periods and properly disposing of them afterward. Enforcement and penalties are identified as key deterrents against non-compliance, stressing the financial and reputational risks of failing to adhere to regulations. Lastly, ethical considerations are mentioned, reminding organizations of the moral implications and responsibilities they bear in managing data.

The evolving regulatory landscape is a dynamic and complex environment that organizations must navigate to ensure compliance with data governance requirements. Staying informed, adapting to changes, and implementing robust data governance practices are essential to mitigate legal and reputational risks.

12.7 Enhanced Focus on Data Literacy

An enhanced focus on data literacy reflects the recognition that data has become a fundamental asset for organizations, and individuals at all levels within those organizations need to be proficient in understanding and effectively using data.

- **Definition of Data Literacy**: Data literacy refers to the ability to read, interpret, analyze, and communicate data. It encompasses a range of skills, including data analysis, data visualization, statistical reasoning, and data-driven decision-making (Tableau 2023).
- **Importance in the Data-Driven Era**: In today's data-driven business environment, data literacy is a critical skill for employees at all levels, from executives to frontline staff. It empowers individuals to make informed decisions based on data insights (DataCamp 2023b).

- **Decision-Making Empowerment:** Data literacy enables individuals to be more self-reliant in accessing and interpreting data, reducing the reliance on data specialists and allowing decision-makers to interact directly with data (TechRepublic 2023).
- **Data-Driven Culture:** Organizations that prioritize data literacy cultivate a data-driven culture where data is used as a strategic asset. This culture promotes data-driven decision-making and innovation (Integrate.io 2023).
- **Training and Education:** Many organizations invest in data literacy training and education programs to upskill their workforce. These programs often include courses on data analysis tools, data visualization, and statistical concepts (TechRepublic 2023).
- **Data Visualization:** Data literacy includes the ability to create and interpret data visualizations effectively. This skill is essential for conveying data insights in a clear and understandable manner (Tableau 2023).
- **Data Ethics:** Data literacy also encompasses an understanding of data ethics, including the responsible and ethical use of data, privacy considerations, and compliance with data protection regulations (Integrate.io 2023).
- **Cross-Functional Collaboration:** Data literacy encourages cross-functional collaboration within organizations. When individuals from different departments can speak the language of data, they can collaborate more effectively on data-driven projects (DataCamp 2023b).
- **Executive Data Literacy:** Executives and leaders play a crucial role in fostering a data-driven culture. Therefore, executive data literacy is essential for setting the tone and priorities for data initiatives (TechRepublic 2023).
- **Data-Driven Innovation:** Data-literate individuals are more likely to identify opportunities for innovation and optimization through data analysis. This can lead to new products, services, and business models (Integrate.io 2023).
- **Measuring Data Literacy:** Organizations may use assessments and metrics to measure the data literacy levels of their employees and identify areas where additional training or support is needed (Tableau 2023).
- **Continuous Learning:** Data literacy is an evolving skill set as data technologies and best practices change. Therefore, individuals and organizations must engage in continuous learning to stay current (DataCamp 2023b).

Figure 12.6, delves into the critical components that underline the increasing emphasis on data literacy in modern organizations. This figure outlines the foundational definition of data literacy, establishing what constitutes competence in understanding and using data effectively. It underscores the importance of data literacy in the data-driven era, highlighting its role in empowering decision-making and fostering a culture that values data-driven insights. Decision-making empowerment is emphasized as a direct benefit of improved data literacy, enabling individuals at all levels of an organization to make informed decisions based on accurate data analysis. The figure also discusses the need for training and education to enhance data literacy, suggesting that continual learning and development are key to maintaining a competitive edge. Data visualization is mentioned as a vital skill, enabling better

12.7 Enhanced Focus on Data Literacy

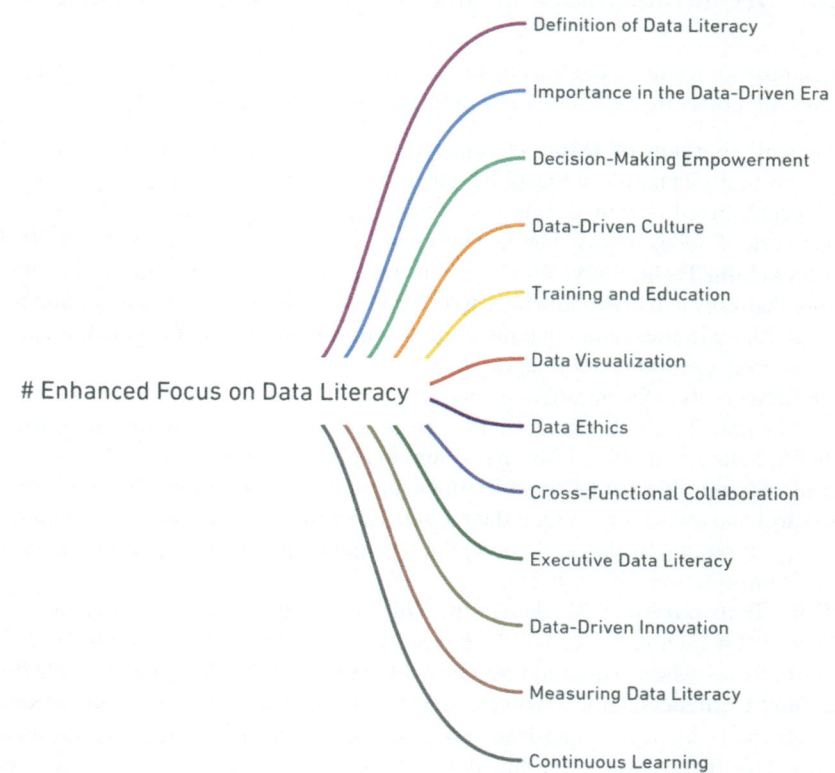

Fig. 12.6 Enhanced focus on data literacy. (Created by the author)

interpretation and communication of data insights. Data ethics are considered crucial, ensuring that data is used responsibly and ethically. Cross-functional collaboration is showcased as essential, promoting an integrative approach to data literacy that spans various departments and disciplines. Executive data literacy is highlighted, stressing the importance of leaders understanding and championing data-driven strategies within their organizations. Data-driven innovation is identified as an outcome of widespread data literacy, leading to new ideas and improvements in processes. Measuring data literacy is shown as necessary for assessing progress and identifying areas needing enhancement. Lastly, continuous learning is emphasized, reflecting the need for ongoing education and adaptation to new data practices and technologies.

An enhanced focus on data literacy reflects the growing recognition that data is a strategic asset that should be accessible and understandable to all individuals within an organization. Data literacy empowers employees to make data-driven decisions, fosters a data-driven culture, and contributes to innovation and competitiveness.

12.8 Decentralization and Blockchain in Data Governance

Decentralization and blockchain technology are innovative approaches to data governance that offer increased security, transparency, and control over data.

- **Decentralization of Data:** Decentralization refers to the distribution of data across multiple nodes or locations rather than storing it in a central repository. In a decentralized system, data is distributed, duplicated, and synchronized across a network of nodes, which can be computers, servers, or devices (Frontiers 2023).
- **Blockchain Technology:** Blockchain is a specific form of decentralized technology that uses a distributed ledger to record transactions securely and immutably. Each block in the chain contains a set of transactions, and once added, it cannot be altered, ensuring data integrity (SpringerLink 2023a).
- **Data Security:** Decentralization and blockchain enhance data security by reducing the risk of a single point of failure or attack. Data is stored redundantly across multiple nodes, making it more resilient to cyber threats (Frontiers 2023).
- **Data Ownership and Control:** Blockchain allows individuals to have greater control and ownership of their data. Users can grant or revoke access to their data using cryptographic keys, enabling them to maintain control over who can access their information (IEEE 2023).
- **Data Transparency:** Blockchain provides transparency as all transactions are recorded in a publicly accessible ledger. This transparency is especially valuable in scenarios where trust and accountability are essential (SpringerLink 2023a).
- **Smart Contracts:** Blockchain can execute self-executing contracts called smart contracts, which automate business processes when predefined conditions are met. This feature can streamline data governance processes and reduce the need for intermediaries (IEEE 2023).
- **Immutable Data:** Once data is recorded on a blockchain, it becomes immutable and cannot be altered or deleted. This feature is crucial for maintaining data integrity and ensuring an accurate historical record (Frontiers 2023).
- **Data Traceability:** Blockchain's ledger provides a comprehensive history of data transactions. Users can trace the origin and changes to data, enhancing auditability and compliance (SpringerLink 2023a).
- **Privacy Considerations:** While blockchain offers transparency, privacy concerns also arise. Some blockchain implementations, like private or permissioned blockchains, provide privacy features to protect sensitive data (IEEE 2023).
- **Use Cases:** Blockchain and decentralization are applied in various data governance use cases, including supply chain management, identity verification, healthcare data sharing, and financial transactions (Frontiers 2023).
- **Challenges:** Implementing blockchain and decentralization in data governance comes with challenges such as scalability, energy consumption (in the case of some blockchains like Bitcoin), and regulatory compliance (SpringerLink 2023a).
- **Interoperability:** Ensuring interoperability between different blockchain networks and with existing systems is crucial for widespread adoption (IEEE 2023).

12.8 Decentralization and Blockchain in Data Governance

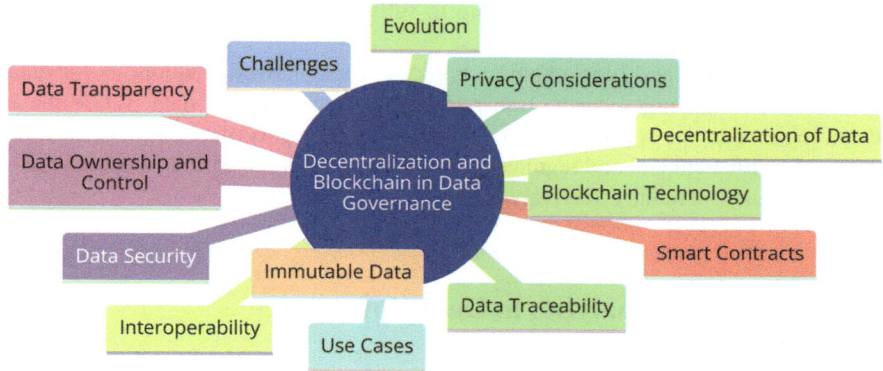

Fig. 12.7 Decentralization and blockchain in data governance map. (Created by the author)

- **Evolution:** Blockchain and decentralized technologies continue to evolve. New consensus mechanisms, scalability solutions, and privacy enhancements are being developed to address current limitations (SpringerLink 2023a).

Figure 12.7, explores how blockchain technology facilitates the decentralization of data governance. This figure emphasizes the importance of blockchain in ensuring data transparency, which strengthens trust among stakeholders by making data operations visible and verifiable. Privacy considerations are crucial, as blockchain can offer enhanced data privacy through encryption and pseudonymization techniques. The decentralization of data highlights the shift from centralized to distributed models of data management, enhancing security and reducing single points of failure. Data ownership and control are depicted as significant benefits, as blockchain enables individuals to maintain control over their data without the need for intermediaries. Data security is bolstered by blockchain's immutable nature, where data cannot be altered once recorded, providing a robust defense against tampering and fraud. Interoperability across different blockchain systems is identified as vital for broader adoption and functional efficacy in diverse applications. Immutable data is a core feature of blockchain that ensures the integrity of data records, making it highly reliable for audit trails and historical data analysis. Data traceability is enhanced through blockchain, allowing the tracking of data origins and changes over time, which is particularly useful in supply chains and regulatory compliance. Use cases for blockchain in data governance are diverse, ranging from enhancing supply chain transparency to securing health records and automating compliance via smart contracts. Smart contracts are explained as self-executing contracts with the terms of the agreement directly written into code, which can automate processes and ensure compliance without human intervention. The figure also discusses the evolution of blockchain technology in data governance, noting its ongoing development and integration into various industries. Finally, challenges such as scalability, energy consumption, and integration with existing systems are acknowledged as current limitations that need addressing to fully realize blockchain's potential in data governance.

Decentralization and blockchain technology offer innovative solutions for data governance, enhancing security, transparency, and control over data. While these technologies present opportunities for improving data governance practices, they also come with technical and regulatory challenges that organizations must navigate to leverage their benefits effectively.

12.9 Predictive Analytics in Data Governance

Predictive analytics in data governance involves using advanced analytical techniques to anticipate and mitigate data-related issues, enhance data quality, and improve decision-making processes. Predictive analytics leverages historical and real-time data to predict future outcomes or identify patterns, trends, and potential issues. In data governance, it focuses on using predictive models to anticipate data quality problems, compliance violations, and data-related risks.

- **Data Quality Improvement:** Predictive analytics can identify data quality issues before they escalate. For example, it can predict data entry errors, missing values, or inconsistencies in data sources and suggest corrective actions (DATAVERSITY 2023a).
- **Anomaly Detection:** Predictive models can detect anomalies or deviations from expected data patterns. This is valuable for identifying data breaches, fraud, or irregularities in data that may indicate a problem (IMD 2023).
- **Compliance Monitoring:** Predictive analytics can help organizations proactively monitor and ensure compliance with data protection regulations. It can detect potential violations and trigger alerts or actions to address them (IBM 2023).
- **Data Matching and Deduplication:** Predictive models can be used to match and deduplicate records in large datasets. This is important for maintaining data accuracy and consistency (Collibra 2023).
- **Customer Behavior Prediction:** In customer-centric industries, predictive analytics is used to anticipate customer behavior, preferences, and needs. This information can inform data governance strategies and marketing efforts (IMD 2023).
- **Data Governance Workflow Optimization:** Predictive analytics can optimize data governance workflows by predicting resource requirements, workload distribution, and potential bottlenecks (Deloitte 2023).
- **Predictive Maintenance:** In industries with equipment and machinery, predictive analytics can be used for predictive maintenance. It anticipates when equipment is likely to fail, helping organizations plan maintenance activities and prevent costly downtime (IBM 2023).

12.9 Predictive Analytics in Data Governance

- **Resource Allocation:** Predictive models can assist in resource allocation, such as allocating data governance personnel to areas with the highest risk of data issues (DATAVERSITY 2023a).
- **Machine Learning Models:** Machine learning algorithms, including regression, classification, and clustering, are commonly used in predictive analytics for data governance (Collibra 2023).
- **Data Preparation:** High-quality predictive analytics relies on clean and well-prepared data. Data governance plays a role in ensuring data is suitable for predictive modeling (IMD 2023).
- **Continuous Improvement:** Predictive models require continuous monitoring and updating to remain accurate and relevant. Data governance processes should include mechanisms for model maintenance (DATAVERSITY 2023a).
- **Interpretability:** Understanding the results of predictive analytics is crucial. Data governance should ensure that predictions and insights are transparent and interpretable to facilitate decision-making (IBM 2023).
- **Ethical Considerations:** Predictive analytics raises ethical considerations, especially when making predictions about individuals or sensitive data. Data governance must include ethical guidelines for model development and usage (Deloitte 2023).
- **Data Security:** Predictive models should be protected against unauthorized access and data breaches. Data governance should enforce security measures to safeguard model data and results (Collibra 2023).

Figure 12.8, titled "Predictive Analytics in Data Governance," maps the integration of predictive analytics within the scope of data governance, emphasizing its transformative impact across various aspects. The figure highlights how predictive analytics is pivotal in data governance workflow optimization, enabling more efficient and effective processes by predicting future trends and needs. Data security is reinforced through predictive analytics by forecasting potential security breaches and vulnerabilities, allowing for preemptive actions. Customer behavior prediction is a significant application, where analytics are used to forecast consumer trends and

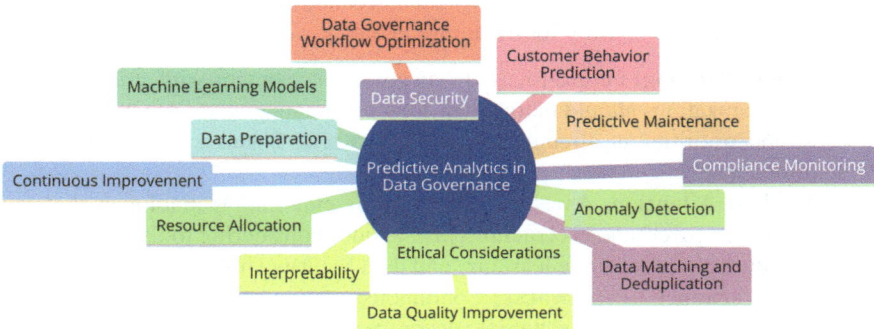

Fig. 12.8 Predictive analytics in data governance. (Created by the author)

preferences, thus enhancing customer engagement and personalized marketing. Predictive maintenance utilizes predictive analytics to foresee equipment failures and schedule maintenance, thus minimizing downtime and operational costs. Compliance monitoring through predictive models helps organizations stay ahead of regulatory requirements by predicting compliance risks. Anomaly detection is crucial in identifying unusual patterns that may indicate fraud, security breaches, or operational inefficiencies. Data matching and deduplication are improved by predictive analytics, which helps in recognizing and merging similar records and removing duplicates, thereby enhancing data quality and accessibility. Ethical considerations are underscored, focusing on the responsible use of predictive analytics to avoid biases and ensure fairness in automated decisions. Data quality improvement benefits from predictive analytics by identifying areas where data integrity can be enhanced, guiding efforts to cleanse and refine data. Interoperability across different data systems and models is facilitated through predictive analytics by enabling seamless data integration and communication. Resource allocation is optimized using predictive insights to better manage and distribute organizational resources. Continuous improvement is a theme throughout, with predictive analytics driving ongoing enhancements in data governance practices. Machine learning models and data preparation are foundational elements, supporting predictive analytics by providing the necessary algorithms and clean data for effective analysis.

Predictive analytics is a powerful tool in data governance, enabling organizations to proactively address data quality issues, compliance violations, and risks. It can lead to more informed decision-making, cost savings, and improved operational efficiency when integrated into data governance processes.

12.10 Preparing for the Future

Preparing for the future in the context of data governance involves anticipating and adapting to emerging trends, technologies, and challenges in the data landscape. It is about ensuring that an organization's data governance strategy remains relevant, agile, and effective in the face of evolving data-related factors. Here are key points to understand about preparing for the future in data governance:

- **Strategic Planning:** Future-focused data governance begins with strategic planning. Organizations should define their long-term data governance goals, objectives, and priorities in alignment with their overall business strategy (Hitachi Solutions 2023).
- **Technology Adoption:** Assessing and adopting emerging technologies that can enhance data governance is crucial. This includes considering tools for data analytics, automation, machine learning, and data security (Informatica 2021).
- **Data Governance Frameworks:** Continuously refine and update the organization's data governance framework to accommodate changing regulatory requirements, industry standards, and best practices (Atlan 2023b).

12.10 Preparing for the Future

- **Data Governance Roles:** Define and assign data governance roles and responsibilities within the organization. Ensure that data stewards, data owners, and data governance teams are equipped to handle future challenges (DATAVERSITY 2023c).
- **Data Governance Policies:** Regularly review and update data governance policies and procedures to address evolving data privacy regulations, data security threats, and compliance requirements (Deloitte 2021).
- **Data Quality Improvement:** Implement strategies for ongoing data quality improvement. Utilize data profiling, cleansing, and enrichment techniques to maintain high-quality data (Atlan 2023b).
- **Data Privacy and Ethics:** Stay informed about evolving data privacy regulations and ethical considerations. Ensure that data governance practices prioritize data privacy and ethical data handling (Informatica 2021).
- **Data Security:** Strengthen data security measures to protect against evolving cybersecurity threats. Implement encryption, access controls, and threat detection systems to safeguard data assets (Hitachi Solutions 2023).
- **Data Integration:** Invest in data integration solutions that facilitate the seamless exchange of data between different systems and platforms, including on-premises and cloud environments (Deloitte 2021).
- **Data Catalogs:** Implement data catalogs and metadata management solutions to maintain a centralized inventory of data assets. This helps with data discovery and usage tracking (Atlan, 2023).
- **Data Governance Training:** Provide ongoing training and education to data governance teams and employees to ensure they are equipped with the latest knowledge and skills (Informatica 2021).
- **Cloud Data Governance:** Adapt data governance practices to encompass cloud data storage and services. Consider cloud-specific governance challenges and solutions (DATAVERSITY 2023c).
- **Scalability:** Ensure that data governance practices and technologies can scale to handle the growing volume and complexity of data generated by the organization (Hitachi Solutions 2023).
- **Data Monetization:** Explore opportunities for data monetization by responsibly sharing or selling data assets to partners or third parties while ensuring compliance with regulations (Atlan 2023b).
- **Continuous Improvement:** Embrace a culture of continuous improvement in data governance. Regularly assess performance, gather feedback, and adjust strategies accordingly (Informatica 2021).
- **Collaboration:** Encourage collaboration between data governance teams, IT departments, and business units to ensure alignment with organizational goals and future needs (Deloitte 2021).
- **Monitoring and Metrics:** Establish key performance indicators (KPIs) and monitoring processes to track the effectiveness of data governance initiatives and make data-driven improvements (Hitachi Solutions 2023).

Fig. 12.9 Preparing for the future in data governance, keys map. (Created by the author)

Figure 12.9, titled "Preparing for the Future in Data Governance," outlines the key areas of focus for organizations to ensure their data governance strategies are future-proof. This figure emphasizes the importance of data security and the ethical handling of data to uphold privacy standards. Continuous improvement is highlighted as essential, with organizations needing to adapt and evolve their data governance practices in response to technological changes and new business requirements. Monitoring and metrics are identified as crucial for assessing the effectiveness of data governance initiatives and making data-driven decisions. Strategic planning is central, allowing organizations to foresee future challenges and opportunities in data governance. Data governance training, roles, and policies are depicted as fundamental components, ensuring that all personnel are equipped with the knowledge and responsibilities necessary to uphold data governance standards. Collaboration across departments and with external partners is shown as vital for a cohesive data governance strategy. Data catalogs are important tools for managing metadata, facilitating easier data discovery and usability. Scalability concerns are addressed, ensuring that data governance frameworks can handle increasing volumes of data. Technology adoption, particularly in cloud data governance, is shown to play a pivotal role in modernizing and strengthening data governance structures. Data integration and the ability to manage diverse data sources seamlessly are critical for comprehensive data oversight. Data monetization is discussed, emphasizing the potential to extract economic value from effectively governed data. Lastly, data quality improvement remains a core objective, ensuring that the data used for decision-making is accurate and reliable, thus supporting overall business objectives and compliance with regulatory standards.

Preparing for the future in data governance involves a proactive and forward-looking approach to managing data assets and data-related processes. By staying adaptable, informed, and innovative, organizations can position themselves to navigate future challenges and leverage data as a strategic asset for success.

12.11 Conclusion: The Evolving Landscape of Data Governance

The future of data governance holds both challenges and exciting opportunities for organizations. This chapter delves into the emerging trends and dynamics that will shape the landscape of data governance in the years to come. Here, we draw conclusions based on the analysis of the key components of this chapter:

- **Artificial Intelligence and Machine Learning:** The rise of AI and ML presents organizations with powerful tools for data analysis and automation. Harnessing these technologies within data governance can lead to more accurate decision-making and proactive issue identification (BARC 2023; LightsOnData 2023).
- **Data Ethics:** Data ethics is no longer a peripheral concern but a central pillar of data governance. Organizations must prioritize ethical data handling and transparency to earn trust and comply with evolving regulations (DATAVERSITY 2024; TechTarget 2023).
- **Big Data and IoT:** The impact of Big Data and IoT continues to grow, introducing vast volumes of data. Effective data governance strategies are essential to harness the potential of these data sources while managing their complexity (TechTarget 2023; LightsOnData 2023).
- **Cloud Governance:** As organizations migrate to the cloud, cloud governance becomes paramount. Ensuring data security, privacy, and compliance in cloud environments is essential for successful data governance (BARC 2023; LightsOnData 2023).
- **Evolving Regulatory Landscape:** The regulatory landscape is in constant flux with new data protection and privacy regulations emerging globally. Organizations must adapt their data governance practices to stay compliant and avoid penalties (DATAVERSITY 2024; TechTarget 2023).
- **Data Literacy:** Enhanced data literacy across the organization empowers individuals to make data-driven decisions. This emphasis on data literacy supports a data-driven culture and better utilization of data assets (BARC 2023; LightsOnData 2023).
- **Decentralization and Blockchain:** Decentralization and blockchain offer innovative solutions for data governance, enhancing data security, transparency, and control. However, their implementation requires careful consideration of use cases and scalability (DATAVERSITY 2024; LightsOnData 2023).
- **Predictive Analytics:** Predictive analytics is a valuable tool for anticipating data quality issues, compliance violations, and other data-related challenges. It enables proactive problem-solving and resource allocation (TechTarget 2023; DATAVERSITY 2024).
- **Preparing for the Future:** Organizations must adopt a forward-looking approach to data governance. This involves strategic planning, technology adoption, compliance readiness, and a commitment to continuous improvement (BARC 2023; TechTarget 2023).

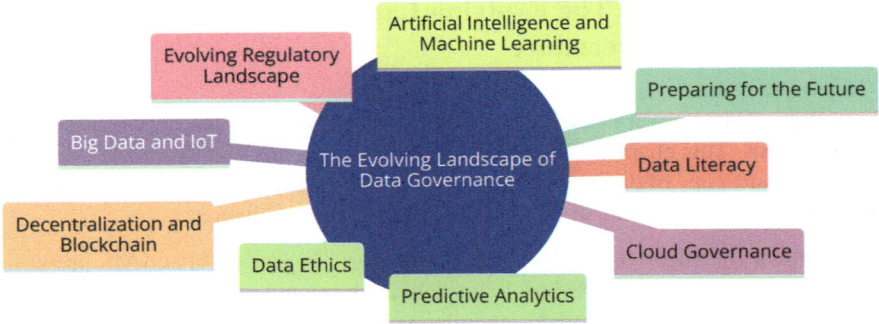

Fig. 12.10 The evolving landscape of data governance. (Created by the author)

Figure 12.10, titled "The Evolving Landscape of Data Governance," illustrates the dynamic nature of data governance, highlighting several key factors driving its evolution. The figure emphasizes the impact of big data and the Internet of Things (IoT) on data governance, noting the increasing volume and complexity of data these technologies generate, necessitating advanced governance strategies. Artificial Intelligence (AI) and Machine Learning (ML) are identified as transformative forces, enhancing the automation and efficiency of data governance processes. Decentralization and blockchain technology are shown as increasingly influential, offering new ways to secure and manage data transparently and immutably. Data ethics is central to the evolving landscape, stressing the need for responsible data management practices that uphold individual rights and societal values. Predictive analytics is highlighted for its role in foreseeing trends and potential issues, allowing organizations to proactively adjust their governance practices. The evolving regulatory landscape is recognized as a critical driver, with new and changing regulations requiring adaptable governance frameworks. Data literacy is underscored as essential, with a need for ongoing education and training to empower all organizational levels to handle data competently. Cloud governance is identified as vital due to the widespread adoption of cloud technologies, necessitating specialized strategies to manage and secure cloud-based data. Finally, the figure addresses the importance of preparing for the future, advocating for forward-thinking approaches that anticipate changes in technology, business processes, and regulatory environments.

The future of data governance is characterized by a dynamic interplay of technological advancements, evolving regulations, ethical considerations, and a growing emphasis on data literacy. Organizations that embrace these trends and adapt their data governance strategies accordingly will be better positioned to harness the full potential of their data, make informed decisions, and maintain a competitive edge in an increasingly data-centric world. Data governance is no longer a static framework but a dynamic and essential element of organizational success.

References

Al-Jarrah OY, Yoo PD, Muhaidat S, Karagiannidis GK, Taha K (2015) Efficient machine learning for big data: a review. Big Data Res 2(3):87–93. Elsevier

Arcesium (2024) Navigating the evolving regulatory landscape with data governance solutions. Arcesium.

Atlan (2023a) Data ethics unveiled: principles & frameworks explored. Atlan

Atlan (2023b) The top 12 data governance trends in 2024. Atlan

BARC (2023) Navigating 2024's data landscape: key trends and developments in data management. BARC

BigID (2023) Cloud data Governance: overview and best practices. BigID

Collibra (2023) The importance of predictive model Governance. Collibra

Data Governance. The definitive guide (E Eryurek, U Gilad, V Lakshmanan, A Kibunguchy-Grant, J Ashdown, eds). Released March 2021. O'Reilly Media

DataCamp (2023a) An introduction to data ethics: what is the ethical use of data? DataCamp

DataCamp (2023b) Closing the data literacy gap: key insights from the state of data literacy 2023 report. DataCamp

DATAVERSITY (2023a) A guide to predictive data analytics (making decisions for the future). DATAVERSITY

DATAVERSITY (2023b) Data governance in the cloud. DATAVERSITY

DATAVERSITY (2023c) The future of data Governance: balancing data governance and data management. DATAVERSITY

DATAVERSITY (2024) Data governance trends in 2024. DATAVERSITY

Deloitte (2021) The future of data Governance in a data-rich world. Deloitte UK

Deloitte (2023) Predictive analytics in government. Deloitte Insights

Frontiers (2023) Decentralized network Governance: Blockchain technology and the future of regulation. Frontiers

Gartner (2022) How IoT impacts data and analytics. Gartner

Hitachi Solutions (2023) Data governance: How to prepare for the future. Hitachi Solutions

IBM (2023) What is predictive analytics? IBM

IEEE (2023) Fine-grained data rights governance in blockchain-based cloud-edge communications. IEEE

IMD (2023) What is predictive analytics? Importance, benefits, & examples. IMD

Imperva (2023) Cloud governance: Framework & model principles. Imperva

Indium Software (2023) Big data's impact on IoT: opportunities and challenges in analytics. Indium Software

Informatica (2021) Data intelligence is the future of data Governance. Informatica

Integrate.io (2023) Why data literacy is essential for a data-driven future. Integrate.io

ISBA (2023) The importance of data ethics: why businesses must take it seriously. ISBA

LightsOnData (2023) In: LightsOnData (ed) Data governance in 2024

Ma J, Gao J, Wu Y, Zhou J (2018) Data quality assessment for data governance: application of machine learning. J Data Inform Qual (JDIQ) 10(4):16. ACM

McKinsey & Company (2023a) Data ethics: what it means and what it takes. McKinsey & Company

McKinsey & Company (2023b) Governance and regulation as generative AI advances. McKinsey & Company

McKinsey & Company (2023c) Putting data ethics into practice. McKinsey & Company

Miller T, Brown G (2018) Artificial intelligence in data governance: from theory to practice. J Data Sci Anal 5(2):102–115. Springer

Protiviti. (2023) Technology for the evolving data privacy regulatory landscape. Protiviti

PwC (2023) Cloud governance on risks and controls. PwC

Red Hat (2023) What is cloud governance? , Red Hat

SAS (2023) IoT success depends on data governance, security, and privacy. SAS

ScienceDirect (2019) IoT-Gov: a structured framework for internet of things governance. ScienceDirect
SpringerLink (2023a) Blockchain and institutions: trust and (De)centralization. SpringerLink
SpringerLink (2023b) Privacy and security challenges and opportunities for IoT technologies. SpringerLink
Tableau (2023) Data literacy explained: definition, examples & more. Tableau
Tealium (2024) Derisking data: regulatory milestones shaping the use of AI and data in 2024. Tealium
TechRepublic (2023) How data literacy is important to data governance. TechRepublic
TechTarget (2023) Data governance strategies for today's evolving IT landscape. TechTarget
Wamba SF, Akter S, Edwards A, Chopin G, Gnanzou D (2015) How 'big data' can make big impact: findings from a systematic review and a longitudinal case study. Int J Prod Econ 165:234–246. Elsevier

Chapter 13
Establishing a Data Governance Culture: Change Management and Leadership

Contents

13.1	Introduction to Data Governance Culture.	392
13.2	The Role of Leadership in Data Governance.	393
13.3	Change Management Principles in Data Governance.	396
13.4	Communicating the Value of Data Governance.	398
13.5	Building Data Governance into Organizational DNA.	401
13.6	Training and Empowerment.	404
13.7	Developing Data Governance Champions.	407
13.8	Incentivizing and Rewarding Compliance.	409
13.9	Overcoming Cultural Barriers.	412
13.10	Continuous Improvement and Adaptation.	414
13.11	Conclusion.	416
References.		417

Abstract This chapter delves into the vital role that organizational culture plays in the successful implementation of data governance. The chapter argues that data governance is not merely a technical initiative but also a cultural shift within an organization, requiring leadership commitment, strategic change management, and widespread stakeholder engagement.

The chapter begins by emphasizing the importance of leadership in driving data governance. Leaders are tasked with setting the vision, allocating resources, and championing the initiatives across the organization. They are instrumental in shaping a data governance team that includes data stewards and other key roles, crucial for the oversight and execution of data governance policies.

Change management principles are detailed as essential for integrating new data governance practices within an organization. The chapter provides a thorough examination of strategies for managing the human aspects of change, such as understanding the need for change, setting clear objectives, engaging stakeholders, and effectively communicating the change. It highlights the necessity of training and skill development to empower employees, enabling them to support and sustain data governance initiatives.

One of the critical focuses of the chapter is on the methods for communicating the value of data governance. It underscores the need to articulate how data governance aligns with broader business goals, emphasizing benefits such as improved data quality, compliance, risk mitigation, and operational efficiency.

Moreover, the chapter discusses the development of a data governance culture as an integral part of the organizational ethos. It explores practical steps for embedding data governance into the organization's DNA, including alignment with corporate values, continuous communication, and the implementation of incentives to promote compliance and foster a data-centric organizational culture.

The chapter concludes by addressing the challenges of cultural barriers, proposing strategies to overcome resistance and integrate data governance into everyday business processes. It stresses the importance of continuous improvement and adaptation, ensuring that data governance evolves in response to new challenges and opportunities.

Overall, this chapter provides a comprehensive guide for organizations aiming to cultivate a robust data governance culture, highlighting the essential roles of leadership and change management in achieving sustainable success in data governance initiatives.

Keywords Data Governance Culture · Change Management · Leadership · Training · Empowerment · Data Governance Champions · Communication · Incentives · Compliance · Overcoming Cultural Barriers · Continuous Improvement, Organizational DNA

13.1 Introduction to Data Governance Culture

This chapter explores a fundamental aspect of successful data governance: the establishment of a Data Governance Culture. In today's data-driven landscape, where data is a strategic asset for organizations, fostering a culture that values, respects, and effectively manages data is of paramount importance. This chapter delves into the key components that contribute to creating a data governance culture and underscores its significance in the broader context of data governance.

As we delve into the various facets of Data Governance Culture in the subsequent sections, we will examine the pivotal role of leadership, the principles of change management, strategies for communicating the value of data governance, and methods for embedding data governance into an organization's DNA. We will also explore the importance of training and empowerment, the development of data governance champions, and the use of incentives to promote compliance.

Moreover, this chapter addresses the challenges posed by cultural barriers and provides insights into strategies for overcoming them. It emphasizes the necessity of continuous improvement and adaptation to ensure that the data governance culture remains dynamic and aligned with the evolving needs and goals of the organization.

This chapter serves as a guide for organizations seeking to instill a data governance culture that not only recognizes the importance of data but also embraces data governance as an integral part of the organizational ethos. Such a culture is the cornerstone of effective data management, ethical data practices, and the realization of the full potential of data as a strategic asset.

13.2 The Role of Leadership in Data Governance

In this section, the focus is on the critical role that leadership plays in the successful implementation and sustenance of a data governance program within an organization. Effective data governance requires strong leadership that provides direction, support, and commitment to the principles and practices of data management. Here are key points to understand about the role of leadership in data governance:

- **Setting the Vision and Strategy**: Leaders, especially top-level executives and senior management, are responsible for defining the vision and strategic objectives of data governance. They must articulate why data governance is essential for the organization and how it aligns with broader business goals (Madsen 2022; McKinsey and Company n.d.).
- **Creating a Data Governance Team**: Leadership is responsible for establishing a dedicated data governance team or committee. This team is tasked with designing, implementing, and overseeing data governance policies and practices (Corporate Finance Institute n.d.; Analytics8 n.d.).
- **Resource Allocation**: Leaders allocate the necessary resources, including budget, technology, and personnel, to support the data governance initiative. Adequate resources are vital for the success of data governance efforts (Madsen 2022).
- **Championing Data Governance**: Effective leaders serve as champions of data governance, advocating for its importance and benefits across the organization. They lead by example and communicate the value of data governance to all stakeholders (Analytics8 n.d.).
- **Defining Roles and Responsibilities**: Leaders ensure that roles and responsibilities within the data governance structure are clearly defined. This includes appointing data stewards, data owners, and other key personnel responsible for data management (Corporate Finance Institute n.d.).
- **Decision-Making Authority**: Leaders establish decision-making authority within the data governance framework. They define who has the final say on data-related matters and ensure that decisions are aligned with organizational objectives (Madsen 2022; McKinsey and Company n.d.).
- **Data Governance Policies**: Leadership plays a significant role in developing and approving data governance policies, standards, and guidelines. They ensure

that these policies are in line with regulatory requirements and industry best practices (Corporate Finance Institute n.d.).
- **Cultural Influence**: Leaders have a significant impact on the organizational culture, including the data culture. They promote a culture of data accountability, transparency, and trust (Analytics8 n.d.).
- **Change Management**: Leaders understand that implementing data governance often requires significant organizational change. They champion change management principles to ensure a smooth transition and acceptance of new data governance practices (Madsen 2022).
- **Monitoring and Accountability**: Leadership holds the data governance team and stakeholders accountable for meeting data governance goals and objectives. They establish metrics and key performance indicators (KPIs) to measure progress (McKinsey and Company n.d.).
- **Communication**: Effective communication is a core leadership responsibility. Leaders ensure that communication channels are open, and they regularly update stakeholders on data governance initiatives and achievements (Madsen 2022).
- **Supporting Data Literacy**: Leaders recognize the importance of data literacy and invest in training and education to improve data-related skills across the organization (Corporate Finance Institute n.d.).
- **Sustaining Commitment**: Leadership's ongoing commitment to data governance is crucial. They demonstrate their dedication by prioritizing data governance even as other organizational priorities evolve (Analytics8 n.d.).
- **Adaptation and Improvement**: Leaders encourage continuous improvement in data governance practices. They are open to feedback and adapt the data governance program as needed to address changing business needs and data challenges (McKinsey and Company n.d.).

Figure 13.1, titled "The Role of Leadership in Data Governance," delineates the pivotal roles and responsibilities that leadership holds in the realm of data governance. This figure emphasizes that leadership is crucial in setting the vision and strategy for data governance, guiding the organization towards a cohesive and strategic approach to managing its data assets. Leadership is responsible for creating a data governance team, assembling a group of individuals with the skills and authority to implement the data governance strategy effectively. Resource allocation by leadership ensures that the data governance team has the necessary resources, such as budget, tools, and personnel, to execute their roles effectively. Champions of data governance, leaders advocate for the importance and benefits of data governance within and outside the organization, promoting a culture that values data as a key asset. Defining roles and responsibilities clearly within the data governance framework helps in clarifying expectations and accountability, which is fundamental for the success of data governance initiatives. Leadership is also involved in assigning decision-making authority, ensuring that the right people have the power to make decisions about data usage and management. They are instrumental in crafting and enforcing data governance policies that align with organizational objectives and compliance requirements. The cultural influence of leadership shapes the

13.2 The Role of Leadership in Data Governance

Fig. 13.1 The role of leadership in data governance. (Created by the author)

organization's overall attitude towards data governance, fostering a culture that embraces transparency, accountability, and data-driven decision-making. Change management is another critical area where leadership plays a vital role, as they guide the organization through transitions in data governance practices and technologies. Monitoring and accountability under leadership ensure that data governance practices are observed and that deviations are addressed promptly. Effective communication by leaders ensures that all stakeholders understand the goals, benefits, and expectations of the data governance program. Leadership support for data literacy initiatives is crucial to enhancing the overall data competence of the organization, empowering employees through training and resources. Sustaining commitment to data governance initiatives over the long term requires ongoing leadership engagement and reinforcement. Lastly, leadership must focus on adaptation and improvement, continually assessing and evolving data governance strategies to meet changing business needs and technological advancements.

Leadership plays a pivotal role in data governance by providing vision, direction, and unwavering support for the program. Strong leadership ensures that data governance is integrated into the organization's culture and strategic priorities, leading to effective data management and the realization of the benefits of data as an asset.

13.3 Change Management Principles in Data Governance

Change management principles are essential in the context of data governance to ensure the successful adoption and integration of new data governance practices and processes within an organization. Change management focuses on managing the human side of change, including understanding the psychology of change, communicating effectively, and addressing resistance to change. Here is what you need to know about change management principles in data governance:

- **Understanding the Need for Change**: The first step in change management is recognizing the need for change. This involves identifying the current challenges and limitations in data management and governance practices that require improvement. It is essential to clearly communicate the reasons for change to stakeholders (Change Management Insight n.d.; Data Governance Institute n.d.).
- **Setting Clear Objectives**: Define clear and measurable objectives for the data governance initiative. What specific outcomes are you aiming to achieve with improved data governance? These objectives provide a roadmap for the change effort (DataGalaxy n.d.; Polestar LLP n.d.).
- **Engaging Stakeholders**: Involving stakeholders from various departments and levels of the organization is crucial. Engage them early in the process to gather input, address concerns, and build support for the changes (BDO Digital n.d.; Polestar LLP n.d.).
- **Creating a Change Team**: Establish a dedicated change management team or committee responsible for planning and executing the change effort. This team should include individuals with expertise in change management principles (Change Management Insight n.d.; DataGalaxy n.d.).
- **Communication Plan**: Develop a comprehensive communication plan that outlines how information about the data governance changes will be disseminated. Regular and transparent communication is key to keeping stakeholders informed and engaged (Polestar LLP n.d.; DataGalaxy n.d.).
- **Training and Skill Development**: Identify the training and skill development needs of employees who will be affected by the changes. Provide training programs and resources to equip them with the necessary knowledge and skills for the new data governance practices (BDO Digital n.d.; Change Management Insight n.d.).
- **Change Champions**: Identify change champions or advocates within the organization who can champion the cause of data governance. These individuals can

13.3 Change Management Principles in Data Governance

help motivate and influence others to embrace the changes (Data Governance Institute n.d.; Polestar LLP n.d.).
- **Addressing Resistance**: Anticipate and address resistance to change. Some individuals may be hesitant or resistant to adopting new data governance practices. Listen to their concerns and provide support and reassurance (DataGalaxy n.d.; Change Management Insight n.d.).
- **Piloting and Testing**: Consider piloting the new data governance practices in a smaller, controlled environment before full implementation. This allows for testing, refinement, and the identification of potential challenges (Data Governance Institute n.d.).
- **Feedback Mechanisms**: Establish feedback mechanisms that allow employees to provide input, share their experiences, and suggest improvements throughout the change process (BDO Digital n.d.; Polestar LLP n.d.).
- **Celebrating Successes**: Recognize and celebrate milestones and successes achieved through the data governance initiative. Positive reinforcement reinforces the importance of the changes (DataGalaxy n.d.; Change Management Insight n.d.).
- **Monitoring and Adaptation**: Continuously monitor the progress of the data governance changes and be prepared to adapt the strategy as needed based on feedback and evolving organizational needs (Polestar LLP n.d.; Data Governance Institute n.d.).
- **Sustaining Change**: The goal is not only to implement changes successfully but also to sustain them over the long term. Ensure that data governance practices become ingrained in the organizational culture (Change Management Insight n.d.; BDO Digital n.d.).
- **Leadership Support**: Leadership plays a critical role in change management. Leaders should visibly support the change effort, model the desired behaviors, and demonstrate commitment to data governance (Polestar LLP n.d.; DataGalaxy n.d.).
- **Documentation**: Document the change management process, including the strategies employed, lessons learned, and best practices. This documentation can inform future change initiatives (BDO Digital n.d.).

Figure 13.2, maps out the essential principles for effectively managing change within data governance frameworks. The figure emphasizes the importance of creating a dedicated change team tasked with overseeing and facilitating change processes and highlights the necessity of strong leadership support to drive and endorse these changes. A communication plan is crucial for ensuring that all stakeholders are informed and engaged throughout the change process. Celebrating successes along the way helps to maintain momentum and morale, showcasing the benefits and positive outcomes of the changes implemented. Training and skill development are fundamental, providing the necessary tools and knowledge to stakeholders to adapt to new systems and practices effectively. Setting clear objectives at the outset of a change initiative ensures alignment and clarity on expected outcomes. Change champions are identified as key influencers who promote and support change efforts,

Fig. 13.2 Change management principles in data governance. (Created by the author)

helping to bridge gaps between different groups and stakeholders. Monitoring and adaptation involve continuous assessment of the change process and making necessary adjustments to ensure the initiative's success. Documentation is critical for maintaining a clear record of changes, decisions, and processes, which supports transparency and accountability. Addressing resistance is a proactive measure to engage skeptics and alleviate concerns, ensuring broad support for the initiative. Feedback mechanisms are essential for gathering input and reactions from those affected by the changes, allowing for responsive and inclusive governance practices. Piloting and testing new processes or tools on a smaller scale before full deployment helps identify potential issues and refine the approach based on real-world feedback. Engaging stakeholders involves their active participation in the change process, ensuring that their insights and needs are considered. Lastly, understanding the need for change is foundational, as it builds a strong case for why changes are necessary, aligning the organization's goals with the strategic benefits of enhanced data governance.

By applying change management principles in data governance, organizations can navigate the complexities of change, minimize disruption, and increase the likelihood of successful implementation and long-term sustainability of data governance practices. It ensures that people are at the center of the change process, leading to improved data management and governance outcomes.

13.4 Communicating the Value of Data Governance

Communicating the value of data governance is a crucial aspect of its successful implementation within an organization. It involves conveying the benefits, importance, and strategic significance of data governance to various stakeholders. Effective communication helps build awareness, gain support, and foster a

13.4 Communicating the Value of Data Governance

data-driven culture. Here is what you need to know about communicating the value of data governance:

- **Identifying Stakeholders**: Start by identifying the key stakeholders within your organization who need to understand the value of data governance. This includes executives, department heads, data practitioners, and other relevant parties (DATAVERSITY n.d.; Collibra n.d.).
- **Tailored Messaging**: Craft tailored messages for different stakeholder groups. The value of data governance may vary depending on roles and responsibilities, so it is essential to communicate in a way that resonates with each group (Informatica 2021).
- **Benefits-Oriented Messaging**: Highlight the tangible benefits of data governance. Focus on how it improves data quality, accuracy, and reliability, leading to better decision-making, reduced operational risks, and enhanced compliance (erwin, Inc. 2021; DATAVERSITY n.d.).
- **Strategic Alignment**: Emphasize how data governance aligns with the organization's strategic goals and objectives. Clearly articulate how it supports business growth, innovation, and competitive advantage (Collibra n.d.; The Data Governance Institute n.d.).
- **Use Cases and Examples**: Provide real-world use cases and examples that demonstrate the positive impact of data governance. Highlight success stories such as how data governance led to cost savings or revenue growth (Informatica 2021; erwin, Inc. 2021).
- **Data-Driven Decision-Making**: Stress the importance of data governance in enabling data-driven decision-making. Explain how it empowers employees at all levels to make informed choices based on high-quality data (DATAVERSITY n.d.).
- **Risk Mitigation**: Communicate how data governance helps mitigate risks related to data breaches, compliance violations, and reputational damage. Highlight the potential consequences of poor data management (The Data Governance Institute n.d.; Collibra n.d.).
- **Regulatory Compliance**: Emphasize how data governance ensures compliance with data privacy regulations, industry standards, and legal requirements. This is especially important in regulated industries (Informatica 2021).
- **Transparency**: Stress the importance of data transparency and trust. Explain how data governance practices enhance transparency by providing clear data lineage, audit trails, and data documentation (erwin Inc. 2021; The Data Governance Institute n.d.).
- **Continuous Improvement**: Communicate that data governance is not a one-time effort but an ongoing commitment to continuous improvement. Highlight the organization's dedication to evolving data governance practices (DATAVERSITY n.d.).
- **Leadership Support**: Share endorsements and messages of support from senior leadership. When leaders champion data governance, it reinforces its importance throughout the organization (Informatica 2021; Collibra n.d.).

- **Education and Training**: Offer educational resources and training programs to help employees understand the value of data governance and how to participate effectively (The Data Governance Institute n.d.).
- **Feedback Mechanisms**: Establish channels for stakeholders to provide feedback and ask questions about data governance. Address concerns and clarify misconceptions promptly (erwin Inc. 2021).
- **Metrics and KPIs**: Use key performance indicators (KPIs) and metrics to measure and communicate the impact of data governance. Regularly report on progress and achievements (Collibra n.d.).
- **Two-Way Communication**: Foster two-way communication channels where stakeholders can share their experiences and suggest improvements. Actively listen to feedback (DATAVERSITY n.d.).
- **Cultural Integration**: Communicate that data governance is not a standalone initiative but an integral part of the organizational culture. Encourage a culture of data responsibility and accountability (The Data Governance Institute n.d.).
- **Visual Aids**: Utilize visual aids such as infographics, dashboards, and presentations to simplify complex concepts and make the value of data governance more accessible (Informatica 2021).
- **Consistent Messaging**: Ensure that messaging about data governance is consistent across all communication channels and materials (erwin Inc. 2021).

Figure 13.3, outlines key strategies and elements involved in effectively conveying the significance and benefits of data governance within an organization. This figure stresses the importance of transparency in all communications, ensuring that the processes, policies, and outcomes of data governance are clear and understandable to all stakeholders. Consistent messaging across various platforms and departments helps maintain a unified understanding of data governance goals and practices. Tailored messaging is critical, as it addresses the specific interests and concerns of different stakeholder groups, making the communication more relevant and impactful. Identifying stakeholders accurately is essential for effective communication, as

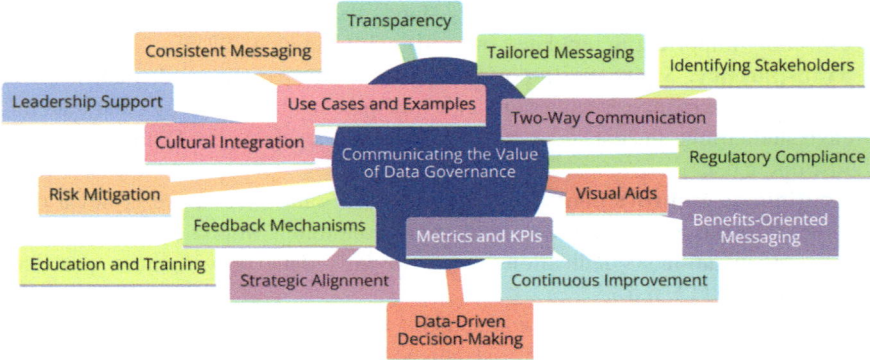

Fig. 13.3 Communicating the value of data governance. (Created by the author)

it allows messages to be directed towards those most affected by or involved in data governance initiatives. Two-way communication is encouraged to foster an environment where feedback is sought and valued, enhancing engagement and alignment. The use of visual aids can simplify complex information and make it more accessible, while benefits-oriented messaging highlights the direct advantages of data governance to the organization and its members. Regulatory compliance is often a key driver for data governance, and communicating how governance actions help meet regulatory requirements can motivate adherence and participation. Metrics and KPIs are vital for quantifying the success and progress of data governance efforts, providing concrete data to support the value proposition. Continuous improvement through regular updates and feedback helps keep the data governance message relevant and forward-thinking. Data-driven decision-making is promoted as a fundamental benefit of robust data governance, enabling more informed, effective, and predictive business strategies. Strategic alignment ensures that the communication strategy is in line with the organization's broader goals and objectives, integrating data governance into the core business processes. Education and training are foundational, equipping stakeholders with the knowledge and skills needed to appreciate and implement data governance principles effectively. Risk mitigation is also a key theme, as data governance significantly reduces the risks associated with data mismanagement and security breaches. Lastly, cultural integration emphasizes embedding data governance into the organizational culture, making it a norm rather than an exception, which can significantly enhance the effectiveness and acceptance of governance initiatives.

Effective communication is essential for creating awareness, building support, and instilling a data-driven mindset within the organization. It helps bridge the gap between technical data governance concepts and the broader organizational understanding of why data governance matters and how it contributes to overall success.

13.5 Building Data Governance into Organizational DNA

Building data governance into an organization's DNA refers to the integration of data governance principles, practices, and values into the core culture and operations of the organization. It signifies that data governance is not just a separate initiative but a fundamental aspect of how the organization operates. Here is what you need to know about building data governance into organizational DNA:

- **Cultural Transformation**: It involves a cultural transformation where data governance becomes a shared mindset and set of behaviors among employees at all levels. Data stewardship, data quality, and responsible data management become ingrained in the organizational culture (Datafloq 2022; McKinsey and Company 2021).
- **Leadership Commitment**: Building data governance into the DNA starts with leadership commitment. Senior executives and leaders must demonstrate their

dedication to data governance and model the desired behaviors (Pythian 2022; McKinsey and Company 2021).
- **Alignment with Values**: Data governance principles should align with the organization's core values and mission. When data governance aligns with what the organization stands for, it becomes more deeply integrated into daily operations (MIT Sloan Management Review 2021).
- **Training and Education**: Invest in training and education programs that help employees understand the importance of data governance and how it supports the organization's goals. Make data literacy a priority (Gartner 2021).
- **Empowerment**: Empower employees to take ownership of data governance responsibilities. Encourage them to identify and address data quality issues, follow data governance policies, and report data-related concerns (Datafloq 2022; McKinsey and Company 2021).
- **Clear Accountability**: Define clear roles and responsibilities for data governance, including data stewards, data owners, and data users. Everyone should know their role in data management (Pythian 2022; Gartner 2021).
- **Communication**: Continuously communicate the importance of data governance and its alignment with organizational objectives. Use regular communication channels to reinforce data governance messages (MIT Sloan Management Review 2021).
- **Integration with Processes**: Embed data governance practices into existing business processes and workflows. For example, data quality checks can be integrated into data entry processes (Gartner 2021).
- **Incentives and Recognition**: Reward and recognize individuals and teams that excel in data governance practices. Incentives can include bonuses, promotions, or public recognition (Datafloq 2022).
- **Monitoring and Reporting**: Establish monitoring mechanisms to track data governance compliance and outcomes. Regularly report on key data governance metrics to keep everyone informed (McKinsey and Company 2021).
- **Feedback Loops**: Create feedback loops that allow employees to provide input and suggestions for improving data governance practices. Act on this feedback to demonstrate responsiveness (Pythian 2022).
- **Scalability**: Ensure that data governance practices are scalable as the organization grows. Scalable processes can adapt to changing data volumes and complexities (Gartner 2021).
- **Documentation**: Document data governance policies, standards, and procedures clearly and make them easily accessible to all employees. Documentation helps establish expectations (Datafloq 2022).
- **Risk Mitigation**: Emphasize how data governance practices mitigate risks associated with data breaches, compliance violations, and reputational damage. Highlight the consequences of data mishandling (MIT Sloan Management Review 2021).
- **Continuous Improvement**: Encourage a culture of continuous improvement in data governance. Regularly assess and update data governance practices to meet evolving needs (McKinsey and Company 2021).

13.5 Building Data Governance into Organizational DNA

- **Data-Driven Decision-Making**: Promote data-driven decision-making at all levels of the organization. Ensure that data is used to inform strategies and tactics (Datafloq 2022).
- **Transparency**: Foster transparency in data management. Provide visibility into data lineage, data quality, and data governance processes (Pythian 2022).
- **Long-Term Perspective**: Building data governance into organizational DNA is a long-term endeavor. It requires persistence and a commitment to sustaining data governance practices over time (MIT Sloan Management Review 2021).

Figure 13.4, elaborates on the core aspects necessary for ingraining data governance deeply within the organizational culture. This figure highlights the pivotal role of leadership commitment in driving the data governance initiative forward, ensuring it is prioritized and supported at the highest levels. Data-driven decision-making is showcased as a critical outcome of effective data governance, emphasizing the ability to make better decisions based on accurate and timely data. Alignment with organizational values ensures that data governance efforts are consistent with the company's mission and ethical standards, facilitating a seamless integration into everyday business practices. The concept of cultural transformation is introduced as essential for changing the organizational mindset to value and utilize data effectively. Continuous improvement is depicted as necessary for keeping data governance practices relevant and effective, adapting to new challenges and opportunities as they arise. A long-term perspective is crucial, focusing on sustainable strategies rather than short-term fixes. Empowerment of employees through access to data and decision-making capabilities is highlighted as key to fostering a sense of ownership and responsibility towards data. Integration with existing business processes ensures that data governance becomes a part of the regular workflow, not an external add-on. Risk mitigation is emphasized as a fundamental benefit, helping to reduce potential losses and increase security. Incentives and recognition are discussed as strategies to motivate and reward employees for effective data governance participation. Clear accountability structures are necessary to ensure that responsibilities are well defined and that individuals understand their roles in the data governance

Fig. 13.4 Building data governance into organizational DNA. (Created by the author)

framework. Feedback loops are important for gathering insights on the effectiveness of data governance initiatives and for making necessary adjustments. Communication, transparency, and documentation are all crucial for maintaining clarity, fostering trust, and ensuring that all stakeholders have the necessary information to engage with data governance practices effectively. Scalability is highlighted to ensure that data governance mechanisms can grow and expand as the organization evolves. Training and education are vital to equip employees with the necessary skills and knowledge to participate in and support data governance initiatives. Finally, monitoring and reporting are essential for tracking progress, measuring success, and identifying areas for improvement in the data governance strategy.

By building data governance into the organizational DNA, organizations can create a culture where data is treated as an asset, and responsible data management is a shared responsibility. This cultural shift leads to improved data quality, better decision-making, and enhanced competitiveness in the data-driven business landscape.

13.6 Training and Empowerment

Training and empowerment are critical components of a successful data governance program. They involve equipping employees with the knowledge, skills, and authority they need to effectively participate in data governance activities. Here is what you need to know about training and empowerment in the context of data governance:

- **Training Programs**: Organizations should develop comprehensive training programs that cover various aspects of data governance. These programs may include data literacy training, data stewardship training, data quality training, and more (Quest 2021; McKinsey and Company 2021).
- **Data Literacy**: Data literacy training is essential for all employees. It ensures that individuals across the organization understand data terminologies, concepts, and the importance of data in decision-making (erwin 2021; Quest 2021).
- **Data Stewardship Training**: Data stewards are individuals responsible for overseeing and managing specific data domains. They require specialized training to perform their roles effectively. This includes understanding data policies, data quality standards, and data lineage (BCG 2021).
- **Data Quality Training**: Training programs should address data quality principles and best practices. Data quality training helps employees recognize data quality issues and take corrective actions (McKinsey and Company 2021; BCG 2021).
- **Customized Training**: Tailor training programs to the specific needs and roles of employees. Not all employees require the same level of data governance knowledge, so customize training content accordingly (Quest 2021).

13.6 Training and Empowerment

- **Regular Training Updates**: Data governance is an evolving field. Regularly update training materials to reflect changes in data governance policies, technologies, and best practices (erwin 2021).
- **Hands-On Learning**: Incorporate hands-on learning experiences and practical exercises into training programs. This helps reinforce learning and allows employees to apply their knowledge in real-world scenarios (BCG 2021).
- **Certifications**: Consider offering data governance certifications or badges to employees who complete training programs successfully. Certifications can motivate employees to engage in data governance activities (Quest 2021).
- **Empowerment**: Empower employees to take ownership of data governance responsibilities. This includes granting them the authority to make data-related decisions within their roles (McKinsey and Company 2021).
- **Clear Guidelines**: Provide clear guidelines and procedures for employees to follow when addressing data governance issues or making data-related decisions. Clarity reduces ambiguity and empowers employees to act confidently (BCG 2021).
- **Access to Tools**: Ensure that employees have access to the necessary tools and technologies to support data governance activities. This may include data quality tools, metadata management systems, and data cataloging platforms (erwin 2021).
- **Feedback Channels**: Establish feedback channels where employees can report data quality issues, data governance concerns, or suggestions for improvement. Act on feedback to demonstrate responsiveness (Quest 2021).
- **Recognition**: Recognize and celebrate employees who excel in data governance efforts. Acknowledgment can include awards, certificates, or public recognition within the organization (McKinsey and Company 2021).
- **Leadership Support**: Leadership should actively support and promote training and empowerment initiatives. When leaders prioritize data governance training, it sends a strong message to the organization (BCG 2021).
- **Continuous Learning Culture**: Foster a culture of continuous learning and improvement. Encourage employees to seek ongoing education and stay updated on data governance trends and practices (erwin 2021).
- **Integration with Onboarding**: Incorporate data governance training into employee onboarding programs. New hires should receive training on data governance principles as part of their orientation (Quest 2021).
- **Measuring Impact**: Implement mechanisms to measure the impact of training and empowerment efforts. Assess how well employees apply their knowledge and skills in data governance activities (BCG 2021).
- **Accessibility**: Ensure that training materials and resources are easily accessible to all employees, including those in remote locations or with disabilities (McKinsey and Company 2021).

Figure 13.5, titled "Training and Empowerment in Data Governance," maps out the essential components of training and empowerment strategies crucial for effective data governance. The figure highlights the importance of integration with onboarding processes, ensuring that new employees receive immediate training in

Fig. 13.5 Training and empowerment keys. (Created by the author)

data governance principles and practices as part of their introduction to the organization. Accessibility to tools and resources is emphasized, enabling employees to apply data governance practices effectively. Regular training updates are essential to keep the workforce current with the latest data governance standards and technologies. Clear guidelines help clarify expectations and provide a structured approach to data governance within the organization. Data literacy is presented as a core goal of training programs, aiming to enhance the overall understanding and competence of employees in using data responsibly and effectively. Feedback channels are important for gauging the effectiveness of training programs and for collecting suggestions for improvement. Certifications may be offered to recognize and validate the data governance skills and knowledge acquired by employees. Hands-on learning experiences are advocated to provide practical skills and real-world applications of data governance principles. Customized training addresses specific needs and roles within the organization, enhancing relevance and effectiveness. Leadership support is crucial in fostering a culture that values and prioritizes data governance training. Data quality training is highlighted as fundamental, focusing on ensuring that employees understand how to manage and maintain the integrity of data. Data stewardship training is also vital, equipping designated staff with the skills to oversee data assets properly. Recognition of achievements in training and empowerment initiatives motivates and rewards employees, reinforcing the value of ongoing education and engagement. A continuous learning culture is advocated, promoting lifelong learning and adaptation in data governance practices. Lastly, measuring the impact of training programs is crucial for assessing their effectiveness and for demonstrating the return on investment in training and development efforts.

Training and empowerment are investments in an organization's data governance capabilities. When employees are well-trained and empowered to participate in data governance, it leads to improved data quality, better decision-making, and a data-driven culture that can drive organizational success.

13.7 Developing Data Governance Champions

Developing data governance champions is a strategic approach to building support and enthusiasm for data governance initiatives within an organization. Data governance champions are individuals who are passionate about data governance and take a proactive role in promoting its principles and practices. Here is what you need to know about developing data governance champions:

- **Identifying Potential Champions**: Start by identifying employees who have a natural interest in data governance, a strong understanding of its importance, and the ability to influence others positively (McKinsey and Company 2021).
- **Leadership and Enthusiasm**: Data governance champions are often leaders or influencers within their respective departments or teams. They possess enthusiasm for data governance and are willing to advocate for its benefits (BCG 2022).
- **Training and Education**: Provide data governance champions with specialized training and education. This may include advanced data governance concepts, communication skills, and change management principles (Precisely 2023).
- **Clear Roles and Responsibilities**: Clearly define the roles and responsibilities of data governance champions. They should know their specific duties, such as promoting data governance, facilitating communication, and assisting with implementation (Deloitte 2022).
- **Cross-Functional Representation**: Ideally, data governance champions should come from various departments and teams within the organization. This ensures a cross-functional approach to data governance advocacy (BCG 2022).
- **Communication Skills**: Data governance champions should have effective communication skills. They should be able to articulate the value of data governance, address concerns, and engage with stakeholders effectively (McKinsey and Company 2021).
- **Advocacy and Promotion**: Champions actively advocate for data governance by promoting its principles and practices to their peers and colleagues. They can conduct workshops, presentations, and awareness campaigns (Precisely 2023).
- **Problem Solving**: Champions should be capable problem solvers who can address data-related issues and challenges within their departments. They serve as a valuable resource for finding solutions (Deloitte 2022).
- **Support and Resources**: Provide champions with the necessary support and resources to carry out their advocacy roles effectively. This includes access to data governance tools, training materials, and ongoing guidance (BCG 2022).
- **Collaboration**: Encourage collaboration among data governance champions. They can share insights, best practices, and success stories, fostering a community of practice (McKinsey and Company 2021).
- **Feedback Channels**: Establish feedback channels for champions to report on the progress and challenges related to data governance initiatives. Their feedback can inform program improvements (Precisely 2023).

- **Recognition**: Recognize and reward data governance champions for their contributions. Publicly acknowledge their efforts and highlight their successes within the organization (Deloitte 2022).
- **Continuous Learning**: Data governance is an evolving field. Encourage champions to stay updated on industry trends and emerging best practices through continuous learning and professional development (McKinsey and Company 2021).
- **Leadership Support**: Ensure that senior leadership actively supports and values the role of data governance champions. Leadership endorsement adds credibility to their advocacy efforts (BCG 2022).
- **Measurement of Impact**: Implement mechanisms to measure the impact of data governance champions. Assess the extent to which they have influenced data governance adoption and improved data-related outcomes (Precisely 2023).
- **Sustainability**: Data governance champions should play a sustainable role in the organization's data governance efforts. Their advocacy should continue over the long term (Deloitte 2022).

Figure 13.6, titled "Developing Data Governance Champions," outlines the essential elements involved in cultivating leaders within the organization who can drive and advocate for effective data governance practices. This figure emphasizes the importance of identifying potential champions who demonstrate the enthusiasm, skill, and influence necessary to promote data governance initiatives effectively. Collaboration across various departments is crucial, fostering a cooperative environment that supports the widespread adoption and implementation of data governance policies. Cross-functional representation ensures that all areas of the organization are involved and that data governance champions can address and understand diverse needs and perspectives. Feedback channels are highlighted as necessary for maintaining open lines of communication between data governance champions and the rest of the organization, facilitating timely updates and adjustments based on staff input and evolving requirements. Advocacy and promotion by

Fig. 13.6 Developing data governance champions map. (Created by the author)

champions help raise awareness and underline the significance of data governance within the organization. Effective communication skills are fundamental for champions, enabling them to convey the importance, benefits, and requirements of data governance clearly and persuasively. Training and education are essential, equipping champions with the knowledge and skills needed to lead data governance efforts confidently. Support and resources from senior management and the organization at large are critical, providing the tools and backing needed for champions to perform their roles effectively. Clear roles and responsibilities must be defined to avoid confusion and ensure that champions know what is expected of them and what they are authorized to do. Recognition of the efforts and successes of data governance champions motivates them and others to continue engaging with data governance initiatives. Sustainability of efforts is important, ensuring that data governance remains a permanent and evolving aspect of the organizational culture. Leadership support underpins the effectiveness of champions, providing the authority and backing necessary to drive change and maintain momentum. Problem-solving skills are vital, enabling champions to address challenges as they arise and find effective solutions. Continuous learning is promoted, encouraging champions to stay updated on the latest data governance trends, technologies, and best practices. Finally, the measurement of impact is crucial for assessing the effectiveness of data governance champions, demonstrating the value of their work, and identifying areas for improvement. This comprehensive approach ensures that data governance champions are well-prepared, supported, and effective in their roles, driving the successful integration of data governance into the organization's DNA.

Developing data governance champions is an effective strategy for driving data governance adoption and ensuring that it becomes an integral part of the organizational culture. These champions serve as advocates, educators, and problem solvers, contributing to improved data quality and data-driven decision-making across the organization.

13.8 Incentivizing and Rewarding Compliance

Incentivizing and rewarding compliance is a key strategy to promote and sustain data governance practices within an organization. It involves recognizing and rewarding individuals and teams for adhering to data governance policies, standards, and best practices. Here is what you need to know about incentivizing and rewarding compliance in the context of data governance:

- **Alignment with Goals**: Ensure that the incentives and rewards align with the organization's strategic goals and objectives. Data governance compliance should contribute to broader organizational success (McKinsey and Company 2021).

- **Recognition**: Recognize and celebrate individuals or teams that consistently comply with data governance policies. Publicly acknowledge their efforts and achievements (JDSupra 2023).
- **Tangible Rewards**: Consider offering tangible rewards such as bonuses, promotions, or financial incentives to individuals or teams that excel in data governance compliance. These rewards provide direct incentives for compliance (BDO Digital 2021).
- **Non-Financial Rewards**: Non-financial rewards can also be effective. These may include certificates, badges, or special recognition within the organization (Deloitte 2022).
- **Performance Metrics**: Establish key performance indicators (KPIs) related to data governance compliance. Link these metrics to performance evaluations and rewards (Microsoft Learn 2024).
- **Timely Recognition**: Ensure that recognition and rewards are timely. Immediate acknowledgment of compliance achievements reinforces desired behaviors (JDSupra 2023).
- **Healthy Competition**: Foster healthy competition by creating friendly competitions or challenges related to data governance compliance. Recognize top performers (McKinsey and Company 2021).
- **Peer Recognition**: Encourage peer recognition by allowing colleagues to nominate and acknowledge individuals or teams for their data governance efforts (BDO Digital 2021).
- **Training Opportunities**: Offer training and professional development opportunities as rewards for compliance. Employees can gain additional skills and knowledge in exchange for their commitment to data governance (Deloitte 2022).
- **Career Advancement**: Consider data governance compliance as a factor in career advancement decisions. Individuals who consistently demonstrate data governance competence can be considered for promotions or leadership roles (Microsoft Learn 2024).
- **Feedback Mechanisms**: Establish feedback mechanisms that allow employees to provide input on the types of incentives and rewards that motivate them the most. Tailor rewards to employee preferences (BDO Digital 2021).
- **Long-Term Perspective**: Incentivizing and rewarding compliance should be viewed as a long-term strategy. Consistent rewards over time reinforce a culture of data governance (McKinsey and Company 2021).
- **Fairness and Equity**: Ensure that incentive programs are fair and equitable, with clear criteria for earning rewards. Avoid favoritism or bias in the reward process (JDSupra 2023).
- **Transparency**: Be transparent about the criteria for earning rewards and the process for selecting recipients. Transparency enhances trust in the incentive program (Deloitte 2022).
- **Communication**: Communicate the existence of incentive programs and their importance to data governance. Employees should be aware of the potential rewards for compliance (Microsoft Learn 2024).

13.8 Incentivizing and Rewarding Compliance

- **Evaluation and Adjustment**: Regularly evaluate the effectiveness of the incentive program and adjust as needed. Seek feedback from participants to improve the program (BDO Digital 2021).
- **Combination of Incentives**: Consider a combination of financial and non-financial incentives to appeal to a diverse workforce (JDSupra 2023).
- **Measuring Impact**: Measure the impact of incentivizing and rewarding compliance by tracking changes in data governance compliance rates and overall data quality (Microsoft Learn 2024).

Figure 13.7, titled "Incentivizing and Rewarding Compliance," presents a complex and interconnected diagram illustrating the various factors and mechanisms involved in promoting compliance within an organization. Central to the diagram is "Healthy Competition," surrounded by related concepts such as "Peer Recognition," "Timely Recognition," "Evaluation and Adjustment," "Recognition," "Performance Metrics," "Career Advancement," "Training Opportunities," "Tangible Rewards," "Long-Term Perspective," "Alignment with Goals," "Measuring Impact," "Combination of Incentives," "Non-Financial Rewards," "Fairness and Equity," "Feedback Mechanisms," "Transparency," and "Communication." Each element connects to adjacent ones, suggesting a dynamic and multifaceted approach to motivating and rewarding compliance efforts effectively. This visualization highlights the importance of a balanced approach, incorporating both tangible and intangible rewards, ensuring transparency, and fostering a culture of continuous improvement and recognition within compliance frameworks.

Incentivizing and rewarding compliance can motivate employees to actively participate in data governance initiatives and adhere to data governance principles. It reinforces the importance of responsible data management and contributes to the overall success of data governance programs within the organization.

Fig. 13.7 Incentivizing and rewarding compliance map. (Created by the author)

13.9 Overcoming Cultural Barriers

Overcoming cultural barriers is a critical aspect of establishing a successful data governance culture within an organization. Cultural barriers often involve resistance to change, lack of awareness about data governance, and entrenched behaviors that hinder the adoption of data governance practices. Here is what you need to know about overcoming cultural barriers in data governance:

- **Identifying Cultural Barriers**: Start by identifying the specific cultural barriers that exist within the organization. These may include resistance to sharing data, a culture of data silos, a lack of trust in data quality, or a preference for gut feelings over data-driven decision-making (DATAVERSITY 2022; McDermott and O'Dell 2001).
- **Leadership Commitment**: Senior leadership plays a crucial role in overcoming cultural barriers. Leaders should openly endorse and support data governance initiatives, demonstrating their commitment to the cultural shift (Four Cornerstone 2022; DEV Community 2022).
- **Communication**: Effective communication is essential. Clearly articulate the reasons for implementing data governance, its benefits, and how it aligns with the organization's goals. Address common misconceptions and concerns (DATAVERSITY 2022; Four Cornerstone 2022).
- **Change Management**: Apply change management principles to facilitate cultural change. Create a change management plan that outlines objectives, strategies, and timelines for overcoming cultural barriers (DEV Community 2022; Emerald Insight 2001).
- **Education and Awareness**: Educate employees about data governance and its importance. Raise awareness through training programs, workshops, and internal communications (McDermott and O'Dell 2001; DATAVERSITY 2022).
- **Storytelling**: Use storytelling to illustrate the impact of data governance on real-world scenarios within the organization. Share success stories and case studies to make data governance relatable (Four Cornerstone 2022).
- **Inclusivity**: Involve employees from diverse departments and levels in data governance initiatives. Encourage their input and participation to create a sense of ownership (DEV Community 2022).
- **Cultural Champions**: Identify and empower cultural champions within the organization. These individuals can advocate for data governance, share their positive experiences, and inspire others to embrace change (DATAVERSITY 2022; Emerald Insight 2001).
- **Rewards and Recognition**: Recognize and reward individuals and teams that embrace data governance practices and contribute to cultural change. Celebrate their achievements (Four Cornerstone 2022).
- **Alignment with Values**: Align data governance principles with the organization's core values. Show how data governance supports the organization's mission and culture (McDermott and O'Dell 2001).

13.9 Overcoming Cultural Barriers

- **Transparency**: Be transparent about data governance processes, decision-making, and outcomes. Transparency builds trust and reduces resistance (DEV Community 2022).
- **Continuous Feedback**: Establish feedback mechanisms that allow employees to express concerns and provide input regarding data governance initiatives. Act on feedback to address issues promptly (Four Cornerstone 2022).
- **Training and Development**: Provide training and development opportunities to help employees acquire the skills needed for data governance. Address skill gaps that may contribute to cultural barriers (DATAVERSITY 2022).
- **Pilot Programs**: Implement data governance pilot programs in select departments or teams to demonstrate the benefits of cultural change. Use these pilots as examples to inspire broader adoption (Emerald Insight 2001).
- **Measurement and Metrics**: Define key performance indicators (KPIs) to measure progress in overcoming cultural barriers. Monitor and report on these metrics regularly (DATAVERSITY 2022).
- **Persistence**: Cultural change takes time and persistence. Recognize that overcoming cultural barriers is a gradual process that may require ongoing efforts (DEV Community 2022).
- **Feedback Loops**: Create feedback loops that allow employees to provide input on the cultural change process. Use feedback to adjust strategies and tactics (Four Cornerstone 2022).
- **Celebrate Milestones**: Celebrate milestones and achievements in cultural change. Recognize the organization's progress in embracing data governance (McDermott and O'Dell 2001).

Figure 13.8, titled "Overcoming Cultural Barriers," showcases a visual map designed to illustrate the various strategies and concepts crucial for addressing cultural obstacles within organizations. At the center of this map is "Overcoming Cultural Barriers," surrounded by key thematic areas such as "Storytelling," "Continuous Feedback," "Training and Development," "Feedback Loops," "Pilot Programs," "Cultural Champions," "Persistence," "Leadership Commitment,"

Fig. 13.8 Overcoming cultural barriers map. (Created by the author)

"Rewards and Recognition," "Education and Awareness," "Inclusivity," "Transparency," "Alignment with Values," "Celebrate Milestones," "Identifying Cultural Barriers," "Measurement and Metrics," "Change Management," and "Communication." Each area connects to adjacent concepts, suggesting a dynamic and comprehensive approach to cultural integration. This visual diagram underscores the importance of a multifaceted strategy that incorporates leadership involvement, transparent communication, and consistent recognition, all aimed at fostering a more inclusive and adaptable organizational culture.

Overcoming cultural barriers in data governance is essential for fostering a data-driven culture where data is valued, trusted, and used effectively to drive decision-making. It requires a multifaceted approach that combines leadership commitment, effective communication, education, and a continuous commitment to change management principles.

13.10 Continuous Improvement and Adaptation

Continuous improvement and adaptation are essential principles in the context of data governance. These principles emphasize the importance of ongoing refinement and flexibility in data governance practices to ensure their effectiveness and relevance over time. Here is what you need to know about continuous improvement and adaptation in data governance:

- **Iterative Process**: Data governance should be viewed as an iterative process that evolves over time. It is not a one-time project but an ongoing effort to enhance data management practices (Alation 2022; Atlan 2022).
- **Feedback-Driven**: Continuous improvement relies on feedback from various stakeholders, including data users, data stewards, and business units. Collect feedback to identify areas for improvement (ISACA 2020; Change Management Insight 2022).
- **KPIs and Metrics**: Establish key performance indicators (KPIs) and metrics to measure the effectiveness of data governance initiatives. Regularly assess these metrics to gauge progress (Alation 2022; Atlan 2022).
- **Benchmarking**: Compare data governance practices with industry benchmarks and best practices. Identify gaps and opportunities for improvement based on benchmarking results (ISACA 2020).
- **Adaptive Governance**: Embrace adaptive governance practices that allow data governance policies and procedures to adapt to changing business needs, technological advancements, and regulatory requirements (SpringerLink 2023).
- **Continuous Learning**: Encourage a culture of continuous learning and development among data governance professionals. Stay updated on industry trends, innovative technologies, and emerging best practices (Change Management Insight 2022; Atlan 2022).

13.10 Continuous Improvement and Adaptation

- **Documentation**: Maintain up-to-date documentation of data governance policies, standards, and procedures. Document changes and revisions to ensure transparency and compliance (ISACA 2020).
- **Change Management**: Implement effective change management processes to facilitate the adoption of new data governance practices. Communicate changes clearly and provide support to affected stakeholders (Change Management Insight 2022).
- **Stakeholder Engagement**: Engage stakeholders in the decision-making process for data governance improvements. Collaborate with business units to identify their evolving data needs (Alation 2022).
- **Pilot Programs**: Use pilot programs to test and refine new data governance initiatives before implementing them organization-wide. Pilots provide valuable insights and reduce risks (Atlan 2022).
- **Data Quality Monitoring**: Continuously monitor data quality and address data quality issues promptly. Implement data quality improvement initiatives based on monitoring results (SpringerLink 2023).
- **Technology Updates**: Keep data governance technology and tools up to date. Leverage technological advancements to enhance data governance capabilities (Alation 2022).
- **Regulatory Compliance**: Stay informed about changing regulatory requirements related to data governance, data privacy, and data security. Adapt policies and practices to remain compliant (ISACA 2020).
- **Feedback Loops**: Establish feedback loops that allow data users to report data-related issues and suggest improvements. Act on feedback to drive continuous enhancements (Change Management Insight 2022).
- **Data Governance Committee**: Maintain a data governance committee or steering group responsible for reviewing and approving changes to data governance policies and practices (Atlan 2022).
- **Cultural Evolution**: Recognize that cultural change is an integral part of continuous improvement. Encourage a data-driven culture where employees value and contribute to data governance (SpringerLink 2023).
- **Risk Management**: Continuously assess data governance risks and develop risk mitigation strategies. Adapt risk management practices to changing circumstances (ISACA 2020).
- **Flexibility**: Data governance practices should be flexible and adaptable to accommodate diverse data sources, formats, and use cases. Avoid rigid approaches that hinder agility (Alation 2022).
- **Feedback Integration**: Actively integrate feedback into the improvement process. Show stakeholders that their input is valued and contributes to positive changes (Change Management Insight 2022).
- **Documentation and Knowledge Sharing**: Document lessons learned and share knowledge within the organization. Ensure that best practices and improvements are communicated effectively (ISACA 2020).

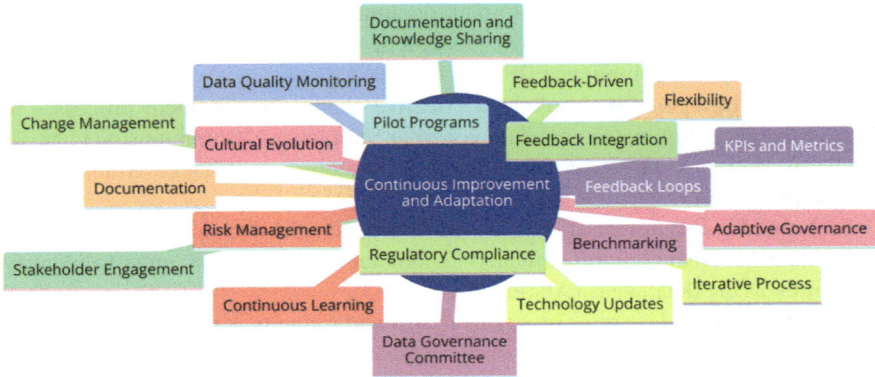

Fig. 13.9 Continuous improvement and adaptation map. (Created by the author)

Figure 13.9, depicts a detailed visual map designed to emphasize strategies and concepts integral to fostering an environment of ongoing enhancement and adaptability within organizations. At the center, "Continuous Improvement and Adaptation" links to various related elements, including "Pilot Programs," "Regulatory Compliance," "Data Governance Committee," "Technology Updates," "Benchmarking," "Feedback Loops," "Feedback Integration," "Feedback-Driven," "Flexibility," "KPIs and Metrics," "Adaptive Governance," "Iterative Process," "Documentation and Knowledge Sharing," "Data Quality Monitoring," "Cultural Evolution," "Change Management," "Documentation," "Risk Management," "Stakeholder Engagement," and "Continuous Learning." These connections illustrate the comprehensive and interconnected approach required to maintain continual growth and responsiveness within an organization's structure and processes. The map highlights the importance of a cohesive strategy that incorporates feedback mechanisms, regulatory adherence, and stakeholder involvement, aimed at ensuring the organization remains forward-looking and adaptable to changes.

Continuous improvement and adaptation in data governance ensure that data remains a valuable and reliable asset for the organization. By staying agile and responsive to changing circumstances, data governance practices can evolve to meet the evolving needs of the business.

13.11 Conclusion

In conclusion, effective data governance is a multifaceted and dynamic endeavor that encompasses various key principles and practices. It serves as the foundation for responsible data management, data quality assurance, and data-driven decision-making within organizations. Throughout this discussion, we have explored essential aspects of data governance, including:

1. **Data Governance Culture**: The establishment of a data governance culture involves leadership commitment, change management, and fostering a data-driven mindset among employees. It requires clear communication of the value of data governance and the alignment of data practices with organizational goals.
2. **Leadership's Role**: Leadership plays a pivotal role in championing data governance initiatives, providing the necessary resources, and setting the tone for data stewardship and compliance.
3. **Change Management**: Effective change management principles are crucial for overcoming resistance to data governance initiatives and ensuring successful adoption across the organization.
4. **Communication and Awareness**: Clear communication and education efforts are vital for ensuring that employees understand the importance of data governance and their roles in its implementation.
5. **Data Governance Champions**: Identifying and empowering data governance champions can drive advocacy and support for data governance practices at various levels of the organization.
6. **Incentivizing and Rewarding Compliance**: Incentives and rewards can motivate individuals and teams to comply with data governance policies and contribute to a culture of responsible data management.
7. **Overcoming Cultural Barriers**: Addressing cultural barriers involves promoting a data-driven culture, fostering collaboration, and addressing resistance to change.
8. **Continuous Improvement and Adaptation**: Embracing continuous improvement and adaptation ensures that data governance practices remain effective and relevant in the face of evolving business needs, technology advancements, and regulatory changes.

These principles collectively contribute to the development of a robust data governance framework that enhances data quality, fosters trust in data, and facilitates informed decision-making. Data governance is an ongoing journey, and organizations that prioritize these principles are better equipped to harness the full potential of their data assets, driving innovation and competitiveness in today's data-driven landscape. As organizations continue to recognize the strategic importance of data, the principles outlined here will serve as a guide for establishing and sustaining effective data governance practices.

References

Alation (2022) How to ensure continuous improvement with data governance. Alation
Analytics8 (n.d.) Defining data governance roles & responsibilities. Analytics8
Atlan (2022) How to improve data governance? Steps, tips & template. Atlan
BCG (2021) A show-don't-tell approach to data governance. Boston Consulting Group
BCG (2022) Any company can become a resilient data champion. Boston Consulting Group
BDO Digital (2021) The comprehensive guide to data governance. BDO

BDO Digital (n.d.) The comprehensive guide to data governance. BDO Digital
Change Management Insight (2022) The process of change management in data governance. Change Management Insight
Change Management Insight (n.d.) The process of change management in data governance. Change Management Insight
Collibra (n.d.) Top 6 best practices of data governance. Collibra
Corporate Finance Institute (n.d.) Data governance: overview, role, importance, goals. Corporate Finance Institute
Data Governance Institute (n.d.) In: Data Governance Institute (ed) Implementing change management
Datafloq (2022) Building a robust data governance framework for organizational success
DataGalaxy (n.d.) Executive sponsorship and change management in data governance
DATAVERSITY (2022) Data governance challenges (and how to fix them). DATAVERSITY
DATAVERSITY (n.d.) Demonstrating the value of data governance. DATAVERSITY
Deloitte (2022) Deploying effective data governance to achieve key business priorities. Deloitte Insights
DEV Community (2022) Best practices for creating a robust data governance framework. DEV Community
Emerald Insight (2001) Overcoming cultural barriers to sharing knowledge. J Knowl Manage. MCB UP Ltd
erwin Inc (2021, March 25) The value of data governance and how to quantify it. Quest Software
Four Cornerstone (2022) How to overcome big data cultural barriers. Four Cornerstone LLC
Gartner (2021) Building a comprehensive data governance program. Gartner
Informatica (2021, September 8) The business value of data governance: describing "What's in it for me?". Informatica
ISACA (2020) Improving data governance and management processes. ISACA
JDSupra (2023) Incentivizing compliance. Thomas Fox - Compliance Evangelist
Madsen L (2022) Dear Laura: what role should leadership play in data governance? DATAVERSITY
McDermott R, O'Dell C (2001) Overcoming cultural barriers to sharing knowledge. J Knowl Manage. MCB UP Ltd.
McKinsey & Company (2021) Designing data governance that delivers value. McKinsey & Company
McKinsey & Company (n.d.) Designing data governance that delivers value. McKinsey & Company
Microsoft Learn (2024) Data, compliance, and governance. Microsoft
MIT Sloan Management Review (2021) Data governance in the 21st-century organization. MIT Sloan Management Review
Polestar LLP (n.d.) Mastering change management for optimal data governance success. Polestar LLP
Precisely (2023) Data governance business case: Here's how to build one. Precisely
Pythian (2022) Building organizational models for better data governance. Pythian
Quest (2021) Data empowerment. Quest Software
SpringerLink (2023) Adaptive governance for a resilient digital society. SpringerLink
The Data Governance Institute (n.d.) DGI data governance framework

Chapter 14
Measuring the Impact of Data Governance: Metrics and Key Performance Indicators

Contents

14.1	Introduction to Measuring Data Governance Impact.	420
14.2	Defining Relevant Metrics and KPIs.	421
14.3	Metrics for Data Quality.	423
14.4	Compliance and Risk Management Metrics.	426
14.5	Metrics for Data Usage and Business Impact.	429
14.6	Tracking and Reporting Mechanisms.	431
14.7	Balancing Quantitative and Qualitative Measures.	434
14.8	Case Studies: Measuring Success in Data Governance.	437
14.9	Continuous Improvement Through Metrics.	438
14.10	Conclusion.	441
References.		443

Abstract This chapter delves into the crucial practice of assessing the effectiveness of data governance strategies through well-defined metrics and KPIs. This chapter outlines the importance of aligning these metrics with organizational goals to effectively track progress, manage risks, ensure data quality, and, ultimately, realize the business value of data assets.

The chapter begins by highlighting the necessity of understanding organizational goals, which guides the selection of relevant metrics and KPIs. It stresses the importance of identifying specific data governance objectives and key data domains crucial to the organization's operations, helping tailor metrics that truly reflect performance in critical areas.

It then explores various categories of data governance metrics, such as data quality metrics (e.g., the percentage of data records with missing values), data compliance metrics (e.g., the compliance rate in responding to data access requests), data usage metrics (e.g., the number of reports generated from a central data repository), and data impact metrics (e.g., revenue growth attributed to data-driven marketing campaigns).

A significant portion of the chapter is dedicated to the practical steps of setting up these metrics, including defining baseline measurements, determining measurement frequency, assigning responsibility, setting targets, documenting measurement methods, and regularly reviewing and refining these metrics to align with evolving business needs.

The narrative emphasizes the integration of data governance initiatives with these metrics, the continuous tracking of progress, and the use of the results to inform and adjust data governance strategies. It also discusses the importance of celebrating successes when targets are met or exceeded, reinforcing the value of data governance throughout the organization.

Moreover, the chapter underscores the balance between quantitative measures (e.g., numerical data accuracy rates) and qualitative assessments (e.g., stakeholder satisfaction and cultural impact), advocating for a holistic view of data governance's effectiveness.

In conclusion, this chapter provides a comprehensive framework for organizations seeking to measure and enhance the effectiveness of their data governance efforts. It makes the case that systematic, metric-based evaluation is essential not only for accountability but also for continuously improving data governance practices to support strategic business objectives.

Keywords Data Governance · Metrics · Key Performance Indicators (KPIs) · Data management · Data quality · Compliance · Risk management · Data usage · Business impact · Continuous improvement · Reporting mechanisms · Quantitative measures · Qualitative measures

14.1 Introduction to Measuring Data Governance Impact

This chapter delves into the crucial aspect of measuring the impact of data governance using metrics and key performance indicators (KPIs). Understanding the effectiveness of data governance initiatives is essential for organizations to assess their data management practices, ensure data quality, manage risks, and realize the business value of their data assets. In this section, we will explore the fundamental importance of measuring the impact of data governance. It serves as the foundation for assessing the success of data governance programs, making informed decisions, and continuously improving data management practices. Effective measurement allows organizations to gain insights into various aspects of data governance, from data quality and compliance to business impact.

14.2 Defining Relevant Metrics and KPIs

Defining relevant metrics and key performance indicators (KPIs) is a critical step in measuring the impact of data governance. These metrics and KPIs provide a clear and quantifiable way to assess the effectiveness of data governance initiatives. Here is what you need to know about defining relevant metrics and KPIs, along with examples:

Understand Organizational Goals Start by understanding the strategic objectives and goals of your organization. Data governance should align with these goals. For example, if an organization's goal is to increase customer satisfaction, relevant metrics may focus on data quality related to customer data (Secureframe 2023).

Identify Data Governance Objectives Define the specific objectives of your data governance program. These objectives should be clear and measurable. For instance, an objective might be to improve data accuracy (DATAVERSITY 2022a).

Select Key Data Domains Identify key data domains or areas that are critical to your organization. These domains represent the types of data that are most important for achieving your objectives. Examples include customer data, financial data, and product data (Arndt 2023).

Determine Data Governance Categories Categorize your metrics and KPIs into different data governance categories. Common categories include data quality, data compliance, data usage, and data impact. Here are real examples for each category:

- **Data Quality Metrics**: Metrics for data quality include accuracy, completeness, consistency, timeliness, and reliability. For instance, a metric might be the percentage of data records with missing values, aiming to reduce missing values in customer contact information by 20% within the next quarter (Secureframe 2023; Datafloq 2023).
- **Data Compliance Metrics**: Metrics might include the number of data access requests handled within the specified timeframe, targeting a 95% compliance rate in responding to data access requests within 48 hours (Arndt 2023).
- **Data Usage Metrics**: An example metric could be the number of reports generated from a central data repository, with a KPI to increase the utilization of the central data repository by 15% over the next 6 months (DATAVERSITY 2022a).
- **Data Impact Metrics**: Metrics could include revenue growth attributed to data-driven marketing campaigns, with a goal to achieve a 10% increase in revenue attributed to data-driven marketing efforts in the next fiscal year (Experian 2023).

Set Baseline Measurements Before implementing data governance initiatives, establish baseline measurements for your selected metrics and KPIs. Baseline data provides a starting point for assessing improvement (Secureframe 2023).

Define Measurement Frequency Determine how often you will measure the selected metrics and KPIs. Some metrics may be measured daily, weekly, monthly, or quarterly depending on their nature (Experian 2023).

Assign Responsibility Clearly define who within the organization is responsible for collecting, tracking, and reporting on the metrics and KPIs. Assign accountability to relevant teams or individuals (Datafloq 2023).

Establish Targets Set specific targets or benchmarks for each metric and KPI. Targets should be realistic and achievable. For example, if you aim to improve data accuracy, you might set a target of achieving 98% data accuracy within a year (DATAVERSITY 2022a).

Document Measurement Methods Document the methods and tools used for measuring each metric. Ensure that measurement methods are consistent and reliable (Arndt 2023).

Review and Refine Regularly review your metrics and KPIs to ensure their continued relevance and alignment with organizational goals. Refine them as needed based on changing priorities or business conditions (Experian 2023).

Communication Communicate the selected metrics and KPIs to relevant stakeholders within the organization. Ensure that all stakeholders understand the significance of these measurements in assessing data governance impact (Secureframe 2023).

Implement Data Governance Initiatives With your metrics and KPIs in place, implement data governance initiatives aimed at improving data quality, compliance, usage, and impact (Datafloq 2023).

Measure Progress Continuously measure and track progress against the defined metrics and KPIs. Compare current measurements to baseline data and targets (DATAVERSITY 2022a).

Adjust Data Governance Strategies Use the measurement results to make informed decisions and adjustments to your data governance strategies. If progress is not on track to meet targets, consider modifications to your initiatives (Arndt 2023).

Celebrate Success Recognize and celebrate achievements when you meet or exceed your targets. Share success stories to motivate teams and reinforce the importance of data governance (Secureframe 2023).

Figure 14.1 is designed to guide organizations in the establishment of effective metrics and key performance indicators for data governance. This figure visually outlines the process of identifying, defining, and utilizing metrics and KPIs that align with an organization's goals and data governance objectives. Starting with

14.3 Metrics for Data Quality

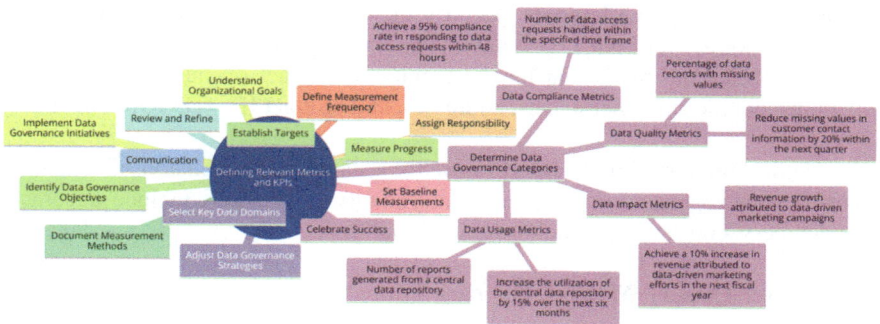

Fig. 14.1 Defining relevant metrics and KPIs map. (Created by the author)

"Understand organizational goals" and "Identify data governance objectives," the map emphasizes the initial steps in aligning data governance efforts with broader business objectives. Key data domains are highlighted as focal points for metric development, followed by categorizing these metrics into meaningful groups such as data quality, data compliance, data usage, and data impact. For each category, specific examples of metrics and KPIs are provided—such as reducing missing values in data records for data quality, and increasing the number of data access requests met within specified timeframes for data compliance. The map further details setting baseline measurements, defining measurement frequencies, and assigning responsibilities for tracking these metrics. It culminates in the celebration of successes, which not only acknowledges achievements but also motivates ongoing compliance and improvement. This structured approach ensures that metrics and KPIs are not only reflective of data governance needs but are also actionable and directly tied to enhancing organizational performance and compliance.

By following these steps and defining relevant metrics and KPIs, organizations can effectively measure the impact of their data governance efforts, make data-driven decisions, and drive continuous improvement in data management practices.

14.3 Metrics for Data Quality

Metrics for data quality are essential for assessing the accuracy, completeness, consistency, and reliability of data within an organization. These metrics provide quantifiable measures of data quality, allowing organizations to identify areas that require improvement. Here are step-by-step insights into defining and using metrics for data quality, along with examples:

Define Data Quality Objectives Start by defining the specific data quality objectives for your organization. These objectives should align with your overall data governance goals. For example, an objective could be to ensure that customer contact information is accurate and up to date (Arndt 2023; Secureframe 2023).

Identify Critical Data Elements Identify the critical data elements (CDEs) that are most important for your organization's operations and decision-making. CDEs are the specific data fields that have a significant impact on business processes. Examples include customer names, addresses, and product IDs (DiGGrowth 2023).

Select Data Quality Dimensions Choose the data quality dimensions that are relevant to your organization. Common data quality dimensions include accuracy, completeness, consistency, timeliness, and reliability.

Define Data Quality Metrics For each critical data element and data quality dimension, define specific data quality metrics. Here are real examples for each dimension:

- **Accuracy**: Percentage of incorrect or inconsistent data values in customer addresses. Accuracy measures the correctness of data by comparing it with a known source or standard, ensuring data values are accurate and free from errors (Secureframe 2023; DATAVERSITY 2022b).
- **Completeness**: Percentage of missing data in customer profiles. This metric evaluates whether all necessary data elements are present within a dataset. High completeness indicates that the dataset is fully populated with the required information (Experian 2023; DiGGrowth 2023).
- **Consistency**: Number of data conflicts between different data sources. Consistency ensures uniformity and coherence across different datasets, eliminating discrepancies and contradictions (DATAVERSITY 2022b; Datapine 2023).
- **Timeliness**: Average time taken to update customer information after a change request. Timeliness measures how up-to-date the data is, ensuring it reflects the most recent information available (DiGGrowth 2023; Secureframe 2023).
- **Reliability**: Percentage of duplicate records in the customer database. Reliability assesses the stability of data over time, ensuring data remains consistent and trustworthy (Experian 2023; Datapine 2023).

Establish Data Quality Baseline Before implementing data quality improvement initiatives, establish a baseline measurement for each metric. Baseline data serves as a reference point for tracking improvements (Secureframe 2023).

Measurement Frequency Determine how often you will measure data quality metrics. Some metrics may be measured continuously while others may be assessed periodically (e.g., monthly or quarterly) (Experian 2023).

Data Quality Tools Implement data quality tools and processes to automate the measurement and monitoring of data quality metrics. These tools can help identify issues in real time (DiGGrowth 2023).

Data Quality Reports Generate regular data quality reports that provide insights into the current state of data quality. These reports should highlight areas that require attention and provide actionable recommendations (Datapine 2023).

14.3 Metrics for Data Quality

Data Quality Governance Assign responsibility for data quality governance. Data stewards and data quality teams should be responsible for monitoring and improving data quality (Arndt 2023).

Data Quality Improvement Initiatives Based on the measurement results, initiate data quality improvement projects to address identified issues. These projects may involve data cleansing, standardization, and validation efforts (Secureframe 2023).

Continuous Monitoring Continuously monitor data quality metrics to track progress and ensure that data quality remains at the desired level. Regularly compare measurements to the established targets (DATAVERSITY 2022b).

Feedback and Iteration Use feedback from data users and stakeholders to refine data quality metrics and measurement processes. Continuous improvement is essential to maintaining high data quality standards (Experian 2023).

Reporting and Communication Communicate data quality improvements and successes to relevant stakeholders to demonstrate the value of data quality initiatives (Datapine 2023).

Figure 14.2 showcases a comprehensive visual map structured to guide organizations in defining and monitoring effective data quality metrics. This map organizes metrics across key data quality dimensions including timeliness, completeness, uniqueness, consistency, and accuracy. Starting from establishing a data quality baseline, the map branches out into selecting relevant data quality dimensions. For timeliness, metrics included focus on the average time taken to update customer information after a change request and striving to minimize the update time to as quickly as within a single business day. Completeness is assessed by monitoring the extent of missing data in customer profiles and aiming to maintain high-profile completeness. Uniqueness is evaluated by examining the extent of duplicate records in

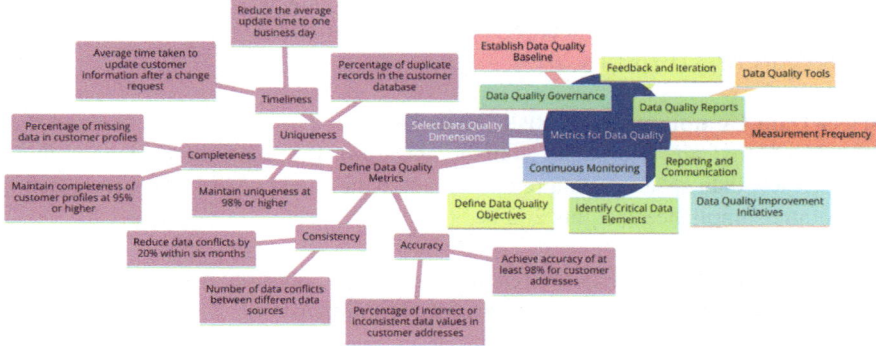

Fig. 14.2 Metrics for data quality. (Created by the author)

the customer database, with a goal to ensure high uniqueness. Consistency metrics include efforts to reduce data conflicts over a specified period and tracking the frequency of data conflicts between sources. Accuracy is centered on ensuring high accuracy for customer addresses and monitoring the occurrence of incorrect or inconsistent data values in those addresses. The map also integrates processes such as defining data quality metrics, identifying critical data elements, and initiating data quality improvement efforts. These are linked to tools and practices including data quality tools, data quality reports, continuous monitoring, feedback and iteration, and establishing a regular measurement frequency. This emphasizes a dynamic, responsive approach to managing data quality, ensuring the metrics are practical and aligned with the organization's continuous improvement goals.

Using well-defined data quality metrics, organizations can systematically assess and improve the quality of their data, leading to more reliable and trustworthy data assets for informed decision-making.

14.4 Compliance and Risk Management Metrics

Compliance and risk management metrics are essential for ensuring that an organization adheres to data-related regulations, standards, and policies while mitigating data-related risks. These metrics help measure an organization's compliance efforts and its ability to manage data-related risks effectively. Here are step-by-step insights into defining and using compliance and risk management metrics, along with examples:

Identify Applicable Regulations and Standards Start by identifying the data-related regulations, industry standards, and internal policies that are applicable to your organization. These could include GDPR, HIPAA, ISO 27001, or specific industry regulations (Secureframe 2023).

Define Compliance and Risk Management Objectives Clearly define the compliance and risk management objectives that your organization aims to achieve. For example, objectives could include ensuring data privacy compliance or reducing data security risks (Experian 2023).

Select Data Compliance and Risk Dimensions Choose the compliance and risk dimensions that are relevant to your organization. Common dimensions include data privacy, data security, data retention, and data access control (OneTrust 2023; Corporate Compliance Insights 2023).

Define Compliance Metrics For each compliance dimension, define specific compliance metrics. Here are real examples for each dimension:

14.4 Compliance and Risk Management Metrics

- **Data Privacy**: Percentage of customer consent obtained for data processing. This metric measures compliance with privacy regulations by ensuring that consent is obtained from data subjects. An example target is achieving a consent rate of at least 90% (Experian 2023).
- **Data Security**: Number of data breaches reported in a year. This metric tracks the incidence of security breaches, aiming to reduce the number of breaches by 20% compared to the previous year (Secureframe 2023).
- **Data Retention**: Percentage of data records retained beyond their expiration date. Ensuring no data records are retained past their expiration date helps maintain compliance with data retention policies (OneTrust 2023).
- **Data Access Control**: Percentage of unauthorized access attempts. This metric measures the effectiveness of access control mechanisms in preventing unauthorized access to data, with a target of maintaining unauthorized access attempts below 5% of total access attempts (Corporate Compliance Insights 2023).

Establish Compliance Baseline Before implementing compliance initiatives, establish a baseline measurement for each compliance metric. Baseline data provides a reference point for tracking improvements (Experian 2023).

Measurement Frequency Determine how often you will measure compliance metrics. Some metrics may be measured continuously while others may be assessed periodically (e.g., monthly or quarterly) (OneTrust 2023).

Compliance Monitoring Tools Implement compliance monitoring tools and processes to automate the measurement and monitoring of compliance metrics. These tools can help detect compliance violations in real time (Corporate Compliance Insights 2023).

Compliance Reports Generate regular compliance reports that provide insights into the organization's compliance status. These reports should highlight areas of noncompliance and include remediation recommendations (Secureframe 2023).

Risk Assessment and Mitigation Conduct regular risk assessments to identify potential data-related risks and vulnerabilities. Develop risk mitigation strategies and track the effectiveness of risk reduction efforts (Experian 2023).

Incident Response Plans Develop incident response plans and metrics to measure the effectiveness of responding to data breaches and incidents. Metrics could include incident response times and resolution rates (OneTrust 2023).

Continuous Monitoring and Audit Continuously monitor compliance and risk management metrics to track progress and ensure ongoing compliance. Regular audits may also be conducted to validate compliance (Corporate Compliance Insights 2023).

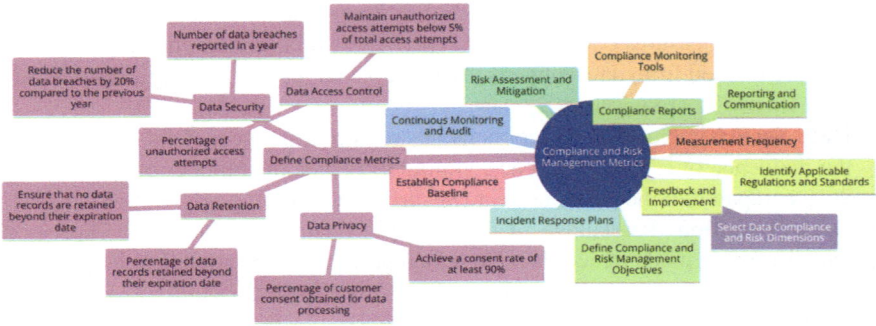

Fig. 14.3 Compliance and risk management metrics map. (Created by the author)

Feedback and Improvement Use feedback from compliance audits and risk assessments to refine compliance metrics and risk management strategies. Continuous improvement is crucial in the ever-evolving regulatory landscape (Secureframe 2023).

Reporting and Communication Communicate compliance achievements and risk management efforts to relevant stakeholders, including regulatory bodies and executive leadership (Experian 2023).

Figure 14.3, titled "Compliance and risk management metrics," provides a detailed visual guide designed to help organizations identify and monitor essential metrics for effective compliance and risk management. This map centrally features "Compliance metrics and risk management," expanding into various key aspects like Data security, Data access control, Data retention, Data privacy, and Compliance management. For Data security, the focus is on reducing the number of data breaches compared to the previous year and monitoring the percentage of unauthorized access attempts. Data Access Control aims to maintain unauthorized access attempts below a minimal threshold, while Data Retention emphasizes ensuring no data remains retained beyond its expiration date and measuring the extent of data records retained past their expiration date. Data privacy measures include evaluating the percentage of customer consent obtained for data processing, with a goal to achieve a high consent rate. Compliance management encompasses elements such as Compliance monitoring tools, Reporting and communication, Measurement frequency, and Feedback and improvement, highlighting the necessity for continuous monitoring and audits, risk assessments and mitigation, establishing compliance baselines, and developing incident response plans. Interconnected elements demonstrate the complex and dynamic nature of compliance and risk management that necessitates continuous monitoring, rigorous assessments, and adaptable response mechanisms to align with changing regulations and standards. This structured approach supports a comprehensive strategy to maintain and enhance compliance and risk management across organizations.

Using well-defined compliance and risk management metrics, organizations can demonstrate their commitment to compliance, manage data-related risks effectively, and ensure data handling practices align with regulatory requirements and industry standards.

14.5 Metrics for Data Usage and Business Impact

Metrics for data usage and business impact are crucial for organizations to assess how effectively they are leveraging their data assets to drive business value. These metrics provide insights into how data is being utilized, its impact on business processes, and the return on investment in data-related initiatives. Here are step-by-step insights into defining and using metrics for data usage and business impact, along with examples:

Define Data Usage and Business Impact Objectives Begin by defining clear objectives for data usage and its impact on the business. These objectives should align with the organization's strategic goals. For example, objectives may include improving sales conversion rates through data-driven marketing (Informatica 2023; McKinsey 2023).

Identify Key Data Usage Scenarios Identify key data usage scenarios or use cases that are critical to achieving your business objectives. These scenarios represent how data is used in specific business processes. Examples include customer segmentation, demand forecasting, and product recommendations (Informatica 2023; DATAVERSITY 2022a).

Select Data Usage and Business Impact Dimensions Choose the data usage and business impact dimensions that are relevant to your organization. Common dimensions include revenue generation, cost reduction, customer satisfaction, and operational efficiency (Arndt 2023; DATAVERSITY 2022a).

Define Data Usage Metrics For each data usage scenario, define specific data usage metrics. Here are real examples of different data usage dimensions:
- **Revenue Generation**: Metric: Increase in sales revenue attributed to data-driven marketing campaigns. Calculation: (Revenue from data-driven campaigns—Baseline revenue)/Baseline revenue. Target: Achieve a 15% increase in revenue through data-driven campaigns (Informatica 2023; DATAVERSITY 2022a).
- **Cost Reduction**: Metric: Percentage reduction in operational costs through data-driven process optimization. Calculation: (Cost savings achieved/Total operational costs) * 100%. Target: Achieve a 10% cost reduction through data-driven process optimization (McKinsey 2023; Experian 2023).

- **Customer Satisfaction**: Metric: Net Promoter Score (NPS) based on customer feedback and engagement data. Calculation: NPS score. Target: Maintain an NPS score of 8 or higher (Informatica 2023).
- **Operational Efficiency**: Metric: Percentage reduction in process cycle time through data-driven automation. Calculation: (Cycle time reduction/Baseline cycle time) * 100%. Target: Reduce process cycle time by 20% through data-driven automation (DATAVERSITY 2022a; Experian 2023).

Establish Baseline Measurements Before implementing data usage initiatives, establish baseline measurements for each metric. Baseline data serves as a reference point for tracking improvements (McKinsey 2023; Informatica 2023).

Measurement Frequency Determine how often you will measure data usage and business impact metrics. Some metrics may be measured continuously while others may be assessed periodically (e.g., monthly or quarterly) (Experian 2023).

Data Usage Analytics Tools Implement data usage analytics tools and processes to track and measure data usage scenarios. These tools help monitor data utilization in real time (Informatica 2023).

Data Impact Reports Generate regular data impact reports that provide insights into how data usage is affecting business processes and outcomes. These reports should highlight successful use cases and areas for improvement (McKinsey 2023; DATAVERSITY 2022a).

Business Impact Assessment Conduct business impact assessments to evaluate the overall contribution of data usage to business objectives. Assessments may involve surveys, interviews, and data analysis (Informatica 2023).

Continuous Improvement Continuously monitor data usage and business impact metrics to track progress and identify opportunities for improvement. Adjust data usage strategies based on performance (DATAVERSITY 2022a; McKinsey 2023).

Feedback and Optimization Use feedback from business impact assessments and user feedback to optimize data usage scenarios and strategies. Continuous optimization is key to maximizing business impact (Informatica 2023).

Reporting and Communication Communicate the positive business impacts achieved through data usage to relevant stakeholders, including senior leadership and decision-makers (Experian 2023).

Figure 14.4, titled "Metrics for data usage and business impact," presents a detailed visual map structured to assist organizations in defining and tracking the effectiveness of data utilization and its impact on business performance. This map is organized around the central theme of aligning metrics with strategic business goals, enhancing both operational efficiency and customer satisfaction. The visualization begins with "Define objectives," which is crucial for setting the framework

14.6 Tracking and Reporting Mechanisms

Fig. 14.4 Metrics for data usage and business impact. (Created by the author)

of what businesses aim to achieve with their data strategies. This leads to "Analytics tools," which support the analysis and interpretation of data to improve decision-making processes such as customer segmentation, demand forecasting, and product recommendations. Key metrics highlighted include the Net promoter score for satisfaction, reductions in operational costs, and increases in sales revenue, showcasing a direct link between data-driven strategies and business outcomes. Additional components of the map include "Business impact assessment" and "Identify key scenarios," which help businesses understand the broader impact of their data utilization on various aspects of operations and strategic planning. "Measurement frequency" and "Data impact reports" are included to ensure ongoing monitoring and reporting of results. The map also emphasizes continuous improvement through "Feedback and optimization," suggesting that businesses should regularly assess and refine their data strategies based on performance metrics. This loop of measurement, reporting, and feedback ensures that data usage remains closely tied to achieving business objectives such as cost reduction, revenue generation, and improvement in customer satisfaction. "Communication of impacts to stakeholders" is a final element, highlighting the importance of transparency and the sharing of results with key stakeholders to maintain support for data-driven initiatives. This comprehensive approach not only supports a dynamic adaptation to business needs but also fosters a culture of continuous development and improvement in utilizing data for strategic advantage.

Using well-defined metrics for data usage and business impact, organizations can effectively measure the value of their data assets and make informed decisions to optimize data-driven strategies for business success.

14.6 Tracking and Reporting Mechanisms

Tracking and reporting mechanisms in the context of data governance refer to the processes and tools used to monitor, measure, and report on various aspects of data management, quality, compliance, and usage. These mechanisms are essential for maintaining data governance practices and ensuring that data-related goals and objectives are met. Here is what you need to know about tracking and reporting mechanisms:

Purpose of Tracking and Reporting The primary purpose of tracking and reporting mechanisms in data governance is to provide visibility into the status and performance of data-related initiatives and processes. This includes monitoring data quality, compliance with regulations, data usage, and the achievement of data governance goals (McKinsey 2023).

Components of Tracking and Reporting Mechanisms Tracking and reporting mechanisms typically consist of the following components:

- **Data Metrics:** Metrics and key performance indicators (KPIs) related to data quality, data usage, compliance, and other relevant aspects (Secureframe 2023).
- **Data Monitoring Tools:** Software tools and solutions that automate the monitoring of data and data-related processes (Twilio Segment 2023).
- **Data Reporting Tools:** Tools for generating reports and dashboards that present data-related insights to stakeholders (DATAVERSITY 2022b).
- **Data Governance Dashboard:** A centralized dashboard that provides a consolidated view of data governance activities and performance (Secureframe 2023).
- **Alerting Systems:** Systems that trigger alerts or notifications when predefined thresholds or anomalies are detected (DATAVERSITY 2022b).
- **Data Auditing:** Regular audits and assessments of data quality and compliance to identify issues (Twilio Segment 2023).
- **Documentation and Documentation Management:** Maintaining records of data governance policies, procedures, and activities (McKinsey 2023).

Key Functions Tracking and reporting mechanisms perform several key functions including:

- **Monitoring:** Continuously monitoring data-related processes and activities (McKinsey 2023).
- **Measurement:** Measuring data quality, compliance, usage, and other relevant metrics (Secureframe 2023).
- **Reporting:** Generating regular reports and dashboards to communicate insights (DATAVERSITY 2022b).
- **Alerting:** Providing real-time or near-real-time alerts when issues or anomalies are detected (DATAVERSITY 2022b).
- **Auditing:** Conducting data audits to assess compliance and quality (Twilio Segment 2023).
- **Documentation:** Maintaining a documented record of data governance activities and changes (McKinsey 2023).

Examples of Tracking and Reporting Here are some examples of what tracking and reporting mechanisms can monitor and report on:

14.6 Tracking and Reporting Mechanisms

- **Data quality metrics:** Tracking data accuracy, completeness, consistency, and timeliness (Secureframe 2023).
- **Compliance metrics:** Ensuring adherence to data privacy regulations, industry standards, and internal policies (Twilio Segment 2023).
- **Data usage metrics:** Measuring the impact of data on business processes and outcomes (DATAVERSITY 2022b).
- **Data governance progress:** Reporting on the implementation of data governance policies and procedures (McKinsey 2023).
- **Data access and permissions:** Monitoring who has access to what data and whether access is appropriate (DATAVERSITY 2022b).
- **Data lineage:** Visualizing the flow of data across systems and processes (Twilio Segment 2023).
- **Data security incidents:** Reporting on data breaches or security vulnerabilities (Secureframe 2023).

Benefits Effective tracking and reporting mechanisms offer several benefits including:

- Improved data quality and reliability (Secureframe 2023).
- Timely detection and mitigation of compliance issues (DATAVERSITY 2022b).
- Enhanced visibility into data usage and its impact on business outcomes (McKinsey 2023).
- Informed decision-making based on data-driven insights (Twilio Segment 2023).
- Accountability and transparency in data governance activities (Secureframe 2023).
- Proactive identification of data-related risks and opportunities for improvement (McKinsey 2023).

Continuous Improvement Organizations should continuously refine their tracking and reporting mechanisms to adapt to evolving data governance requirements, technology changes, and business needs. Feedback from stakeholders should be used to enhance reporting effectiveness (McKinsey 2023; Twilio Segment 2023).

Figure 14.5, titled "Tracking and reporting mechanisms," illustrates a comprehensive visual map designed to guide organizations in the effective tracking and reporting of data usage, quality, and compliance. This map serves as a strategic tool for enhancing organizational transparency and accountability through meticulous data management and monitoring. Central to the diagram is "Auditing," linked to essential practices like "Data auditing," "Governance dashboard," and various monitoring tools that support the tracking of data metrics related to quality, usage, and compliance. These elements help ensure that data management practices adhere to set standards and regulations. Branching from the central auditing focus are sections dedicated to "Documentation management," "Visibility into data initiatives," and "Monitoring data quality and compliance." These branches highlight the importance of maintaining detailed records, improving data visibility, and rigorously monitoring data integrity and regulatory compliance. The map also emphasizes "Continuous improvement" and "Adapt to changes," suggesting that effective data governance requires ongoing evaluation and adaptation to new challenges and

Fig. 14.5 Tracking and reporting mechanisms map. (Created by the author)

information. This dynamic approach is supported by feedback loops from stakeholders, which inform adjustments and enhancements in reporting effectiveness. Further elements like "Alerting systems," "Tracking and reporting mechanisms," and "Risk identification" stress the need for proactive measures in data management, ensuring that potential issues are identified and addressed promptly to mitigate risks. Overall, Fig. 14.5 presents a structured yet adaptable approach to data tracking and reporting, underlining the critical roles of measurement, documentation, and continuous refinement in fostering a culture of accountability and transparency within data governance frameworks. This ensures that all levels of the organization are informed and engaged in maintaining high standards of data management and compliance.

Tracking and reporting mechanisms are integral to effective data governance. They provide the means to monitor, measure, and report on various aspects of data management, ensuring that data is of high quality, compliant, and used to drive business success.

14.7 Balancing Quantitative and Qualitative Measures

Balancing quantitative and qualitative measures in the context of data governance is crucial for gaining a comprehensive understanding of data-related processes, quality, and impact on the organization. Both quantitative and qualitative measures offer unique insights, and finding the right balance between them ensures a more comprehensive approach to data governance. Here is what you need to know about balancing these measures:

Quantitative Measures Quantitative measures involve using numerical data and metrics to assess various aspects of data governance. These measures are typically objective and can be precisely quantified. Examples include data accuracy percent-

14.7 Balancing Quantitative and Qualitative Measures

ages, compliance scores, data usage statistics, and data quality metrics like error rates and completeness percentages (Laney 2018; Martins and Belo 2017).

Qualitative Measures Qualitative measures, on the other hand, involve subjective assessments that are often descriptive or narrative in nature. They provide insights into the "how" and "why" of data governance practices and outcomes. Qualitative measures include factors like user feedback, data governance culture assessments, and narratives explaining compliance challenges or success stories (Pigni et al. 2016).

Importance of Balancing Balancing quantitative and qualitative measures is essential for the following reasons:

- **Comprehensive Understanding:** Qualitative measures provide context and insights that quantitative metrics alone may not capture. Combining both types of measures ensures a more comprehensive understanding of data governance (Bensberg 2003).
- **Human Perspective:** Qualitative measures consider the human element, including user experiences, perceptions, and the cultural aspects of data governance. They shed light on issues like data user satisfaction and organizational alignment (Earley 2020).
- **Holistic Assessment:** By balancing quantitative and qualitative measures, organizations can assess both the hard measurable aspects (quantitative) and the softer human-centered aspects (qualitative) of data governance (Henderson 2015).
- **Decision Support:** Qualitative insights can inform decision-making by providing explanations for trends or anomalies identified through quantitative measures. They help answer questions like "Why is data quality declining?" or "What cultural factors affect compliance?" (Subramanian 2017).

Examples of Balancing Measures Here are some examples of how organizations can balance quantitative and qualitative measures in data governance:

- **Quantitative:** Calculate data quality scores based on accuracy, completeness, and consistency. Track compliance metrics such as the number of regulatory violations (Uttamchandani 2020).
- **Qualitative:** Conduct surveys or interviews with data users to gather their feedback on data quality and ease of data access. Collect narratives from compliance teams on challenges faced during compliance efforts (Toomanian et al. 2011).

Data Governance Dashboard A data governance dashboard is a valuable tool for presenting a balanced view of quantitative and qualitative measures. It can include charts and graphs displaying quantitative metrics alongside sections for qualitative insights, user feedback, and narratives (Laney 2018).

Continuous Improvement The balance between quantitative and qualitative measures may evolve over time as data governance maturity increases. Continuous feedback and adjustment are necessary to ensure that the balance remains effective and aligned with organizational goals (Pigni et al. 2016).

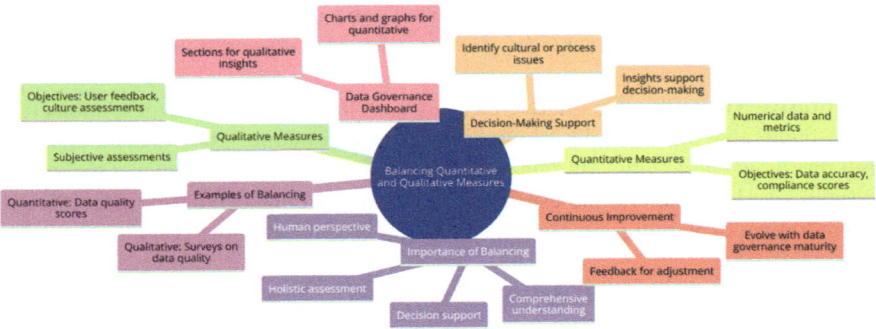

Fig. 14.6 Balancing quantitative and qualitative measures map. (Created by the author)

Decision-Making Support The combined insights from both types of measures can support data governance decision-making. For example, if quantitative measures show a decrease in data quality, qualitative insights can help identify cultural or process issues contributing to the decline and inform improvement strategies (Henderson 2015; Subramanian 2017).

Figure 14.6, titled "Balancing quantitative and qualitative measures," depicts a detailed visual map designed to assist organizations in integrating both quantitative and qualitative approaches to enhance decision-making and data governance. Central to the map is the concept of "Data governance dashboard," which serves as a focal point for balancing diverse metrics. The map branches into several key areas highlighting the integration of different data types and assessment methods. "Quantitative measures" and "Qualitative measures" are depicted as complementary components, with quantitative measures focusing on data accuracy, compliance scores, and numeric data and metrics. In contrast, qualitative measures emphasize user feedback, cultural assessments, and subjective evaluations. "Decision-making support" emerges as a crucial theme, with the dashboard enabling leaders to utilize insights from both data types to support strategic decisions. "Continuous improvement" is emphasized through feedback mechanisms that allow for adjustments based on comprehensive understanding garnered from both quantitative and qualitative data. Additional elements such as "Charts and graphs for quantitative insights" and "Sections for qualitative insights" illustrate the methods for presenting data that support clear communication and informed decision-making. "Examples of balancing" and "Importance of balancing" stress the need for a holistic view, incorporating the human perspective to ensure a more comprehensive assessment. "Feedback for adjustment" and "Evolve with data governance maturity" suggest an adaptive approach, indicating that the balance between these measures should evolve as the organization's data governance matures. This map not only guides the structuring of a governance dashboard but also advocates for a dynamic and balanced approach to using quantitative and qualitative data for enhanced decision-making and governance effectiveness.

Balancing quantitative and qualitative measures in data governance allows organizations to take a more holistic and informed approach to managing their data assets. It ensures that both the technical and human aspects of data governance are considered, leading to more effective data management and governance practices.

14.8 Case Studies: Measuring Success in Data Governance

This section provides insights into how five prominent organizations have implemented data governance to drive success and contribute toward their business outcomes. These case studies illustrate the diverse approaches to data governance and the significant benefits it can bring to an organization.

Airbnb Airbnb used data literacy to promote data-driven decision-making by launching an in-house educational initiative called "Data University," aimed at boosting data literacy across all departments (Atlan 2022a, b; EPJ Data Science 2023). This program empowered employees to understand, interpret, and use data effectively, leading to democratized data and scaled decision-making while ensuring responsible data use.

GE Aviation GE Aviation centralized its scattered data sources, making them more accessible and reliable through their Self-Service Data (SSD) initiative. They established a dedicated data team for user enablement, tooling, and data product deployment, along with a Database Admin team for data governance (Atlan 2022a, b). This approach accelerated decision-making processes and improved safety and operational efficiency by instilling a sense of data ownership among employees.

Wells Fargo Wells Fargo's data governance strategy focused on creating a single source of truth to enhance data accuracy and reliability. They centralized data from multiple sources to create a unified, trustworthy source that reduced discrepancies and improved consistency (Sage Journals 2022; Atlan 2022a, b). This strategy streamlined data management and allowed for more accurate reporting and analysis, thereby enhancing decision-making across the organization.

CSE Insurance CSE Insurance transformed its data culture to improve data management by implementing robust data governance policies and procedures. This included ensuring regulatory compliance, maintaining data quality, protecting data security and privacy, and eliminating data silos (Atlan 2022a, b). These efforts led to better data analytics, streamlined documentation, and significant cost savings for the company.

Uber Uber set up a flexible approach to data governance to handle its large and diverse data sets. This approach included establishing clear data governance policies, implementing data quality measures, and ensuring compliance with data protection regulations (Atlan 2022a, b). Uber's data governance framework enabled them to manage data effectively, leading to improved decision-making and operational efficiency.

Each of these resources provides a detailed look into the strategic implementations of data governance across different industries, highlighting the tangible benefits and methodologies for measuring success. These examples underline the crucial role data governance plays in ensuring data quality, compliance, operational efficiency, and strategic decision-making within organizations.

14.9 Continuous Improvement Through Metrics

Continuous improvement through metrics is a fundamental concept in data governance and broader organizational management. It involves the ongoing process of using data-driven metrics and insights to identify areas for improvement, make data-related decisions, and enhance data governance practices. Here is what you need to know about continuous improvement through metrics:

Purpose of Continuous Improvement Through Metrics The primary purpose of continuous improvement through metrics in data governance is to:

- Identify areas where data governance practices can be enhanced (McKinsey & Company 2019).
- Track the performance and effectiveness of data governance initiatives (Laney 2018).
- Ensure that data governance aligns with organizational goals and objectives (Frontiers 2022).
- Facilitate informed decision-making based on data-driven insights (Journal of Big Data 2021).

Key Steps in Continuous Improvement The process of continuous improvement through metrics typically involves the following steps:

1. **Data Collection:** Gather relevant data and metrics related to data quality, compliance, data usage, and other aspects of data governance (Journal of Big Data 2021).
2. **Data Analysis:** Analyze the collected data to identify trends, anomalies, and areas for improvement. This may involve statistical analysis, data visualization, and trend analysis (McKinsey & Company 2019).
3. **Benchmarking:** Compare current metrics to industry benchmarks or best practices to assess performance relative to peers (Laney 2018).
4. **Goal Setting:** Establish specific improvement goals and objectives based on data insights. These goals should be aligned with the organization's strategic priorities (Frontiers 2022).

14.9 Continuous Improvement Through Metrics

5. **Action Planning:** Develop action plans and strategies to address identified areas for improvement. Assign responsibilities and timelines for implementation (McKinsey & Company 2019).
6. **Implementation:** Execute the action plans, which may involve changes in data governance policies, processes, or technologies (Laney 2018).
7. **Monitoring:** Continuously monitor the impact of implemented changes on data governance metrics. Ensure that improvements are sustained (Journal of Big Data 2021).
8. **Feedback Loop:** Collect feedback from stakeholders and data users to assess the effectiveness of improvements and make further adjustments as needed (McKinsey & Company 2019).

Types of Metrics for Continuous Improvement Metrics for continuous improvement in data governance can be categorized into various areas including:

- **Data Quality Metrics:** Assessing data accuracy, completeness, consistency, and timeliness (Journal of Big Data 2021).
- **Compliance Metrics:** Measuring adherence to data privacy regulations, industry standards, and internal policies (Laney 2018).
- **Data Usage Metrics:** Evaluating the impact of data on business processes and outcomes (Frontiers 2022).
- **Operational Metrics:** Tracking the efficiency and effectiveness of data-related processes (McKinsey & Company 2019).
- **User Satisfaction Metrics:** Gathering feedback from data users on their experiences and satisfaction with data access and quality (Journal of Big Data 2021).

Benefits of Continuous Improvement through Metrics Continuous improvement through metrics offers several benefits:

- **Proactive Issue Identification:** Early detection of data quality issues or compliance violations (McKinsey & Company 2019).
- **Informed Decision-Making:** Data-driven insights support better decision-making (Laney 2018).
- **Enhanced Data Governance Practices:** Identifying areas for improvement leads to more effective data governance (Frontiers 2022).
- **Alignment with Business Goals:** Metrics help ensure that data governance aligns with organizational objectives (Journal of Big Data 2021).
- **Accountability and Transparency:** Metrics promote accountability for data-related responsibilities (Laney 2018).

Data Governance Dashboard A data governance dashboard is a valuable tool for visualizing and monitoring metrics related to continuous improvement. It provides a real-time or near-real-time view of key performance indicators (KPIs) and progress toward improvement goals (McKinsey & Company 2019).

Iterative Process Continuous improvement through metrics is an iterative process that evolves over time. As data governance matures and organizational priorities

Fig. 14.7 Continuous improvement through metrics map. (Created by the author)

change, the metrics and improvement initiatives may also evolve to address new challenges and opportunities (Frontiers 2022).

Figure 14.7, titled "Continuous improvement through metrics," presents a detailed visual map designed to guide organizations in leveraging metrics to drive ongoing enhancements in data governance and decision-making processes. Central to the diagram is the theme of "Continuous improvement through metrics," emphasizing the pivotal role that metrics play in fostering organizational evolution and adaptability. The diagram expands into various categories of metrics and associated processes. Types of metrics such as operational, compliance, user satisfaction, data usage, and data quality metrics each contribute to assessing different aspects of organizational performance and governance. The purpose of these metrics is to align with organizational goals to effectively track performance and effectiveness, ultimately enhancing data governance practices. Action planning is a critical component, involving goal setting, data collection, monitoring, and implementation to ensure that metrics are not only collected but also effectively acted upon. The iterative process highlighted in the map suggests that the approach to metrics should evolve with the maturity of data governance and adapt to organizational changes, ensuring that data-driven decision-making remains relevant and effective over time. A central platform, the "Data governance dashboard," facilitates benchmarking and utilizes feedback loops for continual adjustment and improvement, aligning closely with business goals. The benefits derived from this metrics-driven approach include enhanced data governance, proactive issue identification, informed decision-making, and increased accountability and transparency. Overall, the map portrays a dynamic system where continuous improvement is facilitated through the strategic application of various metrics, supported by iterative reviews and adaptability to internal and external changes. This comprehensive framework aids in decision support and ensures that governance strategies are progressively refined to meet evolving business and regulatory demands.

Continuous improvement through metrics is a dynamic and proactive approach to enhancing data governance practices, ensuring data quality, and maximizing the value of data assets within an organization.

14.10 Conclusion

This chapter, titled "Measuring the Impact of Data Governance: Metrics and Key Performance Indicators," has provided a comprehensive exploration of the critical role that metrics and key performance indicators (KPIs) play in the realm of data governance. This chapter has covered a wide range of topics, from the introduction to the measurement of data governance impact to the practical aspects of defining relevant metrics, tracking data quality, compliance, and business impact, and finally, to the crucial concept of continuous improvement through metrics. In this conclusion, we summarize the key insights and takeaways from each section of this chapter.

- The chapter began by highlighting the significance of measuring the impact of data governance.
- It emphasized that effective measurement requires well-defined metrics aligned with organizational goals.
- The next section delved into the process of defining metrics and KPIs tailored to an organization's unique data governance needs.
- It stressed the importance of selecting metrics that align with strategic objectives and key data governance areas.

14.10.1 Metrics for Data Quality

- The focus here was on metrics related to data quality, including accuracy, completeness, consistency, and timeliness.
- Data quality metrics are essential for identifying and addressing data errors and ensuring data reliability.

14.10.2 Compliance and Risk Management Metrics

- This section explored metrics associated with compliance and risk management in data governance.
- Metrics in this area help organizations maintain adherence to data protection regulations and mitigate data-related risks.

14.10.3 Metrics for Data Usage and Business Impact

- Data usage metrics were discussed, emphasizing their role in understanding how data impacts business processes and outcomes.
- These metrics contribute to data-driven decision-making and maximizing the value of data assets.

14.10.4 Tracking and Reporting Mechanisms

- The chapter highlighted the importance of effective tracking and reporting mechanisms for presenting metrics in a clear and actionable manner.
- Data governance dashboards were introduced as tools to facilitate real-time monitoring.

14.10.5 Balancing Quantitative and Qualitative Measures

- Balancing quantitative and qualitative measures was emphasized as essential for a comprehensive view of data governance success.
- Qualitative insights provide context and human perspectives to complement quantitative metrics.

14.10.6 Case Studies: Measuring Success in Data Governance

- This section discussed real-world case studies illustrating how organizations measure success in data governance.
- Case studies highlighted the practical application of metrics and the impact on data governance outcomes.

14.10.7 Continuous Improvement through Metrics

- The chapter concluded with a focus on the iterative nature of data governance improvement through metrics.
- Continuous monitoring, analysis, and adaptation based on metrics were highlighted as best practices.

In conclusion, this chapter has underscored the critical role of metrics and KPIs in data governance. It has shown that measuring success is not only a best practice but a strategic imperative. Effective measurement empowers organizations to ensure data quality, mitigate compliance risks, optimize data usage, foster accountability, and drive continuous improvement. As organizations continue to navigate the complexities of data governance, embracing robust measurement practices will be essential for success in the data-driven landscape.

References

Arndt A (2023) Data governance metrics & KPIs to measure success. Experian. 14 March 2023
Atlan (2022a) 5 data governance examples: case studies, takeaways & more. Atlan
Atlan (2022b) Data governance in insurance: 7 ways it impacts insurers. Atlan
Bensberg F (2003) Data quality management and data governance. J Data Manag 12(4):56–72
Corporate Compliance Insights (2023) 8 crucial compliance KPIs. Corporate Compliance Insights
Datafloq (2023) 35 metrics you should use to monitor data governance. Datafloq
Datapine (2023) Data quality management: see metrics for data control. Datapine
DATAVERSITY (2022a) What are data governance metrics? DATAVERSITY. 21 June 2022
DATAVERSITY (2022b) What is data governance? Definition, types, uses. DATAVERSITY
DiGGrowth (2023) The essential guide to data governance metrics: what to track and why. DiGGrowth
Earley S (2020) Leveraging data for competitive advantage: strategies for monetization and risk management. Inf Manag J 34(2):22–34
EPJ Data Science (2023) A global-scale analysis of the sharing economy model—an Airbnb case study. EPJ Data Science
Experian (2023) Measuring success in data governance. Experian
Frontiers (2022) A balanced scorecard for maximizing data performance. Frontiers in Psychology
Henderson D (2015) Improving organizational performance through data governance. Perform Manag Rev 23(3):19–27
Informatica (2023) Measuring success in data governance. Informatica
Journal of Big Data (2021) Big data quality framework: a holistic approach to continuous quality management. Journal of Big Data
Laney D (2018) Infonomics: how to monetize, manage, and measure information as an asset for competitive advantage. Taylor & Francis
Martins R, Belo O (2017) Balanced scorecard for data performance: a case study. J Bus Anal 5(1):13–21
McKinsey (2023) Designing data governance that delivers value. McKinsey
McKinsey & Company (2019) How continuous improvement can build a competitive edge. McKinsey & Company
OneTrust (2023) Compliance program performance metrics: how to measure compliance. OneTrust
Pigni F, Piccoli G, Watson RT (2016) Digital data streams: creating value from data-driven innovation. MIS Q Exec 15(2):63–72
Sage Journals (2022) Wells Fargo: fall from great to miserable: a case study on corporate governance. Sage Journals
Secureframe (2023) 7 data governance metrics and KPIs every business should track. Secureframe
Subramanian S (2017) Reducing risks through effective data governance. Risk Manag J 45(2):28–36
Toomanian A, Salehi M, Ashuri B (2011) Data governance in large data warehouses. Int J Data Sci 7(3):34–45
Twilio Segment (2023) What is a data governance framework? Examples & models. Twilio Segment
Uttamchandani S (2020) Maximizing data value through governance and analytics. J Data Anal 9(1):15–24

Chapter 15
Legal and Regulatory Considerations in Data Governance

Contents

15.1	Introduction to Legal and Regulatory Aspects	446
15.2	Global Data Protection and Privacy Laws	448
15.3	Navigating Industry-Specific Regulations	449
15.4	Cross-Border Data Transfer and Compliance	451
15.5	Developing a Compliance-Oriented Data Governance Framework	454
15.6	Data Governance in the Context of Legal Discovery and Audits	456
15.7	Cybersecurity Laws and Regulations	458
15.8	Case Studies: Legal and Regulatory Compliance	461
15.9	Future Legal and Regulatory Trends	463
15.10	Conclusion	464
References		465

Abstract This chapter addresses the intricate legal and regulatory frameworks that organizations must navigate to manage data responsibly. The chapter emphasizes the importance of understanding and complying with various laws, regulations, and standards to mitigate legal risks, avoid fines, and maintain a strong reputation.

The chapter begins by exploring different data protection and privacy laws, such as the GDPR in Europe and the CCPA in the United States. These laws dictate how organizations must handle personal data, including obtaining explicit consent for data collection, ensuring data subject rights, and executing data breach notifications. The narrative highlights the complexities involved in adhering to these laws, given their extensive requirements and the severe penalties for non-compliance.

It then discusses industry-specific regulations, like HIPAA in healthcare and PCI DSS in finance, which impose additional data governance obligations tailored to particular sectors' unique risks and needs. These regulations require specialized knowledge and strategies to ensure compliance, and the chapter provides insights into best practices for navigating these industry-specific challenges.

Cross-border data transfer is another focus area, emphasizing the need for organizations to manage international data flows compliantly. The chapter outlines mechanisms like Standard Contractual Clauses and Binding Corporate Rules to secure data transfers outside of jurisdictions with stringent data protection laws.

The narrative shifts to the development of compliance-oriented data governance frameworks, which are essential for organizations to systematically manage their data in line with legal and regulatory requirements. Such frameworks integrate policies, procedures, and controls to facilitate compliance and support efficient data management.

Additionally, the chapter covers the roles of data governance in the context of legal discovery and audits, explaining how proper data governance facilitates the organization's ability to respond to legal requests and audits efficiently and accurately.

In conclusion, this chapter provides a comprehensive overview of the legal and regulatory landscape impacting data governance. It underscores the necessity for organizations to develop robust data governance frameworks that not only comply with current laws and regulations but are also adaptable to future changes. This approach helps organizations protect their data assets, comply with legal obligations, and enhance their overall data governance practices.

Keywords Legal and regulatory considerations · Data governance · GDPR · CCPA · HIPAA · Compliance frameworks · Data protection · Privacy laws · Cybersecurity · Cross-border data transfer · Compliance audits · Data security · Regulatory compliance · Legal discovery · Data breaches · Industry-specific regulations · Future legal trends

15.1 Introduction to Legal and Regulatory Aspects

Legal and regulatory aspects in the context of data governance refer to the complex web of laws, regulations, and standards that govern how organizations collect, use, store, and protect data. These aspects are critical considerations for any organization that handles data, as noncompliance can lead to legal consequences, fines, and damage to reputation. Here is what you should know about the legal and regulatory aspects of data governance:

Data Protection and Privacy Laws Various laws such as the General Data Protection Regulation (GDPR) in Europe (European Union 2016a) and the California Consumer Privacy Act (CCPA) in the United States (California Legislature 2018) dictate how organizations must handle personal data. They require explicit consent for data collection, data subject rights, and data breach notification.

15.1 Introduction to Legal and Regulatory Aspects

Industry-Specific Regulations Some industries have specific regulations governing data governance. For example, the healthcare sector must adhere to the Health Insurance Portability and Accountability Act (HIPAA) (U.S. Congress 1996), while the financial industry follows regulations like the Payment Card Industry Data Security Standard (PCI DSS) (Payment Card Industry Security Standards Council 2018).

Cross-Border Data Transfer When data crosses international borders, additional regulations come into play. Organizations must ensure that data transfers comply with applicable laws such as the EU-U.S. Privacy Shield (U.S. Federal Trade Commission 2016) or Standard Contractual Clauses (SCCs) (European Commission 2004).

Compliance-Oriented Data Governance To meet legal and regulatory requirements, organizations establish compliance-oriented data governance frameworks. These frameworks include policies, procedures, and controls to ensure data is handled in accordance with the law.

Legal Discovery and Audits Legal discovery involves the process of providing relevant data in legal proceedings. Data governance practices must facilitate efficient data retrieval. Audits by regulatory authorities assess an organization's compliance with data protection laws.

Cybersecurity Laws and Regulations Data breaches can result in legal action and regulatory penalties. Organizations must adhere to cybersecurity laws and standards to protect data from breaches (U.S. Congress 2002).

Case Studies Real-world case studies provide examples of organizations successfully navigating legal and regulatory challenges in data governance. These cases offer insights into best practices.

Future Trends As technology and data practices evolve, so do legal and regulatory requirements. Organizations must monitor and adapt to future trends in data governance and compliance.

The Fig. 15.1, highlights the critical legal and regulatory frameworks that organizations must navigate to manage data effectively and compliantly. It covers topics such as data protection and privacy laws, industry-specific regulations, cross-border data transfer, compliance-oriented data governance, legal discovery and audits, cybersecurity laws and regulations, real-world case studies, and future trends in data governance and compliance.

Legal and regulatory aspects are fundamental considerations in data governance. Organizations must develop robust data governance strategies that align with applicable laws, industry regulations, and best practices to protect data, maintain compliance, and mitigate legal risks.

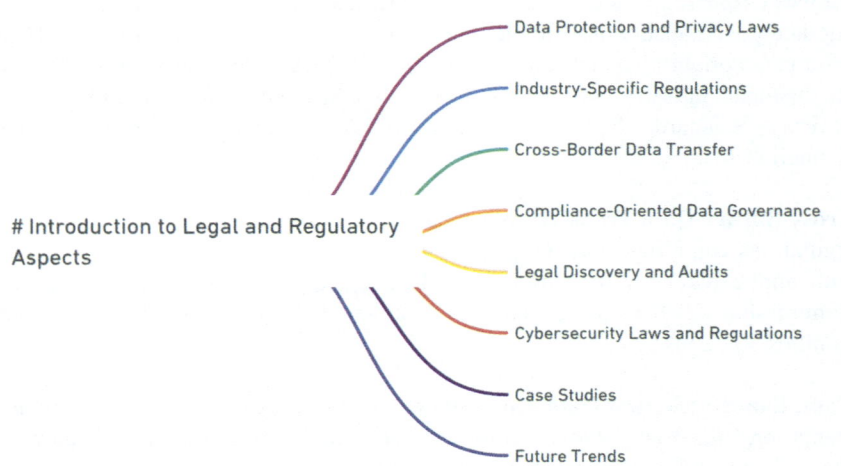

Fig. 15.1 Introduction to legal and regulatory aspects. (Created by the author)

15.2 Global Data Protection and Privacy Laws

Global Data Protection and Privacy Laws refer to a set of regulations and legal frameworks that are designed to protect the privacy and personal data of individuals in the digital age. These laws have become increasingly important as data has become an asset and as concerns about data privacy and security have grown. Here are key points about global data protection and privacy laws:

- **GDPR (General Data Protection Regulation)**: The GDPR is one of the most significant data protection laws globally. It was enacted by the European Union (EU) in 2016 and has extraterritorial reach, meaning it applies to organizations outside the EU that process the data of EU residents. The GDPR grants individuals control over their personal data and imposes strict obligations on organizations regarding data handling, consent, data breach notification, and more (European Union 2016a).
- **CCPA (California Consumer Privacy Act)**: The CCPA is a privacy law in the United States, specifically in California. It grants California residents certain rights over their personal data, including the right to know what data is collected and the right to opt out of data sales. It applies to businesses that meet specific criteria even if they are not located in California (California Legislature 2018).
- **LGPD (Lei Geral de Proteção de Dados)**: The LGPD is the data protection law in Brazil, similar in some respects to the GDPR. It grants individuals rights over their data and imposes obligations on organizations, including the appointment of a Data Processing Officer (DPO) and data breach notification requirements.

- **Data Transfer Mechanisms**: Global data protection laws also address cross-border data transfers. The EU has mechanisms such as Standard Contractual Clauses (SCCs) (European Commission 2004) and the EU-U.S. Privacy Shield (though it has been invalidated) (U.S. Federal Trade Commission 2016) to ensure data protection when data leaves the EU.
- **Data Subject Rights**: These laws typically grant individuals rights over their data, including the right to access, rectify, and delete data, as well as the right to know how their data is used and the right to object to certain data processing.
- **Data Security and Breach Notification**: Data protection laws often require organizations to implement security measures to protect data from breaches. They also mandate data breach notification to authorities and affected individuals in case of a breach (International Organization for Standardization 2013; U.S. Department of Health and Human Services 2000).
- **Penalties and Fines**: Noncompliance with these laws can result in significant fines and penalties. For example, the GDPR allows fines of up to €20 million or 4% of the global annual revenue, whichever is higher (European Union 2016a).
- **Data Protection Authorities**: Most countries with data protection laws have regulatory authorities responsible for enforcing these laws and ensuring compliance.
- **Consent and Transparency**: Transparency in data processing and obtaining informed consent from individuals are fundamental principles of these laws.
- **Impact on Global Business**: Global data protection and privacy laws have a profound impact on businesses that handle personal data. They require organizations to adopt comprehensive data protection strategies and may influence the design of products and services.

Global data protection and privacy laws are a response to the growing importance of personal data in the digital age. They aim to protect individuals' privacy rights and impose obligations on organizations to handle data responsibly and securely. These laws have a far-reaching impact on businesses, and compliance is a top priority for organizations worldwide.

15.3 Navigating Industry-Specific Regulations

Navigating Industry-Specific Regulations refers to the process of understanding, complying with, and managing regulations and standards that are specific to industries. These regulations are in addition to general data protection and privacy laws and are tailored to address unique challenges and requirements of specific sectors. Here is what you should know about navigating industry-specific regulations: Different industries have their own set of regulations and standards that govern how data is handled, stored, and protected. These regulations are designed to address industry-specific challenges and protect sensitive information related to that sector.

Examples of Industry-Specific Regulations
- **Healthcare**: The healthcare industry is subject to regulations such as the Health Insurance Portability and Accountability Act (HIPAA) in the United States. HIPAA mandates strict data security and privacy protections for patient health information (U.S. Congress 1996).
- **Financial Services**: The financial sector is regulated by laws like the Payment Card Industry Data Security Standard (PCI DSS) and the Dodd-Frank Wall Street Reform and Consumer Protection Act. These regulations aim to safeguard financial data and prevent fraud (Payment Card Industry Security Standards Council 2018; U.S. Congress 2010).
- **Pharmaceuticals**: Pharmaceutical companies must adhere to regulations like the Food and Drug Administration's (FDA) Title 21 CFR Part 11, which governs electronic records and electronic signatures (U.S. Food and Drug Administration 1997).
- **Energy**: The energy sector may have regulations related to the protection of critical infrastructure and the handling of sensitive energy-related data, such as the North American Electric Reliability Corporation (NERC) Critical Infrastructure Protection (CIP) Standards (North American Electric Reliability Corporation 2016).

Complexity Industry-specific regulations can be complex and may require specialized knowledge for compliance. They often involve strict data security measures, data retention policies, and reporting requirements.

Compliance Challenges Organizations operating in regulated industries must invest in compliance efforts to ensure they meet the specific requirements of these regulations. Noncompliance can lead to legal consequences and reputational damage.

Cross-Industry Impact In some cases, industry-specific regulations may have implications beyond the sector they directly govern. For example, the GDPR has a broad impact and applies to organizations globally, including those in healthcare, finance, and other industries.

Integration with Data Governance To effectively navigate industry-specific regulations, organizations need to integrate compliance efforts with their broader data governance strategies. This includes implementing policies and procedures, data protection measures, and regular audits.

Data Governance Officers Many organizations appoint Data Protection Officers (DPOs) or Compliance Officers responsible for ensuring compliance with industry-specific regulations. These officers play a key role in monitoring and enforcing compliance.

Constant Monitoring As regulations evolve, organizations must stay up to date with changes and adapt their data governance practices accordingly. This often requires ongoing monitoring and adjustment of policies and procedures.

15.4 Cross-Border Data Transfer and Compliance

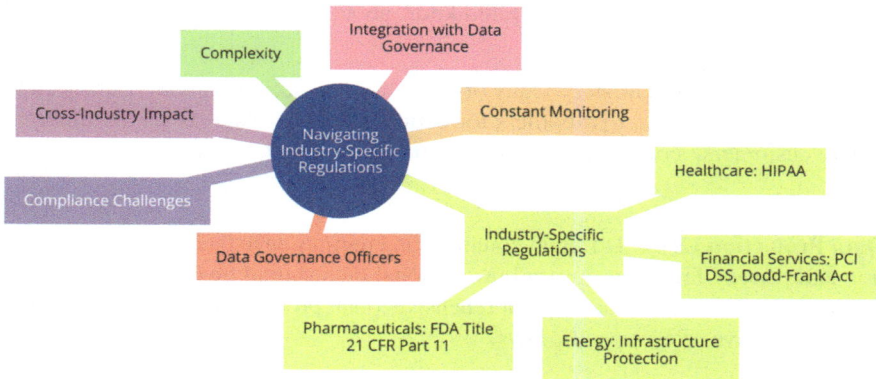

Fig. 15.2 Navigating industry-specific regulations map. (Created by the author)

Figure 15.2, titled "Navigating industry-specific regulations," illustrates the complexities and multifaceted nature of regulatory compliance across different industries. At the center of this diagram is the core concept of navigating these regulations, highlighted by a blue circle. From this central point, various interconnected elements branch out, emphasizing the key areas of focus and concern in industry-specific regulatory landscapes. These include the complexity of the regulations themselves, their cross-industry impacts, the specific challenges of compliance, and the roles of data governance officers in ensuring adherence. The map also points to the importance of integrating regulatory compliance within broader data governance frameworks and the necessity for constant monitoring to stay compliant. Specific regulations mentioned are HIPAA for healthcare, PCI DSS and the Dodd-Frank Act for financial services, FDA Title 21 CFR Part 11 for pharmaceuticals, and infrastructure protection standards for the energy sector. Each element in the diagram is visually connected, underscoring the interconnected nature of these aspects and their collective impact on organizational data governance strategies.

Navigating industry-specific regulations is a crucial aspect of data governance for organizations operating in regulated sectors. These regulations are tailored to address industry-specific risks and requirements, and compliance is essential to avoid legal and financial penalties. Integrating compliance efforts into overall data governance strategies is key to effectively managing industry-specific regulations.

15.4 Cross-Border Data Transfer and Compliance

Cross-Border Data Transfer and Compliance refer to the challenges and requirements associated with transferring personal data across international borders while ensuring compliance with data protection and privacy laws. Data transfer between countries or regions can be subject to strict regulations to protect individuals' privacy rights. Here is what you should know about cross-border data transfer and compliance:

International Data Transfer Many organizations, especially global ones, need to transfer personal data across borders for various purposes, such as serving international customers, collaborating with partners, or centralizing data processing centers.

Data Protection Laws Different countries and regions have their own data protection and privacy laws that govern how personal data should be handled and transferred. Some countries have comprehensive data protection regulations, while others may have sector-specific laws.

EU GDPR and Data Transfers The General Data Protection Regulation (GDPR) in the European Union (EU) sets stringent requirements for the transfer of personal data outside the EU and European Economic Area (EEA). Transfers to countries without an "adequacy decision" require additional safeguards (European Union 2016a).

Mechanisms for Data Transfer To facilitate cross-border data transfers, various mechanisms and safeguards can be used, including:

- **Standard Contractual Clauses (SCCs)**: These are standardized contractual clauses that organizations can use to ensure adequate data protection when transferring data to non-adequate countries (European Commission 2004).
- **Binding Corporate Rules (BCRs)**: These are internal rules for multinational organizations that define how personal data should be protected when transferred between their entities.
- **Data Transfer Agreements**: Organizations can enter into data transfer agreements with third parties that include specific data protection provisions.
- **Privacy Shield**: This was a framework that allowed U.S. organizations to self-certify compliance with EU data protection requirements. However, it has been invalidated (U.S. Federal Trade Commission 2016).
- **Adequacy Decisions**: Some countries or regions may be granted an "adequacy decision" by the EU, indicating that their data protection laws are equivalent to GDPR standards. Data transfers to countries with adequacy decisions are unrestricted.

Data Minimization and Purpose Limitation Organizations must ensure that transferred data is limited to what is necessary for the intended purpose and that it is not used for purposes incompatible with the original data collection.

Data Localization Laws Some countries have data localization laws that require certain types of data to be stored within their borders. Compliance with these laws may affect cross-border data transfer strategies.

15.4 Cross-Border Data Transfer and Compliance

Compliance Audits Organizations may need to conduct regular audits and assessments to ensure that data transfers comply with relevant laws and mechanisms.

Privacy Impact Assessments (PIAs) PIAs can help organizations assess the potential risks and privacy implications of cross-border data transfers and implement necessary safeguards.

Future Developments Data protection laws and regulations related to cross-border data transfer are subject to change. Organizations must stay informed about developments in international data protection to maintain compliance.

Figure 15.3, outlines the intricate network of factors that influence the management of international data exchange under strict compliance frameworks. At the heart of this diagram is the central theme of cross-border data transfer and compliance, highlighting the core challenge of navigating this complex area. The diagram branches out to detail various components critical to managing data across borders. It includes sections on Data protection laws and Compliance audits, which are foundational to ensuring that data transfers meet the established privacy standards. The visual also details Mechanisms for data transfer, such as Data transfer agreements, Standard contractual clauses (SCCs), and Binding corporate rules (BCRs), which are structural methods enabling organizations to transfer data internationally legally. Another branch touches on the Privacy shield, noted as invalidated, representing previous attempts to streamline compliance between different regulatory jurisdictions like the EU and the US. Adequacy decisions, which assess if countries outside the EU offer sufficient levels of data protection to permit easier data flows from the EU, are also featured. The map points to Future developments and Data localization

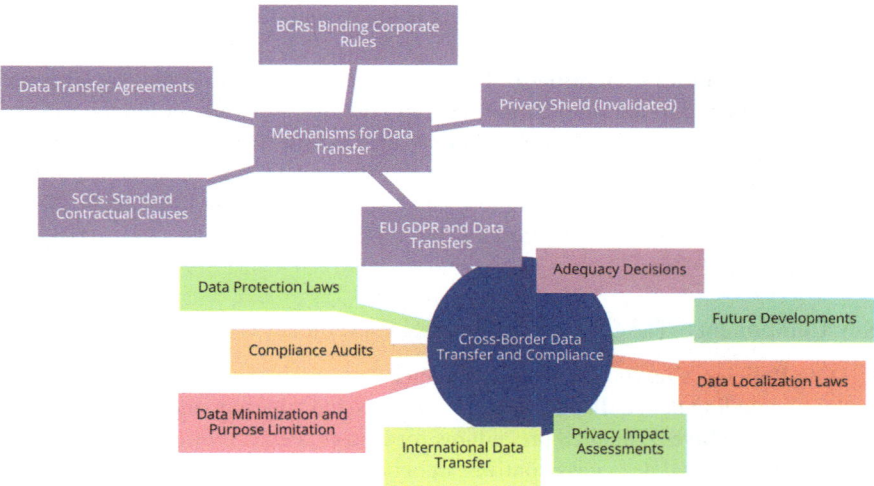

Fig. 15.3 Cross-border data transfer and compliance map. (Created by the author)

laws, indicating the evolving legal landscape and the trend in some regions to mandate local data processing and storage. Finally, aspects such as International data transfer and Privacy impact assessments emphasize the practical steps in managing cross-border data flows, ensuring that each transfer is scrutinized for privacy risks and complies with global standards. Together, these elements illustrate a comprehensive approach to managing and safeguarding personal data across international boundaries, highlighting the various facets of the regulatory environment impacting cross-border data transfer.

Cross-border data transfer and compliance involve navigating a complex landscape of data protection laws and mechanisms to ensure that personal data is transferred securely and in compliance with relevant regulations. It requires a combination of legal, technical, and organizational measures to safeguard individuals' privacy rights during data transfers.

15.5 Developing a Compliance-Oriented Data Governance Framework

Developing a Compliance-Oriented Data Governance Framework involves creating a structured and comprehensive approach to data governance that prioritizes compliance with relevant laws, regulations, and standards. Such a framework ensures that an organization's data practices align with legal requirements and industry-specific regulations. Here is what you should know about developing a compliance-oriented data governance framework:

Foundation of Data Governance A compliance-oriented data governance framework is built upon the foundational principles of data governance. It includes defining data ownership, data stewardship, data policies, and data procedures.

Legal and Regulatory Alignment The framework aligns with applicable laws and regulations related to data protection, privacy, industry-specific requirements, and cross-border data transfer. Key regulations include the General Data Protection Regulation (GDPR) in the European Union (European Union 2016a), the California Consumer Privacy Act (CCPA) in the United States (California Legislature 2018), and the Health Insurance Portability and Accountability Act (HIPAA) (U.S. Congress 1996).

Key Components The framework consists of several key components:

- **Data Policies**: Clearly defined data policies that specify how data should be handled, including data collection, storage, processing, sharing, and disposal.
- **Data Procedures**: Detailed procedures for data-related activities to ensure compliance, such as data access controls, data retention, and data breach response.

15.5 Developing a Compliance-Oriented Data Governance Framework

- **Data Governance Roles**: Identification of roles and responsibilities within the organization, including data owners, data stewards, and compliance officers.
- **Data Risk Assessment**: Regular assessments of data-related risks, including privacy risks, security risks, and legal risks, and strategies to mitigate these risks.
- **Data Privacy Impact Assessments (PIAs)**: Conducting PIAs to evaluate and mitigate the privacy implications of data processing activities (United Nations 1990).
- **Data Security Measures**: Implementation of data security measures and controls to protect data from unauthorized access and breaches, in line with standards such as ISO/IEC 27001 (International Organization for Standardization 2013).
- **Data Auditing and Monitoring**: Continuous monitoring of data activities and regular audits to ensure compliance and identify potential issues.
- **Data Training and Awareness**: Training programs and awareness campaigns to educate employees about data governance and compliance requirements.
- **Data Mapping and Inventory**: Developing a comprehensive data map and inventory to understand where data is stored, who has access to it, and how it is used.
- **Data Protection Impact Assessments (DPIAs)**: Conducting DPIAs to assess and mitigate risks to individuals' rights and freedoms when processing personal data as required by some data protection laws (European Union 2016a).
- **Cross-Functional Collaboration**: Collaboration between legal, compliance, IT, and other relevant departments to ensure that data governance aligns with legal requirements.
- **Documentation and Record-Keeping**: Maintaining detailed records of data governance activities, including data processing activities, data protection measures, and compliance efforts.
- **Change Management**: Implementing change management processes to adapt the data governance framework as laws and regulations evolve.
- **Regular Audits and Assessments**: Conducting regular audits and assessments to evaluate the effectiveness of the framework and identify areas for improvement.
- **Incident Response Plan**: Developing a data breach incident response plan to address data breaches promptly and in compliance with breach notification requirements (California Legislature 2018).
- **Continuous Improvement**: The framework includes mechanisms for continuous improvement and adaptation to changing compliance requirements.
- **Reporting and Accountability**: Defining reporting structures and accountability mechanisms to ensure that compliance obligations are met.

A compliance-oriented data governance framework is essential for organizations to meet legal and regulatory requirements while effectively managing data. It provides a structured approach to data governance, focusing on alignment with compliance standards, risk mitigation, and ongoing monitoring and improvement. Such a framework helps organizations build trust with stakeholders and protect individuals' privacy rights.

15.6 Data Governance in the Context of Legal Discovery and Audits

Data Governance in the Context of Legal Discovery and Audits refers to the management and governance of data to ensure that it is discoverable, preserved, and compliant with legal requirements when the organization is involved in legal proceedings, investigations, or audits. This aspect of data governance is critical for organizations to respond effectively to legal requests, protect their interests, and demonstrate compliance with relevant laws and regulations. Here is what you should know about data governance in the context of legal discovery and audits:

Legal Discovery Legal discovery, also known as eDiscovery, is the process by which organizations collect, preserve, review, and produce electronically stored information (ESI) in response to legal requests such as subpoenas, litigation, or regulatory investigations. Effective data governance ensures that data is organized, accessible, and preserved in a manner that complies with legal requirements.

Importance of Data Governance Effective data governance is essential for legal discovery because it ensures that data is organized, accessible, and preserved in a manner that complies with legal requirements. It helps organizations identify and locate relevant data quickly and efficiently.

Key Aspects of Data Governance for Legal Discovery
- **Data Classification**: Classifying data based on its sensitivity and relevance to legal matters. This includes identifying personally identifiable information (PII), sensitive documents, and relevant communication records.
- **Data Retention Policies**: Implementing data retention policies that specify how long data should be retained, ensuring that data is not prematurely deleted or destroyed, which could lead to legal sanctions (Federal Rules of Civil Procedure 2015).
- **Data Preservation**: Putting in place mechanisms to preserve data that may be relevant to ongoing or anticipated legal matters. This includes suspending routine data deletion processes and ensuring data integrity during preservation.
- **Data Access Controls**: Implementing strict access controls to limit who can access and modify data that may be subject to legal discovery. This prevents unauthorized alterations or deletions (U.S. Department of Health and Human Services 2000).
- **Audit Trails**: Maintaining comprehensive audit trails that record who accessed, modified, or deleted data, when these actions occurred, and for what purpose (International Organization for Standardization 2013).
- **Legal Hold**: Implementing legal holds on data that is subject to legal proceedings to prevent data destruction. Legal holds must be properly documented and monitored.

15.6 Data Governance in the Context of Legal Discovery and Audits

- **Data Search and Retrieval**: Establishing search capabilities to quickly locate and retrieve relevant data during legal discovery. This includes metadata indexing and advanced search tools.
- **Data Privacy Considerations**: Ensuring that data processing during legal discovery complies with data privacy laws and regulations, including redacting sensitive information as necessary.
- **Data Security Measures**: Implementing data security measures to protect sensitive legal information from unauthorized access or breaches.
- **Documentation and Reporting**: Documenting all actions taken during the legal discovery process, including data collection, preservation, and production. Reporting may be required to demonstrate compliance with legal requirements.

Audits In addition to legal discovery, organizations may be subject to audits by regulatory bodies or internal audit teams. Data governance plays a crucial role in providing auditors with evidence of compliance with data protection, privacy, and security standards.

Cross-Functional Collaboration Effective data governance in the context of legal discovery and audits requires collaboration between legal, compliance, IT, and data management teams to ensure that data practices align with legal and regulatory requirements.

Third-Party Tools Many organizations use specialized eDiscovery and audit management software to streamline the process and ensure compliance (The Sedona Conference 2018).

Figure 15.4, titled "Data governance in legal discovery and audits," provides a detailed overview of the essential components and considerations that organizations must manage when integrating data governance with legal processes. Centered on the concept of data governance within the realm of legal discovery and audits, the diagram highlights various key areas connected to this central theme. It starts with the Importance of data governance, emphasizing how foundational data governance practices underpin all other aspects of legal readiness and compliance. From there, it connects to Cross-functional collaboration, which is crucial for effective data governance, ensuring that various departments work together seamlessly in the face of legal challenges. The diagram also features Third-party tools and Legal discovery, indicating tools that aid in managing data specifically for legal processes, and the broader concept of audits, which are regular and systematic checks to ensure compliance and governance standards are met. Specific processes such as Data retention policies, Data preservation, and Audit trails are marked out as essential for maintaining historical data integrity and accessibility. These are critical for any legal audit or discovery process, where the ability to retrieve and verify data history is paramount. Data search and retrieval and Data access controls are noted as operational aspects of data governance, ensuring that data can be efficiently located and accessed under controlled conditions. Similarly, Data security measures, Data privacy considerations, and Data classification outline the security and categorization

Fig. 15.4 Data governance in legal discovery and audits map. (Created by the author)

strategies that protect and organize data according to sensitivity and relevance, crucial for legal holds and compliance with privacy laws. Lastly, Documentation and reporting encapsulate the need for maintaining clear records of data governance activities and audit trails, which are essential for demonstrating compliance during legal reviews. This map visually ties together these various strands, illustrating how they contribute to a comprehensive strategy for managing data governance in the context of legal discovery and audits. Each branch not only defines a specific area of focus but also shows how interconnected and vital each is to ensuring that an organization meets its legal and regulatory obligations effectively.

Data governance in the context of legal discovery and audits is a critical component of overall data governance. It ensures that organizations can effectively respond to legal requests, protect their data, and demonstrate compliance with legal requirements and standards. A well-implemented data governance framework simplifies the process of identifying, preserving, and producing relevant data, reducing the risk of legal and regulatory sanctions.

15.7 Cybersecurity Laws and Regulations

Cybersecurity laws and regulations are legal frameworks that govern and enforce the protection of digital assets, information systems, and data from cyber threats, attacks, and breaches. These laws are essential for promoting cybersecurity practices, ensuring the privacy and security of sensitive data, and holding organizations

15.7 Cybersecurity Laws and Regulations

accountable for safeguarding their digital assets. Here is what you should know about cybersecurity laws and regulations:

Purpose Cybersecurity laws and regulations aim to:

- Protect sensitive data: Laws require organizations to implement measures to safeguard personal and confidential information.
- Promote best practices: They establish standards and guidelines for cybersecurity best practices.
- Prevent cybercrimes: Laws define cybercrimes such as hacking, data breaches, and identity theft and prescribe penalties for offenders.
- Ensure data breach notification: Some laws require organizations to notify affected individuals and authorities in the event of a data breach.
- Regulate critical infrastructure: Certain regulations focus on securing critical infrastructure sectors like energy, finance, and healthcare.

Key Elements of Cybersecurity Laws and Regulations
- **Data Protection**: Laws often include provisions for protecting sensitive data such as personally identifiable information (PII) and financial data (U.S. Congress 1996; California Legislature 2018).
- **Data Breach Notification**: Some regulations require organizations to notify affected individuals and relevant authorities in the event of a data breach (U.S. Congress 2015).
- **Security Standards**: Laws establish cybersecurity standards and requirements that organizations must follow, including encryption, access controls, and security audits (International Organization for Standardization 2013).
- **Compliance and Auditing**: Organizations may be required to undergo cybersecurity audits and demonstrate compliance with legal requirements.
- **Penalties**: Legal frameworks typically outline penalties for noncompliance, which may include fines, legal actions, and sanctions.
- **Cross-Border Data Transfer**: Regulations address the transfer of data across borders, imposing restrictions on cross-border data flows to protect data privacy.
- **Regulation of Critical Infrastructure**: Some laws focus on securing critical infrastructure sectors such as energy, transportation, and healthcare due to their importance to national security (European Union 2016b).
- **International Considerations**: Laws may have international implications, especially in the context of data transfer and the extraterritorial reach of certain regulations.

Examples of Cybersecurity Laws and Regulations
- **General Data Protection Regulation (GDPR)**: Applies to the European Union (EU) and regulates the processing of personal data, including stringent data protection requirements and data breach notification obligations (European Union 2016b).
- **California Consumer Privacy Act (CCPA)**: Provides California residents with rights regarding their personal information and requires businesses to disclose data practices and provide opt-out options (California Legislature 2018).

- **Health Insurance Portability and Accountability Act (HIPAA)**: Regulates the protection of health information in the United States, requiring healthcare organizations to implement security measures (U.S. Congress 1996).
- **Cybersecurity Information Sharing Act (CISA)**: Encourages the sharing of cybersecurity threat information between the government and private sector in the United States (U.S. Congress 2015).
- **NIS Directive**: Applies to EU member states and establishes security requirements for operators of essential services and digital service providers (European Union 2016b).

Enforcement Enforcement of cybersecurity laws and regulations may involve government agencies, regulatory bodies, and legal authorities. Non-compliance can result in fines, legal actions, and reputational damage.

Evolution Cybersecurity laws are continuously evolving to keep pace with emerging cyber threats and technologies. New regulations may be enacted to address specific issues such as the security of Internet of Things (IoT) devices or critical infrastructure protection.

Global Impact Many cybersecurity laws and regulations have a global impact because they apply to organizations that handle data or conduct business internationally.

Figure 15.5, titled "Cybersecurity laws and regulations," visually encapsulates the comprehensive framework of elements that make up the regulatory landscape surrounding cybersecurity. This map centrally features the overarching category of Cybersecurity laws and regulations, surrounded by various branches that delve into different aspects of cybersecurity governance. The core of the diagram stresses the

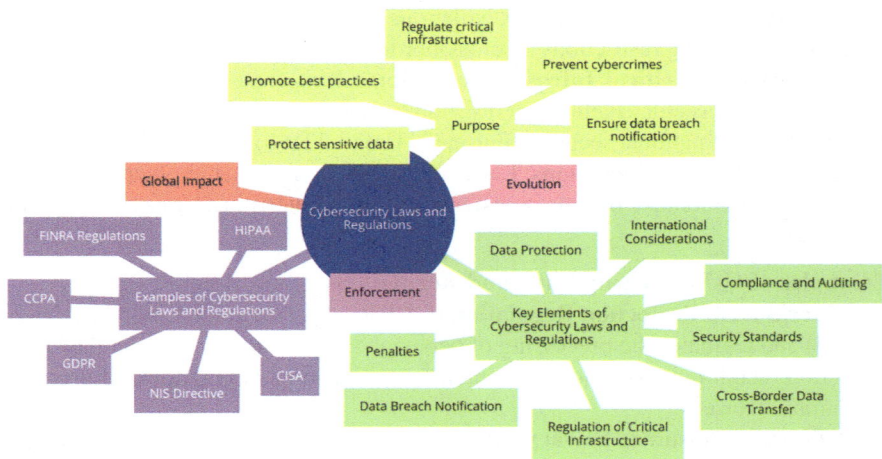

Fig. 15.5 Cybersecurity laws and regulations map. (Created by the author)

Purpose of these laws and regulations, which includes promoting best practices, protecting sensitive data, preventing cybercrimes, ensuring data breach notifications, and regulating critical infrastructure. These goals underscore the intent behind cybersecurity regulations: to safeguard data and systems from unauthorized access and attacks while ensuring that organizations have protocols in place for when breaches occur. Branching out from the purpose are areas like Enforcement, which includes penalties for noncompliance and mechanisms for data breach notification, highlighting the actions taken to ensure adherence to laws. This part of the diagram connects to various specific examples of cybersecurity laws and regulations like HIPAA, which protects health information; CCPA and GDPR, which govern consumer data privacy; and the NIS Directive along with CISA, which focuses on improving national cybersecurity and protecting infrastructure. Further branching explores the Evolution of these regulations, pointing to the dynamic nature of cybersecurity laws as they adapt to new challenges posed by technological advancements and increasingly sophisticated cyber threats. This segment ties into International Considerations and Compliance and Auditing, indicating the global impact and the need for consistent application and verification of security measures across borders. Security Standards and Regulation of Critical Infrastructure are also depicted as foundational components, addressing the specifics of what is required from organizations to safeguard against threats and the special considerations needed for the systems critical to national and economic security. Overall, this map not only outlines the specific areas covered by cybersecurity laws but also interconnects these aspects to demonstrate the comprehensive and interconnected approach required to effectively manage and enforce cybersecurity in the modern digital world.

Cybersecurity laws and regulations are crucial for promoting cybersecurity, protecting sensitive data, and holding organizations accountable for their cybersecurity practices. They vary by district and industry, but their common goal is to enhance cybersecurity measures, mitigate cyber risks, and ensure data privacy and security. Organizations must stay informed about relevant laws and comply with their requirements to mitigate legal and financial risks.

15.8 Case Studies: Legal and Regulatory Compliance

Real-world case studies in legal and regulatory compliance cover a wide range of topics, from data protection and privacy laws to cybersecurity and industry-specific regulations. Here are some key examples:

General Data Protection Regulation (GDPR) Enacted in the European Union in May 2018, GDPR is a privacy law that sets standards for storing user or customer details online, such as IP addresses, device IDs, and home addresses. It applies to all companies doing business in Europe, including those not based in the EU. GDPR is not static; it undergoes updates to address evolving issues, like the clarification on cookie walls in 2020. Companies need to continually monitor GDPR updates to remain compliant, as even accidental oversights can result in substantial fines (European Union 2016a).

California Consumer Privacy Act (CCPA) Effective from January 2020, the CCPA allows consumers to understand and control the data collected about them. This law applies to commercial companies meeting certain criteria, like processing data of over 50,000 California residents annually or generating more than $25 million in yearly revenue. CCPA also undergoes periodic updates, such as the data minimization regulations introduced in 2022, which require private consumer data to be collected, used, and shared only as necessary and proportionate to the purpose for which it was collected (California Legislature 2018).

Payment Card Industry Data Security Standard (PCI DSS) This is a regulatory compliance framework for online merchants to ensure secure processing and storage of credit and debit card numbers. It includes guidelines like maintaining a working security system, encrypting transmitted data, and conducting regular process and system testing. While not a law, PCI DSS compliance is often required by payment and merchant service providers, and noncompliance can lead to significant fines (Payment Card Industry Security Standards Council 2018).

Health Insurance Portability and Accountability Act (HIPAA) HIPAA is a compliance framework designed to protect data in healthcare and medical settings, especially electronically stored data. Businesses in the healthcare sector must have robust data protection protocols and comply with IT security guidelines to protect patient information (U.S. Congress 1996).

Sarbanes-Oxley Act (SOX) Introduced in 2002, SOX aims to prevent financial fraud by corporations by strengthening transparency in corporate accounting and implementing new regulatory and compliance requirements. It includes standards for external auditors and mandates that all public businesses store records and messages for at least 5 years (U.S. Congress 2002).

Case Studies by Deloitte Deloitte provides insights on the challenges and principles of regulating emerging technologies like AI, big data analytics, and the Internet of Things. These case studies highlight the balance regulators must strike between fostering innovation and enforcing regulations to protect consumers and ensure fair markets (Deloitte 2020).

Compliance Case Studies by Kroll Kroll's Financial Services Compliance and Regulation team conducts diverse types of compliance reviews and support for firms globally. This includes health checks, deep-dive thematic assessments, and regulatory reviews. One example is the redesigning of a Compliance Monitoring Program for a private-equity-backed trust and corporate service provider, which involved creating a risk-focused and easy-to-use program for clear reporting and management (Kroll 2021).

These examples illustrate the diverse and dynamic nature of legal and regulatory compliance in various industries and regions. They underscore the importance of staying informed and adapting to regulatory changes to ensure compliance and avoid penalties.

15.9 Future Legal and Regulatory Trends

Future legal and regulatory trends in data governance, particularly for 2024 and beyond, indicate several evolving focuses. Key trends identified include:

Shift Left in Data Governance Organizations are expected to implement data governance and security measures earlier in the data journey. This "Shift Left" approach aims at improving data quality and compliance with increasing regulations regarding data privacy and security, thereby mitigating risks of data exposure before it reaches cloud data warehouses (Eryurek et al. 2021).

AI and Machine Learning in Data Management The integration of AI and machine learning in data governance is becoming increasingly prominent. These technologies automate data processing tasks, support predictive analytics, provide personalized user experiences, and enable scalability in handling large data sets. AI and machine learning are not only enhancing data management processes but also introducing new challenges in data security and ethical use (Lee and Darbellay 2021).

Data Privacy and Compliance with Emerging Regulations Adhering to a variety of data privacy laws such as GDPR, CCPA, and HIPAA remains a complex challenge. Organizations need to stay vigilant about regulatory changes and update their policies accordingly. This includes data classification, risk management, and ensuring consumer rights are protected (European Union 2016a; California Legislature 2018; U.S. Congress 1996).

Data Quality Management High-quality data is essential for reliable and useful insights. Organizations focus on data validation, enrichment, monitoring, and collaboration between business and IT to maintain high data quality standards (Eryurek et al. 2021).

Data Lineage and Metadata Management Understanding data flow and managing metadata are crucial for data transparency and usability. Tools for visualizing data lineage and automated metadata collection are becoming more important in this context (Eryurek et al. 2021).

Integration of Cloud-based Solutions The adoption of cloud computing in data governance offers scalability, flexibility, collaboration, security, and cost-effectiveness. Cloud solutions are increasingly being used to manage and leverage data assets effectively (Lee and Darbellay 2021).

Decentralized Data Governance Models There is a trend towards decentralizing data governance, giving more control and responsibility to individual departments. This approach aligns governance practices with specific business objectives and encourages cross-functional collaboration (Data Governance in AI, FinTech and LegalTech 2021).

Real-time Data Governance The capability for real-time governance allows organizations to respond immediately to changes, enhancing agility and competitiveness. This includes continuous monitoring of data quality, security, compliance, and the integration of governance tools with operational systems (Eryurek et al. 2021).

Data as a Service (DaaS) DaaS centralizes data management while enhancing accessibility and customization of data access. This approach helps in maintaining consistent data quality and managing costs more efficiently (Eryurek et al. 2021).

Increased Focus on Data Ethics Ethical considerations in data usage are gaining importance. Organizations are developing ethical guidelines and policies, focusing on transparency and accountability to build trust and align with societal norms and expectations (Lee and Darbellay 2021).

These trends indicate a dynamic future for data governance, where technological advancements, regulatory compliance, and ethical considerations will play significant roles in shaping strategies and practices.

15.10 Conclusion

This chapter, "Legal and Regulatory Considerations in Data Governance," comprehensively addresses the multifaceted aspects of legal and regulatory frameworks that influence data governance. The chapter encompasses various themes, each critical to understanding and implementing effective data governance in today's complex and evolving digital landscape:

1. **Introduction to Legal and Regulatory Aspects**: This section sets the stage for understanding the importance of legal and regulatory considerations in data governance, emphasizing the need for alignment with legal requirements and regulatory standards to ensure data is managed responsibly and ethically.
2. **Global Data Protection and Privacy Laws**: It delves into various international laws like GDPR and CCPA, highlighting the need for organizations to comply with diverse and sometimes conflicting privacy regulations across different districts.

3. **Navigating Industry-Specific Regulations**: This part focuses on the unique regulatory challenges and requirements in different industries, stressing the importance of tailored data governance strategies to meet specific sectoral needs.
4. **Cross-Border Data Transfer and Compliance**: This section deals with the complexities of data transfer across borders, underscoring the challenges in complying with different regional laws and the importance of securing data in transit and at rest.
5. **Developing a Compliance-Oriented Data Governance Framework**: It outlines strategies for building a data governance framework that not only complies with legal requirements but also embeds compliance into the fabric of organizational data management practices.
6. **Data Governance in the Context of Legal Discovery and Audits**: This part emphasizes the role of data governance in legal contexts, such as in responding to audits and legal discovery requests, highlighting the need for efficient data retrieval and management processes.
7. **Cybersecurity Laws and Regulations**: The chapter discusses the critical role of cybersecurity laws in data governance, stressing the need for robust security measures to protect data against breaches and cyber threats.
8. **Case Studies: Legal and Regulatory Compliance**: Real-world examples illustrate the practical application of legal and regulatory principles in data governance, providing insights into how organizations navigate these complex requirements.
9. **Future Legal and Regulatory Trends**: It looks forward to emerging trends in data governance, such as the increasing role of AI and machine learning, the evolution of privacy laws, and the shift towards decentralized and real-time data governance models.

This chapter offers a thorough exploration of the legal and regulatory landscape of data governance. It underscores the necessity for organizations to stay informed and agile in their data governance approaches, adapting to ongoing changes in legal, regulatory, and technological realms. The chapter serves as a vital guide for organizations seeking to navigate the intricate web of data governance requirements while harnessing data as a strategic asset.

References

California Legislature. (2018). California Consumer Privacy Act (CCPA). California Civil Code, Title 1.81.5

Data Governance in AI, FinTech and LegalTech (2021) Law and regulation in the financial sector by Joseph Lee (Editor). Aline Darbellay

Deloitte (2020) Case Studies on Regulatory Compliance

Eryurek E et al (2021) Data governance: the definitive guide: people, processes, and tools to operationalize data trustworthiness. O'Reilly Media

European Commission (2004) Standard Contractual Clauses (SCCs) for the transfer of personal data to third countries. Off J Eur Union L 119/89

European Union (2016a) General Data Protection Regulation (GDPR). Off J Eur Union L 119/1

European Union (2016b) Directive (EU) 2016/1148 concerning measures for a high common level of security of network and information systems across the Union (NIS Directive). Off J Eur Union L 194/1

Federal Rules of Civil Procedure (2015) Rule 26: Duty to Disclose; General Provisions Governing Discovery

International Organization for Standardization (2013) ISO/IEC 27001: information security management. ISO

Kroll (2021) Compliance monitoring program design

Lee J, Darbellay A (eds) (2021) Data Governance in AI, FinTech and LegalTech: law and regulation in the financial sector. Edward Elgar Publishing

North American Electric Reliability Corporation (2016) Critical Infrastructure Protection (CIP) Standards

Payment Card Industry Security Standards Council (2018) Payment Card Industry Data Security Standard (PCI DSS)

The Sedona Conference (2018) The Sedona Principles: best practices, recommendations & principles for addressing electronic document production

U.S. Congress (1996) Health Insurance Portability and Accountability Act (HIPAA). Public Law 104–191

U.S. Congress (2002) Sarbanes-Oxley Act of 2002 (SOX). Public Law 107–204

U.S. Congress (2010) Dodd-Frank Wall Street Reform and Consumer Protection Act. Public Law 111–203

U.S. Congress (2015) Cybersecurity Information Sharing Act (CISA). Public Law 114–113

U.S. Department of Health and Human Services (2000) HIPAA Privacy Rule

U.S. Federal Trade Commission (2016) EU-U.S. Privacy Shield Framework

U.S. Food and Drug Administration (1997) Title 21 CFR Part 11: Electronic Records; Electronic Signatures

United Nations (1990) Guidelines for the regulation of computerized personal data files

Chapter 16
Data Stewardship and Ownership: Best Practices

Contents

16.1	Introduction to Data Stewardship and Ownership.	468
16.2	Defining Roles and Responsibilities.	470
16.3	Best Practices in Data Stewardship.	471
16.4	Establishing Data Ownership.	472
16.5	Collaboration Between Stewards and Owners.	474
16.6	Training and Empowerment.	476
16.7	Accountability and Performance Measurement.	478
16.8	Case Studies: Effective Stewardship and Ownership.	480
16.9	Overcoming Common Challenges.	482
16.10	Conclusion.	484
References.		485

Abstract This chapter provides a comprehensive exploration of the roles and responsibilities crucial to effective data management within organizations. It emphasizes the importance of clearly defined roles, accountability, and collaboration in fostering robust data governance frameworks.

The chapter begins by defining data stewardship and ownership, explaining that stewards act as custodians of data, ensuring its accuracy, accessibility, and alignment with business goals and regulatory requirements. Data owners, usually senior personnel, hold ultimate accountability for data assets and make critical decisions regarding their use and protection.

Key best practices discussed include the clear definition of roles to prevent overlap and confusion, and the establishment of accountability mechanisms to ensure data is managed according to both internal standards and external regulations. The chapter stresses the importance of strategic involvement from data stewards, who should not only manage but also strategize on data utilization for organizational benefits.

Additionally, the chapter addresses the importance of quality and rigor in data management, advocating for regular audits and user feedback to maintain high data standards. Collaboration and transparency are highlighted as essential for effective data stewardship, necessitating open lines of communication across departments and clear data access protocols.

The cultural aspect of data governance is also discussed, with a focus on aligning data stewardship practices with organizational culture to facilitate smooth implementation and acceptance of data policies. Community building within and outside the organization is recommended to foster a supportive network for data stewards, promoting continuous learning and professional development.

In summary, this chapter outlines the foundational elements of effective data stewardship and ownership, providing guidance on implementing practices that enhance data governance. It underscores the need for strategic planning, quality management, and a collaborative and culturally aligned approach to ensure that data serves as a strategic asset for the organization.

Keywords Data Stewardship · Data Ownership · Best Practices · Data Management · Accountability · Data Quality · Collaboration · Strategic Planning · Role Definition · Training · Empowerment · Compliance · Performance Measurement · Data Security · Organizational Culture · Stakeholder Engagement

16.1 Introduction to Data Stewardship and Ownership

Data Stewardship and Ownership are critical components in effective data governance, playing a vital role in ensuring that an organization's data is managed responsibly and efficiently. Best practices in this area involve a combination of strategic, operational, and collaborative approaches:

Defining Roles and Responsibilities Data stewards act as central authorities on business data, collaborating across functions to ensure data is fit for all purposes within the organization. Their responsibilities include documenting data meaning, stipulating data quality expectations, and setting rules for data usage and management (Plotkin 2014; Redman 2008).

Emphasizing Accountability Data stewards are accountable for ensuring data is managed according to both internal and external rules, including industry guidelines and various regulations like GDPR and CCPA. They should also promote safe and creative uses of data, acting as data ambassadors within the organization (Plotkin 2014; Loshin 2012).

16.1 Introduction to Data Stewardship and Ownership

Strategic Involvement Data stewards should engage in high-level strategic thinking, looking for innovative ways data can serve the organization. This includes maintaining data relevance for decision-making and resolving the root causes of data-related issues (Dyché 2015).

Quality and Rigor Maintaining high standards of data accuracy and quality is essential. Regular audits and feedback from users help ensure these standards, with data stewards playing a key role in identifying, escalating, and fixing data quality issues (Plotkin 2014; Otto 2021).

Collaboration and Transparency Effective data stewardship requires collaboration with users and other stakeholders. Data stewards should facilitate easy access to information about data and its proper use, holding educational sessions and being available for communication (Redman 2008).

Respecting Organizational Culture Data stewards often work to change the way people do their jobs, so considering corporate culture across the data governance spectrum is crucial. This means understanding and aligning with the organization's values, beliefs, and norms (Dyché 2015).

Community Building Forming a data stewardship community both internally and externally is beneficial. Internally, it allows data stewards to collaborate on data standards, policies, and best practices. Externally, it enables them to view their roles in a broader context and engage in professional development opportunities (Plotkin 2014; Otto 2021).

Figure 16.1, illustrates the multifaceted aspects involved in managing and overseeing data within an organization. At the center of the diagram is the primary concept of data stewardship and ownership, emphasizing the critical role of accountability and proper management of data assets. The diagram branches out to highlight the formation of a data stewardship community, focusing on community building and the importance of forming groups dedicated to managing data. This includes defining roles and responsibilities, which is essential for clarity and effective data governance. Setting rules for data usage ensures that data is handled consistently and appropriately, supported by documenting data meaning and quality, which aids in maintaining high standards. Emphasizing accountability is a key aspect, ensuring that those responsible for data are clearly identified and held to account for managing data accurately and effectively. This extends to quality and rigor, which involves regular audits and user feedback to uphold high standards of data accuracy. Collaboration and transparency are vital components, promoting internal and external collaboration, and facilitating access to data information. This includes working closely with users and stakeholders to ensure data is used effectively and safely, reflecting the organizational culture and aligning with the organization's values. Strategic involvement is another critical area, which involves

Fig. 16.1 Data stewardship and ownership map. (Created by the author)

maintaining data relevance through high-level strategic thinking and ensuring data management practices align with broader organizational goals. Respecting organizational culture means considering corporate culture in data practices, which is crucial for ensuring that data governance aligns with the overall mission and values of the organization. Overall, this map visually represents the comprehensive approach required for effective data stewardship and ownership, highlighting the interconnected roles, responsibilities, and practices necessary to manage data assets responsibly and strategically within an organization.

Data stewardship and ownership are about more than just managing data; they encompass a comprehensive approach that integrates strategic planning, quality management, collaborative practices, and cultural alignment to ensure that data serves as an asset for the organization.

16.2 Defining Roles and Responsibilities

Defining roles and responsibilities in the context of data stewardship and ownership involves establishing clear and specific duties for individuals who manage and govern data within an organization. This includes:

- **Data Stewards**: Responsible for overseeing the quality, accessibility, and lifecycle of data. They ensure data meets organizational standards and compliance requirements, and they often act as a liaison between IT and business units (Plotkin 2014; Loshin 2012).
- **Data Owners**: Typically senior-level personnel who have ultimate accountability for data assets. They make decisions about data access, sharing, and security, ensuring that data is used effectively and responsibly in alignment with business goals and legal requirements (Ladley 2012; DAMA International 2017).

These roles are crucial in maintaining the integrity, security, and usability of an organization's data, thus supporting informed decision-making and strategic planning. Effective communication and collaboration between data stewards and owners are essential for successful data governance.

16.3 Best Practices in Data Stewardship

Best Practices in Data Stewardship are essential for ensuring the effective management and use of data within organizations. Here are some key practices:

1. **Define Clear Roles and Responsibilities**: Assign specific duties to data stewards, ensuring they know what is expected of them. This includes overseeing data quality, accessibility, and compliance with data policies and regulations (Plotkin 2014; DAMA International 2017).
2. **Promote Data Stewardship Training and Awareness**: Ensure that data stewards are well-trained and understand the significance of their role. This involves regular training on data management practices, data governance policies, and the latest trends in data technology (Loshin 2012).
3. **Implement Data Governance Frameworks**: Data stewards should operate within a structured data governance framework. This framework should clearly outline data standards, policies, and procedures, ensuring a consistent approach to data management across the organization (Ladley 2012).
4. **Focus on Data Quality**: Data stewards should prioritize the accuracy, completeness, and reliability of data. Implement processes for regular data quality checks and establish protocols for correcting data issues (Plotkin 2014; Otto 2021).
5. **Collaboration and Communication**: Encourage data stewards to work closely with other departments, including IT, legal, and business units, to ensure that data management aligns with organizational goals and complies with legal standards (Loshin 2012).
6. **Manage Data Security and Privacy**: Data stewards should be involved in ensuring that data is securely stored and handled, complying with relevant data protection laws such as GDPR and HIPAA (Plotkin 2014).
7. **Use Technology to Enhance Data Stewardship**: Leverage data management tools and technologies to streamline data stewardship tasks. This includes using data cataloging tools, metadata management systems, and data quality software (DAMA International 2017).
8. **Regular Review and Improvement**: Continuously assess and improve data stewardship practices. This could involve regular audits of data management activities and updating policies as needed based on evolving data governance needs (Plotkin 2014; Otto 2021).

Fig. 16.2 Best practices in data stewardship. (Created by the author)

9. **Encourage a Data-Driven Culture**: Data stewards should advocate for a data-driven culture within the organization, where data is recognized as an asset and used to inform decision-making (Loshin 2012).
10. **Support Data Compliance and Ethical Use**: Ensure that data stewards are equipped to navigate the complex landscape of data compliance and ethics. This involves understanding the legal implications of data use and promoting ethical data practices (Ladley 2012; Otto 2021).

The Fig. 16.2, outlines essential best practices in data stewardship, emphasizing the need for clear roles, continuous training, and a robust governance framework to manage data effectively within organizations. Key practices include assigning specific responsibilities to data stewards, promoting awareness and training, implementing governance frameworks, focusing on data quality, fostering collaboration, ensuring data security and privacy, utilizing technology, regularly reviewing, and improving practices, advocating for a data-driven culture, and supporting compliance and ethical data use. These practices are crucial for the effective management and utilization of data, ensuring that it serves as an asset in informed decision-making and aligns with organizational goals and legal standards. Implementing these best practices can significantly enhance the integrity, accessibility, and security of data, thereby supporting the organization's overall success and compliance efforts.

16.4 Establishing Data Ownership

Establishing data ownership is a critical aspect of data governance and management in any organization. It involves identifying and assigning responsibility to specific individuals or departments for the control and management of data assets. Key aspects of establishing data ownership include:

16.4 Establishing Data Ownership

1. **Identification of Data Owners**: Data owners are usually identified based on their role, expertise, and authority within the organization. They are typically senior-level executives or managers who have the authority to make decisions about the use and management of specific data sets (Ladley 2012; DAMA International 2017).
2. **Defining Ownership Roles**: Clearly defining the roles and responsibilities of data owners is crucial. This includes decision-making authority regarding access to data, data sharing, data quality, and compliance with relevant laws and regulations (Plotkin 2014; Otto 2021).
3. **Accountability and Responsibility**: Data owners are accountable for the data's accuracy, privacy, security, and compliance with regulatory requirements. They are responsible for establishing data policies, standards, and procedures, and for ensuring that these are followed across the organization (Redman 2008).
4. **Collaboration with Data Stewards**: Data owners often work closely with data stewards who are responsible for the day-to-day management and maintenance of data. Data stewards support the data owners by implementing policies, managing data quality, and providing access to data (Plotkin 2014; DAMA International 2017).
5. **Data Lifecycle Management**: Data owners oversee the entire lifecycle of their data assets from creation and acquisition to archiving and disposal. This includes understanding how data is used, who uses it, and how it is maintained and protected throughout its lifecycle (Ladley 2012; Otto 2021).
6. **Training and Communication**: Ensuring that data owners and all relevant staff are trained in data governance policies and practices is essential. Effective communication about data ownership roles and responsibilities throughout the organization is also crucial (Plotkin 2014).
7. **Aligning with Organizational Goals**: Data ownership should align with the overall business strategy and goals of the organization. This alignment ensures that data management practices support business objectives and add value to the organization (Ladley 2012).
8. **Legal and Regulatory Compliance**: Data owners must ensure that data is managed in compliance with relevant laws, regulations, and industry standards. This is particularly important for sensitive or personal data, where compliance is not just best practice but a legal requirement (Redman 2008).

Figure 16.3, provides a visual representation of the key components necessary for defining and implementing data ownership within an organization. At the center is the core concept of establishing data ownership, surrounded by various critical elements that contribute to this objective. The map includes the Identification of Data Owners, which is crucial for clearly assigning responsibility for specific data assets. Defining Ownership Roles follows, ensuring that the responsibilities and expectations for data owners are clearly articulated. Accountability and Responsibility highlight the importance of holding data owners accountable for the management and security of their data. This accountability is supported by Training and Communication, which emphasizes the need for ongoing education and clear

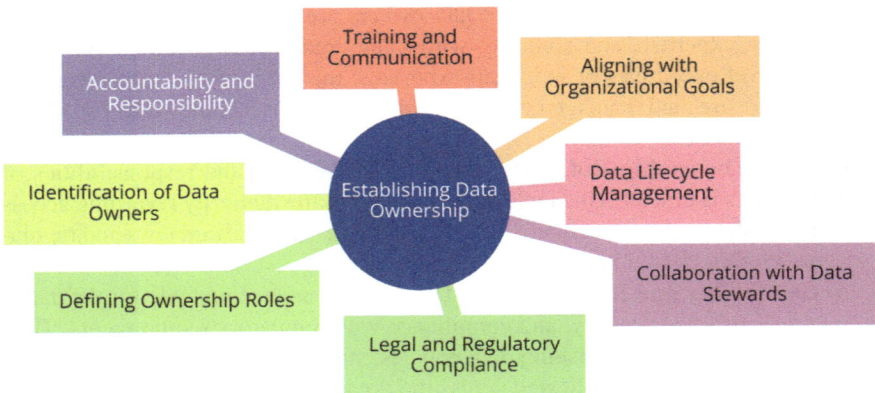

Fig. 16.3 Establishing data ownership map. (Created by the author)

communication regarding data ownership roles and responsibilities. Aligning with Organizational Goals ensures that data ownership practices are in sync with the broader strategic objectives of the organization. Data Lifecycle Management is included to stress the importance of overseeing data throughout its entire lifecycle, from creation to deletion. Collaboration with Data Stewards points to the necessity of data owners working closely with data stewards to ensure data integrity and compliance. Legal and Regulatory Compliance underscores the requirement for data ownership practices to meet all relevant legal and regulatory standards. Together, these elements form a comprehensive framework for establishing data ownership, ensuring that data is managed responsibly, aligned with organizational goals, and compliant with legal standards. This integrated approach is critical for effective data governance and the protection of organizational data assets.

Establishing data ownership involves a structured approach to assigning responsibility and accountability for data within an organization, ensuring that data is managed effectively, securely, and in compliance with legal and regulatory requirements.

16.5 Collaboration Between Stewards and Owners

Collaboration between data stewards and data owners is a fundamental component of effective data governance. This collaboration is essential for ensuring that data is managed efficiently and aligned with the organization's strategic goals. Here are key aspects of this collaboration:

1. **Clear Role Definition and Communication**: Both parties need a clear understanding of their roles. Data owners typically have decision-making authority over data, while data stewards handle the day-to-day management. Effective

16.5 Collaboration Between Stewards and Owners

communication channels between them are vital to ensure a shared understanding of responsibilities and objectives (Ladley 2012; Plotkin 2014).
2. **Joint Strategy Development**: Stewards and owners should collaborate on developing data strategies, ensuring that these strategies align with broader business goals and data governance policies (DAMA International 2017).
3. **Policy Development and Implementation**: Data owners and stewards should work together to develop and implement data-related policies, including data quality, privacy, and security policies (Otto 2021).
4. **Data Quality Management**: Data stewards often lead efforts in data quality management, but they need to work closely with data owners to define quality metrics and standards that align with business needs (Plotkin 2014).
5. **Compliance and Risk Management**: Both groups need to collaborate on compliance issues, ensuring that data practices adhere to regulatory requirements. Data stewards can help data owners understand the implications of various regulations and assist in implementing compliant processes (Redman 2008).
6. **Change Management**: When changes are made to data governance policies or practices, stewards and owners should collaborate to manage these changes, communicating impacts and new procedures across the organization (Ladley 2012).
7. **Training and Advocacy**: Data owners and stewards can jointly advocate for the importance of data governance and provide training and resources to other employees, fostering a data-aware culture within the organization (DAMA International 2017).
8. **Problem-Solving and Decision Making**: Data stewards often identify issues in data management, but effective resolution usually requires involvement and decisions from data owners, particularly when issues have broader business implications (Plotkin 2014).
9. **Feedback Loops**: Regular meetings and feedback loops between data owners and stewards help in continuously improving data governance practices and addressing any emerging issues promptly (Otto 2021).
10. **Technology and Tools Utilization**: Collaboration on the selection and use of data management tools and technologies ensures these tools are aligned with the organization's data governance strategy and objectives (Redman 2008).

Figure 16.4, titled "Collaboration Between Stewards and Owners," highlights the essential components required for effective teamwork between data stewards and data owners within an organization. At the center of the diagram is the key concept of collaboration between stewards and owners, from which various critical elements branch out. The diagram emphasizes Clear Role Definition and Communication, ensuring that both stewards and owners understand their responsibilities and can communicate effectively. Training and Advocacy are crucial for equipping these roles with the necessary skills and promoting the importance of their collaboration. Problem-Solving and Decision Making is a vital aspect, as it highlights the need for collaborative approaches to address challenges and make informed decisions regarding data management. Compliance and Risk Management underline the

Fig. 16.4 Collaboration between stewards and owners map. (Created by the author)

importance of ensuring that data practices meet regulatory requirements and mitigate risks. Feedback Loops are necessary to maintain continuous improvement and adaptation in data governance practices. Data Quality Management is highlighted to ensure that collaboration efforts maintain high standards of data accuracy and reliability. Policy Development and Implementation stress the need for jointly developed policies that guide data management practices, while Change Management addresses the need to manage organizational changes effectively that impact data governance. Technology and Tools Utilization underscores the importance of leveraging appropriate tools and technologies to facilitate collaboration and improve data management processes. Joint Strategy Development is emphasized to ensure that stewards and owners work together towards common goals and strategic objectives. This map visually represents the interconnected elements that contribute to successful collaboration between data stewards and owners, ensuring effective data governance and the achievement of organizational data management goals.

Effective collaboration between data stewards and owners is crucial for the successful governance of data, ensuring that it is managed as an asset that supports the organization's objectives and compliance requirements.

16.6 Training and Empowerment

Training and empowerment in the context of data stewardship and data governance are crucial for fostering a data-centric culture within an organization. This involves equipping individuals with the necessary knowledge, skills, and authority to effectively manage and utilize data. Key elements include:

1. **Comprehensive Training Programs**: Implementing training programs that cover the fundamentals of data governance, data quality, data security, and relevant regulations (like GDPR or HIPAA) is essential. These programs should be

16.6 Training and Empowerment

tailored to distinct roles within the organization, including data stewards, data owners, and other employees who handle data (Plotkin 2014; DAMA International 2017).

2. **Continuous Learning and Development**: The data landscape is constantly evolving, so ongoing education and professional development opportunities are crucial. This could include workshops, webinars, conferences, and online courses to keep staff updated on the latest trends and best practices in data management (Loshin 2012).
3. **Empowering Data Stewards and Owners**: Empowerment involves granting data stewards and owners the authority and resources they need to make decisions and implement changes related to data governance. This includes giving them a say in setting policies and involving them in strategic discussions about data (Ladley 2012; Plotkin 2014).
4. **Encouraging a Data-Driven Culture**: Training and empowerment should aim to cultivate a data-driven culture within the organization. Employees at all levels should be encouraged to use data in their decision-making processes and understand the value of high-quality data (Redman 2008).
5. **Tools and Technology Proficiency**: Ensuring that staff are proficient in using data governance tools and technologies is vital. Training should include practical aspects of using these tools effectively for data management, quality control, and compliance (DAMA International 2017).
6. **Soft Skills Development**: Besides technical skills, soft skills like communication, problem-solving, and collaboration are vital for data stewards and owners. These skills help in effectively communicating data governance policies and collaborating with different departments (Plotkin 2014).
7. **Feedback Mechanisms**: Implement feedback loops where employees can share their experiences, challenges, and suggestions regarding data governance. This feedback can help in refining training programs and governance strategies (Otto 2021).
8. **Recognition and Incentives**: Recognizing and rewarding employees who effectively implement data governance practices can motivate others. Incentives can include recognition in meetings, awards, or career advancement opportunities (Loshin 2012).

Figure 16.5, titled "Training and Empowerment," provides a comprehensive overview of the essential elements required to effectively train and empower data stewards and owners within an organization. At the center of the diagram is the core concept of training and empowerment, with various critical components branching out to support this initiative. Empowering Data Stewards and Owners is highlighted as the central objective, ensuring these roles have the knowledge and authority to manage data effectively. Recognition and Incentives emphasize the importance of acknowledging and rewarding individuals' contributions to data governance, thereby motivating continued excellence and engagement. Encouraging a Data-Driven Culture underscores the need to foster an organizational environment that values and utilizes data for decision-making. Comprehensive Training Programs are

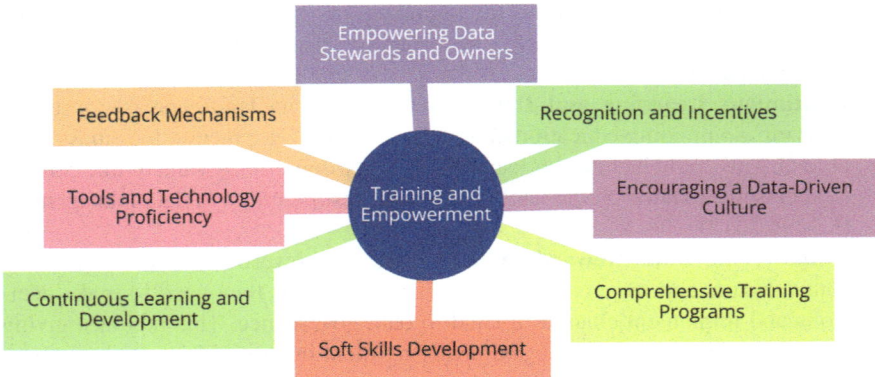

Fig. 16.5 Training and empowerment map. (Created by the author)

essential for providing structured and thorough education on data governance principles, tools, and practices. Feedback Mechanisms ensure continuous improvement by allowing data stewards and owners to receive and act on constructive feedback. Tools and Technology Proficiency highlights the importance of training in the specific tools and technologies used for data management, ensuring that stewards and owners can effectively utilize these resources. Continuous Learning and Development stress the importance of ongoing education to keep up with the evolving landscape of data governance. Soft Skills Development is also crucial, as it includes training in areas such as communication, teamwork, and problem-solving, which are essential for effective collaboration and data management. Together, these elements create a robust framework for training and empowering data stewards and owners, ensuring they are well-equipped to manage data assets effectively and contribute to the organization's data governance goals. This map visually ties these various components together, illustrating how they collectively support a comprehensive approach to training and empowerment in data governance.

Training and empowerment in data governance are about equipping individuals with the skills and authority they need to effectively manage data, while also fostering an organizational culture that values and utilizes data for strategic decision-making.

16.7 Accountability and Performance Measurement

Accountability and performance measurement are vital components of data governance, ensuring that data management activities align with organizational objectives and comply with relevant standards and regulations. Key aspects include:

1. **Defining Clear Accountability:** This involves assigning specific individuals (data stewards, data owners) clear responsibilities for data quality, security, com-

16.7 Accountability and Performance Measurement

pliance, and usage. Each role should have defined expectations and responsibilities to ensure that everyone knows who is accountable for various aspects of data management (Plotkin 2014; DAMA International 2017).

2. **Performance Metrics and KPIs:** Establishing key performance indicators (KPIs) and metrics to measure the effectiveness of data governance is crucial. These metrics might include data quality scores, compliance levels, the number of data incidents, or the speed of data processing. They help in assessing how well the data governance program is performing against its goals (Ladley 2012; Redman 2008).
3. **Regular Monitoring and Reporting:** Implementing a process for regular monitoring of these metrics and reporting on them to senior management ensures ongoing oversight of the data governance program (Plotkin 2014; Loshin 2012).
4. **Continuous Improvement:** Using the insights gained from performance measurement, organizations should continually refine and improve their data governance processes. This might involve adjusting policies, providing additional training, or implementing innovative technologies (Otto 2021).
5. **Transparency in Reporting:** Transparency in reporting on data governance performance helps build trust within the organization and with external stakeholders. It demonstrates a commitment to effective data management and regulatory compliance (Plotkin 2014; DAMA International 2017).
6. **Linking to Business Outcomes:** Ideally, performance metrics should be linked to broader business outcomes to demonstrate the value of data governance in supporting organizational objectives (Redman 2008).
7. **Feedback Mechanisms:** Establishing channels for feedback from various stakeholders, including data users, IT staff, and business unit leaders, can provide valuable insights into the effectiveness of data governance practices (Loshin 2012).
8. **Compliance Audits:** Regular audits can help ensure that data governance practices comply with internal policies and external regulations. These audits can also identify areas for improvement (Ladley 2012).

Fig. 16.6 Accountability and performance measurement map. (Created by the author)

Figure 16.6, visually outlines the key components necessary for establishing and maintaining accountability and measuring performance within an organization. At the center of the diagram is the main concept of accountability and performance measurement, surrounded by several interconnected elements that support this objective. Defining Clear Accountability is emphasized, ensuring that roles and responsibilities are clearly articulated and understood. Feedback Mechanisms are crucial for providing ongoing, constructive feedback that drives improvement and accountability. Linking to Business Outcomes highlights the importance of connecting performance metrics to tangible business results, ensuring that efforts are aligned with organizational goals. Performance Metrics and KPIs are essential for quantitatively measuring performance and progress towards these goals. Transparency in Reporting underscores the need for clear and open communication about performance metrics, fostering trust and clarity within the organization. Regular Monitoring and Reporting emphasize the importance of consistent oversight and documentation to track performance over time. Compliance Audits are necessary to ensure that practices meet regulatory standards and internal policies, reinforcing accountability. Continuous Improvement is a key element, focusing on the ongoing enhancement of processes and performance based on the insights gained from performance measurement and feedback. Together, these elements form a comprehensive framework for accountability and performance measurement, ensuring that organizational activities are aligned with goals, transparently tracked, and continually improved. This map visually ties these components together, illustrating how they collectively support effective accountability and performance management in an organization.

Accountability and performance measurement in data governance are about ensuring that data is managed effectively and in alignment with the strategic goals of the organization. By measuring performance and holding individuals accountable, organizations can enhance the value they derive from their data assets.

16.8 Case Studies: Effective Stewardship and Ownership

Real-world case studies in effective data stewardship and ownership highlight various approaches and strategies that organizations have adopted to manage their data assets successfully.

One such example is the Data Stewardship Program implemented by the Ministry of Children and Youth Services for the Government of Ontario, Canada. This program focused on building staff capacity at various levels to effectively use data, thereby improving the Ministry's culture around data. The program was developed to address several key questions, including the universal involvement in data work, changing attitudes and behaviors towards data, addressing fears related to data, and establishing accountability for achieving program outcomes. The Ministry conducted an Information Management Maturity Assessment, leading to a recognition of their data management capabilities being in an immature state and the need for

16.8 Case Studies: Effective Stewardship and Ownership

evidence-informed decision-making. The program included components like Individual Area Action Plans, general meetings, skill-building sessions, and communities of practice, all aimed at enhancing data stewardship and governance within the Ministry (Ministry of Children and Youth Services, Government of Ontario, Canada).

Another perspective on effective data stewardship is provided by Cognopia, which emphasizes the critical role of data stewards as central authorities on business data. Data stewards in any data-driven business are essential for holding together the definition, documentation, improvement, and use of data under their stewardship. Their responsibilities include documenting acceptable use and privacy policies for data, ensuring data quality, and documenting the provenance of data. These roles are integral in preventing misunderstanding and enabling organizations to make better decisions more quickly. The guide by Cognopia also outlines diverse types of data stewards, such as data object stewards, business data stewards, and process data stewards, each with specific roles and responsibilities tailored to the organization's structure and data governance goals (Cognopia).

The Role of Data Stewards in Baltimore, Maryland's Blueprint 2.0: This case focuses on the Baltimore County Public Schools (BCPS), the 25th largest school district in the U.S., which implemented a data governance program as part of its strategic plan, Blueprint 2.0. The program aimed to leverage technology and data to enhance teaching, learning, and organizational effectiveness. Data stewards played a crucial role in overseeing data quality, usage, and reliability within their respective areas, including student information systems, special education, and finance. They were responsible for defining data collection, reporting, and quality assurance processes within their subject areas, establishing access rights and security levels, and providing training to ensure compliance with governance policies. The impact of data stewards was significant in ensuring data integrity and quality, making data accessible and actionable, empowering decision-making, and streamlining data requests across the organization (Baltimore County Public Schools).

Freddie Mac's Data Stewardship Model: Freddie Mac operationalized a new Data Stewardship Model to enhance its data governance. The model was developed in three phases: Ready (laying the groundwork and establishing the program), Set (identifying and working with stewards and stakeholders), and Go (execution of the program). A key component of the model's success was engaging with stakeholders across the organization to understand their needs and challenges. This approach led to the creation of a roadmap for evolving into a robust data organization, identifying key areas for improvement such as Data Governance, Data Quality, and Master Data Management. The model emphasized branding to communicate the program's value, setting data standards through collaborative working groups, and ensuring roles and accountabilities within data governance. This comprehensive approach has made a substantial difference in promoting a data-centric culture and addressing the people, culture, change, and communication aspects critical to effective data governance (Freddie Mac).

These case studies underscore the importance of well-structured data stewardship programs in enhancing the value of data as a strategic enterprise asset, fostering a culture of responsible data use, and ensuring compliance with regulations. They also highlight the need for continuous engagement with stakeholders, setting clear standards, and ensuring roles and accountabilities in data governance initiatives.

16.9 Overcoming Common Challenges

Overcoming usual challenges in data governance and management often involves addressing issues related to data quality, stakeholder engagement, technology integration, and adherence to policies and regulations. Here are some strategies for overcoming these challenges:

1. **Improving Data Quality:** Establish clear data quality standards and conduct regular audits. Implement tools for data cleaning and validation to maintain high-quality data. Ensuring data accuracy, completeness, and consistency is crucial for effective decision-making (Plotkin 2014).
2. **Securing Stakeholder Buy-in:** Engage stakeholders early in the data governance process. Clearly communicate the benefits of data governance to all levels of the organization, emphasizing how it supports business objectives. Building a strong business case and demonstrating quick wins can help in gaining stakeholder support (Dallas and Lubrano 2018).
3. **Managing Data Silos:** Break down data silos by promoting cross-departmental collaboration. Implementing a centralized data management system can help unify disparate data sources, making data more accessible and usable across the organization (Plotkin 2014).
4. **Addressing Lack of Data Culture:** Foster a data-driven culture through training and awareness programs. Encourage data literacy and the use of data in decision-making across the organization. Leadership support is critical in promoting a culture that values data (Plotkin 2014; Dallas and Lubrano 2018).
5. **Ensuring Regulatory Compliance:** Stay informed about relevant data regulations and ensure compliance through regular training and updates to data policies. Implementing a compliance monitoring program can help in maintaining adherence to legal requirements (Redman 2008).
6. **Handling Technology Challenges:** Choose the right technology solutions that align with your data governance goals. Ensure that your data governance framework is flexible enough to adapt to innovative technologies. Regularly review and update technology tools to keep pace with advancements (Plotkin 2014).
7. **Resource Constraints:** Prioritize data governance initiatives and seek executive support for necessary resources. Consider leveraging external expertise if internal resources are limited. Effective resource allocation and management are essential for the success of data governance programs (Dallas and Lubrano 2018).
8. **Change Management:** Effectively manage the change process by setting realistic expectations, communicating changes clearly, and providing support during the transition. Change management strategies should include stakeholder engagement, training, and continuous feedback loops (Plotkin 2014).

16.9 Overcoming Common Challenges

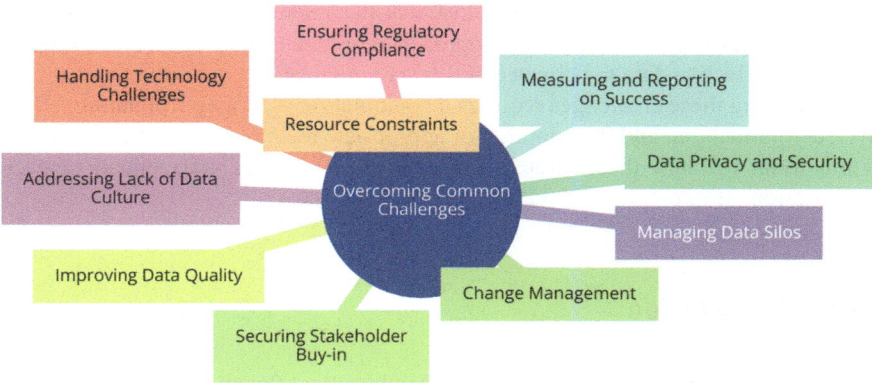

Fig. 16.7 Overcoming common challenges map. (Created by the author)

9. **Data Privacy and Security:** Implement robust data security measures and privacy policies. Regularly update these measures to counter evolving cyber threats and maintain compliance with privacy laws. Protecting sensitive information is critical for maintaining trust and avoiding legal issues (Redman 2008).
10. **Measuring and Reporting on Success:** Develop clear metrics to measure the success of data governance initiatives and regularly report these to stakeholders to demonstrate progress and value. Metrics should align with business goals and provide insights into the effectiveness of data governance practices (Plotkin 2014).

Figure 16.7, provides a visual representation of the various hurdles organizations face in data governance and the strategies to address them. The central theme of overcoming common challenges is surrounded by several key components, each representing a significant area that requires attention and action. Handling Technology Challenges is highlighted as a critical area, focusing on addressing issues related to technological advancements and integrations. Addressing Lack of Data Culture emphasizes the need to foster a culture that values and understands the importance of data. Improving Data Quality is crucial for ensuring that the data used in decision-making is accurate, complete, and reliable. Securing Stakeholder Buy-In stresses the importance of gaining support from key stakeholders to ensure the successful implementation and sustainability of data governance initiatives. Change Management is necessary for effectively managing organizational changes that impact data governance practices. Managing Data Silos points to the need to break down isolated data storage practices to ensure data is accessible and integrated across the organization. Data Privacy and Security are critical for protecting sensitive information and complying with legal requirements. Measuring and Reporting on Success involves tracking the effectiveness of data governance initiatives and communicating outcomes to stakeholders. Ensuring Regulatory Compliance focuses on adhering to relevant laws and regulations, which is essential for avoiding legal penalties and maintaining organizational integrity. Resource Constraints highlight the challenges related to limited resources and the need to prioritize and allocate them effectively. Overall, this map illustrates the interconnected nature of these challenges and the comprehensive approach required to address them, ensuring robust and effective data governance within an organization.

Addressing these challenges requires a combination of strategic planning, effective communication, and the adoption of appropriate technologies and processes. It is important to tailor these strategies to the specific needs and context of each organization.

16.10 Conclusion

The chapter of "Data Stewardship and Ownership: Best Practices" provides a comprehensive guide to the effective management and governance of data within organizations. The chapter covers various aspects of data stewardship and ownership, emphasizing the importance of best practices in these areas.

1. **Introduction to Data Stewardship and Ownership**: This section sets the stage for the chapter, highlighting the significance of data stewardship and ownership in modern organizations and the impact of these roles on data governance.
2. **Defining Roles and Responsibilities**: It delves into the specific roles and responsibilities of data stewards and owners, detailing the scope and limits of each role in the context of data governance.
3. **Best Practices in Data Stewardship**: The chapter discusses key best practices in data stewardship, including maintaining data quality, ensuring compliance, and fostering a culture of data-centric decision-making.
4. **Establishing Data Ownership**: This part emphasizes the importance of clearly defining data ownership within an organization, detailing how it contributes to effective data governance and overall business strategy.
5. **Collaboration Between Stewards and Owners**: It highlights the necessity of collaboration between data stewards and owners, focusing on the constructive collaboration required to manage data effectively and align it with business goals.
6. **Training and Empowerment**: The chapter underscores the importance of training and empowerment in building a data-literate workforce, capable of handling data responsibly and effectively.
7. **Accountability and Performance Measurement**: This section outlines the importance of establishing accountability in data governance and ways to measure the performance of data stewardship and ownership initiatives.
8. **Case Studies: Effective Stewardship and Ownership**: Real-world case studies are presented to illustrate the application of best practices in data stewardship and ownership, providing practical insights into successful data governance strategies.
9. **Overcoming Common Challenges**: The chapter concludes with a discussion on usual challenges in data governance and strategies to overcome them, including managing data silos, securing stakeholder buy-in, and ensuring regulatory compliance.

This chapter serves as a valuable resource for organizations seeking to enhance their data governance practices. It emphasizes the critical role of data stewards and owners in managing data as a strategic asset and provides practical insights for implementing effective data governance strategies.

References

Baltimore County Public Schools (BCPS) (n.d.) Blueprint 2.0: Data Governance Program
Cognopia (n.d.) The role of data stewards in data-driven businesses
Dallas G, Lubrano M (2018) Governance, stewardship and sustainability: theory, practice and evidence, 2nd edn. Routledge
DAMA International (2017) The DAMA Guide to the Data Management Body of Knowledge (DAMA-DMBOK). Dent Technics Publications
Dyché J (2015) The new IT: how technology leaders are enabling business strategy in the digital age. McGraw-Hill Education
Freddie Mac (n.d.) Data stewardship model: enhancing data governance
Ladley J (2012) Data governance: how to design, deploy, and sustain an effective data governance program. Elsevier
Loshin D (2012) Business intelligence: the savvy Manager's guide. Morgan Kaufmann
Ministry of Children and Youth Services, Government of Ontario, Canada (n.d.) Data Stewardship Program
Otto B (2021) Designing data governance: conceptual foundations and practical application. Springer
Plotkin D (2014) Data stewardship: an actionable guide to effective data management and data governance. Morgan Kaufmann
Redman TC (2008) Data driven: profiting from your most important business asset. Harvard Business Review Press

Chapter 17
Data Governance Maturity Models: Assessing and Enhancing Your Program

Contents

17.1	Introduction to Data Governance Maturity Models.	488
17.2	Overview of Common Data Governance Maturity Models.	490
17.3	Assessing Your Current Maturity Level.	491
17.4	Developing a Roadmap for Maturity Advancement.	493
17.5	Key Factors Influencing Data Governance Maturity.	496
17.6	Aligning Maturity Improvement with Business Objectives.	500
17.7	Case Studies: Maturity Model Implementation.	503
17.8	Challenges in Advancing Maturity.	504
17.9	Continuous Improvement in Data Governance.	506
17.10	Conclusion.	508
References.		509

Abstract This chapter provides an in-depth exploration of frameworks that help organizations assess and incrementally improve their data governance capabilities. These maturity models enable a structured approach to evaluate current practices, identify areas for improvement, and guide strategic enhancements to achieve higher levels of data governance maturity.

The chapter begins by introducing the concept of data governance maturity models, which delineate various stages of maturity an organization can achieve—from initial, ad-hoc practices to optimized and continuously improving processes. These models are instrumental for organizations to understand their current state, set clear improvement goals, and align their data governance efforts with overarching business objectives.

A detailed overview of several well-known maturity models is provided, including the Data Management Maturity (DMM) model, the Capability Maturity Model Integration (CMMI), and the Information Governance Reference Model (IGRM). Each model offers a unique framework and is tailored to specific organizational needs and industry standards, helping businesses navigate the complexities of data governance.

The process of assessing an organization's current maturity level is described as a crucial step, involving the selection of a suitable model, defining assessment criteria, and conducting evaluations to pinpoint the organization's position within the maturity spectrum. This assessment aids in identifying strengths and weaknesses in current data governance practices.

Furthermore, the chapter discusses the development of a roadmap for maturity advancement. This roadmap is based on the findings from the maturity assessment and includes setting specific, measurable goals, prioritizing improvement initiatives, and allocating necessary resources. The roadmap is a dynamic tool that guides organizations through the necessary steps to elevate their data governance practices.

Key factors influencing data governance maturity are also explored, such as executive leadership, organizational culture, regulatory requirements, and technological infrastructure. These factors can either facilitate or hinder the progress of data governance maturity and must be carefully managed to support advancement.

Real-world case studies are included to illustrate the practical application of data governance maturity models in various organizational contexts. These examples highlight the challenges faced and the strategies implemented by organizations to enhance their data governance frameworks.

In summary, this chapter provides a comprehensive guide on using data governance maturity models to systematically assess and improve data governance initiatives. It underscores the importance of continuous evaluation, strategic planning, and alignment with business objectives to effectively manage and utilize data as a strategic asset.

Keywords Data governance maturity models · Data management · Organizational assessment · Continuous improvement · Capability maturity model integration · Data quality · Compliance · Business Alignment · Data management maturity model · Strategic planning · Organizational development · Process improvement · Data stewardship · Risk management · Information governance

17.1 Introduction to Data Governance Maturity Models

Data Governance Maturity Models are frameworks or models used to assess and enhance the maturity level of a data governance program within an organization. These models help organizations understand where they currently stand in terms of data governance and provide a roadmap for improving their data management practices.

17.1 Introduction to Data Governance Maturity Models

What Is a Data Governance Maturity Model? A Data Governance Maturity Model is a structured framework that outlines various levels or stages of maturity an organization can achieve in its data governance efforts. These models typically consist of several stages, often ranging from ad-hoc or initial stages to more advanced and mature stages, helping organizations visualize their growth path in data governance. The Data Management Maturity (DMM) model from the CMMI Institute is a prime example of such frameworks, defining clear maturity pathways tailored for data governance (CMMI Institute 2014).

Why Use a Data Governance Maturity Model? Data governance is essential for organizations to manage data effectively, ensure data quality, compliance, and support data-driven decision-making. Maturity models help organizations:

- Assess their current state of data governance (Mosley 2017)
- Identify strengths and weaknesses in their data management practices
- Set clear goals and priorities for improvement
- Measure progress over time
- Align data governance efforts with business objectives (Enterprise Data Management Council 2020a)

Common Stages in Data Governance Maturity Models While the specific stages and criteria may vary, common stages in data governance maturity models often include:

- Ad-Hoc: No formal data governance practices; data management is decentralized.
- Initial/Chaotic: Initial efforts to establish data governance, but no standardized processes.
- Defined: Basic data governance policies and processes are defined but not consistently applied.
- Managed: Data governance processes are consistently followed, and roles and responsibilities are well-defined.
- Optimized: Continuous improvement and optimization of data governance practices.

Assessment and Enhancement Organizations use data governance maturity models to assess their current stage and identify gaps. Once weaknesses are identified, they can work on enhancing their data governance program by implementing best practices, refining policies, improving data quality, and ensuring compliance with data regulations (Chrissis et al. 2003).

Choosing the Right Model Organizations should choose a data governance maturity model that aligns with their industry goals and existing data governance practices. Commonly used models include the Data Management Maturity (DMM) model, the Capability Maturity Model Integration (CMMI), and industry-specific models. IBM's Information Governance Solutions provide methodologies and tools that support the adoption of these models (IBM Corporation 2010).

Continuous Improvement Data governance is an ongoing process, and organizations should regularly reassess their maturity level and make necessary adjustments to enhance their program continually.

Data Governance Maturity Models provide organizations with a structured approach to assess, improve, and mature their data governance practices. By using such models, organizations can better manage their data assets, ensure data quality, and support their overall business objectives.

17.2 Overview of Common Data Governance Maturity Models

This section provides insights into the various frameworks that organizations can use to assess and enhance their data governance practices. Below are some well-known data governance maturity models:

Data Management Maturity (DMM) Model
The DMM model, developed by the CMMI Institute, is one of the most widely used frameworks for assessing data governance maturity. It defines specific maturity levels, each representing increasingly sophisticated data management practices. This model is particularly useful for organizations looking to systematically improve their data governance processes (CMMI Institute 2014).

Capability Maturity Model Integration (CMMI)
CMMI is a broader framework used for assessing and improving process maturity across various domains, including data governance. Organizations can adapt CMMI to assess their data governance practices by defining specific maturity levels and criteria for data-related processes (Chrissis et al. 2003).

Information Governance Reference Model (IGRM)
Developed by EDRM, the IGRM focuses on information governance practices and maturity. It outlines various components of information governance, such as data stewardship, data lifecycle management, and compliance, providing a maturity model for each component. This model helps organizations understand and improve their information governance structures (EDRM 2012).

Enterprise Data Management Council (EDMC) Framework
The EDM Council offers a comprehensive framework for data governance and management, focusing particularly on the financial services industry. The DCAM framework outlines maturity levels and best practices for areas like data governance, data quality, and data architecture within the context of financial services (Enterprise Data Management Council 2020b).

17.3 Assessing Your Current Maturity Level

Data Management Body of Knowledge (DAMA-DMBOK)
Edited by Mosley, the DAMA-DMBOK serves as an essential guide for data management principles, including data governance. This body of knowledge provides foundational insights for organizations seeking to establish a robust data governance framework that aligns with recognized best practices (Mosley 2017).

IBM Information Governance Solutions
IBM offers methodologies and tools that support the adoption of data governance maturity models. Their solutions are designed to help organizations implement and optimize their data governance practices effectively, ensuring alignment with business objectives and compliance requirements (IBM Corporation 2010).

Data Governance Best Practices Frameworks
Some organizations and consulting firms have developed their proprietary data governance maturity models and best practice frameworks. These frameworks often draw from industry standards and practices to create customized models that suit specific industries or organizational needs.

Industry-Specific Models
In addition to generic models, certain industries may have specific data governance maturity models tailored to their unique requirements and regulations.

For example, healthcare organizations may have models that focus on data governance in the context of healthcare data, and similar adaptations exist for other industries.

Customized Models
Some organizations create their own data governance maturity models tailored to their specific goals, culture, and existing practices.

Customized models allow for a highly personalized approach to data governance maturity assessment and enhancement.

When choosing a data governance maturity model, organizations should consider their industry, the scope of their data governance program, and their specific objectives. Regardless of the model used, the key is to assess the current state of data governance, identify areas for improvement, and establish a roadmap for enhancing data governance practices over time.

17.3 Assessing Your Current Maturity Level

Assessing your current maturity level in the context of data governance or any other discipline is a crucial step in understanding where your organization stands in terms of best practices and capabilities. Here is what you need to know about assessing your current maturity level:

Purpose of Assessment

The primary purpose of assessing your current maturity level is to gain a clear understanding of how well your organization manages its data assets. This assessment helps identify strengths, pinpoint weaknesses, and set clear, strategic directions for improvement. It can also facilitate the alignment of data governance with business objectives, ensuring that data management efforts directly support organizational goals (Atlan 2023a).

Selecting a Maturity Model

To begin the assessment, choose an appropriate maturity model that aligns with your organization's specific needs and objectives. The "Data Governance Maturity Model: A Roadmap to Optimizing Your Data Initiatives and Driving Business Value" by CastorDoc provides a comprehensive guide that can help organizations tailor their assessment according to their operational contexts and data governance challenges (CastorDoc 2023).

Define Assessment Criteria

Once a model is selected, define specific criteria or indicators for each maturity level within the model. This includes measurable standards that cover various aspects of data governance such as data quality, data access, data stewardship, and compliance. These criteria serve as benchmarks to evaluate your current practices against established best practices (DATAVERSITY 2023a).

Data Collection and Evaluation

Gather relevant data and evaluate your current practices against the predefined criteria. This step may involve surveys, interviews, document reviews, and performance analysis. The goal is to objectively determine which maturity level your organization currently occupies and identify key areas for improvement.

Maturity Levels

Maturity models typically consist of multiple levels or stages, such as ad-hoc, initial, defined, managed, and optimized. Each level represents a different degree of maturity in the area being assessed.

Scoring and Assessment Results

Assign scores based on the collected data to quantify the maturity level. This scoring will help in creating a detailed assessment report that outlines current capabilities, gaps, and potential areas for improvement.

Identify Gaps and Prioritize

Analyze the assessment results to identify significant gaps in your data governance practices. Prioritize these gaps based on their impact on organizational performance and compliance requirements. This step is crucial for setting targeted improvement goals that will yield the most significant benefits.

Create a Roadmap for Improvement

Using the insights gained from the assessment, develop a strategic roadmap that outlines specific actions needed to enhance maturity levels. This plan should include short-term and long-term goals, assigned responsibilities, and timelines for achieving each milestone.

Regular Reassessment
Data governance is a dynamic field that requires ongoing attention and adaptation. Regularly reassess your maturity level to ensure continuous improvement and alignment with evolving business needs and technological advancements.

Communication and Stakeholder Buy-In
Share the assessment results and improvement roadmap with relevant stakeholders within your organization. Engage leadership and teams in the process to secure buy-in and support for data governance improvement initiatives.

Assessing your current maturity level is a foundational step in the journey to enhance data governance or any other aspect of your organization's operations. It provides a baseline for improvement efforts, helps set realistic goals, and ensures that resources are allocated where they will have the most significant impact on your organization's success.

17.4 Developing a Roadmap for Maturity Advancement

Developing a roadmap for maturity advancement is a crucial step in the process of improving an organization's capabilities, whether it is in the context of data governance or any other area. Here is what you need to know about creating a roadmap for maturity advancement:

1. **Assessment Findings as a Starting Point**
   Begin by reviewing the findings from your initial data governance maturity assessment. This review should highlight areas needing improvement and the strengths to leverage. Utilizing tools and frameworks like those outlined in "Data Governance Maturity Model: A Roadmap to Optimizing Your Data Initiatives and Driving Business Value" by CastorDoc can provide structured guidance in this evaluation process (CastorDoc 2023).
2. **Set Clear Objectives and Goals**
   Define clear, measurable objectives for your data governance program that align with your organization's broader business goals. McKinsey & Company's guide "Designing Data Governance that Delivers Value" provides insights into aligning data governance initiatives with strategic business outcomes, emphasizing the importance of clear objectives to drive organizational buy-in and success (McKinsey & Company 2023a). Ensure that your objectives are SMART: Specific, Measurable, Achievable, Relevant, and Time-bound.
3. **Prioritize Improvement Initiatives**
   Not all data governance activities have equal impact. Prioritize initiatives based on their potential to address critical gaps and deliver quick wins. DATAVERSITY's article, "The Importance of a Data Governance Roadmap," discusses strategies for prioritizing data governance efforts that can help

organizations focus on the most impactful areas first (DATAVERSITY 2023b). Consider factors like potential risks, benefits, resource availability, and dependencies.

4. **Create a Detailed Plan**

   Develop a detailed action plan for each data governance initiative. This plan should include specific tasks, responsible parties, deadlines, and required resources. Atlan's "10 Steps to Create an Effective Data Governance Roadmap" offers a practical approach to structuring these plans, ensuring that each step is actionable and aligned with the overall data governance strategy (Atlan 2023b). Assign responsibilities to individuals or teams, and establish timelines and milestones for each task.

5. **Allocate Resources**

   Ensure that each initiative within the roadmap has the necessary resources, including budget, personnel, and technology. Adequate resourcing is critical for the successful implementation of data governance improvements.

   Sequencing and Dependencies.

   Consider the sequence in which initiatives should be executed. Some improvements may depend on others being completed first. Identify dependencies between initiatives and plan accordingly to ensure a smooth execution process.

6. **Risk Management**

   Assess potential risks and challenges associated with each initiative. Develop mitigation plans and contingency strategies to address these risks.

   Risk management should be an integral part of your roadmap.

7. **Monitoring and Measurement**

   Define key performance indicators (KPIs) and metrics that will allow you to track progress and measure the success of your maturity advancement efforts.

   Regularly monitor and assess your initiatives against these KPIs.

8. **Communication and Stakeholder Engagement**

   Clearly communicate the roadmap and its objectives to all relevant stakeholders, including senior leadership, project teams, and those affected by the changes.

   Engage stakeholders throughout the process to maintain alignment and support.

9. **Iterative Approach**

   Understand that maturity advancement is an ongoing process. Your roadmap may evolve as you progress, and new insights and challenges emerge.

   Be prepared to adapt your plan as needed to address changing circumstances or priorities.

10. **Documentation and Documentation**

    Keep thorough documentation of your roadmap, including objectives, plans, timelines, progress reports, and outcomes. This documentation is essential for accountability and future reference.

11. **Celebrate Achievements**

    Recognize and celebrate milestones and achievements along the way. Positive reinforcement can boost team morale and motivation.

17.4 Developing a Roadmap for Maturity Advancement

Fig. 17.1 Roadmap for maturity advancement. (Created by the author)

Figure 17.1, titled "Developing a Roadmap for Maturity Advancement," outlines the critical steps and considerations necessary for progressing an organization's data governance maturity. At the center is the primary concept of developing a roadmap for maturity advancement, with various key components branching out to support this objective. Starting with Assessment Findings as a Starting Point, the diagram emphasizes the importance of using current assessments to identify areas for improvement. Prioritizing Improvement Initiatives ensures that the most critical areas are addressed first, based on their impact and feasibility. Monitoring and Measurement are crucial for tracking progress and ensuring that initiatives are moving towards the desired goals. Risk Management highlights the need to identify and mitigate potential risks that could hinder maturity advancement. Sequencing and Dependencies focus on the logical order of initiatives, ensuring that dependencies are managed effectively to avoid bottlenecks. An Iterative Approach suggests that the process should be flexible and adaptive, allowing for adjustments as new insights and challenges emerge. Setting Clear Objectives and Goals is essential for providing direction and measurable targets for maturity advancement efforts. Creating a Detailed Plan outlines the steps and actions required to achieve these goals. Documentation and Documentation emphasize the importance of keeping thorough records of plans, processes, and progress, which supports transparency and accountability. Communication and Stakeholder Engagement ensure that all relevant parties are informed and involved, fostering support and collaboration. Allocating Resources is necessary to provide the financial, human, and technological resources needed to implement the plan effectively. Celebrating Achievements recognizes and rewards milestones and successes, which helps maintain momentum and motivation. Together, these elements create a comprehensive framework for developing a roadmap for maturity advancement, ensuring that efforts are well-planned, monitored, and aligned with organizational goals. This map visually integrates these components, illustrating how they collectively support the progression of data governance maturity within an organization.

Developing a roadmap for maturity advancement provides a structured and organized approach to achieving your goals and enhancing your organization's capabilities. It serves as a guiding document that helps ensure that your efforts are aligned with your strategic objectives and that you make measurable progress toward maturity.

17.5 Key Factors Influencing Data Governance Maturity

Data governance maturity is influenced by a combination of factors that impact an organization's ability to effectively manage its data assets. These factors can either facilitate or hinder the development and advancement of data governance practices. Here are key factors that influence data governance maturity:

1. **Executive Leadership and Commitment**
    Strong executive support is paramount for driving data governance initiatives forward. The analysis by Duranton et al. in "Is Your Company Gaining Momentum in Data?" highlights how leadership commitment influences the pace and effectiveness of adopting data-driven strategies across the organization (Duranton et al. 2021).
    Organizational Culture
2. An organizational culture that values and utilizes data effectively is more likely to advance in its data governance maturity. The McKinsey & Company report, "Designing Data Governance that Delivers Value," stresses the importance of culture in embedding data governance into business processes and decision-making (McKinsey & Company 2023b).
3. **Regulatory and Compliance Requirements**
    Adherence to regulatory standards significantly drives the implementation of robust data governance frameworks. The "Data Governance: A Conceptual Framework, Structured Review, and Research Agenda" from ScienceDirect provides an overview of how regulatory pressures shape data governance practices, urging organizations to maintain compliance and avoid legal pitfalls (ScienceDirect 2019).
4. **Data Complexity and Volume**
    The complexity and volume of an organization's data can influence its data governance maturity. Managing large volumes of data or dealing with complex data structures may require more advanced governance practices.
5. **Data Quality and Integrity**
    The integrity and reliability of data are foundational to effective data governance. Keith D. Foote's work, "Creating a Data Maturity Model: What, Why, How," discusses the role of data quality in achieving higher maturity levels, emphasizing that well-governed data leads to trustworthy business insights (DATAVERSITY 2023c).

17.5 Key Factors Influencing Data Governance Maturity

6. **Technological Infrastructure**
   Advanced technological tools and solutions facilitate efficient data governance by enabling better data integration, security, and analytics. Informatica's "Data Governance Framework: 4 Pillars for Success" outlines how technology underpins all critical aspects of data governance, providing the necessary support to ensure scalability and responsiveness (Informatica 2023).
7. **Data Stewardship and Roles**
   The presence of dedicated data stewards and clearly defined roles and responsibilities for data management contribute to higher maturity. Stewards are responsible for data quality, governance, and data asset management.
8. **Data Governance Framework**
   The existence and implementation of a well-defined data governance framework or model, such as those discussed earlier, provide a structured approach for organizations to follow.
9. **Education and Training**
   The level of education and training provided to staff on data governance principles and best practices can significantly impact maturity. A knowledgeable workforce is essential.
10. **Data Privacy and Security Concerns**
    Concerns about data privacy and security breaches can drive organizations to establish more robust data governance practices to protect sensitive information.
11. **Data Governance Metrics and Measurement**
    The establishment of key performance indicators (KPIs) and metrics to measure data governance effectiveness is crucial. Data-driven insights into governance maturity help guide improvements.
12. **Change Management Capabilities**
    An organization's ability to manage change and effectively implement new data governance processes and policies is essential for advancement.
13. **Budget and Resource Allocation**
    Adequate budget and resource allocation for data governance initiatives are critical. Without sufficient resources, it can be challenging to make progress.
14. **Business Value and ROI**
    Demonstrating the business value and return on investment (ROI) of data governance initiatives can help maintain leadership support and secure additional resources.
15. **External Factors and Market Competition**
    Market dynamics and competition can influence data governance maturity. Organizations in highly competitive industries may invest more in data governance to gain a competitive edge.
16. **Size and Complexity of the Organization**
    The size and complexity of an organization can determine the scope and scale of data governance efforts. Larger organizations or those with diverse data sources often require more comprehensive governance.

17. **Maturity Model Selection**

The choice of a data governance maturity model and its alignment with organizational needs can influence the maturity path.

18. **Data Governance Champion**

Having a dedicated data governance champion or advocate who promotes and oversees data governance initiatives can be a driving force in maturity advancement.

Figure 17.2, titled "Key Factors Influencing Data Governance Maturity," visually represents the various elements that significantly impact the maturity of data governance within an organization. The central theme is key factors influencing data governance maturity, from which several critical components extend. Executive Leadership and Commitment highlight the importance of having strong support and involvement from top management to drive data governance initiatives. Organizational Culture underscores the need for a culture that values and understands the significance of data governance. Regulatory and Compliance Requirements emphasize the necessity of meeting legal and regulatory standards, which can significantly influence data governance practices. Data Complexity and Volume reflect the challenges associated with managing large and intricate datasets. Data Quality and Integrity focuses on maintaining high standards of data accuracy and reliability, which are essential for effective data governance. Technological Infrastructure points to the importance of having robust and appropriate technology to support data management activities. Data Stewardship and Roles emphasize the need for clearly defined roles and responsibilities in managing data assets. The Data Governance Framework is crucial for providing a structured approach to implementing data governance. Education and Training highlight the importance of ongoing learning and development to equip individuals with the necessary skills for data governance. Data Privacy and Security Concerns address the need to protect sensitive information and comply with privacy regulations. Data Governance Metrics and Measurement are essential for tracking the progress and effectiveness of data governance initiatives. Change Management Capabilities stress the ability to manage organizational changes that impact data governance. Budget and Resource Allocation underline the necessity of providing adequate resources to support data governance activities. Business Value and ROI focus on demonstrating the financial and strategic benefits of data governance to gain support and investment. External Factors and Market Competition reflect the influence of industry trends and competitive pressures on data governance practices. Size and Complexity of the Organization highlight how the scale and intricacy of an organization affect its data governance needs. Maturity Model Selection points to the importance of choosing the right framework or model to guide the maturity process. Data Governance Champion emphasizes the role of a dedicated individual or team in leading and advocating for data governance within the organization. This map illustrates how these interconnected factors collectively influence the maturity of data governance, providing a comprehensive view of the elements that organizations must consider to develop and advance their data governance capabilities effectively.

17.6 Aligning Maturity Improvement with Business Objectives

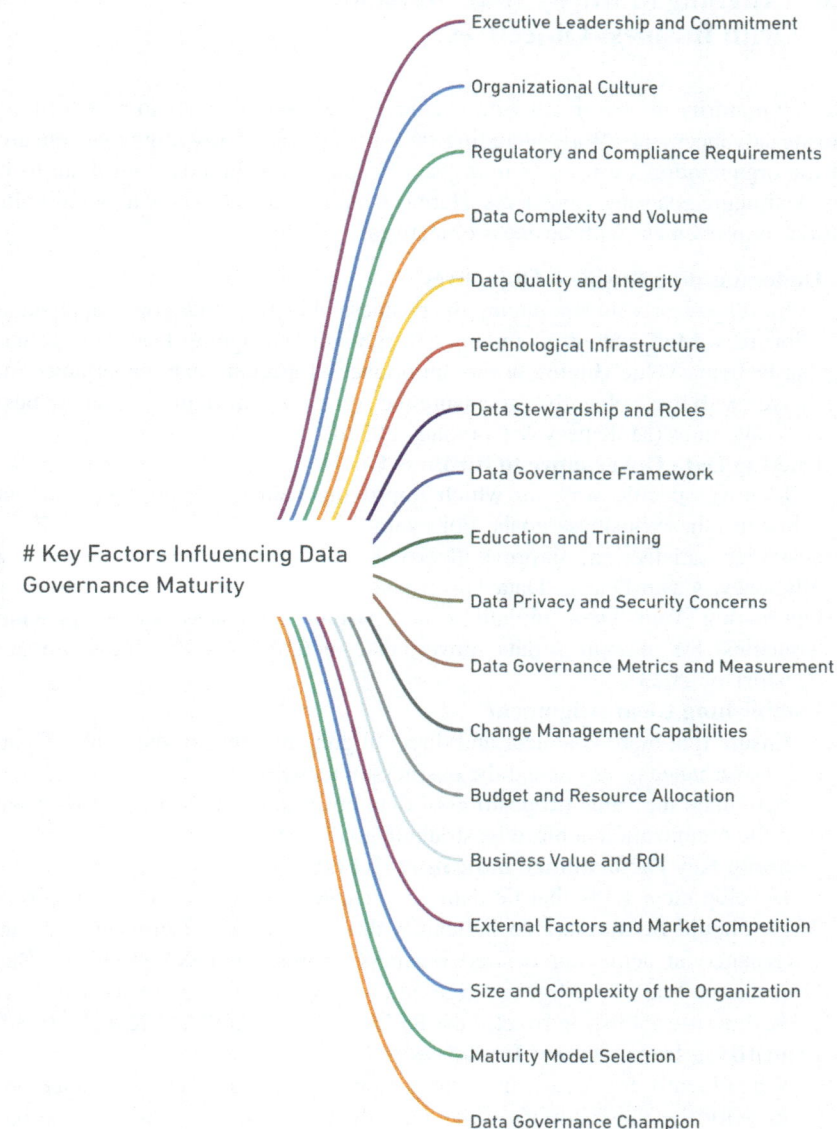

Fig. 17.2 Key factors influencing data governance maturity. (Created by the author)

Organizations need to consider these factors and tailor their data governance strategies accordingly. By addressing these key factors, organizations can enhance their data governance maturity and effectively manage their data assets.

17.6 Aligning Maturity Improvement with Business Objectives

Aligning maturity improvement with business objectives is a critical aspect of any maturity advancement effort, including data governance. This alignment ensures that the organization's efforts to improve its capabilities directly contribute to its overall strategic goals and objectives. Here is what you need to know about aligning maturity improvement with business objectives:

1. **Understanding Business Objectives**
   Start by clearly understanding the business objectives that your organization prioritizes. McKinsey & Company's insights in "Designing Data Governance that Delivers Value" highlight the importance of aligning data governance initiatives with these objectives to ensure they directly contribute to strategic business outcomes (McKinsey & Company 2023d).
2. **Linking Data Governance to Business Goals**
   Identify specific ways in which improving data governance can support achieving these business goals. For example, better data quality can enhance customer satisfaction, improve decision-making, and increase operational efficiency. CastorDoc's "Data Governance Maturity Model: A Roadmap to Optimizing Your Data Initiatives and Driving Business Value" provides strategies for mapping data governance activities to business impacts (CastorDoc 2023).
3. **Establishing Clear Alignment**
   Ensure that there is a clear and direct alignment between the goals of your data governance program and the business objectives.
   Articulate the value proposition of data governance in terms of how it will help the organization achieve its strategic goals.
4. **Defining Key Performance Indicators (KPIs)**
   Develop clear KPIs that tie data governance efforts to business outcomes. These should be measurable indicators that reflect the effectiveness of data governance in achieving desired business results. DATAVERSITY's "Data Governance Success: Aligning Your Strategy with Business Objectives" discusses how to establish and track these KPIs effectively (DATAVERSITY 2023d).
5. **Prioritizing Improvement Initiatives**
   When identifying areas for improvement within your data governance program, prioritize those initiatives that have the most significant potential impact on business objectives.
   Focus resources and efforts on initiatives that directly align with strategic priorities.
6. **Resource Allocation**
   Ensure that resources—budget, personnel, technology—are allocated in alignment with the prioritized governance activities that drive the most significant business value. IBM's guide, "A Step-by-Step Guide to Setting Up a Data

17.6 Aligning Maturity Improvement with Business Objectives

Governance Program," offers a practical approach to resource planning and management in data governance initiatives (IBM 2023).

7. **Communication and Collaboration**

   Foster collaboration between data governance teams and business units. Effective communication and collaboration ensure that data governance efforts are integrated into business processes.

   Involve business stakeholders in data governance decision-making and implementation.

8. **Regular Monitoring and Reporting**

   Continuously monitor and report on the progress of data governance initiatives and their impact on business objectives.

   Share success stories and demonstrate how data governance is contributing to the organization's success.

9. **Adaptation and Flexibility**

   Be prepared to adapt and adjust data governance priorities and initiatives as business objectives evolve.

   Flexibility is key to ensuring that data governance remains relevant and aligned with changing business needs.

10. **Demonstrating ROI**

    Establish mechanisms to track the return on investment (ROI) of data governance initiatives. Demonstrating a positive ROI can help secure ongoing support from executives and stakeholders.

11. **Feedback and Continuous Improvement**

    Solicit feedback from business units and stakeholders to understand their evolving needs and challenges.

    Use this feedback to continuously refine and improve data governance practices to better serve business objectives.

12. **Executive Sponsorship**

    Secure ongoing executive sponsorship and support for data governance initiatives to maintain alignment with business objectives.

Figure 17.3, titled "Aligning Maturity Improvement with Business Objectives," visually outlines the essential elements for ensuring that data governance maturity initiatives are in harmony with the broader business goals of an organization. The central theme of aligning maturity improvement with business objectives is connected to various critical components that support this alignment. Understanding Business Objectives is foundational, emphasizing the need to clearly grasp the overall goals and strategies of the organization. Linking Data Governance to Business Goals ensures that data governance initiatives directly contribute to achieving these goals. Establishing Clear Alignment highlights the importance of ensuring that all data governance activities are consistently aligned with business objectives. Defining Key Performance Indicators (KPIs) is crucial for setting measurable targets that reflect both data governance and business priorities. Prioritizing Improvement Initiatives involves selecting and focusing on the most impactful areas for maturity advancement that will drive business success. Resource Allocation underscores the necessity of providing the appropriate resources to support these

Fig. 17.3 How to align maturity improvement with business objectives map. (Created by the author)

initiatives. Communication and Collaboration are vital for ensuring that all stakeholders are informed and engaged in the maturity improvement process. Regular Monitoring and Reporting emphasize the importance of continuous oversight and documentation to track progress and ensure transparency. Adaptation and Flexibility are essential for allowing the maturity improvement process to adjust to new challenges and insights. Demonstrating ROI focuses on showing the financial and strategic returns from maturity improvement efforts to gain and maintain support. Feedback and Continuous Improvement highlight the need for ongoing evaluation and refinement of data governance practices based on performance feedback. Executive Sponsorship stresses the importance of having strong leadership support to drive and sustain maturity improvement initiatives. This map visually integrates these elements, illustrating how they collectively ensure that efforts to improve data governance maturity are strategically aligned with and supportive of the organization's broader business objectives.

Aligning maturity improvement efforts with business objectives is essential for ensuring that data governance and related initiatives are not seen as isolated technical endeavors but as strategic enablers of business success. This alignment enhances the chances of achieving tangible business benefits through improved data management and governance practices.

17.7 Case Studies: Maturity Model Implementation

There are several case studies that highlight the implementation of Data Governance Maturity Models in different organizations:

IBM Data Governance Council Maturity Model serves as a valuable resource for understanding the implementation of a maturity model in data governance. This guide outlines a framework for evaluating and designing data governance programs across various maturity levels, covering key areas such as organizational structures, stewardship, policy, value creation, risk management, and more.

DATAVERSITY provides insights into creating a Data Maturity Model, highlighting its importance in transitioning organizations from minimal to maximal use of data. This approach involves assessing an organization's current data use, identifying gaps, and promoting data maturity through strategic planning and behavior patterns that maximize data utilization. Moreover, DATAVERSITY discusses the critical role of Data Governance in supporting data maturity, noting that governance programs can significantly improve an organization's data maturity by establishing a common language, promoting coherent coordination, and facilitating more effective teamwork around data usage and accessibility.

Ally Financial's Data Management Maturity Model Implementation Ally Financial, as described in another DATAVERSITY online seminar, was an early adopter of the Data Management Maturity (DMM) model. They conducted a broad-based evaluation of their data management practices and created a strategy and sequence plan for improvements. This implementation, which included data governance and a robust data quality program, led to a "Satisfactory" rating in their latest regulatory audit. The case study illustrates how Ally leveraged the DMM model for comprehensive improvements in data management and governance.

Data Governance Maturity Assessment Tool A design science approach" by Philippe Marchildon, Simon Bourdeau, Pierre Hadaya, and Aldrin Labissière, is indeed a real study published in the journal Projectics/Proyéctica/Projectique in 2018. It details a comprehensive approach toward assessing the maturity level of data governance within organizations using a design science research methodology. The authors developed a Data Governance Maturity (DGM) assessment tool, comprising 11 dimensions and 72 questions, aimed at helping organizations evaluate their data governance practices against best practices, identify gaps, and guide the development, deployment, or improvement of their Data Governance Framework (DGF). This study makes a significant contribution to the field by operationalizing the concept of data governance maturity, providing a practical tool for organizations to assess and prioritize their data governance initiatives. It addresses a gap in the existing literature and practice by offering a means to evaluate an organization's level of maturity in data governance comprehensively. The DGM assessment tool is aligned with existing data governance maturity frameworks and methodologies, making it a valuable resource for organizations seeking to improve their data governance practices.

These case studies provide valuable insights into the practical aspects of implementing data governance maturity models, demonstrating the challenges and strategies involved in such endeavors. They highlight the importance of leadership buy-in, effective communication, and the adaptation of governance models to specific organizational contexts.

17.8 Challenges in Advancing Maturity

Advancing the maturity of any organization, especially in the context of data governance or management, presents a set of usual challenges:

1. **Lack of Executive Support and Buy-In**: One of the primary challenges is securing ongoing executive support. McKinsey & Company's research "Data analytics: Accelerating maturity in the US public sector" underscores the necessity of top-level buy-in for the success of data initiatives, highlighting strategies to engage senior leadership effectively (McKinsey 2023b).
2. **Cultural Resistance**: Organizational culture often resists changes necessary for advancing maturity. DATAVERSITY's "Data Governance Challenges (and How to Fix Them)" offers solutions for addressing resistance, such as engaging stakeholders early and demonstrating the tangible benefits of enhanced data governance (DATAVERSITY 2023d).
3. **Data Quality Issues**: Poor data quality significantly hampers progress. The "Data Governance Maturity Model" by CastorDoc provides a roadmap for systematically improving data quality as part of advancing governance maturity, emphasizing the role of quality management practices in achieving higher maturity levels (CastorDoc 2023).
4. **Integration of Data Across Silos**: The technical and political challenges of integrating siloed data can be formidable. IT Governance Docs' "Data Governance Maturity Model" discusses methodologies for breaking down silos and integrating data effectively to foster a unified governance framework (IT Governance Docs 2023).
5. **Resource Constraints**: Advancing maturity often requires significant investment in terms of time, money, and human resources. Organizations may struggle to allocate these resources effectively, especially if the immediate benefits are not clear.

17.8 Challenges in Advancing Maturity

6. **Lack of Clear Strategy or Roadmap**: Developing and adhering to a clear strategy or roadmap for advancing maturity is crucial. Without this, efforts can become disjointed and ineffective.
7. **Keeping Pace with Technological Changes**: The rapid pace of technological advancement means that data governance strategies and tools can quickly become outdated. Organizations need to stay informed and adaptable to keep up.
8. **Measuring Progress and Value**: Demonstrating the value and progress of data governance initiatives can be challenging. Clear metrics and regular reporting are necessary to maintain support and funding.
9. **Regulatory Compliance**: Keeping up with changing regulations and ensuring compliance can be a moving target, adding complexity to data governance efforts.
10. **Talent and Skill Shortages**: The specialized nature of data governance means that it can be challenging to find and retain the right talent with the necessary skills.

Figure 17.4, titled "Challenges in Advancing Maturity," illustrates the various obstacles organizations may encounter when trying to improve their data governance maturity. The central theme of challenges in advancing maturity is connected to several critical factors that can impede progress. Lack of Executive Support and Buy-In highlights the difficulty of advancing maturity without strong leadership

Fig. 17.4 Challenges in advancing maturity. (Created by the author)

endorsement and commitment. Cultural Resistance points to the challenge of overcoming ingrained behaviors and mindsets that resist change. Data Quality Issues emphasize the problems related to ensuring data accuracy, completeness, and reliability, which are foundational to effective data governance. Integration of Data Across Silos addresses the difficulty of unifying disparate data sources to create a cohesive data environment. Resource Constraints underline the limitations related to budget, personnel, and technology that can hinder maturity initiatives. Lack of Clear Strategy or Roadmap highlights the need for a well-defined plan to guide maturity advancement efforts. Keeping Pace with Technological Changes focuses on the challenge of adapting to rapid technological advancements and integrating new tools and systems. Measuring Progress and Value underscores the importance of tracking the effectiveness and benefits of maturity initiatives. Regulatory Compliance points to the necessity of adhering to legal and regulatory standards, which can be complex and demanding. Talent and Skill Shortages highlight the difficulty of finding and retaining individuals with the necessary skills and expertise to drive maturity improvement. This map visually integrates these elements, illustrating how these interconnected challenges collectively impact the ability to advance data governance maturity within an organization. Each branch highlights a specific area of concern, providing a comprehensive view of the potential obstacles to achieving higher maturity levels.

Each of these challenges requires a tailored approach, often involving a combination of leadership commitment, cultural change, strategic planning, and investment in resources and training. Addressing these challenges effectively is key to advancing the maturity of an organization's data governance or management capabilities.

17.9 Continuous Improvement in Data Governance

Continuous improvement in data governance refers to the ongoing process of enhancing data governance practices and policies within an organization. This approach is essential to ensure that data governance keeps pace with changes in business requirements, technology, and regulatory landscapes. Key elements of continuous improvement in data governance include:

1. **Regular Review and Updating of Policies and Standards**: As business environments and technologies evolve, it's critical to continuously review and update data governance policies and standards. Alation's guide, "How to Ensure Continuous Improvement With Data Governance," discusses methods to keep data governance practices aligned with current business requirements and technological advancements (Alation 2023).
2. **Feedback Mechanisms**: Implementing robust feedback mechanisms from stakeholders, including end-users, data managers, and executives, is vital. These inputs can highlight areas of success and those needing improvement.

17.9 Continuous Improvement in Data Governance

Atlan's "How to Improve Data Governance? Steps, Tips & Template" provides practical advice on setting up effective feedback loops that inform continuous improvement strategies (Atlan 2023c).

3. **Performance Metrics and Monitoring**: Establishing and monitoring performance metrics is crucial for evaluating the effectiveness of data governance initiatives. Analytics8's publication, "How to Improve Data Quality with Effective Data Governance," outlines how to use metrics to track improvements in data quality, a key aspect of governance that directly affects business outcomes (Analytics8 2023).
4. **Stakeholder Engagement**: Maintaining continuous engagement with stakeholders ensures that data governance efforts are well-supported and aligned with business needs. The document "The Journey to Data Governance Brilliance: Embracing Continuous Improvement" by Data Governance Platforms highlights strategies for engaging stakeholders in a manner that fosters a culture of continuous improvement (Data Governance Platforms 2023).
5. **Adapting to Regulatory Changes**: Keeping abreast of regulatory changes and ensuring that data governance policies and practices comply with these changes is a critical aspect of continuous improvement.
6. **Investing in Training and Awareness**: Regular training and awareness programs for staff at all levels can help to embed a culture of data governance within the organization.
7. **Leveraging Technology and Tools**: Utilizing the latest tools and technologies to streamline data governance processes can enhance efficiency and effectiveness.
8. **Incident Management and Learning from Mistakes**: Implementing robust processes for handling data governance incidents and learning from these incidents is key to improving practices over time.
9. **Benchmarking and Best Practices**: Comparing an organization's data governance practices with industry benchmarks and best practices can provide insights into areas of improvement.
10. **Agility and Flexibility**: Being agile and flexible in adapting data governance practices to meet new challenges and opportunities as they arise is crucial for continuous improvement.

Figure 17.5 illustrates the various components and strategies essential for maintaining and enhancing data governance practices within an organization. The central theme of continuous improvement in data governance is surrounded by several key elements that support this objective. Regular Review and Updating of Policies and Standards emphasize the need for constant evaluation and revision of data governance policies to ensure they remain relevant and effective. Investing in Training and Awareness highlights the importance of educating staff and promoting awareness about data governance practices and principles. Benchmarking and Best Practices involve comparing current practices with industry standards and adopting proven strategies to improve data governance. Leveraging Technology and Tools underscores the significance of using advanced tools and technologies to enhance

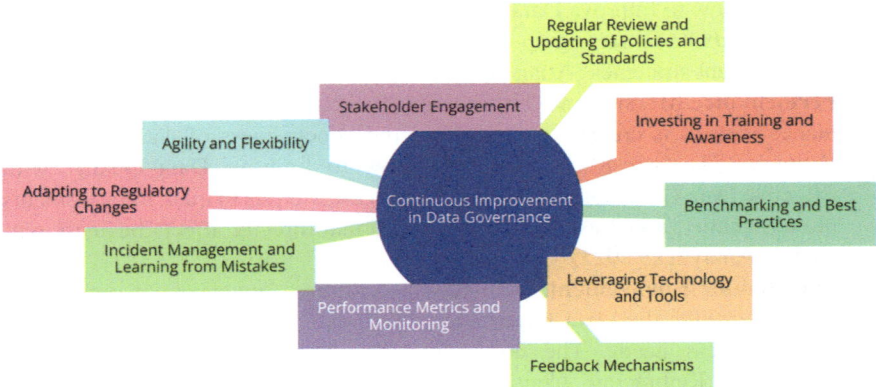

Fig. 17.5 Continuous improvement in data governance map. (Created by the author)

data management and governance processes. Feedback Mechanisms ensure that continuous improvement is guided by input from various stakeholders, allowing for adjustments and refinements based on real-world experiences. Performance Metrics and Monitoring focus on tracking the effectiveness of data governance initiatives through quantitative measures. Incident Management and Learning from Mistakes highlight the importance of addressing data governance issues promptly and using these experiences to prevent future occurrences. Adapting to Regulatory Changes stresses the need to stay compliant with evolving legal and regulatory requirements. Agility and Flexibility emphasize the ability to respond quickly to changes and new challenges in the data governance landscape. Stakeholder Engagement is crucial for ensuring that all relevant parties are involved in and committed to continuous improvement efforts. This map visually integrates these elements, illustrating how they collectively support a dynamic and evolving approach to data governance. Each branch represents a specific area of focus, providing a comprehensive view of the strategies needed to achieve continuous improvement in data governance within an organization.

Continuous improvement in data governance is about creating a dynamic, responsive approach to managing an organization's data assets, ensuring they remain a valuable and secure resource that supports the organization's goals and needs.

17.10 Conclusion

In concluding the chapter on Data Governance Maturity Models, the journey towards improving data governance in an organization is both complex and multifaceted. This chapter has provided a comprehensive overview, covering various aspects of data governance maturity:

1. **Introduction to Data Governance Maturity Models**: This section highlighted the importance of understanding data governance maturity models as frameworks that help organizations assess and enhance their data governance programs.
2. **Overview of Common Data Governance Maturity Models**: Different models were explored, providing insights into various approaches organizations can adopt to assess and structure their data governance efforts.
3. **Assessing Your Current Maturity Level**: This part emphasized the importance of accurately assessing the current state of an organization's data governance maturity as a starting point for improvement.
4. **Developing a Roadmap for Maturity Advancement**: The chapter underscored the need for a strategic roadmap tailored to an organization's specific goals and maturity level, guiding the progression towards higher levels of data governance maturity.
5. **Key Factors Influencing Data Governance Maturity**: Critical factors that impact the maturity of data governance, such as leadership commitment, stakeholder engagement, and resource allocation, were discussed.
6. **Aligning Maturity Improvement with Business Objectives**: This section stressed the importance of aligning data governance maturity improvement efforts with overall business objectives to ensure relevance and value.
7. **Case Studies: Maturity Model Implementation**: Real-world examples provided practical insights into the implementation of data governance maturity models, highlighting challenges and successes experienced by various organizations.
8. **Challenges in Advancing Maturity**: The chapter identified and explored usual challenges faced by organizations in advancing their data governance maturity, such as cultural resistance, resource constraints, and technological changes.
9. **Continuous Improvement in Data Governance**: Finally, the emphasis was on the need for ongoing improvement in data governance practices, highlighting the dynamic nature of data governance and the continuous effort required to maintain and enhance maturity over time.

Overall, the chapter provides a thorough understanding of the intricacies involved in developing, assessing, and enhancing data governance within an organization. It underscores the necessity of a strategic approach, continuous assessment, and adaptation to the evolving business and technological landscapes to effectively manage and govern data assets.

References

Alation (2023) How to ensure continuous improvement with data governance. Alation
Analytics8 (2023) How to improve data quality with effective data governance. Analytics8
Atlan (2023a) Data governance readiness assessment: a step-by-step guide. Atlan
Atlan (2023b) 10 steps to create an effective data governance roadmap. Atlan
Atlan (2023c) How to improve data governance? Steps, tips & template. Atlan

CastorDoc (2023) Data governance maturity model: a roadmap to optimizing your data initiatives and driving business value. CastorDoc
Chrissis MB, Konrad M, Shrum S (2003) CMMI: guidelines for process integration and product improvement. Addison-Wesley Professional
CMMI Institute (2014) Data management maturity (DMM) model. Software Engineering Institute, Carnegie Mellon University
Data Governance Platforms (2023) The journey to data governance brilliance: embracing continuous improvement. Data Governance Platforms
DATAVERSITY (2023a) Data governance and the maturity assessment model. DATAVERSITY
DATAVERSITY (2023b) The importance of a data governance roadmap. DATAVERSITY
DATAVERSITY (2023c) Data governance success: aligning your strategy with business objectives. DATAVERSITY
DATAVERSITY (2023d) Data governance challenges (and how to fix them). DATAVERSITY
Duranton S et al (2021) Is your company gaining momentum in data? BCG
EDRM (2012) Information governance reference model (IGRM). EDRM
Enterprise Data Management Council (2020a) EDM council data management capability assessment model (DCAM). EDM Council
Enterprise Data Management Council (2020b) DCAM: data management capability assessment model. EDM Council
IBM (2023) A step-by-step guide to setting up a data governance program. IBM
IBM Corporation (2010) IBM information governance solutions. IBM
Informatica (2023) Data governance framework: 4 pillars for success. Informatica
IT Governance Docs (2023) Data governance maturity model. IT Governance Docs
McKinsey & Company (2023a) Designing data governance that delivers value. McKinsey & Company
McKinsey & Company (2023b) Data analytics: accelerating maturity in the US public sector. McKinsey
Mosley M (ed) (2017) DAMA-DMBOK: data management body of knowledge, 2nd edn. DAMA International
ScienceDirect (2019) Data governance: a conceptual framework, structured review, and research agenda. ScienceDirect

Chapter 18
Conclusion: The Evolving Landscape of Data Governance

Contents

18.1	Reflection on the Journey of Data Governance	514
18.2	The Current State of Data Governance	516
18.3	Major Takeaways from the Book	517
18.4	The Future of Data Governance	518
18.5	Preparing for Ongoing Changes	519
18.6	The Role of Leadership and Culture	519
18.7	Final Thoughts on Building a Resilient Data Governance Framework	520
18.8	Encouraging a Community of Practice	521
18.9	Closing Remarks	522
References		523

Abstract This chapter serves as a comprehensive synthesis of the themes discussed throughout the book, highlighting the dynamic and ever-changing nature of data governance. This chapter encapsulates the significant aspects of data governance, emphasizing how it has become an indispensable part of modern organizational management due to the increasing reliance on data-driven decision-making and the need for compliance with evolving regulations.

The chapter begins by revisiting the critical role of data privacy regulations like GDPR and CCPA, which have reshaped how organizations manage, protect, and utilize data. It discusses the increased importance of data security measures in the wake of rising cyber threats and the ways organizations are strengthening their defenses to protect sensitive information.

Further, the chapter touches on the significance of maintaining high data quality and the integration of advanced tools like data catalogs and metadata management to enhance the understanding and usability of data assets. It also examines the growing incorporation of AI and machine learning in governance practices, which are helping organizations automate processes and ensure consistency and compliance across their data management systems.

The shift toward cloud and hybrid data environments is identified as a pivotal trend, presenting both challenges and opportunities in data governance. This migration necessitates adaptable governance frameworks that can operate effectively in increasingly complex IT landscapes.

Another focal point of the chapter is the discussion of various data governance frameworks and the best practices that have emerged to aid organizations in crafting effective governance strategies. It acknowledges the rising awareness of data ethics, prompting organizations to consider the broader implications of their data practices, particularly concerning AI and data handling.

The chapter also delves into the maturity of data governance programs, noting that organizations typically evolve from basic to more sophisticated governance practices. This maturation process is crucial for organizations aiming to leverage data governance as a strategic asset, enhancing operational efficiency, compliance, and decision-making capabilities.

In conclusion, the chapter reflects on the journey of data governance, stressing its status as a critical strategic function that transcends mere technical necessity. It highlights the need for continuous adaptation and improvement in governance strategies to keep pace with technological advancements and regulatory changes. This dynamic field requires ongoing attention, flexibility, and a proactive approach to harness the benefits of data governance fully.

Overall, this chapter provides a reflective summary of the evolving landscape of data governance, reinforcing the essential nature of robust data governance frameworks in supporting the strategic goals of modern organizations. It serves as both a conclusion and a call to action for organizations to continually evaluate and enhance their data governance practices in response to an increasingly data-driven world.

Keywords Data governance · Data privacy regulations · Data security · Data quality · Metadata management · AI governance · Cloud environments · Data ethics · Data governance tools · Data governance frameworks · Regulatory compliance · Technological trends · Data-driven decisions · Organizational change

In this chapter of this book about the evolving landscape of data governance, we discuss the current state of data governance practices, challenges, and trends in the world of data management. Data governance refers to the processes, policies, and controls put in place by organizations to ensure that data is managed, used, and protected effectively and in compliance with regulations. Here are some key topics:

1. **Data Privacy Regulations**: Data governance has been heavily influenced by data privacy regulations like GDPR (General Data Protection Regulation) and CCPA (California Consumer Privacy Act) in recent years. Organizations have had to adapt their data governance practices to ensure compliance with these regulations.
2. **Data Security**: With the increasing frequency and sophistication of data breaches, data security has become a paramount concern in data governance.

Organizations are adopting stricter security measures and encryption techniques to protect sensitive data.
3. **Data Quality**: Maintaining high data quality is crucial for making informed business decisions. Data governance efforts often include data quality assessments, cleansing, and data stewardship programs.
4. **Data Catalogs and Metadata Management**: Managing metadata and creating data catalogs have gained prominence in data governance. These tools help organizations discover and understand their data assets better.
5. **AI and Machine Learning Governance**: As AI and machine learning become more integrated into business processes, there is a need for governance around the development, deployment, and monitoring of AI models to ensure fairness, transparency, and compliance.
6. **Cloud and Hybrid Environments**: The migration to cloud-based data storage and processing has introduced new challenges for data governance. Organizations need to adapt their governance practices to work seamlessly in cloud and hybrid environments.
7. **Data Governance Frameworks**: Various data governance frameworks and best practices have emerged to help organizations establish effective data governance programs, such as the DAMA (Data Management Association) framework and the Data Governance Institute's framework.
8. **Data Ethics**: With increased awareness of the ethical implications of data use, organizations are integrating ethical considerations into their data governance strategies, including responsible AI and data handling.
9. **Data Governance Tools**: The market for data governance tools and platforms has grown, offering solutions for data lineage, data cataloging, data stewardship, and more.
10. **Data Governance Maturity**: Organizations are continually assessing and improving their data governance maturity. They often start with basic data governance practices and gradually mature into more advanced and comprehensive programs.

Figure 18.1 illustrates the dynamic and rapidly changing nature of data governance. This landscape is shaped by various factors, including emerging regulations, technological advancements, and the increasing importance of ethical considerations. Organizations are now more focused on ensuring data privacy, strengthening security measures, and maintaining high data quality to make informed decisions. The integration of artificial intelligence (AI) and machine learning (ML) into business processes necessitates robust governance to ensure fairness and compliance. Additionally, the shift towards cloud and hybrid environments introduces new challenges that require adaptable governance practices. Frameworks and tools have become essential for establishing effective governance strategies, facilitating data management, and promoting ethical data use. As organizations navigate these complexities, the maturity of their data governance programs continues to evolve, highlighting the ongoing journey toward achieving excellence in data management and

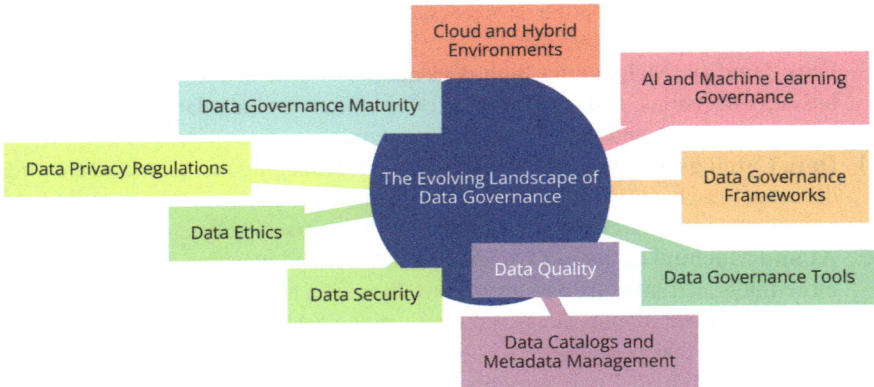

Fig. 18.1 The evolving landscape of data governance. (Created by the author)

governance. This dynamic field requires continuous attention and adaptation to meet the demands of an increasingly data-driven world.

The landscape of data governance is rapidly evolving, shaped by emerging regulations, technological advancements, and the growing importance of ethical considerations. Organizations are increasingly focusing on ensuring data privacy, bolstering security measures, and maintaining high data quality to make informed decisions.

The integration of AI and machine learning into business processes necessitates robust governance to ensure fairness and compliance. Moreover, the shift towards cloud and hybrid environments poses new challenges, requiring adaptable governance practices. Frameworks and tools have become indispensable for establishing effective governance strategies, facilitating data management, and fostering ethical data use.

As organizations navigate through these complexities, the maturity of their data governance programs continues to evolve, highlighting the ongoing journey toward achieving excellence in data management and governance.

This dynamic field requires continuous attention and adaptation to meet the demands of an increasingly data-driven world.

18.1 Reflection on the Journey of Data Governance

"Reflection on the Journey of Data Governance" refers to a retrospective or introspective examination of an organization's experiences, challenges, and progress in the field of data governance. This reflection may take various forms, including reports, articles, presentations, or discussions, and it typically aims to draw lessons from the data governance journey. Here are some general insights into what such a reflection might encompass:

1. **Historical Perspective**: A reflection on the journey of data governance may begin by looking back at the organization's historical approach to data management (Alhassan et al. 2018). This could include the reasons for initiating data governance efforts, key milestones, and initial challenges.
2. **Achievements and Progress**: The reflection would highlight the achievements and progress made in the field of data governance. This might include improvements in data quality, compliance with regulations, increased data transparency, and better data-driven decision-making (Ross and Beath 2019).
3. **Challenges and Obstacles**: It is essential to acknowledge and analyze the challenges and obstacles encountered during the data governance journey. These could include resistance to change, resource constraints, technology limitations, and organizational culture issues (Janssen et al. 2012).
4. **Lessons Learned**: Reflecting on the journey of data governance involves identifying and documenting lessons learned from both successes and failures (Liebowitz 2013). These lessons can inform future data governance initiatives and help avoid repeating past mistakes.
5. **Impact on Business**: An assessment of how data governance has positively impacted the organization's business processes, customer relationships, and overall performance (Ross and Beath 2019) is typically included. This could involve case studies or data-driven evidence of improved outcomes.
6. **Regulatory Compliance**: If applicable, the reflection may discuss how data governance efforts have ensured compliance with data privacy and security regulations, such as GDPR, CCPA, HIPAA, etc. (Voigt and von dem Bussche 2017).
7. **Data Culture**: An organization's data culture plays a significant role in the success of data governance. Reflecting on changes in the data culture, including increased data awareness, data literacy, and data ownership, can be crucial (Floridi 2013).
8. **Future Directions**: The reflection may conclude by outlining the organization's plans and aspirations for data governance (Aaronson 2018). This could involve setting new goals, exploring emerging trends, and considering advancements in data technology.
9. **Stakeholder Feedback**: Gathering feedback from key stakeholders, such as data stewards, business units, and IT teams, can provide valuable insights into the effectiveness of data governance efforts.
10. **Continuous Improvement**: Recognizing that data governance is an ongoing process, the reflection may emphasize the importance of continuous improvement and adaptation to evolving data challenges and opportunities.

Overall, a reflection on the journey of data governance serves to assess the past, inform the present, and shape the future of an organization's data governance efforts. It helps organizations refine their strategies and make data governance an integral part of their business operations.

18.2 The Current State of Data Governance

The current state of data governance reflects an increasingly complex and pivotal area in the world of data management and information technology. Here are some key aspects:

1. **Rising Importance**: Data governance has become more crucial as organizations recognize the value of data as a strategic asset (Janssen et al. 2012. This importance is driven by the need for better data quality, compliance with regulations, and efficient data management. Example: Major financial institutions have established centralized data governance teams to ensure data is utilized effectively for decision-making and strategy formulation.
2. **Regulatory Compliance**: With the introduction of regulations like the General Data Protection Regulation (GDPR) in Europe, California Consumer Privacy Act (CCPA), and others, data governance plays a vital role in ensuring compliance (Aaronson 2018). These regulations mandate how data is collected, stored, processed, and shared, making data governance a necessity for legal compliance. Example: Companies like Facebook and Google have had to significantly revamp their data governance policies to comply with GDPR, affecting how they handle user data globally.
3. **Integration of AI and ML**: Artificial intelligence (AI) and machine learning (ML) are increasingly being integrated into data governance frameworks. These technologies help in automating data governance processes, analyzing large datasets, and providing insights for better decision-making (Ross and Beath 2019). Example: Retail giants like Amazon use AI to automate data categorization and quality checks, streamlining their data governance processes.
4. **Data Quality and Integrity**: Ensuring the quality and integrity of data is a core focus. This involves establishing processes and standards to ensure that data is accurate, consistent, and reliable (Liebowitz 2013). Example: Healthcare providers, such as hospitals, implement rigorous data quality protocols to ensure patient information's accuracy, which is vital for treatment and care.
5. **Data Privacy and Security**: Protecting sensitive data from breaches and unauthorized access is a critical aspect of data governance. This involves implementing robust security measures and privacy policies (Floridi 2013). Example: After high-profile data breaches, companies like Equifax have strengthened their data governance frameworks to protect sensitive customer information more effectively.
6. **Data Literacy**: Organizations are focusing on improving data literacy among employees to ensure that everyone understands the importance of data governance and how to handle data responsibly (Liebowitz 2013). Example: Global consulting firms like Deloitte offer training programs to their employees to enhance understanding and adherence to data governance policies.
7. **Tools and Technologies**: The market for data governance tools is growing, with solutions offering capabilities in data cataloging, quality control, compliance management, and more (Talburt et al. 2020). Example: Financial services

companies are increasingly adopting platforms like Collibra for data governance, which helps in managing data compliance, quality, and lineage.
8. **Collaboration and Culture**: Effective data governance requires a shift in organizational culture, emphasizing collaboration across different departments and levels. This collaborative culture fosters more effective data governance practices (Ross and Beath 2019). Example: Tech companies like Microsoft encourage a collaborative culture where IT and business units work together for effective data governance.
9. **Cloud and Hybrid Environments**: With the shift towards cloud and hybrid IT environments, data governance strategies are adapting to manage data across on-premises and cloud platforms. This shift necessitates new governance models that can function effectively in these diverse environments (Aaronson 2018). Example: Enterprises like IBM have transitioned to cloud-based data governance solutions to manage data across diverse environments, both on-premises and in the cloud.
10. **Ethical Considerations**: Ethical considerations, including how data is used and for what purposes, are increasingly part of the data governance conversation, especially in sectors like healthcare and finance. These considerations ensure that data governance also aligns with ethical standards (Floridi 2013). Example: AI research and deployment companies, like DeepMind, are incorporating ethical data governance practices to ensure responsible use of data, especially in sensitive fields like AI ethics.

These examples demonstrate how various industries and companies are responding to and shaping the current trends in data governance.

18.3 Major Takeaways from the Book

The book focuses on several key aspects of data governance:

1. **Overview and Historical Evolution**: It covers the definition, scope, and historical development of data governance, emphasizing its growing relevance in the modern digital landscape.
2. **Key Objectives and Benefits**: The book discusses the primary goals of data governance, such as ensuring data quality, regulatory compliance, and enhancing decision-making. It also outlines the benefits like operational efficiency and improved customer trust.
3. **Pillars of Data Governance**: Key pillars such as data quality, security, privacy, and efficient data management are examined in detail.
4. **Challenges and Misconceptions**: It addresses usual challenges in implementing data governance and debunks prevalent myths, highlighting the strategic, not just technical, nature of data governance.
5. **Aligning with Business Objectives**: The book stresses the importance of aligning data governance with business goals to ensure its effectiveness and relevance.

6. **Success Stories**: Real-world case studies are presented to illustrate the transformative impact and tangible benefits of effective data governance in various organizations.
7. **Implementation Strategies**: It offers insights into overcoming resistance to change, addressing data quality issues, and the vital role of technology and tools in supporting data governance efforts.

Major takeaways include the historical evolution and current relevance of data governance, key objectives such as ensuring data quality and regulatory compliance, and the benefits of implementing data governance like increased operational efficiency and improved customer trust. It also discusses the challenges in maintaining data quality, the importance of data security and privacy, efficient data management strategies, and the significance of data governance in modern organizations. Real-world success stories and case studies are presented to illustrate the effective implementation and impact of data governance. The book provides a comprehensive understanding of data governance and its critical role in contemporary organizational management.

18.4 The Future of Data Governance

The future of data governance is likely to be shaped by several key trends:

1. **Increased Automation and AI Integration**: Data governance processes are expected to become more automated with advanced AI and machine learning integration, leading to more efficient data management and decision-making (Ross and Beath 2019).
2. **Enhanced Focus on Privacy and Security**: As data breaches and privacy concerns continue to rise, data governance will increasingly focus on robust security measures and privacy regulations (Aaronson 2018).
3. **Greater Emphasis on Data Ethics**: Ethical considerations will become a central aspect of data governance, addressing how data is used, particularly in sensitive sectors like healthcare and finance (Floridi 2013).
4. **Cloud-Based Governance Models**: The shift towards cloud computing will see more cloud-based data governance models offering flexibility and scalability, which are essential for managing data across diverse environments (Aaronson 2018).
5. **Data Literacy Across Organizations**: There will be a greater emphasis on data literacy, ensuring all organizational levels understand the importance of data governance (Liebowitz 2013).
6. **Regulatory Evolution**: Data governance will continue to evolve in response to changing global regulations, requiring organizations to stay agile and compliant (Janssen et al. 2012).

Real-world examples include companies like Google and Amazon investing heavily in AI for data management and European Union's GDPR influencing global data governance practices.

18.5 Preparing for Ongoing Changes

Preparing for ongoing changes in any field, especially in areas like technology, data governance, or organizational management, involves several key strategies:

1. **Continuous Learning and Adaptation**: Stay updated with the latest trends and developments in your field. This could involve regular training, attending workshops, or keeping up with industry publications (Ross and Beath 2019).
2. **Flexibility and Resilience**: Develop a mindset and organizational culture that embraces change rather than resists it. This includes being open to innovative ideas and approaches (Floridi 2013).
3. **Scenario Planning**: Anticipate potential future scenarios and challenges. This can help in creating flexible strategies that can adapt to various situations (Janssen et al. 2012).
4. **Investing in Technology and Skills**: Keeping technology infrastructure and skills updated is crucial for dealing with changes, especially in fields driven by rapid technological advancements (Talburt et al. 2020).
5. **Building a Collaborative Environment**: Encourage collaboration and communication within teams and across departments to ensure a cohesive response to changes (Ross and Beath 2019).
6. **Risk Management**: Identify and assess potential risks associated with changes and develop plans to mitigate them (Aaronson 2018).

For instance, companies like Amazon and Google constantly update their algorithms and infrastructure to stay ahead in the fast-paced tech industry, demonstrating a commitment to continuous learning and adaptation. Similarly, businesses worldwide had to quickly adapt to remote working models during the COVID-19 pandemic, highlighting flexibility and resilience.

18.6 The Role of Leadership and Culture

The role of leadership and culture in any organization is pivotal. Leadership sets the tone and direction, influencing how goals are pursued and values upheld. Effective leaders inspire, motivate, and guide their teams, fostering a positive and productive work environment (Plotkin 2013). They also play a crucial role in decision-making and strategizing. Culture embodies the shared values, beliefs, and practices within an organization. It shapes the way people interact, work, and approach challenges. A strong, positive culture can enhance job satisfaction, collaboration, and overall performance (O'Reilly Media 2017). Leaders must champion and strategically support Data Governance initiatives while organizational culture needs to evolve to embrace data literacy, collaboration, and a shared understanding of the value of Data Governance (Redman 2008). Together, leadership and culture are fundamental in shaping the ethos of an organization, influencing its success, adaptability, and resilience. They are instrumental in driving change, promoting innovation, and maintaining a healthy work environment (Finlay 2018). Furthermore, the role of leadership in ensuring compliance with regulations such as GDPR, and fostering a culture that upholds ethical standards, is paramount (Voigt and von dem Bussche 2017).

Demonstrative examples include companies like Apple, where leadership under Steve Jobs established a culture of innovation and excellence, or Google, known for its open and collaborative culture fostered by its leadership.

18.7 Final Thoughts on Building a Resilient Data Governance Framework

Building a resilient data governance framework involves several key components:

1. **Strong Leadership Commitment**: Leadership must be actively involved and committed to data governance. They set the vision and ensure alignment with organizational goals (Ross and Beath 2019).
2. **Cultural Integration**: Data governance should be integrated into the organizational culture. This means promoting data awareness and literacy across all levels (Floridi 2013).
3. **Flexible and Scalable Framework**: The framework should be adaptable to changes in technology, business processes, and regulatory environments (Aaronson 2018).
4. **Effective Communication**: Clear communication channels and regular updates are crucial for engaging stakeholders and ensuring everyone is on the same page (Ross and Beath 2019).
5. **Regular Training and Education**: Continuous education and training programs help in keeping the team updated and competent (Janssen et al. 2012).
6. **Robust Technology Infrastructure**: "Invest in the right tools and technologies to support data governance processes (Talburt et al. 2020).
7. **Comprehensive Policies and Procedures**: Develop and enforce clear policies and procedures for data management, privacy, and security (Floridi 2013).
8. **Monitoring and Continuous Improvement**: Regularly monitor the effectiveness of data governance initiatives and be open to adjustments and improvements (Ross and Beath 2019).

Building a resilient data governance framework requires a comprehensive approach that combines strong leadership, a supportive culture, adaptable structures, clear communication, ongoing education, robust technology, and continuous monitoring and improvement.

18.8 Encouraging a Community of Practice

Encouraging a Community of Practice is an essential aspect of successful data governance, as it fosters collaboration, knowledge sharing, and a culture of responsible data management among employees.

Encouraging a Community of Practice

1. **Define Objectives**
   - Clearly articulate the goals and objectives of the data governance initiative to the organization (Liebowitz 2013).

2. **Leadership Support**
   - Secure executive sponsorship to demonstrate the importance of data governance (Alhassan et al. 2018).

3. **Education and Training**
   - Provide training sessions and resources to help employees understand data governance principles (Liebowitz 2013).

4. **Communication**
   - Foster open communication channels to discuss data governance issues, challenges, and successes (Floridi 2013).

5. **Cross-Functional Teams**
   - Form cross-functional teams with members from different departments to collaborate on data-related projects (Janssen et al. 2012).

6. **Knowledge Sharing**
   - Encourage the sharing of best practices, case studies, and success stories related to data governance (Alhassan et al. 2018).

7. **Feedback Mechanisms**
   - "Create feedback loops to gather input and suggestions from employees regarding data governance improvements" (Janssen et al. 2012).

8. **Recognition and Rewards**
   - Acknowledge and reward individuals or teams for their contributions to data governance and best practices (Liebowitz 2013).

9. **Community Events**
   - Organize regular meetings, workshops, or webinars to bring the community together for learning and networking (Floridi 2013).

10. **Documentation**
    - Maintain a central repository of data governance documentation, resources, and templates for easy access (Alhassan et al. 2018).

By encouraging a Community of Practice, organizations can harness the power of data to make informed decisions, ensure data integrity, and drive business success while fostering a culture of collaboration and responsibility among employees.

18.9 Closing Remarks

As we conclude this journey through the multifaceted world of Data Governance, it is important to reflect on the critical role that data plays in today's organizations. This book has traversed the expansive landscape of Data Governance, underscoring its significance in shaping the modern digital ecosystem. From ensuring regulatory compliance and enhancing decision-making capabilities to fostering operational efficiency and driving innovation, the realm of Data Governance is both dynamic and indispensable.

The evolution of Data Governance from a technical concern to a strategic priority reflects the growing recognition of data as a vital organizational asset. This transformation necessitates a comprehensive approach, integrating people, processes, and technology. We have explored the challenges in implementing effective Data Governance frameworks, including technological complexities, cultural resistances, and the need for organizational alignment.

A key takeaway from this exploration is the necessity for organizations to be agile and adaptable. The ever-evolving regulatory landscapes, technological advancements, and changing business needs demand a flexible and proactive approach to Data Governance. Balancing the need for advanced technological solutions with considerations of cost, scalability, and integration with existing systems is paramount.

Furthermore, the role of leadership and the importance of fostering a data-centric culture within organizations cannot be overstated. Leaders must champion and strategically support Data Governance initiatives, while organizational culture needs to evolve to embrace data literacy, collaboration, and a shared understanding of the value of Data Governance.

As we step into the future, the significance of Data Governance will only amplify. The increasing volume, velocity, and variety of data, coupled with escalating cybersecurity threats and stringent regulatory requirements, will continue to challenge organizations. However, these challenges also present opportunities for innovation, growth, and competitive advantage.

In closing, remember that Data Governance is not a destination but a journey. It requires continuous improvement, adaptability, and commitment. The principles and practices discussed in this book are not just theoretical concepts but practical tools to navigate the complex and ever-changing world of data. Embrace these challenges, leverage the opportunities, and let Data Governance be the beacon that guides your organization toward digital resilience and strategic excellence.

References

Aaronson S (2018) Data is different: why the world needs a new approach to governing cross-border data flows, CIGI papers no. 197. Centre for International Governance Innovation

Alhassan I, Sammon D, Daly M (2018) Data governance activities: a comparison between scientific and practice-oriented literature. J Enterp Inf Manag (Emerald Publishing Limited) 31:300

Farkas J, Shrestha R, Akram H (2020) Security and privacy issues in IoT devices and sensor networks. Springer

Finlay S (2018) Artificial intelligence and machine learning for business. Relativistic Publishing

Floridi L (2013) The ethics of information. Oxford University Press

Giordano AD (2014) Data governance: how to design, deploy, and sustain an effective data governance program. IBM Press

Janssen M, Charalabidis Y, Zuiderwijk A (2012) Benefits, adoption barriers and myths of open data and open government. Inf Syst Manag (Taylor & Francis) 29:258

Liebowitz J (2013) Data governance: creating value from information assets. CRC Press

O'Reilly Media (2017) Data governance: the definitive guide. O'Reilly Media

Plotkin D (2013) Data stewardship: an actionable guide to effective data management and data governance. Elsevier

Redman TC (2008) Data driven: profiting from your most important business asset. Harvard Business Review Press

Ross JW, Beath CM (2019) Designing and executing digital strategies. Management Information Systems Quarterly, Association for Information Systems

Talburt et al (2020) Data curation: foundations, principles, and practice. Springer

Voigt P, von dem Bussche A (2017) The EU general data protection regulation (GDPR): a practical guide. Springer International Publishing

References

Aaronson S (2018) Data is different: why the world needs a new approach to governing cross-border data flows, CIGI papers no. 197. Centre for International Governance Innovation

Abedjan Z, Golab L, Naumann F (2015) Profiling relational data: a survey. VLDB J 24(4):557–581

Abraham R, Schneider J, Vom Brocke J (2019a) Data governance: a conceptual framework, structured review, and research agenda. Int J Inf Manag 49:424–438

Abraham R, Schneider J, Vom Brocke J (2019b) Data governance: a conceptual framework, structured review, and research agenda. Springer

Accenture (2018) The pulse of enterprise IT: data-driven innovation. Accenture

AHIMA (n.d.) Healthcare data governance: practice brief. AHIMA

Ahmad K, Maabreh M, Ghaly K, Khan J, Qadir J (2022) Developing future human-centered smart cities: critical analysis of smart city security, data management, and ethical challenges. Comput Sci Rev 43:100452

Ahsan M, Tushar FI, Islam M (2022) Machine learning in cybersecurity: techniques and applications. J Cybersecur

Al Jazeera (2018) Cambridge analytica and Facebook: the scandal so far. Al Jazeera

Alation (2020) Data catalog & data governance platform. Alation

Alation (2022) How to ensure continuous improvement with data governance. Alation

Alation (2023a) How to ensure continuous improvement with data governance. Alation

Alation (2023b) Top 6 common data governance challenges (and their solutions!). Alation

Alation (n.d.) Data governance in telecom: what you need to know. Alation

Al-Badi A, Tarhini A, Khan AI (2018) Exploring big data governance frameworks comparing different data governance frameworks to the ISO 8000 standard. Springer

Aldoseri A, Al-Khalifa KN, Hamouda AM (2023) Re-thinking data strategy and integration for artificial intelligence: concepts, opportunities, and challenges. Appl Sci 13(12):7082

Alhassan I, Sammon D (2016) Data governance activities: an analysis of the literature. Springer

Alhassan I, Sammon D, Daly M (2016) Data governance activities: an analysis of the literature. J Decis Syst 25(sup1):64–75

Alhassan I, Sammon D, Daly M (2018a) Data governance activities: a comparison between scientific and practice-oriented literature. J Data Inf Qual 31:300

Alhassan I, Sammon D, Daly M (2018b) Data governance activities: a comparison between scientific and practice-oriented literature. J Enterp Inf Manag (Emerald Publishing Limited) 31:300

Al-Jarrah OY, Yoo PD, Muhaidat S, Karagiannidis GK, Taha K (2015) Efficient machine learning for big data: a review. Big Data Res 2(3):87–93. Elsevier

Almulihi A, Alassery F, Khan W (2022) Implications of healthcare data breaches in the big data era. Health Informatics J

© The Editor(s) (if applicable) and The Author(s), under exclusive license to Springer Nature Switzerland AG 2024
D. Sargiotis, *Data Governance*, https://doi.org/10.1007/978-3-031-67268-2

American Health Information Management Association (AHIMA) (2019) Data Quality Management model. AHIMA
American Journal of Health-System Pharmacy (2021) HIPAA Compliance in Healthcare. American Journal of Health-System Pharmacy
American Journal of Public Health (2019) HIPAA compliance in healthcare. American Journal of Public Health
Analytics8 (2023) How to improve data quality with effective data governance. Analytics8
Analytics8 (n.d.) Defining data governance roles & responsibilities. Analytics8
Anthony David Giordano (2014) Data governance: how to design, deploy, and sustain an effective data governance program. IBM Press
Apache (2021a) Apache NiFi documentation
Apache (2021b) What is Airflow™?
Arndt A (2023) Data governance metrics & KPIs to measure success. Experian
Ashraf N (2022) Corporate governance and data breaches: the role of peer events. J Corp Gov
Askham N (2023) Navigating data mesh and evolving data governance: a practical guide
Astera (n.d.) Data governance in financial services: a complete analysis. Astera
Ataccama (2021) Automating metadata management. Ataccama
Ataccama (2023) How GenAI boosts data governance initiatives. Ataccama
Atlan (2022a) 5 data governance examples: case studies, takeaways & more. Atlan
Atlan (2022b) Data governance in insurance: 7 ways it impacts insurers. Atlan
Atlan (2022c) How to improve data governance? Steps, tips & template. Atlan
Atlan (2023a) How to improve data governance? Steps, tips & template. Atlan
Atlan (2023b) 5 data governance examples: case studies, takeaways & more. Atlan
Atlan (2023c) Data ethics unveiled: principles & frameworks explored. Atlan
Atlan (n.d.-a) Airbnb—data literacy for decision-making. Atlan
Atlan (n.d.-b) GE aviation—centralizing data sources. Atlan
Atlan (n.d.-c) Wells Fargo—single source of truth for governance. Atlan
Atlan. (2023) The top 12 data governance trends in 2024. Atlan
Aura (2023) 17 most common types of cyber attacks & examples (2024)
Bain & Company (2020) Data analytics in retail. Bain & Company Reports
Baltimore County Public Schools (n.d.) Blueprint 2.0: data governance program. Baltimore County Public Schools
BARC (2023) Navigating 2024's data landscape: key trends and developments in data management. BARC
Barlette Y, Baillette P (2022) Big data analytics in turbulent contexts: towards organizational change for enhanced agility. Production Planning & Control
Basukie J, Wang Y, Li S (2020) Big data governance and algorithmic management in sharing economy platforms: a case of ridesharing in emerging markets, vol 161. Technol Forecast Soc Change, p 120310
Batini C, Scannapieco M (2016) Data and information quality: dimensions, principles and techniques. Springer
Batini C, Cappiello C, Francalanci C, Maurino A (2009) Methodologies for data quality assessment and improvement. ACM Comput Surv (CSUR) 41(3):1–52
BCG (2021) A show-don't-tell approach to data governance. Boston Consulting Group
BCG (2022) Any company can become a resilient data champion. Boston Consulting Group
BCG (Boston Consulting Group) (2020) The data-driven transformation. Boston Consulting Group
BDO Digital (2021) The comprehensive guide to data governance. BDO
Benneyan JC, Lloyd RC, Plsek PE (2003) Statistical process control as a tool for research and healthcare improvement. BMJ Qual Saf 12(6):458–464
Bensberg F (2003) Data quality management and data governance. J Data Manag 12(4):56–72
BigID. (2023) Cloud data governance: overview and best practices. BigID
Bipartisan Policy Center (2023) History of the Cambridge Analytica controversy. Bipartisan Policy Center

References

BMC Public Health (2023) Effectiveness of capacity building interventions relevant to public health practice: a systematic review
Bode I, Deneckere A, Rossi G (2020) The role of hydrogen in decarbonising the steel industry. J Clean Prod 247:119115
Bowers S (2011) Scientific data management in the coming decade. Data Sci J 10:12–24
Brickclay (2023) Data governance: implementation, challenges, and solutions. Brickclay
Brookings (2023) Highlights: the GDPR and CCPA as benchmarks for federal privacy legislation
Brous P, Janssen M, Vilminko-Heikkinen R (2016) Coordinating decision-making in data management activities: a systematic review of data governance principles. Springer
Brous P, Janssen M, Herder P (2019) Internet of things adoption for reconfiguring decision-making processes in asset management. Bus Process Manag J 25:495
Brown G, Kyttä M (2018) Key issues and research priorities for public participation GIS (PPGIS): a synthesis based on empirical research. Appl Geogr 95:122–136
Buckman J, Hashim MJ, Woutersen T, Bockstedt J (2019) Fool me twice? Data breach reductions through stricter sanctions. In: The Cambridge handbook of compliance
California Law Review (2020) Understanding the CCPA and its implications
California Legislature (2018) California consumer privacy act (CCPA). California Civil Code, Title 1.81.5
Cambridge University Press (2019) Education and planning: anticipating and responding to skill gaps, changing skill needs and competencies
Cambridge University Press (2024) Building better global data governance
Cannon DM, Kessler L (2007) Danger–corporate data breach! J Corp Acc Finance 18(5):41–49
Carlton M, Levy Y (2017) Mitigating advanced persistent threats through cybersecurity skills. J Inf Syst Secur
CastorDoc (2023) Data governance maturity model: a roadmap to optimizing your data initiatives and driving business value. CastorDoc
CastorDoc (n.d.) Data governance in retail: best practices, challenges, and viable solutions. CastorDoc
Cavusoglu H, Mishra B, Raghunathan S (2004) The effect of internet security breach announcements on market value. Int J Electron Commer 9(1):70–104
Census (2023) How Airbnb democratized their data to empower their employees
Center for Internet Security, 2023
Chand R, Bhushan B, Jain S (2023) Role of blockchain technology in digital currency security. J Cryptogr Eng
Change Management Insight (n.d.) The process of change management in data governance. Change Management Insight
Chen JV, Li HC, Yen DC, Bata KV (2012) Did IT consulting firms gain when their clients were breached? Comput Hum Behav 28(2):456–464
Cheong LK, Chang V (2007) The need for data governance: a case study. Springer
Chrissis MB, Konrad M, Shrum S (2003) CMMI: guidelines for process integration and product improvement. Addison-Wesley Professional
CISA (2023) Incident response plan basics
Cisco (2020) Cisco consumer privacy survey. Cisco
Cisco (2021) Data privacy benchmark study. Cisco
Cisco (2023) What is a cyberattack?
CISecurity (2023) Why employee cybersecurity awareness training is important
Clarkston Consulting (n.d.) The importance of data governance in retail. Clarkston Consulting
CloudCodes (2021) Data governance best practices: top 6 best practices. CloudCodes
CMMI Institute (2014a) Data management maturity (DMM) model. CMMI Institute
CMMI Institute (2014b) Data management maturity (DMM) model. Software Engineering Institute, Carnegie Mellon University
CNIL. (2019). GDPR fines and data breach survey. CNIL
Cognopia (n.d.) The role of data stewards in data-driven businesses. Cognopia

Collibra (2020) Data governance in the cloud. Collibra
Collibra (2021a) Customer stories. Collibra
Collibra (2021b) Data governance council: what is it and why do you need one? Collibra
Collibra (2023) The importance of predictive model governance. Collibra
Collibra (n.d.) Top 6 best practices of data governance. Collibra
Communications of the ACM (2021) Understanding cyber attack methodologies. Communications of the ACM
Computer Law & Security Review (2021a) Investigation and assessment post-breach
Computer Law & Security Review (2021b) Privacy by default in data protection
Confluent (n.d.) Aligning data governance initiatives with business objectives
Consumer Financial Protection Bureau (2019) Cost of a data breach report 2020. Ponemon Institute
ContactPigeon (n.d.) The 24 best eCommerce retail case studies. ContactPigeon
Corporate Compliance Insights (2023) 8 crucial compliance KPIs. Corporate Compliance Insights
Corporate Finance Institute (n.d.) Data governance: overview, role, importance, goals. Corporate Finance Institute
Coursera (2024) Skills gap analysis: a guide to training your teams
Coyle D, Diepeveen S, Wdowin J, Kay L, Tennison J (2020) Informing the global data future: benchmarking data governance frameworks. Data & Policy, Cambridge Core
CrowdStrike (2023a) Data encryption explained
CrowdStrike (2023b) What is Data Loss Prevention (DLP)?
CrowdStrike (2023c) 12 most common types of cyberattacks today
CrowdStrike (2023d) Insider threats and how to identify them
CrowdStrike (2023e) Incident response plan: frameworks and steps
Customer Experience Magazine (2020) The importance of data quality in customer service. Customer Experience Magazine
Cybersecurity and Infrastructure Security Agency (CISA) (2023) Defining insider threats
Cybsafe (2023) 7 reasons why security awareness training is important
Dallas G, Lubrano M (2018) Governance, stewardship, and sustainability: theory, practice, and evidence, 2nd edn. Routledge
DAMA-DMBOK: Data management body of knowledge (2nd ed.). Technics Publications.
DAMA International (2017a) DAMA-DMBOK: data Management body of knowledge. DAMA International
DAMA International (2017b) The DAMA guide to the data management body of knowledge (DAMA-DMBOK). Technics Publications
Data Governance Handbook: A Practical Approach to Building Trust in Data by Wendy Batchelder
Data Governance Platforms (2023) The journey to data governance brilliance: embracing continuous improvement. Data Governance Platforms
Data Management and Governance Services: Simple and effective approaches by Tejasvi Addagada, Barry Lyons
Data Privacy Manager (2022) GDPR fine: Danske Bank fined €1.3 million over non-compliant data deletion processes. Data Privacy Manager
Data Quality: Dimensions, Measurement, Strategy, Management, and Governance by Rupa Mahanti
Databricks (2023) How John Deere uses industrial AI in precision agriculture. Databricks
Databricks (n.d.) What is data governance? A comprehensive guide. Databricks
DataCamp (2023a) An introduction to data ethics: what is the ethical use of data? DataCamp
DataCamp (2023b) Closing the data literacy gap: key Insights from the state of data literacy 2023 report. DataCamp
Datafloq (2023) 35 metrics you should use to monitor data governance. Datafloq
DataGalaxy (2023) Aligning data governance to support business line objectives. DataGalaxy
DataGalaxy (n.d.) Data governance for supporting business objectives. DataGalaxy
DataRobot (2021) AI and machine learning platform. DataRobot
Dataversity (2015) Case study: an inside look at Airbnb's data science journey. Dataversity
DATAVERSITY (2018) The data governance imperative for retailers. DATAVERSITY

DATAVERSITY (2022a) Data governance challenges (and how to fix them). DATAVERSITY
DATAVERSITY (2022b) What is data governance? Definition, types, uses. DATAVERSITY
DATAVERSITY (2022c) What are data governance metrics? DATAVERSITY
Dataversity (2023) Case study: three strategies for data governance success. Dataversity
DATAVERSITY (2023a) Data governance challenges (and how to fix them). DATAVERSITY
DATAVERSITY (2023b) Data governance success: aligning your strategy with business objectives. DATAVERSITY
DATAVERSITY (2023c) The importance of a data governance roadmap. DATAVERSITY
DATAVERSITY (2023d) A guide to predictive data analytics (making decisions for the future). DATAVERSITY
DATAVERSITY (2023e) Case study: three strategies for data governance success. DATAVERSITY
DATAVERSITY (2023f) Data governance best practices. DATAVERSITY
DATAVERSITY (2023h) Data governance metrics: how to measure success. DATAVERSITY
DATAVERSITY (2023i) Data governance roles and responsibilities. DATAVERSITY
DATAVERSITY (2023j) Data governance in the cloud. DATAVERSITY
DATAVERSITY (2023k) The future of data governance: balancing data governance and data Management. DATAVERSITY
DATAVERSITY (2024) Data governance trends in 2024. DATAVERSITY
DATAVERSITY (n.d.) Demonstrating the value of data governance. DATAVERSITY
Davenport TH (2014) Big data at work: dispelling the myths, uncovering the opportunities. Harvard Business Review Press
David Plotkin (2013) Data stewardship: an actionable guide to effective data Management and data governance. Elsevier
De Mauro A, Greco M, Grimaldi M (2015) What is big data? A consensual definition and a review of key research topics. AIP Conf Proc 1648(1):89–93
Delacroix S, Lawrence ND (2019) Bottom-up data trusts: disturbing the 'one size fits all' approach to data governance. Int Data Privacy Law. https://doi.org/10.1093/idpl/ipz014
Deloitte (2019a) Managing risk in Digital transformation. Deloitte
Deloitte (2019b) The culture of data literacy. Deloitte Insights
Deloitte (2020) Case studies on regulatory compliance. Deloitte
Deloitte (2021) The future of data governance in a data-rich world. Deloitte UK
Deloitte (2022) Deploying effective data governance to achieve key business priorities. Deloitte Insights
Deloitte (2023) Predictive analytics in government. Deloitte Insights
Deloitte (n.d.) Case studies on regulatory compliance in emerging technologies. Deloitte
Deloitte US (2023) Blockchain and quantum technologies in digital trust. Deloitte US
DEV Community (2022) Best practices for creating a robust data governance framework. DEV Community
Devane H (2021) The complete guide to data security compliance laws and regulations
Developing and Implementing Policies for Academic Integrity (2021) Oxford University Press
DiGGrowth (2023) The essential guide to data governance metrics: what to track and why. DiGGrowth
Diligent Corporation (n.d.) Regulatory compliance 101: definition, requirements & solutions. Diligent Corporation
DW (2020) H&M fined €35 million for privacy breaches. DW
Dyché J (2015) The new IT: how technology leaders are enabling business strategy in the digital age. McGraw-Hill Education
Earley S (2020) Leveraging data for competitive advantage: strategies for monetization and risk management. Inf Manag J 34(2):22–34
EDRM (2012) Information governance reference model (IGRM). EDRM
Emerald Insight (2001) Overcoming cultural barriers to sharing knowledge. J Knowl Manag. MCB UP Ltd.
Emerald Insight (2021) Big data analysis for decision-making processes: challenges and opportunities for the management of health-care organizations. Emerald Insight

Enforcing Regulation (2021) Oxford University Press

Enterprise Data Management Council (2020) DCAM: data Management capability assessment model. EDM Council

EPJ Data Science (2023) A global-scale analysis of the sharing economy model—an Airbnb case study. EPJ Data Science

Erwin Inc (2021) The value of data governance and how to quantify it. Quest Software

Eryurek E et al (2021) Data governance: the definitive guide: people, processes, and tools to operationalize data trustworthiness. O'Reilly Media

EU General Data Protection Regulation (GDPR) – An Implementation and Compliance Guide by IT Governance Privacy Team

European Commission (2004) Standard contractual clauses (SCCs) for the transfer of personal data to third countries. Official Journal of the European Union, L 119/89

European Commission (2018) 2018 reform of EU data protection rules. European Commission

European Data Protection Board (2020) Hamburg Commissioner Fines H&M 35.3 million euro for data protection violations in service Centre. European Data Protection Board

European Data Protection Board (2021) Guidelines on data protection by design and by default. European Data Protection Board

European Data Protection Board (EDPB) (2021) Guidelines 01/2021 on examples regarding data breach notification. European Data Protection Board (EDPB)

European Union (2016) General data protection regulation (GDPR). Official Journal of the European Union, L 119/1

European Union Agency for Cybersecurity (ENISA) (2023) Physical and environmental security

Experian (2023a) Data governance metrics & KPIs to measure success. Experian

Experian (2023b) Measuring success in data governance. Experian

EY India (2023) Transforming the future of Indian logistics sector. EY

Factom (2020) Harmony Integrates Blockchain Technology

Fan W, Geerts F (2022) Foundations of data quality management. Springer

Farkas J, Shrestha R, Akram H (2020) Security and privacy issues in IoT devices and sensor networks. Springer

Federal Rules of Civil Procedure (2015) Rule 26: duty to disclose; general provisions governing discovery. Federal Rules of Civil Procedure

Ferraiolo DF, Kuhn DR, Chandramouli R (2003) Role-based access control. Artech House

Financial Times (2019) Banking on GDPR: a financial services case study. Financial Times

Finlay S (2018) Artificial intelligence and machine learning for business. Relativistic Publishing

First San Francisco Partners (2016) Agile master data management and data governance in action. First San Francisco Partners

Forbes (2020a) The importance of data cleaning in business analytics. Forbes

Forbes (2020b) The human element of cybersecurity, Forbes

Forrester (2021a) The Forrester wave™: data quality solutions Q2 2021. Forrester

Forrester (2021b) The Forrester wave™: cloud-based data governance solutions, Q1 2021. Forrester

Four Cornerstone (2022) How to overcome big data cultural barriers. Four Cornerstone LLC

Freddie Mac (n.d.) Data stewardship model: enhancing data governance. Freddie Mac

Frontiers (2022) A balanced scorecard for maximizing data performance. Frontiers in Psychology

Frontiers (2023) Decentralized network governance: Blockchain technology and the future of regulation. Frontiers

Furman J, Seamans R (2019) AI and the economy. Innov Policy Econ 19(1):161–191

Gailhofer P, Franke C (2021) Data governance and regulation for sustainable smart cities. Front Environ Sci

Gartner (2020a) Market guide for data loss prevention. Gartner

Gartner (2020b) Predicts 2020: data and analytics strategies—invest, influence, and impact. Gartner

Gartner (2021a) Magic quadrant for data quality solutions. Gartner

Gartner (2021b) The state of data Quality: current practices and evolving trends. Gartner

References

Gartner (2021c) Building a comprehensive data governance program. Gartner
Gartner (2021d) Data governance key initiative overview. Gartner
Gartner (2021e) Data quality solutions. Gartner
Gartner (2021g) Improve data quality for better business outcomes. Gartner
Gartner (2021h) Top 10 data and analytics technology trends for 2021. Gartner
Gartner (2021i) Top security and risk management trends. Gartner
Gartner (2022) How IoT impacts data and analytics. Gartner
GDPR.eu (2020a) GDPR compliance. GDPReu
GDPR.eu (2020b) GDPR fines: lessons learned. GDPR.eu
Ge M, Helfert M (2007) A review of information quality research. Data and Information Quality Workshop
Global Data Governance Project (2021) The global data governance project: 52 case studies. George Washington University
Golightly D, Sloan D, Sime J (2022) The impact of GDPR on data protection and privacy. J Data Protect Privacy
Gordon LA, Loeb MP, Zhou L (2011) The impact of information security breaches: has there been a downward shift in costs? J Comput Secur 19(1):33–56
Gordon LA, Loeb MP, Lucyshyn W, Zhou L (2018) Empirical evidence on the determinants of cybersecurity investments in private sector firms. J Inf Secur 9(2):133–153
Grafanaki S (2017) Regulatory challenges revisited: the cases of infrastructure regulation and data governance. Regulation & Governance
Griffin GW, Holcomb D (2023) Data literacy and skills development. In: Building a data culture: the usage and flow data culture methodologies
Gudivada V, Apon A, Ding J (2017) Data quality considerations for big data and machine learning: going beyond data cleaning and transformations. Int J Adv Softw 10(1):1–20
Gunningham N (2021) Enforcing Environmental Regulation. J Environ Law 23(2):169–193. Oxford University Press
Hamdare S, Kulkarni R, Singh A (2023) Cybersecurity risks in electric vehicle charging systems. J Electr Veh Technol
Harvard Business Review (2018) John Deere bets the farm on AI IoT. Harvard Business Review
Harvard Business Review (2019a) Ensuring data quality in the age of big data. Harvard Business Review
Harvard Business Review (2019b) Why data cleaning plays a vital role in data science. Harvard Business Review
Harvard Business Review (2019c) Why data quality matters for business success. Harvard Business Review
Harvard Business Review (2019d) Drive innovation with better decision-making. HBR
Harvard Business Review (2020a) Creating a data-driven culture. Harvard Business Review
Harvard Business Review (2020b) Data quality and its impact on decision-making. Harvard Business Review
Harvard Business Review (2020c) The business case for data quality. Harvard Business Review
Harvard International Law Journal (2018) The GDPR: impact and implications
Harvard Journal of Law & Technology (2018) Innovating in uncertainty: effective compliance and the GDPR
Harvard Law Review (2020) The evolution of data privacy: the GDPR effect. Harvard Law Review
HashiCorp (2021) Vault: identity-based security for modern applications. HashiCorp
Hassani H, MacFeely S (2023) Driving excellence in official statistics: unleashing the potential of comprehensive digital data governance. Big Data Cogn Comput 7(3):134
Haug A, Zachariassen F, Van Liempd D (2011) The costs of poor data quality. J Ind Eng Manag 4(2):168–193
HBS Online (2020) The advantages of data-driven decision-making. Harvard Business School Online
Healthcare IT News (2020) Standardizing patient records for improved care

Henderson D (2015) Improving organizational performance through data governance. Perform Manag Rev 23(3):19–27

Hiatt JM, Creasey TJ (2012) Change management: the people side of change. Prosci Learning Center Publications

HIMSS (2020) Healthcare data management: three case studies. HIMSS

Hitachi Solutions (2023) Data governance: how to prepare for the future. Hitachi Solutions

Hoeren T, Pinelli C (2020) The California consumer privacy act: a comparative analysis with GDPR. California Law Review

Hrouga M, Sbihi A (2023) Logistics 4.0 for supply chain performance: perspectives from a retailing case study. Bus Process Manag J 29(6):1892–1919. Emerald Publishing Limited

IBM (2020a) Accelerating your journey to AI. IBM

IBM (2020b) Watson knowledge catalog. IBM

IBM (2021) Bankdata case study. IBM

IBM (2023) A step-by-step guide to setting up a data governance program. IBM

IBM (2023a) What is encryption?

IBM (2023b) What is Data Loss Prevention (DLP)?

IBM (2023c) What are insider threats?

IBM (2024) What is data governance? IBM

IBM Security (2023a) What is data privacy?

IBM Security (2023b) Data protection strategy: key components and best practices

IDC (2021) IDC's worldwide internet of things forecast, 2021–2025. IDC

Identity Theft Resource Center (2020) 2019 data breach report. Identity Theft Resource Center

IEEE (2020) Ethical issues related to data privacy and security: why we must balance ethical and legal requirements in the connected world. IEEE Digital Privacy

IEEE (2023) Fine-grained data rights governance in Blockchain-based cloud-edge communications. IEEE

IEEE (2024) Research on big data security and privacy risk governance. IEEE

IEEE Security & Privacy (2021) Risk assessment in cybersecurity. IEEE Security & Privacy

IEEE Transactions on Big Data (2021) Securing big data in the age of AI. IEEE Transactions on Big Data

IEEE Transactions on Dependable and Secure Computing (2021) Encryption techniques for data protection. IEEE Transactions on Dependable and Secure Computing

IMD (2023) What is predictive analytics? Importance, benefits & examples. IMD

Immuta (2020) Data access & control. Immuta

Imperva (2023) What is Data Loss Prevention (DLP)?

Implementing International Environmental Law (2021) Oxford University Press

Indium Software (2023) Big data's impact on IoT: opportunities and challenges in analytics. Indium Software

Informatica (2020a) Axon data governance. Informatica

Informatica (2020b) Enterprise data catalog. Informatica

Informatica (2020c) Master Data Management (MDM) solutions and tools. Informatica

Informatica (2021b) Data intelligence is the future of data governance. Informatica

Informatica (2021c) The business value of data governance: describing "What's in it for me?". Informatica

Informatica (2023a) Data governance framework: 4 pillars for success. Informatica

Informatica (2023b) Measuring success in data governance. Informatica

Information Commissioner's Office (2020) ICO Fines British airways £20m for data breach affecting more than 400000 customers. Information Commissioner's Office

Integrate.io (2023) Why data literacy is essential for a data-driven future. Integrate.io

Integrate.io (2024) Overcoming data governance challenges. Integrate.io

Intellias (n.d.) Big data in the telecom: trends, use cases & case studies. Intellias

International Association for Privacy Professionals (IAPP) (2021) Data lifecycle management and privacy. IAPP

References

International Journal of Electronic Commerce (2020) Security measures in E-commerce. International Journal of Information Management

International Journal of Information Management (2020a) Access control in data security. International Journal of Information Management

International Journal of Information Management (2020b) Data governance and employee training. International Journal of Information Management

International Journal of Information Management (2021a) Data privacy regulations: the implications for data governance. International Journal of Information Management

International Journal of Information Management (2021b) Legal compliance in data breach response. International Journal of Information Management

International Organization for Standardization (2013) ISO/IEC 27001: information security management. ISO

Institute of Data (2023a) The importance of cybersecurity awareness training for employees

ISACA (2020) Practical data security and privacy for GDPR and CCPA

ISACA (2021a) An ethical approach to data privacy protection

ISACA (2021b) Essential cybersecurity components: continuous monitoring, human intelligence and commitment

ISACA (2023) Balancing privacy with businesses' need for data. ISACA

ISBA (2023) The importance of data ethics: why businesses must take it seriously. ISBA

ISO (2011) ISO 8000 data quality and data governance standard. International Organization for Standardization

ISO/IEC (2013) ISO/IEC 27001:2013 information security management systems – requirements. International Organization for Standardization

IT Governance Docs (2023) Data governance maturity model. IT Governance Docs

Janssen M, Charalabidis Y, Zuiderwijk A (2012) Benefits, adoption barriers, and myths of open data and open government. Inf Syst Manag (Taylor & Francis) 29:258

Jarmul K (2023) Practical data privacy: enhancing privacy and security in data

Jay Liebowitz (2013) Data governance: creating value from information assets. CRC Press

JDSupra (2023) Incentivizing compliance. Thomas Fox – Compliance Evangelist

Jin GZ (2018) Artificial intelligence and consumer privacy. National Bureau of Economic Research. https://doi.org/10.3386/w24253

Jochem P, Babrowski S, Fichtner W (2015) Assessing CO_2 emissions of electric vehicles in Germany in 2030. Transp Res A Policy Pract 78:68–83

Journal of Big Data (2021) Big data quality framework: a holistic approach to continuous quality management. Journal of Big Data

Journal of Business Analytics (2021) Predictive analytics in business decision-making. Journal of Business Analytics

Journal of Business Ethics (2020) Ethical considerations in data privacy and security

Journal of Computer and System Sciences (2021) Recent advances in encryption technology

Journal of Cybersecurity (2020) Case studies on data breach incidents

Journal of Cybersecurity and Privacy (2020) Insider threats in cybersecurity

Journal of Cybersecurity and Privacy (2021a) Modern data security challenges

Journal of Cybersecurity and Privacy (2021b) Risk assessment in cybersecurity

Journal of Data and Information Quality (2020a) Best practices in data cleansing

Journal of Data and Information Quality (2020b) The organizational impact of data quality

Journal of Data and Information Quality (2021a) Data consistency in financial services

Journal of Data and Information Quality (2021b) Improving customer data quality in retail

Journal of Data Quality (2020) Best practices in data quality management

Journal of Financial Regulation and Compliance (2021a) Data quality and regulatory compliance in financial services

Journal of Financial Regulation and Compliance (2021b) GDPR adaptation in financial services

Journal of Healthcare Management (2021) Improving patient data quality in healthcare

Journal of Information Security (2020a) Cybersecurity threats and data breach implications

Journal of Information Security (2020b) Developing effective security policies
Journal of Information Security (2021) Testing and updating incident response plans
Journal of Information Systems (2021) Regulatory compliance in data security and privacy
Journal of Network and Computer Applications (2021) Trends in data loss prevention solutions
Kahn BK, Strong DM, Wang RY (2002) Information quality benchmarks: product and service performance. Commun ACM 45(4):184–192
Kahneman D (2011) Thinking, fast and slow. Farrar, Straus and Giroux
Kamilaris A, Kartakoullis A, Prenafeta-Boldú FX (2017) A review on the practice of big data analysis in agriculture. Comput Electron Agric 143:23–37
Karkouch A, Mousannif H, Al Moatassime H, Noel T (2016) Data quality in internet of things: a state-of-the-art survey. J Netw Comput Appl (Elsevier) 73:57
Kerasidou A, Kerasidou C (2023) Data-driven research and healthcare: public trust, data governance and the NHS. BMC Med Ethics 24:51
Khaitan R, Sharma R (2019) The Unified Payments Interface (UPI) success story in India. J Financ Serv Res 56(3):245–264
Khatri V, Brown CV (2010) Designing data governance. Commun ACM 53(1):148–152
Kim HY, Cho JS (2018) Data governance framework for big data implementation with NPS Case Analysis in Korea. Journal of Big Data
Kimball R, Caserta J (2004) The data warehouse ETL toolkit: practical techniques for extracting, cleaning, conforming and delivering data. Wiley
Kimball R, Ross M (2013) The data warehouse toolkit: the definitive guide to dimensional modeling. Wiley
Kirshteyn ME (n.d.) Data governance framework: a comprehensive guide to success
Koltay T (2016) Data governance, data literacy and the management of data quality. Springer
Kotter JP (1996) Leading change. Harvard Business Review Press
KPMG (2020) Guardians of trust: the global chief data officer report. KPMG International Cooperative
KPMG Global (n.d.) Data governance in healthcare. KPMG
Kroll (2021) Compliance monitoring program design. Kroll
Kroll (n.d.) Compliance monitoring program design for financial services. Kroll
Ladley J (2012) Data governance: how to design, deploy, and sustain an effective data governance program. Elsevier
Ladley J (2019a) Data governance: how to design, deploy, and sustain an effective data governance program, 2nd edn. Academic
Ladley J (2019b) Data governance: how to design, deploy, and sustain an effective data governance program. Elsevier
Laney D (2018) Infonomics: how to monetize, manage, and measure information as an asset for competitive advantage. Taylor & Francis
Lee J, Darbellay A (eds) (2021) Data governance in AI, FinTech and LegalTech: law and regulation in the financial sector. Edward Elgar Publishing
Lee YW, Strong DM, Kahn BK, Wang RY (2002) AIMQ: a methodology for information quality assessment. Inf Manag 40(2):133–146
LightsOnData (2023a) Data governance in 2024. LightsOnData
LightsOnData (2023b) Top 10 data governance courses and training. LightsOnData
Loshin D (2012) Business intelligence: the savvy manager's guide. Morgan Kaufmann
Loshin D (2013) The practitioner's guide to data quality improvement. Morgan Kaufmann
Luciano Floridi (2013) The ethics of information. Oxford University Press
Ma J, Gao J, Wu Y, Zhou J (2018) Data quality assessment for data governance: application of machine learning. J Data Inf Qual (JDIQ) 10(4):16. ACM
MacFeely S et al (2022) Informing the global data future: benchmarking data governance frameworks. Cambridge University Press
Madsen L (2022) Dear Laura: what role should leadership play in data governance? DATAVERSITY

References

Mahanti R (2019) Data governance and data Management: contextualizing data governance drivers, technologies, and tools. Kindle Edition

Managing Strategic Change (2021) Oxford University Press

Martins R, Belo O (2017) Balanced scorecard for data performance: a case study. J Bus Anal 5(1):13–21

Master Data Management in Practice: Achieving True Customer MDM by Dalton Cervo, Mark Allen, Jill Dyché

McDermott R, O'Dell C (2001) Overcoming cultural barriers to sharing knowledge. J Knowl Manag (MCB UP Ltd) 5:76

McKinsey (2021) Data ethics: what it means and what it takes

McKinsey (2023a) Designing data governance that delivers value. McKinsey

McKinsey (2023c) Digital logistics and the technology race. McKinsey

McKinsey (2023d) Making collaboration across functions a reality. McKinsey

McKinsey & Company (2019a) How continuous improvement can build a competitive edge. McKinsey & Company

McKinsey & Company (2019b) The role of data in digital transformation. McKinsey & Company

McKinsey & Company (2020) Beyond hiring: how companies are reskilling to address talent gaps. McKinsey & Company

McKinsey & Company (2021) Designing data governance that delivers value. McKinsey & Company

McKinsey & Company (2022) Using skill gap assessments to help future-proof your organization. McKinsey & Company

McKinsey & Company (2023a) Data analytics: accelerating maturity in the US public sector. McKinsey

McKinsey & Company (2023b) Designing data governance that delivers value. McKinsey & Company

McKinsey & Company (2023c) Data ethics: what it means and what it takes. McKinsey & Company

McKinsey & Company (2023d) Governance and Regulation as generative AI advances. McKinsey & Company

McKinsey & Company (2023e) Putting data ethics into practice. McKinsey & Company

McKinsey & Company (n.d.) Designing data governance that delivers value. McKinsey & Company

MDPI (2023) Integration of IoT-enabled technologies and Artificial Intelligence (AI) for smart city scenario: recent advancements and future trends

Microsoft (2023a) What is data loss prevention (DLP)?

Microsoft (2023b) What is insider threat?

Microsoft Learn (2024) Data, compliance, and governance. Microsoft

Microsoft Security (2023a) What is data security?

Microsoft Security (2023b) Stay compliant with Zero Trust security

Microsoft Security (2023c) Microsoft Digital defense report.

Microsoft Security (2023d) What is access control?

Miller T, Brown G (2018) Artificial intelligence in data governance: from theory to practice. J Data Sci Anal 5(2):102–115. Springer

Ministry of Children and Youth Services, Government of Ontario, Canada (n.d.) Data Stewardship Program

MIT Sloan Management Review (2021a) Adapting data governance for evolving business needs. MIT Sloan Management Review

MIT Sloan Management Review (2021b) Building Trust in an era of data breaches. MIT Sloan Management Review

MIT Sloan Management Review (2021c) Data governance in the 21st-century organization. MIT Sloan Management Review

Moilanen J (2021) Standardized data product metadata examples based on real-world published data products

Monte Carlo (2023) The future of data governance: 4 trends to watch out for. Monte Carlo

Mosley M (ed) (2017) DAMA-DMBOK: data Management body of knowledge, 2nd edn. DAMA International
National Institute of Standards and Technology (NIST) (2018) Framework for improving critical infrastructure cybersecurity (Version 1.1)
National Institute of Standards and Technology (NIST) (2023a) Protecting information systems
National Institute of Standards and Technology (NIST) (2023b) Continuous monitoring
Network Security (2020) Effective incident response planning
Network Security (2021) Software and system security updates
Nexocode (2023) Supply chain visibility: the role of real-time data in logistics. Nexocode
Nicho M (2018) Information security governance: a framework. J Inf Secur 9(2):115–132
NordLayer (n.d.) What is regulatory compliance and why is it important?
North American Electric Reliability Corporation (2016) Critical Infrastructure Protection (CIP) Standards
O'Reilly Media (2017) Data governance: the definitive guide. O'ReReilly Media
OECD (2018) Implementing effective ethics standards in government and the civil service. OECD
OECD (2019) Data governance in the public sector. OECD iLibrary
OECD (2021) The path to becoming a data-driven public sector. OECD iLibrary
OECD (2023) Global trends in government innovation. OECD Observatory of Public Sector Innovation
Okera (2020) Data access platform
Okta (2021) Customer success stories
Olson JE (2003a) Data quality. Morgan Kaufmann
Olson JE (2003b) Data quality: the accuracy dimension. Morgan Kaufmann
Omeda (2023) How to achieve cross-functional collaboration in data governance
OneTrust (2020) Privacy management software. OneTrust
OneTrust (2023) Compliance program performance metrics: how to measure compliance. OneTrust
Onna (2023) Why cross-functional collaboration is critical in data governance
Ono T, Lafortune G, Schoenstein M (2013) Health workforce planning in OECD countries: a review of 26 projection models from 18 countries, OECD health working papers, no. 62. OECD Publishing
Otto B (2011) Organizing data governance: findings from the telecommunications industry and consequences for large service providers. Commun Assoc Inf Syst 29:Article 3
Otto B (2021) Designing data governance: conceptual foundations and practical application. Springer
Otto B, Österle H (2016) Data quality management. Springer
OvalEdge (2023) Data governance maturity models and how to measure it. OvalEdge
Oxford Handbook of Management (2021) Oxford Academic
Palo Alto Networks (2023) What is access control?
Paskaleva K, Evans J, Martin C (2017) Data governance in the sustainable smart city. Information Polity
Payment Card Industry Security Standards Council (2018) Payment card industry data Security standard (PCI DSS)
Phillips J, Klein JD (2021) Change management: from theory to practice. TechTrends 67(1):189–197
Pigni F, Piccoli G, Watson RT (2016) Digital data streams: creating value from data-driven innovation. MIS Q Exec 15(2):63–72
Pipino LL, Lee YW, Wang RY (2002) Data quality assessment. Commun ACM 45(4):211–218
Plotkin D (2014) Data stewardship: an actionable guide to effective data Management and data governance. Morgan Kaufmann
Polestar LLP (n.d.) Mastering change management for optimal data governance success. Polestar LLP
Pollack J, Pollack R (2015) The change Management paradigm. Oxford University Press
Precisely. (2023) Data governance business case: here's how to build one. Precisely

References

Protiviti. (2023) Technology for the evolving data privacy regulatory landscape. Protiviti
PwC (2023) Cloud governance on risks and controls. PwC
Pythian. (2022) Building organizational models for better data governance. Pythian
Quest (2021) Data empowerment. Quest Software
Rapid7 (2023) What is security awareness training for employees?
Red Hat (2023) What is cloud governance? Red Hat
Redman TC (2008a) Data driven: profiting from your Most important business asset. Harvard Business Review Press
Redman TC (2008b) Data driven: profiting from your most important business asset. Harvard Business Press
Rosário AT, Raimundo RJ, Cruz SP (2022) Sustainable entrepreneurship: a literature review. Sustainability 14(9):5556
Ross JW, Beath CM (2019) Designing and executing Digital strategies. Management Information Systems Quarterly, Association for Information Systems
Royal Society (2023) Data governance: case studies. Royal Society
Royal Society (2024) Data governance: Case studies. Royal Society
Sage Journals (2022) Wells Fargo: fall from great to miserable: a case study on corporate governance. Sage Journals
Salesforce (2021) Salesforce security and privacy. Salesforce
Sambasivan N, Kapania S, Highfill H, Parthasarathy S, Akrong D, Matthews T, Churchill E (2021) Everyone wants to do the model work, not the data work. In: Data cascades in high-stakes AI. Proceedings of the 2021 CHI conference on human factors in computing systems, pp 1–15
Sarker MRI, Bhattacharjee U, Khan R (2020) Integration of hybrid renewable energy based electric vehicles for efficient power management. Energy Rep 6:175–184
SAS (2023) IoT success depends on data governance, security, and privacy. SAS
SAS Institute Inc (2020) SAS Data Management. SAS Institute Inc
ScienceDirect (2019) IoT-Gov: a structured framework for internet of things governance. ScienceDirect
ScienceDirect (2023) Data governance: a conceptual framework structured review and future research agenda. ScienceDirect
ScienceDirect (2024) Data governance: a conceptual framework, structured review, and research agenda. ScienceDirect
Secureframe (2023) 7 data governance metrics and KPIs every business should track. Secureframe
Sholler D, Ram K, Boettiger C, Katz DS (2021) Enforcing public data archiving policies in academic publishing: A study of ecology journals. Arxiv
Simplilearn (2023) Quantum computing's impact across industries
Singh S, Chana I (2016) Q-aware: quality of service based cloud resource provisioning. Comput Electr Eng 47:138–160
Smallwood RF (2019) Information governance: concepts, strategies, and best practices. Wiley
Smarty (2023) Understanding key data governance players: roles and responsibilities. Smarty
Smithsonian Directive 609 (2019) The CARE principles for indigenous data governance. Data Science Journal
Soares S (2012) Data governance tools: evaluation criteria, big data governance, and alignment with enterprise data management, 1st edn
Soares S (n.d.-a) Data governance tools: evaluation criteria, big data governance, and alignment with enterprise data management
Soares S (n.d.-b) Data governance: creating value from information assets. CRC Press
Soares S (n.d.-c) The chief data officer handbook for data governance
SolverASSIST (n.d.) Data governance best practices for business intelligence. SolverASSIST
Sophos (2023) What is DLP – data loss prevention?
Splunk (2023) Data governance: definitions, components, & best practices. Splunk
SpringerLink (2023) General Data Protection Regulation (GDPR)
SpringerLink (2023a) Adaptive governance for a resilient Digital Society. SpringerLink

SpringerLink (2023b) Blockchain and institutions: trust and (De)centralization. SpringerLink

SpringerLink (2023c) Privacy and security challenges and opportunities for IoT technologies. SpringerLink

Stallings W (2017) Cryptography and network security: principles and practice. Pearson

Subramanian S (2017) Reducing risks through effective data governance. Risk Manag J 45(2):28–36

Supply Chain Management Review (2021) Data quality in supply chain management

Sweeney SM, Hamadeh HK, Abrams N, Adam SJ, Brenner S, Connors DE et al (2023) Case studies for overcoming challenges in using big data in cancer. Cancer Res 83(8):1183–1190

Tableau (2023) Data literacy explained: definition, examples & more. Tableau

Talburt et al (2020) Data curation: foundations, principles, and practice. Springer

Talend (2020) Data catalog

Talend (2023) Data governance success stories

TDWI (2021) Managing data in the cloud: the challenge of complex environments for real-time applications. TDWI

Tealium (2024) Derisking data: regulatory milestones shaping the use of AI and data in 2024. Tealium

Technology Magazine (2023) Navigating the complexity of data governance and AI models

TechRepublic (2023) How data literacy is important to data governance. TechRepublic

TechRepublic (2024) How does data governance affect data security and privacy? TechRepublic

TechTarget (2023) Data governance strategies for today's evolving IT landscape. TechTarget

Thales Group (2021) Data protection regulation around the world

The Data Governance Institute (2020) Data governance best practices. The Data Governance Institute

The Data Literacy Project (2021) Fostering data literacy. The Data Literacy Project

The Sedona Conference (2018) The Sedona principles: best practices, recommendations & principles for addressing electronic document production

Thomas C (2008) Redman, "data driven: profiting from your most important business asset". Harvard Business Review Press

Toomanian A, Salehi M, Ashuri B (2011) Data governance in large data warehouses. Int J Data Sci 7(3):34–45

Torre-Bastida AI, Gil G, Miñón R, Díaz-de-Arcaya J (n.d.) Technological perspective of data governance in data space ecosystems

Twilio Segment (2023) What is a data governance framework? Examples & models. Twilio Segment

U.S. Congress (1996) Health Insurance Portability and Accountability Act (HIPAA). Public Law, pp 104–191

U.S. Congress (2002) Sarbanes-Oxley Act of 2002 (SOX). Public Law, pp 107–204

U.S. Congress (2010). Dodd-Frank wall street reform and consumer protection act. Public Law, pp 111–203

U.S. Congress (2015) Cybersecurity information sharing act (CISA). Public Law, pp 114–113

U.S. Department of Health and Human Services (2000). HIPAA Privacy Rule

U.S. Federal Trade Commission (2016) EU-U.S. Privacy shield framework

U.S. Food and Drug Administration (1997) Title 21 CFR part 11: electronic records; Electronic signatures

Udemy (2023) Data governance – the complete course for beginners

United Nations (1990) Guidelines for the regulation of computerized personal data files. United Nations

Uttamchandani S (2020) Maximizing data value through governance and analytics. J Data Anal 9(1):15–24

Velotix (2023) Data governance in 2023: trends, insights, and analysis. Velotix

Voigt P, von dem Bussche A (2017) The EU general data protection regulation (GDPR): a practical guide. Springer International Publishing

References

Wamba SF, Akter S, Edwards A, Chopin G, Gnanzou D (2015) How 'big data' can make big impact: findings from a systematic review and a longitudinal case study. Int J Product Econ 165:234–246

Wang RY, Lee YW, Pipino LL, Strong DM (1998) Manage your information as a product. Sloan Manag Rev 39(4):95–105

Watson J, Whiting M, Krieg C, Butcher R, Carr PA, Simpson L (2020) COVID-19 self-testing in high-risk populations and remote locations: results and lessons learnt. J Glob Health 10(2):020340

Widianingsih I, Morrell E (2007) Participatory governance in Indonesia: a case study of community-based environmental initiatives. Environ Dev Sustain 9(4):409–425

World Economic Forum (2021) How digital transformation is driving economic change

World Economic Forum (2023a) Quantum computing governance and regulation. World Economic Forum

World Economic Forum (2023b) Top 10 emerging technologies of 2023. World Economic Forum

XenonStack (2020) Metadata management tools and best practices. XenonStack

Y42 (2023) 5 common challenges for cross-functional data governance collaboration

Zagoudis S (2020) Defining purpose, scope, and objectives in a data governance policy. DATAVERSITY

Zaharia M, Chowdhury M, Franklin MJ, Shenker S, Stoica I (2016) Spark: cluster computing with working sets. In: Proceedings of the 2nd USENIX conference on hot topics in cloud computing

Zhang C, Wang Z, Wang Z (2019) A review on the research and applications of agricultural robotics: development, key technologies, and challenges. Precis Agric 20(4):918–940

Zheng Z, Xie S, Dai H, Chen X, Wang H (2018) Blockchain challenges and opportunities: a survey. Int J Web Grid Serv 14(4):352–375

Zyskind G, Nathan O, Pentland A (2015) Decentralizing privacy: using Blockchain to protect personal data. In: 2015 IEEE security and privacy workshops. IEEE, pp 180–184

SPRINGER NATURE

GPSR Compliance

The European Union's (EU) General Product Safety Regulation (GPSR) is a set of rules that requires consumer products to be safe and our obligations to ensure this.

If you have any concerns about our products, you can contact us on ProductSafety@springernature.com

In case Publisher is established outside the EU, the EU authorized representative is:

Springer Nature Customer Service Center GmbH
Europaplatz 3
69115 Heidelberg, Germany

The manufacturer's authorised representative in the EU is Springer Nature Customer Service Centre GmbH, Europaplatz 3, 69115 Heidelberg, Germany. If you have any concerns regarding our products, please contact ProductSafety@springernature.com

Printed and bound by CPI Group (UK) Ltd, Croydon, CR0 4YY
25/03/2026
02078170-0007